MATHEMATICAL PHYSICS AND APPLIED MATHEMATICS

Editors:

M. FLATO, *Université de Dijon, Dijon, France*

R. RĄCZKA, *Institute for Nuclear Research, Warsaw, Poland*

with the collaboration of:

M. GUENIN, *Institut de Physique Théorique, Geneva, Switzerland*

D. STERNHEIMER, *Collège de France, Paris, France*

VOLUME 2

MATHEMATICAL PHYSICS AND PHYSICAL MATHEMATICS

MATHEMATICAL PHYSICS AND PHYSICAL MATHEMATICS

PROCEEDINGS OF THE INTERNATIONAL SYMPOSIUM
ORGANIZED BY THE MATHEMATICAL INSTITUTE
OF THE POLISH ACADEMY OF SCIENCES,
THE INSTITUTE FOR NUCLEAR RESEARCH
AND UNIVERSITY OF WARSAW,
HELD IN WARSAW 25–30 MARCH, 1974

Edited by

KRZYSZTOF MAURIN
University of Warsaw

and

RYSZARD RĄCZKA
Institute for Nuclear Research

D. REIDEL PUBLISHING COMPANY
DORDRECHT–HOLLAND/BOSTON–U.S.A.

PWN–POLISH SCIENTIFIC PUBLISHERS
WARSAW

Library of Congress Cataloging in Publication Data
Main entry under title:

Mathematical physics and physical mathematics.

1. Mathematical physics — Congresses. I. Maurin, Krzysztof, ed. II. Rączka, Ryszard, ed. III. Polska Akademia Nauk. Instytut Matematyczny. IV. Warsaw. Instytut Badań Jądrowych. V. Warsaw. Uniwersytet.
QC19.2.M37 530.1'5 74-34289
ISBN 90-277-0537-2

Distributors for Albania, Bulgaria, Chinese People's Republic, Czechoslovakia, Cuba, German Democratic Republic, Hungary, Korean People's Democratic Republic, Mongolia, Poland, Rumania, Democratic Republic of Vietnam, the U.S.S.R., and Yugoslavia
ARS POLONA — RUCH
Krakowskie Przedmieście 7, 00–068 Warszawa, Poland

Distributors for the U.S.A., Canada and Mexico
D. REIDEL PUBLISHING COMPANY, INC.
Lincoln Building, 160 Old Derby Street, Hingham, Mass. 02043, U.S.A.

Distributors for all other countries
D. REIDEL PUBLISHING COMPANY
P.O. Box 17, Dordrecht, Holland

Printed in Poland by D.R.P.

TABLE OF CONTENTS

PART THREE – GROUP REPRESENTATION IN QUANTUM THEORY

PART FOUR – QUANTUM MECHANICS AND PARTICLE PHYSICS

PART FIVE – QUANTUM STATISTICAL PHYSICS

FOREWORD

For a number of years now a need has been felt (once again) to draw together physicists employing advanced modern mathematical apparatus, 'mathematical physicists', and mathematicians to whom the subject-matter of physics is not alien and for whom physics is a source of sound problems, (for symmetry we use a neologism) 'physical mathematicians'. This need has led to the organization of the first large-scale International Symposium on Mathematical Physics ('Recent Progress in Mathematical Physics') which was held March 25–30, 1974, in the large 'Swedish' hotel 'Forum', then not yet officially opened. (This remark of a geographical-geometrical nature is required in order to understand the 'inner history' of the opening address!) This Symposium brought together leading investigators from all over the world (see list of participants) who undertook the far from easy task of reporting, in a manner accessible to the 'educated layman', the principal results and methods from the field on which they were experts. We thank them for sending the manuscripts of their beautiful lectures 'on schedule'.

Such a large undertaking could not, obviously, be realized without strong patronage. The patrons were:

1. Institute of Mathematics, Polish Academy of Sciences (the directors, Professors Z. Olech and Z. Semadeni, displayed great understanding of the need for such a conference);

2. Institute for Nuclear Research, Polish Academy of Sciences (the director, Professor R. Żelazny, organized enormous assistance for the Symposium, for which we express our heartfelt gratitude to him);

3. University of Warsaw (in the persons of the Rector, Professor Z. Rybicki, and Professor J. Werle, the then director of the Institute of Theoretical Physics).

But our world rests not on mighty ones. Our Symposium could not have taken place without the onerous and (seemingly – as is evident from this foreword) thankless work of untitled people necessary to keep such a machine in motion and to provide the requisite facilities for the participants. These people were workers of the Department of Mathematical Methods of Physics at the University of Warsaw, of the Foreign Department of the Institute for Nuclear Research, and of the Elementary Particles Laboratory of the Institute for Nuclear Research.

KRZYSZTOF MAURIN and RYSZARD RĄCZKA

OPENING ADDRESS

After these serious words* allow me to make some remarks which perhaps are not very serious but are of vital importance.

1. *Historical Remarks.* As you may have guessed, this hotel was built by Swedes. Now, why exactly by Swedes and why is the First Conference on Mathematical Physics taking place exactly in this spot? Warsaw was burned and sacked by Swedish armies at least twice, in 1655–1674 and in 1701–1709 ('Northern War'), and this is precisely the spot where the Swedish military camp was situated. As he lay dying on March 25, 1674, the Swedish commander-in-chief – a good Protestant of course (I have forgotten his name; perhaps our Swedish colleagues can help me?) – called out to his fellow countrymen: "Let the Hotel 'Forum' be erected in 1974 as a sign of our expiation and let the First International Conference on Mathematical Physics be convened there."

2. *Philosophical Remarks.* A gulf between mathematics and physics has evolved since the 'twenties and is still deepening. This was unthinkable to former generations and is unimaginable to the average person. What we call mathematical physics is the last desperate attempt to throw a bridge across this abyss.

Every physicist has some smattering of mathematics, but not every mathematician has become somewhat acquainted with physics during his university years (at any rate, in Poland). I have noticed that every physicist (even an experimental physicist) cherishes a deep, unconscious love for mathematics. This love is revealed, for instance, when with great pleasure he introduces some mathematical notions during his lectures.

Every mathematician, on the other hand, secretly admires physicists and feels happiest when his theorem finds application in physics. This admiration was once expressed by my teacher, the great Polish topolo-

* The Symposium was officially opened by Professor Z. Olech and Professor R. Żelazny.

gist Karol Borsuk when he made a melancholy remark to me: "Well, we prove our theorems—physicists change the world".

In his beautiful essay *The Unreasonable Effectiveness of Mathematics in Natural Sciences*, Eugene Wigner stresses that the chief role of mathematics in physics consists not in its being an instrument (e.g. computations) but in being the language of physics. And an admirably adequate language indeed!

This was expressed symbolically by Eddington and Jeans in their statement "God is a mathematician".

Michael Reed and Barry Simon put it in a slightly different way in their excellent book *Methods of Modern Mathematical Physics*: "When a successful mathematical model is created for a physical phenomenon it is natural to think of physical quantities in terms of the mathematical objects which represent them and to interpret similar or secondary phenomena in terms of the same model."

I need not specify what a mathematical physicist is; there are many here. We shall learn by experience during the next six days what mathematical physics is.

3. *Organizational* (*Technical*) *Remarks.* We have not divided this conference into sections on purpose, in order to emphasize the unity of the discipline. On this account, too, we have asked our reporters not to be overly technical so that ordinary mortals of good will might understand just a little of what will be said here. Or at least that we might have the pleasant illusion that we do understand. Man's vested right is to understand and only to be misinformed.

Our program is perhaps a bit overloaded, but mathematicians and physicists—, of course, mathematical physicists—are free people. You need not attend all the sessions.

KRZYSZTOF MAURIN

LIST OF PARTICIPANTS

ADAMCZYK, A., *Institute for Nuclear Research*, Hoża 69, 00-681 Warszawa, Poland.

ALBEVARIO, S., *Institute of Mathematics, University of Oslo*, Blindern, 1000 Oslo 3, Norway.

AMES, W., *College of Engineering, University of Iowa*, Iowa City, Ia. 52240, USA.

ANDERSON, B., *Institute of Theoretical Physics, University of Lund*, Solvegatan 14, 22362 Lund, Sweden.

ANDERSON, S., *Institute of Theoretical Physics*, Fack, 40220 Goteborg 5, Sweden.

ANTOINE, J. P., *Institut de Physique Theorique 2, Chemin du Cyclotron, Université de Louvain*, 1348 Louvain-la-Neuve, France.

ARODZ, H., *Department of Theoretical Physics, Jagellonian University*, Reymonta 4, 30-059 Kraków, Poland.

BAŁABAN, T., *Institute of Mathematics, Warsaw University*, PKiN, Warszawa, Poland.

BARUT, A. O., *Department of Physics, University of Colorado*, Boulder, Co. 80304, USA.

BAŻAŃSKI, S., *Institute of Theoretical Physics, Warsaw University*, Hoża 69, 00-681 Warszawa, Poland.

BECHLER, A., *Institute of Theoretical Physics, Warsaw University*, Hoża 69, 00-681 Warszawa, Poland.

BIAŁYNICKA-BIRULA, Z., *Institute of Physics, Polish Academy of Sciences*, Al. Lotników 32/46, 02-668 Warszawa, Poland.

BIAŁYNICKI-BIRULA, I., *Institute of Theoretical Physics, Warsaw University*, Hoża 69, 00-681 Warszawa, Poland.

BÖHM, A., *Center for Particle Theory, University of Texas*, Austin, Tx. 78712, USA.

BONA, P., *Department of Theoretical Physics, University of Bratislava, Mlynska Dolina*, 81631 Bratislava, Czechoslovakia.

BORCHERS, H., *Institut für Theoretische Physik, Universität Göttingen*, Bunsenstrasse 9, 3400 Göttingen, Federal Republic of Germany.

BROJAN, J., *Institute of Theoretical Physics, Warsaw University*, Hoża 69, 00-681 Warszawa, Poland.

BROS, J., *Department de Physique Theorique, Centre d'Etudes Nucleaires*, Saclay, B. P. No. 2, 91190 Gif-sur-Yvette, France.

BRZESKI, A., *Institute of Theoretical Physics, University of Wrocław*, Cybulskiego 36, 50-205 Wrocław, Poland.

BUDINI, P., *International Centre for Theoretical Physics*, Miramare-Grinano, P. O. Box 586, 34100 Trieste, Italy.

BUGAJSKA, K., *Institute of Physics, Silesian University*, Uniwersytecka 4, 40-007 Katowice, Poland.

BUGAJSKI, S., *Institute of Physics, Silesian University*, Uniwersytecka 4, 40-007 Katowice, Poland.

BURZYŃSKI, A., *Department of Theoretical Physics, Jagellonian University*, Reymonta 4, 30-059 Kraków, Poland.

CASTELL, L., *Max-Planck-Institut*, Riemerschmidstrasse 7, Postfach 1529, 813 Starnberg, Federal Republic of Germany.

CEGŁA, W., *Institute of Theoretical Physics, University of Wrocław*, Cybulskiego 36, 50-205 Wrocław, Poland.

CHIANG CHINN CHANN, *Center for Particle Theory, University of Texas*, Austin, Tx. 78712, USA.

CIEPLAK, M., *Institute of Theoretical Physics, Warsaw University*, Hoża 69, 00-681 Warszawa, Poland.

CIESIELSKI, Z., *Institute of Mathematics, Polish Academy of Sciences*, Śniadeckich 8, 00-656 Warszawa, Poland.

CONNES, A., *Centre de Physique Theorique, C.N.R.S.*, 31 Ch. J. Aiguier, 13274 Marseille 9, CEDEX 2, France.

CRONSTROM, C., *Research Institute for Theoretical Physics, University of Helsinki*, Siltavuorenpenger 20B, 00170 Helsinki 17, Finland.

DIN, A., *Institute of Theoretical Physics*, Fack, 40220 Goteborg 5, Sweden.

DOEBNER, H., *Institut für Theoretische Physik, Technische Universität Clausthal*, Leibnizstrasse ?, 3392 Clausthal-Zellerfeld, Federal Republic of Germany.

DOPPLICHER, S., *Istituto di Fisica, Universita di Roma*, Piazzale delle Science 5, 00185 Roma, Italy.

DÜRR, H., *Max-Planck-Institut für Physik und Astrophysik*, Fohringer Ring 6, 8000 München, Federal Republic of Germany.

DZIEMBOWSKI, Z., *Institute of Theoretical Physics, Warsaw University*, Hoża 69, 00-681 Warszawa, Poland.

FLATO, M., *Labo. Phys.-Math., Fac. des Sciences Mirande, Université de Dijon*, Campus Universitaire, 21000 Dijon, France.

BUDINI, P., *International Centre for Theoretical Physics*, Miramare-Grinano, P. O. Box 586, 34100 Trieste, Italy.

FRY, M. P., *Institut für Theoretische Physik, Universität Graz*, Universitätsplatz 5, 8010 Graz, Austria.

GARBACZEWSKI, P., *Institute of Theoretical Physics, University of Wrocław*, Cybulskiego 36, 50-205 Wrocław, Poland.

GAWĘDZKI, K., *Department of Mathematical Physics, Warsaw University*, Hoża 74, 00-682 Warszawa, Poland.

GEORGE, C., *Faculté des Sciences, Service de Chimie Physique 2, Université de Bruxelles*, Av. F. D. Roosevelt 50, 1050 Bruxelles, Belgium.

GHIRARDELLO, L., *Istituto di Fisica, Universita di Milano*, Via Celoria 16, 20133 Milano, Italy.

GLIMM, J., *Courant Institute of Mathematical Sciences*, New York University, New York, N.Y. 10012, USA.

GORINI, V., *Istituto di Fisica, Universita di Milano*, Via Celoria 16, 20133 Milano, Italy.

GUENIN, M., *Department de Physique Theorique, Université de Genève*, Boulevard d'Yvoy 32, 1211 Genève 4, Switzerland.

GUERRA, F., *Istituto di Fisica, Universita di Salerno*, Via Vernei 42, 84100 Salerno, Italy.

HANCKOWIAK, J., *Institute of Theoretical Physics, University of Wrocław*, Cybulskiego 36, 50-205 Wrocław, Poland.

HARTKAMPER, A., *Fachbereich Physik, Universität Trier-Kaiserslautern*, Pfaffenbergstrasse 95, 6750 Kaiserslautern, Federal Republic of Germany.

HEGERFELDT, G., *Institut für Theoretische Physik, Universität Göttingen*, Bunsenstrasse 9, 3400 Göttingen, Federal Republic of Germany.

HENNING, H., *Institut für Theoretische Physik, Technische Universität Clausthal*, Leibnizstrasse, 3392 Clausthal-Zellerfeld, Federal Republic of Germany.

HEPP, K., *Seminar für Theoretische Physik, Eidgenossische Technische Hochschule*, Honggerberg, 8049 Zurich, Switzerland.

HOEGFORS, C., *Institute of Theoretical Physics*, Fack, 40220 Goteborg 5, Sweden.

HOEGH-KROHN, R., *Institute of Mathematics, University of Oslo*, Blindern, 1000 Oslo 3, Norway.

IAGOLITZER, D., *Department de Physique Theorique, Centre d'Etudes Nucleaires*, Saclay, B. P. No. 2, 91190 Gif-sur-Yvette, France.

INFELD, E., *Institute for Nuclear Research*, Hoża 69, 00-681 Warszawa, Poland.

INGARDEN, R., *Institute of Physics, Nicholas Copernicus University*, Grudziądzka 5, 87-100 Toruń, Poland.

IWIŃSKI, Z., *Institute of Theoretical Physics, Warsaw University*, Hoża 69, 00-681 Warszawa, Poland.

JADCZYK, A., *Institute of Theoretical Physics, University of Wrocław*, Cybulskiego 36, 50-205 Wrocław, Poland.

JAFFE, A., *Physics Department, Harvard University*, Cambridge, Ma. 02138, USA.

JAKUBIEC, A., *Department of Mathematical Physics, Warsaw University*, Hoża 74, 00-682 Warszawa, Poland.

JAMIOŁKOWSKI, A., *Institute of Physics, Nicholas Copernicus University*, Grudziądzka 5, 87-100 Toruń, Poland.

JANCEWICZ, B., *Institute of Theoretical Physics, University of Wrocław*, Cybulskiego36, 50-205 Wrocław, Poland.

JENDRZEJEWSKI, J., *Institute of Theoretical Physics, University of Wrocław*, Cybulskiego 36, 50-205 Wrocław, Poland.

JUREWICZ, A., *Institute for Nuclear Research*, Hoża 69, 00-681 Warszawa, Poland.

KAPUŚCIK, E., *Theoretical Physics Department, Institute of Nuclear Physics*, Radzikowskiego 152, 31-342 Kraków, Poland.

KARWOWSKI, W., *Institute of Theoretical Physics, University of Wrocław*, Cybulskiego 36, 50-205 Wrocław, Poland.

KASTRUP, H., *Institut für Theoretische Physik, Technische Hochschule*, Schinkelstrasse 2, 5100 Aachen, Federal Republic of Germany.

KEHLERT, E., *Kobenhavns Universitets, Matematiske Institut*, Universitetsparken 5, 2100 Kobenhavn, Denmark.

KIHLSBERG, A., *Institute of Theoretical Physics*, Fack, 40220 Goteborg 5, Sweden.

KIJOWSKI, J., *Department of Mathematical Physics, Warsaw University*, Hoża 74, 00-682 Warszawa, Poland.

KISYŃSKI, J., *Institute of Mathematics, Warsaw University*, PKiN, Warszawa, Poland.

KLAUDER, J., *Bell Laboratories*, 600 Mountain Avenue, Murray Hill, N.J. 07974, USA.

KOLLER, K., *Deutsches Elektronen-Synchrotron*, Notkestieg 1, 2000 Hamburg, Federal Republic of Germany.

KOMOROWSKI, J., *Department of Mathematical Physics, Warsaw University*, Hoża 74, 00-682 Warszawa, Poland.

KOSSAKOWSKI, A., *Institute of Physics, Nicholas Copernicus University*, Grudziądzka 5, 87-100 Toruń, Poland.

KOTECKY, R., *Institute of Physics, Czechoslovak Academy of Sciences*, Na Slovance 2, 0116 Praha, Czechoslovakia.

KRÓLIKOWSKI, W., *Institute of Theoretical Physics, Warsaw University*, Hoża 69, 00-681 Warszawa, Poland.

KULIKOWSKA, T., *Institute for Nuclear Research*, 05-400 Otwock-Świerk, Poland.

KUPCZYŃSKI, M., *Institute of Theoretical Physics, Warsaw University*, Hoża 69, 00-681 Warszawa, Poland.

KUSZELL, A., *Institute for Nuclear Research*, Hoża 69, 00-681 Warszawa, Poland.

LASSNER, G., *Sektion Mathematik, Karl-Marx-Universität*, Karl-Marx-Platz, 701 Leipzig, German Democratic Republic.

LEGATOWICZ, A., *Institute for Nuclear Research*, Hoża 69, 00-681 Warszawa, Poland.

LICHNEROWICZ, A., *Aire de Physique Mathematique, College de France*, 11, Place Marcellin-Berthelot, 75005 Paris, France.

LUKIERSKI, J., *Institute of Theoretical Physics, University of Wrocław*, Cybulskiego 36, 50-205 Wrocław, Poland.

LUNDBERG, L. E., *Nordisk Inst. Teoret. Atomfysik, Blegdamsvej* 17, 2100 Kobenhavn, Denmark.

ŁAWRYNOWICZ, J., *Institute of Mathematics, Polish Academy of Sciences*, Śniadeckich 8, 00-656 Warszawa, Poland.

ŁOPUSZAŃSKI, J., *Institute of Theoretical Physics, University of Wrocław*, Cybulskiego 36, 50-205 Wrocław, Poland.

ŁUKASZUK, L., *Institute for Nuclear Research*, Hoża 69, 00-681 Warszawa, Poland.

MAĆKOWIAK, J., *Institute of Physics, Nicholas Copernicus University*, Grudziądzka 5, 87-100 Toruń, Poland.

MAJEWSKI, W., *Institute of Physics, Polish Academy of Sciences*, Al. Lotników 32/46, 02-668 Warszawa, Poland.

MAURIN, K., *Department of Mathematical Physics, Warsaw University*, Hoża 74, 00-682 Warszawa, Poland.

MAYNE, F., *Faculté des Sciences, Service de Chimie Physique 2, Université de Bruxelles*, Av. F. D. Roosevelt 50, 1050 Bruxelles, Belgium.

MICHEL, L., *Institute des Hautes Etudes Scientifiques*, Route de Chartres, 91440 Bures-sur-Yvette, France.

MICKELSSON, J., *University of Jyvaskyla, Department of Mathematics*, Sammonkatu 6, 40100 Jyvaskyla, Finland.

MIELNIK, B., *Institute of Theoretical Physics, Warsaw University*, Hoża 69, 00-681 Warszawa, Poland.

MIHUL, E., *Department of Theoretical Physics, University of Bucharest*, Str. Academiei 14, Bucharest, Rumania.

MIKA, J., *Institute for Nuclear Research*, 05-400 Otwock-Świerk, Poland.

MOSTOWSKI, J., *Institute of Physics, Polish Academy of Sciences*, Al. Lotników 32/45, 02-668 Warszawa, Poland.

MOZRZYMAS, *Institute of Theoretical Physics, University of Wrocław*, Cybulskiego 36, 50-205 Wrocław, Poland.

NAGEL, B., *Institute of Theoretical Physics, Royal Institute Tech.*, Lindstedtsvagen 15, 10044 Stockholm, Sweden.

NAPIÓRKOWSKI, K., *Department of Mathematical Physics, Warsaw University*, Hoża 74, 00-682 Warszawa, Poland.

NAPIÓRKOWSKI, M., *Department of Mathematical Physics, Warsaw University*, Hoża 74, 00-682 Warszawa, Poland.

NEUMANN, H., *Institut für Theoretische Physik, Universität Marburg*, Renthof 7, 3550 Marburg, Federal Republic of Germany.

NIEDERLE, I., *Czechoslovak Academy of Sciences*, Na Slovance 2, 0116 Praha, Czechoslovakia.

NILSSON, J., *Institute of Theoretical Physics*, Fack, 40220 Goteborg 5, Sweden.

OLSZEWSKI, J., *Theoretical Physics Department, Jagellonian University*, Reymonta 4, 30-059 Kraków, Poland.

O'RAIFEARTAIGH, L., *School of Theoretical Physics, Dublin Institute for Advanced Studies*, 10, Burlington Rd., Dublin 4, Ireland.

OZIEWICZ, Z., *Institute of Theoretical Physics, University of Wrocław*, Cybulskiego 36, 50-205 Wrocław, Poland.

PAWLIK, B., *Department of Mathematical Physics, Warsaw University*, Hoża 74, 00-682 Warszawa, Poland.

PECZKIS, M., *Institute of Theoretical Physics, Warsaw University*, Hoża 69, 00-681 Warszawa, Poland.

PIASECKI, J., *Institute of Theoretical Physics, Warsaw University*, Hoża 69, 00-681 Warszawa, Poland.

PINDOR, M., *Institute of Theoretical Physics, Warsaw University*, Hoża 69, 00-681 Warszawa, Poland.

PIRON, C., *Department de Physique Theorique, Université de Genève*, Boulevard d'Yvoy 32, 1211 Genève 4, Switzerland.

PISKOREK, A., *Department of Mathematical Physics, Warsaw University*, Hoża 74, 00-682 Warszawa, Poland.

PRUSKI, S., *Institute of Physics, Nicholas Copernicus University*, Grudziądzka 5, 87-100 Toruń, Poland.

PUSZ, W., *Department of Mathematical Physics, Warsaw University*, Hoża 74, 00-682 Warszawa, Poland.

RAYSKI, J., *Theoretical Physics Department, Jagellonian University*, Reymonta 4, 30-059 Kraków, Poland.

RAYSKI, J., jr., *Theoretical Physics Department, Jagellonian University*, Reymonta 4, 30-059 Kraków, Poland.

RĄCZKA, R., *Institute for Nuclear Research*, Hoża 69, 00-681 Warszawa, Poland.

REBIESA, P., *Theoretical Physics Department, Jagellonian University*, Reymonta 4, 30-059 Kraków, Poland.

REEH, H., *Institut für Theoretische Physik, Universität Göttingen*, Bunsenstrasse 9, 3400 Göttingen, Federal Republic of Germany.

REK, Z., *Institute for Nuclear Research*, Hoża 69, 00-681 Warszawa, Poland.

ROEPSTORFF, G., *II. Institut für Theoretische Physik, Universität Hamburg*, Luruper Chaussee 149, 2000 Hamburg, Federal Republic of Germany.

RÜHL, W., *Fachbereich Physik, Universität Trier-Kaiserslautern*, Pfaffenbergstrasse 95, 6750 Kaiserslautern, Federal Republic of Germany.

RZARZEWSKI, K., *Institute of Theoretical Physics, Warsaw University*, Hoża 69, 00-681 Warszawa, Poland.

SADOWSKI, P., *Department of Mathematical Physics, Warsaw University*, Hoża 74, 00-682 Warszawa, Poland.

SAWICKI, M., *Institute of Theoretical Physics, Warsaw University*, Hoża 69, 00-681 Warszawa, Poland.

SCHRADER, W., *Institut für Theoretische Physik, Freien Universität Berlin*, Arnimallee 3, 1000 Berlin 33, West Berlin.

SEGAL, I., *Department of Mathematics, Massachusetts Institute of Technology*, Cambridge, Ma. 02139, USA.

SEILER, R., *Institut für Theoretische Physik, Freien Universität Berlin*, Arnimallee 3, 1000 Berlin 33, West Berlin.

SENATORSKI, A., *Institute for Nuclear Research*, Hoża 69, 00-681 Warszawa, Poland.

SEWERYŃSKI, M., *Institute for Nuclear Research*, Hoża 69, 00-681 Warszawa, Poland.

SIEGEL, W., *Theoretical Physics Department, Jagellonian University*, Reymonta 4, 30-059 Kraków, Poland.

SIMMS, D., *Mathematics Department, Trinity College*, Dublin 2, Ireland.

SIMON, J., *Laboratoire Phys.-Mathematique, Faculté des Sciences Mirande, Université de Dijon, Campus Universitaire*, 21000 Dijon, France.

SKAGERSTAM, B., *Institute of Theoretical Physics*, Fack, 40220 Goteborg 5, Sweden.

SKORUPSKI, A., *Institute for Nuclear Research*, Hoża 69, 00-681 Warszawa, Poland.

SNELLMAN, H., *Institute of Theoretical Physics, Royal Institute of Technology*, Lindstedtsvagen 15, 10044 Stockholm, Sweden.

SOKOŁOWSKI, L., *Theoretical Physics Department, Jagellonian University*, Reymonta 4, 30-059 Kraków, Poland.

SOMMER, G., *Institut für Theoretische Physik, Universität Bielefeld*, Herforder Str. 28, 4800 Bielefeld, Federal Republic of Germany.

STERNHEIMER, D., *Centre de Physique Theorique, C. N. R. S.*, 31, Ch. J. Aiguier 13274 Marseille 9, CEDEX 2, France.

STRASSBURGER, A., *Department of Mathematical Physics, Warsaw University*, Hoża 74, 00-682 Warszawa, Poland.

STREIT, L., *Institut für Theoretische Physik, Universität Bielefeld*, Herforder Str. 28, 4800 Bielefeld, Federal Republic of Germany.

STROCCHI, F., *Istituto di Fisica, Scuola Normale Superiore*, Piazza dei Cavalieri 7, 56100 Pisa, Italy.

STROM, S., *Institut of Theoretical Physics*, Fack, 40220 Goteborg 5, Sweden.

SUCKEWER, S., *Institute for Nuclear Research*, Hoża 69, 00-681 Warszawa, Poland.

SUDARSHAN, C., *Center for Particle Theory, University of Texas*, Austin, Tx. 78712, USA.

SZCZYRBA, I., *Department of Mathematical Physics, Warsaw University*, Hoża 74, 00-682 Warszawa, Poland.

SZYMACHA, A., *Institute of Theoretical Physics, Warsaw University*, Hoża 69, 00-681 Warszawa, Poland.

SZYMAŃSKI, Z., *Institute for Nuclear Research*, Hoża 69, 00-681 Warszawa, Poland.

ŚWIĘCKI, M., *Institute for Nuclear Research*, Hoża 69, 00-681 Warszawa, Poland.

TAFEL, J., *Institute of Theoretical Physics, Warsaw University*, Hoża 69, 00-681 Warszawa, Poland.

TATUR, S., *Institute of Theoretical Physics, Warsaw University*, Hoża 69, 00-681 Warszawa, Poland.

THOM, R., *Institut des Hautes Etudes Scientifiques*, Route des Chartres, 91440 Bures-sur-Yvette, France.

TODOROV, I. I., *Institute of Physics, Bulgarian Academy of Sciences*, Boul. Lenin 72, Sofia 13, Bulgaria.

TRAUTMAN, A., *Institute of Theoretical Physics, Warsaw University*, Hoża 69, 00-681 Warszawa, Poland.

TRAUTMAN, R., *Institute of Physics, Polish Academy of Sciences*, Al. Lotników 32/46, 02-668 Warszawa, Poland.

TURKO, L., *Institute of Theoretical Physics, University of Wrocław*, Cybulskiego 36, 50-205 Wrocław, Poland.

TURSKI, Ł., *Institute of Theoretical Physics, Warsaw University*, Hoża 69, 00-681 Warszawa, Poland.

UHLENBROCK, D., *Institut für Theoretische Physik, Freien Universität Berlin*, Arnimallee 3, 1000 Berlin 33, West Berlin.

UHLMANN, A., *Sektion Physik, Karl-Marx-Universität*, Karl-Marx-Platz, 701 Leipzig, German Democratic Republic.

URBAŃSKI, P., *Department of Mathematical Physics, Warsaw University*, Hoża 74, 00-682 Warszawa, Poland.

VELO, G., *Istituto di Fisica, Universita di Bologna*, Vie Irnerio 46, 40126 Bologna, Italy.

VOROS, A., *Department de Physique Theorique, Centre d'Etudes Nucleaires*, Saclay, B. P. No. 2, 91190 Gif-sur-Yvette, France.

WAWRZYŃCZYK, A., *Department of Mathematical Physics, Warsaw University*, Hoża 74, 00-682 Warszawa, Poland.

WEHRL, A., *Institut für Theoretische Physik, Universität Wien*, Bolzmanngasse 5, 1090 Wien, Austria.

WERLE, J., *Institute of Theoretical Physics, Warsaw University*, Hoża 69, 00-681 Warszawa, Poland.

WILK, G., *Institute for Nuclear Research*, Hoża 69, 00-681 Warszawa, Poland.

WIZIMIRSKI, Z., *Department of Mathemaitcal Physics, Warsaw University*, Hoża 74, 00-682 Warszawa, Poland.

WODKIFWICZ, K., *Institute of Theoretical Physics, Warsaw University*, Hoża 69, 00-681 Waiszawa, Poland.

WORONOWICZ, L., *Department of Mathematical Physics, Warsaw University*, Hoża 74, 00-682 Warszawa, Poland.

WRZECIONKO, J., *Institute for Nuclear Research*, Hoża 69, 00-681 Warszawa, Poland.

WYCECH, S., *Institute for Nuclear Research*, Hoża 69, 00-681 Warszawa, Poland.

YNGVASON, J., *Institut für Theoretische Physik, Universität Göttingen*, Bunsenstrasse 9, 3400 Göttingen, Federal Republic of Germany.

ZEIDLER, E., *Sektion Mathematik, Karl-Marx-Universität*, Karl-Marx-Platz, 701 Leipzig, German Democratic Republic.

ŻAKOWICZ, W., *Institute for Nuclear Research*, Hoża 69, 00-681 Warszawa, Poland.

ŻELAZNY, R., *Institute for Nuclear Research*, 05-400 Otwock-Świerk, Poland.

PART ONE

QUANTUM FIELD THEORY

THE VALUE AND THE SCOPE OF QUANTUM FIELD THEORY

JOHN R. KLAUDER

Bell Laboratories, Murray Hill, New Jersey 07974, U.S.A.

It should be apparent to the members of this distinguished audience that the last decade has seen a rapid growth and expansion of various disciplines in mathematical physics and especially those that are rooted in the fascinating problems associated with the quantum theory of an infinite number of degrees of freedom. This very conference is testimony to the vitality and conviction felt among those who study mathematical physics, and while this gathering is not solely confined to quantum field theory, there is little doubt in my mind that directly or indirectly quantum field theory provides a principal force of attraction which draws the attendees to such a meeting. This attractive force is strong enough to draw those whose primary interest is explaining the observed natural phenomena of the elementary particle world, as well as those whose interest lies in fundamental mathematical questions – and of course the entire spectrum of people whose interests lie in between.

From my way of thinking, the quantum theory of fields interpreted in its broad sense represents the most profound and comprehensive theoretical framework for the description of physical reality that has been advanced so far. Successful applications of the methods developed in the general theory have been made in solid state, liquid helium, superconductivity, critical phenomena, etc. Yet corresponding successes in the realm of elementary particle physics are for the most part conspicuously absent. Periodically, enthusiasm among elementary particle physicists runs high as new ideas are announced, analyzed and assimilated; and in fact this process is actively occurring at the present time as characterized by the catch words of non-Abelian gauge theories, spontaneous symmetry breaking, unified theories of weak and electromagnetic interactions, asymptotic freedom – to name just a few. Nevertheless, in spite of more than 40 years of study, of myriads of models and methods, we still know remarkably little about theories of genuine physical interest and how they fit in the general scheme of things.

Personally, I feel that the needs of this field can be profitably served from a suitably balanced position based on mathematics and physics. Here 'suitably balanced' suggests a proper mix on the one hand of physical motivation, insight and heuristic proposals and on the other hand of mathematical rigor and independent developments, and especially with a joint regard for logical structure and consistency. Classic examples of mutual enrichment of mathematics and physics are: Newton's invention of calculus for mechanics; Maxwell's completion of the wave equation; the development of non-Euclidean geometry and tensor calculus, and its cross fertilization with general realtivity; the introduction of Hilbert space and the great stimulus for the development of operator theory given by quantum mechanics; the Heaviside calculus and Dirac δ-function as preludes to the development of distribution theory; the application of group theory, algebraic techniques and especially C^*-algebraic approaches in quantum theory; the parallel development of the probability theory of stochastic variables having its own physical roots and its intimate connection with quantum theory; and the developments carried out in a relativistic framework under the general heading of axiomatic quantum field theory which were stimulated by problems in conventional quantum field theory. One must marvel at the progress achieved by such cross fertilization, but recognize that many problems still remain especially in the area of quantum field theory.

Faced with an unknown object of interest all kinds of information are potentially relevant. Consider the classic example of the blind people who try to describe an elephant based on their individual impressions, or of a paleontologist who attempts to reconstruct an extinct ancestor of the elephant from various bits and pieces at his disposal. In so doing consistency of individual elements is essential in order to piece together a larger structure. And in the same way mathematical consistency of individual elements is essential in order to piece together any meaningful physical picture. In view of our present lack of knowledge, I feel that quantum field theory should be pursued and studied in a diverse family of models so as to create a sufficiently broad collection of consistent individual elements. In this view quantum field theory is 'simply' the quantum theory of an infinite number of degrees of freedom interacting in essentially any fashion subject only to a few general quantum mechanical principles. Needless to say some models have more value than others in that they may be closer to reality or may shed insight of a theoretical nature, but of course such differences are to be expected.

Even though the scope of quantum field theory is taken as exceedingly broad, it is fitting and proper that the standards of development and interpretation be set at the highest possible level. In particular I hold that the *arena* in which quantum field theory takes place is a (separable) Hilbert space with a positive-definite metric, which is nothing more, as I see it, than the vision of our founding fathers (e.g., Heisenberg, Jordan, Dirac, Wigner, von Neumann, Pauli, and Yukawa, to name a few). This elegant arena may have been somewhat eroded under periodic attacks, but it has weathered the ravages of time remarkably well. To encompass manifestly covariant descriptions of higher spin massless gauge fields, a little indefiniteness may be needed in the Hilbert space metric. However, I feel that the appeal to indefinite metrics to avoid ultraviolet divergence difficulties – no matter how noble the intentions and elegant the mathematics may be – can compromise fundamental issues involved in the physics of ultraviolet problems. Perhaps it isn't the mathematics that needs a life-giving transfusion but the physics!

The successes of the traditional approach to quantum field theory are of course thoroughly documented, so much so in fact that the methods of the traditional approach often take on the aura of dogma. In one form these articles of faith read as follows: Adopt a complete set of kinematical variables and choose a Hamiltonian defined thereon suggested by an appropriate classical (or at least c-number) theory. For a system with an infinite number of degrees of freedom use an approximating sequence of systems each of a finite number of degrees of freedom, and proceed as above. This is the essence of the statement, and of course variations, elaborations and extensions abound. Prominent among these are path integral approaches, generating functionals, functional differential formulations, real and imaginary time formulations, regularization and renormalization, all capped off with perturbative or perhaps nonperturbative calculational techniques.

Unfortunately, so few problems can really be solved completely that approximations are introduced (e.g., truncating the coupled Green's function equations) and consistency of the original system of equations becomes very hard to test. Simple, soluble systems are evidently valuable since they serve to demonstrate consistency of the methods even if they do not correspond to physical reality. Traditional wisdom asserts that models are to be defined through methods of the traditional approach in which the approximating sequence is generated by truncating the interaction term to a finite number of degrees of freedom, and each such system is then defined in the natural way using an irreducible operator representation at each stage (hereafter called the 'standard way'). For the sake of an analogy I should like to develop, let us regard the free theory as our Sun and theories approachable in the standard way as other stars in our galaxy, the Milky Way. The simplest model in our galaxy is our Sun, namely just the free theory, and of course there are also many important models of considerable complexity. In the interest of clarity it may be helpful to comment on a few primitive examples in our galaxy that have some simple yet significant properties.

Perhaps the easiest way to characterize any model is to give a Hamiltonian that is meaningful classically and which would serve as a natural starting point in proposing a formal quantum Hamiltonian with which to begin. In the relations that follow all sums cover an infinite range.

Simple Models in the Milky Way

Primitive stages of stellar development

(1) $$\mathscr{H} = \tfrac{1}{2}\sum [P_n^2 + m^2(Q_n - c)^2 + \lambda Q_n^4].$$

Remarks. Product problem; exhibits inequivalent canonical commutation relation (CCR) representations for distinct m, λ, c values.

(2) $$\mathscr{H} = \tfrac{1}{2}\sum [P_n^2 + (Q_n - Q_{n+1})^2 + m^2 Q_n^2].$$

Remarks. Prototype of spacial coupling (momentum space cutoff); implies free field is globally inequivalent for distinct m.

(3) $$\mathscr{H} = \tfrac{1}{2}\sum [P_n^2 + (k_n^2 + m^2)Q_n^2].$$

Remarks. Prototype of spacial coupling (configuration space cutoff); field equivalent for distinct m provided $\sum (1+k_n^4)^{-1} < \infty$, otherwise inequivalent; implies free fields of distinct mass are locally equivalent for space dimension $s \leqslant 3$ but not for $s \geqslant 4$.

(4) $$\mathscr{H} = \tfrac{1}{2}\sum [P_n^2 + (Q_n + g\psi^\dagger\psi)^2] + \mu\psi^\dagger\psi.$$

Remarks. Prototype of static nucleon $\left(\text{e.g., } \psi^\dagger\psi = \begin{bmatrix} 1 & 0 \\ 0 & 0 \end{bmatrix}\right)$; field involves direct sum of inequivalent CCR representations; expansion of quadratic term exhibits infinite mass renormalization of nucleon mass due to self energy corrections from prototype Yukawa interactions; also elementary prototype example of field representations that arise in the infrared treatment of Blanchard, and of Kulish and Faddeev.

An important feature that has appeared in the above discussions pertains to the field operator representation, or more precisely to the representation of the CCR. Just as with the rotation group, the choice among inequivalent representations is important, and in general it is not clear *ab initio* how to pick the relevant representation since the proper choice depends on the Hamiltonian. Given that the kinematics is indeed interwoven with the dynamics, the successes of the Green's function approach may be understood since the dynamical solution and representation problem are both solved simultaneously (even if one does not know that it is happening!). It is important to understand, however, that even this method must cope with an infinite number of degrees of freedom, a problem that is treated in the Green's function method by implicit use of the standard way.

Let me return to my stellar analogy. The stars in our galaxy, the Milky Way, have been defined in the standard way and are, in a sense, centered about our Sun, the free theory. Turn off the coupling of the nonlinearity and one passes through the galaxy from model to model finally reaching the Sun at zero coupling. Traditional dogma would say that our galaxy is the universe; there is no more, and we are the center of it all. Embedded as we are deep in our own galaxy it is hard to challenge that concept. The great Polish astronomer Nicholas Copernicus faced similar preconceptions in his own day, and 500 years after his birth it is time to challenge them once again in a more modern context. I hold that there are indeed galaxies of models separate from the Milky Way that are so remote they cannot be approached in the standard way. If models are represented as stars, then how does one tell whether a speck of light is a star in our own galaxy or represents the light from a distant galaxy? One proven way in astronomy and elsewhere is to choose a very special star, e.g., a model of distinguished and exceptional symmetry, that admits a solution so that the distance to it and to the associated stars of its galaxy can be determined. The other models of a distant galaxy are determined in a *revised*, or *non*standard way that is based on the operator representation dictated by the distinguished star and by sequences composed of finite deviations from that particular star.

The key to establishing the existence of a distant galaxy is the finding of a model that cannot be approached in the standard way. With this in mind, let us focus on two distinguished stars of exceptional symmetry.

Simple Models in Distant Galaxies

Models of exceptional symmetry

(1) $\mathscr{H} = \frac{1}{2}\sum (P_n^2+m^2Q_n^2)+\lambda\left(\sum Q_n^2\right)^2.$

Remarks. High symmetry (rotational symmetry); like one version of the spherical model in statistical mechanics, unapproachable in the standard way unless $\lambda \sim$ (number of degrees of freedom)$^{-1}$ which leads to *free* theory with possible shifted mass; symmetry and nonfree theory *demand* reducible CCR representation, and time zero field not cyclic; implications: $\mathscr{H}$ involves other operators and is *not* given as above; path integral *not* given with above Hamiltonian; variant formulations inappropriate; the straitjacket of irreducible CCR representations is too tight – in short, 'Eq. (1)' is already in error! Yet soluble and nonfree. (Solution easily extends to $(\sum Q_n^2)^{p/2}$, $p > 2$, interaction, but no new insight results.)

Such unorthodox behavior does not occur for the interaction Q_1^4 or $\sum Q_n^4$ that belong to the galaxy of the free theory and it means that we are dealing with a *new* galaxy, the RS-galaxy, named for its defining model (RS = rotational symmetry). (An important additional characterizing feature of this and other distant galaxies is given below.) The interaction $(\sum Q_n^2)^2+Q_1^2Q_2^2$ represents just one of the numerous models that make up the RS-galaxy.

In the context in which the free term involves the field in the form $\sum Q_n^2$, such as we have been considering so far, it is interesting to observe one fact about the present case. Consider the class of quartic potentials for which

$$\sum C_{klmn}Q_kQ_lQ_mQ_n \leqslant k\left(\sum Q_n^2\right)^2$$

for some finite k, and for the least such k define

$$v \equiv \frac{\sum C_{klmn} Q_k Q_l Q_m Q_n}{k(\sum Q_n^2)^2} \leqslant 1 .$$

As the Q-values vary, the quantity v is an indicator of where the potential is weak or comparable to the (square of the) free term. Of all potentials of this type the RS-potential has $v \equiv 1$ and is thus everywhere maximal. Stated otherwise, any other potential with finite k is dominated by the RS-potential, the distinguished star of the RS-galaxy.

$$(2) \qquad \mathscr{H} = \int \{\tfrac{1}{2}[\pi^2(\underset{\sim}{x}) + m^2\varphi^2(\underset{\sim}{x})] + \lambda\varphi^4(\underset{\sim}{x})\}\, d\underset{\sim}{x}$$

related by:

$$\varphi(\underset{\sim}{x}) = \sum h_n(\underset{\sim}{x}) Q_n, \qquad \pi(\underset{\sim}{x}) = \sum h_n(\underset{\sim}{x}) P_n$$

and

$$C_{klmn} \equiv \int h_k(\underset{\sim}{x}) h_l(\underset{\sim}{x}) h_m(x) h_n(\underset{\sim}{x})\, d\underset{\sim}{x} .$$

Remarks. High symmetry (ultralocal symmetry); standard way of approach yields no nontrivial solution; symmetry and nonfree theory demand no CCR – in fact field has no conjugate momentum although vacuum cyclic for field; implications: $\mathscr{H}$ is *not* given as above; path integral *not* given with above Hamiltonian, etc.; the straitjacket of *any* CCR representation is too tight – 'Eq. (1)' is in error! Yet soluble and nonfree (noncanonical dimensions, etc.) for any number of space dimensions, $s \geqslant 1$. (Solution readily extends to $\int |\varphi(\underset{\sim}{x})|^p d\underset{\sim}{x}$, $p > 2$, interaction, but no new insight results.)

This example suggestively characterizes yet another isolated galaxy, the UL-galaxy (UL = ultralocal). One of the many other models in this galaxy is that given by the interaction $\int \varphi^4(\underset{\sim}{x})\, d\underset{\sim}{x} + Q_1^2 Q_2^2$. Note for this e xample *no* finite k-bound exists, i.e., there exist regions or points o Q-values where

$$\int \varphi^4(\underset{\sim}{x})\, d\underset{\sim}{x} = \sum C_{klmn} Q_k Q_l Q_m Q_n = \infty$$

even though

$$\int \varphi^2(\underset{\sim}{x})\, dx = \sum Q_n^2 < \infty .$$

In summary, we have discussed some examples of our own galaxy, and indicated that there were some models that were unapproachable in the standard way and are therefore to be indentified with distant galaxies.

Where do the covariant theories with Hamiltonian

$$\mathscr{H} = \int \{\tfrac{1}{2}\pi^2(\underset{\sim}{x}) + (\nabla\varphi)^2(\underset{\sim}{x}) + m^2\varphi^2(\underset{\sim}{x})] + \lambda\varphi^4(\underset{\sim}{x})\}\, d\underset{\sim}{x}$$

fit in? For space dimension $s = 1$ and 2 rigorous results of the constructive quantum field theory school say these models may be approached in the standard way and thus represent stars that lie in our galaxy, the Milky Way. For $s = 3$, arguments of renormalized perturbation theory suggest that this model is also in our galaxy (in the halo?). For $s \geqslant 4$, these are nonrenormalizable models and little is known.

The relativistic models for which the space dimension $s \geqslant 4$ may be dismissed as being unphysical, but that just avoids the issues. Such models must have solutions and it is important to understand such models theoretically in order to gain proper perspective on the theories of ultimate physical interest. Indeed the need to understand such theories may be far more pragmatic as the following line of argument suggests.

In order to focus more sharply on the relevance of singular models in the general scheme of things let us generalize the form of the interaction term so as to consider the class of Hamiltonians of the form

$$\mathscr{H} = \int \{\tfrac{1}{2}[\pi^2(\underset{\sim}{x}) + (\nabla\varphi)^2(\underset{\sim}{x}) + m^2\varphi^2(\underset{\sim}{x})] + \lambda\,|\varphi(\underset{\sim}{x})|^p\}\, d\underset{\sim}{x}\,.$$

According to renormalized perturbation theory, the separation into renormalizable and nonrenormalizable theories is determined by the inequalities $p \leqslant 2n/(n-2)$ and $p > 2n/(n-2)$ respectively, where $n = s+1$ is the space-time dimension. More specifically, $p < 2n/(n-2)$ corresponds to the superrenormalizable theories while $p = 2n/(n-2)$ corresponds to the renormalizable theories.[1] Imagine that interactions can be defined for nonintegral p (which seems far less preposterous than nonintegral space-time dimension!) and that solutions can be found for all $p > 2$. Given those solutions, what are the continuity properties in the parameter p? Are the solutions of the renormalizable theories given by the limit $p \uparrow 2n/(n-2)$? Does the limit $p \downarrow 2n/(n-2)$ yield the same solutions or does it yield something else entirely? Does the acceptability of either limit in any way depend on the chosen value of λ?

These are deep questions that cannot be answered at the present time. Are they of any value? Absolutely, at least in my opinion, if for no other reason than to assess stability of solutions. Far more important: It may happen, for example, that the limit $p \downarrow 2n/(n-2)$ yields an acceptable solution for large λ while the limit $p \uparrow 2n/(n-2)$ does not. Is this complete nonsense?

The preceding questions asked in the context of relativistic field theory can also be asked – and answered – for a single degree of freedom system having roughly a similar structure. Consider the Hamiltonian given by

$$\mathscr{H} = P^2 + Q^2 + \frac{\lambda}{|Q|^\alpha}$$

and study the behavior of the solutions, i.e., eigenvalues and eigenfunctions, as a function of $\lambda \geqslant 0$ and α (the analog of p above). Results of such a study: If $\alpha \leqslant 2$, solutions can be found that pass to the free theory (harmonic oscillator) as $\lambda \downarrow 0$, while if $\alpha > 2$, this behavior is *not* possible. Instead, for any $\alpha > 2$, as $\lambda \downarrow 0$ the solutions limit to a specific 'pseudo-free' theory having different eigenvalues and eigenfunctions than those of the harmonic oscillator. Moreover, if $\lambda \geqslant 3/4$, the solutions for $\alpha = 2$ are obtained by continuity from $\alpha \downarrow 2$ and in general not from $\alpha \uparrow 2$.

The moral of this story comes through loud and clear. (1) Interactions may be so singular that the solution is not even continuous in the coupling at the origin, and it is suggestive that this behavior is associated with nonrenormalizable theories. (2) To treat such theories perturbatively a fundamentally different 'unperturbed' theory than the free theory is required. (3) It may well be that for large coupling the renormalizable theories resemble nonrenormalizable theories more nearly than they resemble super-renormalizable theories. The value that stems from enlarging the scope of investigation is evident!

In addition, we are at last in a position to see the real significance of a multitude of galaxies, and the relation they have to the Milky Way. Take the RS or UL models and pass to the limit $\lambda \downarrow 0$ attempting to recover the free theory. What results in each case is *not* the free theory but a pseudo-free theory, different in each case and characteristic of the particular model. In other words, in turning off the coupling the models remain in the distant galaxy; they do not suddenly make a discontinuous transition at zero coupling to our Sun in the Milky Way. Certainly such a discontinuous transition is impossible in the physical world of astronomy just as in the mathematical world of field theory models.

How can I stress the importance of this picture strongly enough! Take a formula at random from the vast literature of quantum theory and look at it in this new light. For example, consider the relation

$$\exp\left(-i\lambda V(-i\delta/\delta j)\right)\exp\left(i\int j\Delta_F j\right)$$

which represents a common starting point for study of many problems. The right-hand factor represents the generating functional for time-ordered vacuum expectation values of the appropriate free theory, while the left-hand factor provides the effects of interaction and acts by way of functional derivatives. Making precise the action of V, V^2, ..., V^p, on such functionals is the goal of renormalized perturbation theory; keeping the operation in exponential form in principle covers the cases of a nonconvergent power series in λ; expanding effectively in powers of $\hbar$ leads to the loop expansion; reinterpreting Δ_F and adding an i to V changes this prescription from real time to imaginary time, e.g., from a Minkowski to a Euclidean formulation. But implicit in all this is the assumption that as $\lambda \downarrow 0$ the free results emerge. *That* basic assumption is what I want to challenge for sufficiently singular potentials! In the singular potential case the left-hand factor formally acts as a nontrivial projector in combination with its semi-group properties. It is the action of that projection operation on the free theory solution that formally gives the properties of the 'pseudo-free' theory. Lastly, and again formally, in the so projected subspace, in which the left-hand factor acts as a semi-group, an expansion in λ may well give an accurate asymptotic series.

Heuristically it is easy to see what potentials may have this unusual behavior as opposed to those that behave more conventionally. Simply introduce a functional Fourier transform on the free factor and carry out the functional derivatives as required in the potential term. There results in the integrand two expressions, $W_0 = \int \varphi\Delta_F^{-1}\varphi$ and $V(\varphi)$, which effectively determine the support properties of the integrand. Formally, if $V < \infty$ whenever $W_0 < \infty$, then the support properties ought to be dictated by the free theory in the limit $\lambda \downarrow 0$. If this is *not* the case and V punches some holes in the support of the free theory (like 'Swiss cheese'), then the limit as $\lambda \downarrow 0$ should not give the free theory but formally one with reduced support. In this view, nonrenormalizability is just a consequence of hard cores in path space which are directly responsible for the holes in the support. Application of this simple formal argument to the covariant scalar field *precisely* yields the previously stated division into renormalizable $(p \leqslant 2n/(n-2))$ and nonrenormalizable $(p > 2n/(n-2))$ interactions.

The behavior qualitatively described above applies rigorously to the two distinguished stars of exceptional symmetry that can be solved and shown to lie in distant galaxies. A similar behavior must be considered as a definite possibility for covariant quantum field theories whenever the interaction is strong enough (which means nonrenormalizable and possibly large-coupling renormalizable). In the course of time as we discuss various models of physical value, it is essential that we keep open the possibility that what we are studying is not a nearby star in our own galaxy but instead a star situated in a remote galaxy, which is in fact unapproachable in the standard way, and, in virtuc of its great distance, may be endowed with hitherto undreamed of properties. What a fantastic new view of nature might emerge if we expand the scope of our investigations to encompass not only our own galaxy but the vast number of galaxies that undoubtedly lie beyond.

In the foregoing, I have presented a very personal view of the value and the scope of quantum field theory, and of course the various topics discussed at this conference have shed light on these and related subjects from still other points of view. Let us hope that the climate remains favorable so that the development of mathematical methods of ever increasing penetrating power into the physical world will continue to grow and flourish.

It is a pleasure to thank H. Ezawa for several stimulating discussions that have helped focus on the issues and shape the discussion in a number of places.

Reference

[1] For $n = 2$ all power law interactions are super-renormalizable. If nonrenormalizable interactions exist for $n = 2$, one candidate may be $\lambda\{\exp[\varphi^4(\underset{\sim}{x})]-1\}$. We shall ignore this issue here and confine attention to $n \geqslant 3$.

CLUSTER EXPANSIONS AND THEIR APPLICATION TO THE n-PARTICLE STRUCTURE OF WEAKLY COUPLED QUANTUM FIELD MODELS

JAMES GLIMM*

Institut des Hautes Etudes Scientifiques, 91440 Bures-sur-Yvette, France

It is a goal of mathematical physics to contribute to a useful dialogue between mathematics and physics. In the case of constructive quantum field theory, the contact with mathematics is based on the new mathematical developments which arise in its mathematical formulation, for example in partial differential equations in an infinite number of variables, in C^*-algebras and in random fields. The contact with physics is based on the idea that the models studied have some relation to the interactions of elementary particles, and to the methods commonly used by physicists to study these particles.

In recent years, the $p(\varphi)_2$ model for selfinteracting boson quantum fields in two space-time dimensions has been brought to a fairly complete state. There are five distinct methods for constructing these field theories. Each of these constructions has its own advantages. The first of these constructions [4] was based on estimates of function space integrals, and compactness of C^*-algebra states. This construction is still the most general. The second construction [5], [6] is based on convergent expansions similar to the high temperature expansions of statistical mechanics. This construction provides the most detailed structure of the solutions. The third construction, due to Nelson and Guerra, Rosen and Simon, is the simplest. It has been presented here by F. Guerra. It is based on the principle of monotone convergence and has the advantage of yielding correlation inequalities and the Lee Yang theorem of statistical mechanics. A fourth construction has been announced by Dobrushin and Minlos, and seems to be very well suited to the formulation of uniqueness questions. Theorem 2 below gives a fifth construction.

In this lecture, I will concentrate on the second construction, via cluster expansions. We consider the action

$$\int_{R^2} \left[\tfrac{1}{2}\nabla\varphi(x)^2 + \tfrac{1}{2}m_0^2\varphi(x)^2 + \lambda P(\varphi(x))\right] dx\,.$$

Suppose λ_0, m_0 are allowed to be complex numbers in the region

$$\text{(1)} \quad \operatorname{Re}\lambda_0 \geqslant 0, \quad \operatorname{Arg} m_0 \text{ small}, \quad |\lambda_0|/|m_0|^2 \ll 1.$$

THEOREM 1 ([5], [6]). For complex λ_0, m_0 as in (1), the $p(\varphi)_2$ model can be constructed, using convergent series expansions. It satisfied the Osterwalder–Schrader and Wightman axioms.

This result has been extended by Spencer [7], who showed that the addition of a large external field to an arbitrary interaction:

$$p(\varphi)+\mu_0\varphi, \quad \mu_0 \gg 1$$

also gives a convergent series expansion for the solution.

The next result is uniqueness. It is due to Eckmann, Magnen and Sénéor.

THEOREM 2 ([3]). The $(\varphi)_2^4$ model, for small coupling, is Borel summable.

This result is the most satisfying form of uniqueness. It states that by the method of Borel summation, the $(\varphi)_2^4$ model can be constructed from its formal perturbation series expansion.

The perturbation series is used for calculation by physicists, and this result provides a theoretical basis to justify the calculations. The proof of Theorem 2 is based on two ideas. The first is a slight enlargement of the region of analyticity in (1), and the second is improved estimates, with strong decay[1] properties, for the Schwinger functions, defined by Theorem 1.

Probably the most important application of the cluster expansion is to the particle structure, since the particles (rather than the fields) are the experimentally observed quantities.

THEOREM 3 ([5], [6]). For λ_0, m_0 real, $\lambda_0/m_0^2 \ll 1$, the $p(\varphi)_2$ model describes particles. The axioms of the Haag–Ruelle theory are satisfied. The n-particle states exist, and an isometric S matrix exists.

Continuing further in this direction, the most important questions concern bound states and the asymptotic completeness. Correlation inequalities have been used to exclude bound states in some cases [5], [6] and variational calculations indicate their presence in other cases

[5], [6]. A more detailed analysis of bound states, based on properties of the Bethe–Salpeter kernel for small coupling, has been made by Spencer [8], and Spencer and Zirilli [9].

References

* Supported in part by the National Science Foundation, Grant NSF-GP-24003.

[1] Strong cluster decay properties in statistical mechanics are given by M. Duneau, D. Iagolnitzer and B. Souillard [1].

Bibliography

[1] M. Duneau, D. Iagolnitzer and B. Souillard, *Commun. Math. Phys.* **31** (1973) 191.
[2] M. Duneau, D. Iagolnitzer and B. Souillard, *Commun. Math. Phys.* **35** (1974) 307.
[3] J.-P. Eckmann, J. Magnen and R. Sénéor, *Commun. Math. Phys.* **39** (1975) 251.
[4] J. Glimm and A. Jaffe, *Acta Math.* **125** (1970) 203.
[5] J. Glimm, A. Jaffe and T. Spencer, *Annals of Math.* **100** (1974) 585.
[6] J. Glimm, A. Jaffe and T. Spencer, in: *Constructive Quantum Field Theory* (edited by G. Velo and A. S. Wightman), Lecture Notes in Physics, Springer Verlag, Berlin 1973.
[7] T. Spencer, 'The Mass Gap for the $p(\varphi)_2$ Quantum Field Model with a Strong External Field', *Commun. Math. Phys.* **39** (1974) 63.
[8] T. Spencer, *Commun. Math. Phys.* **44** (1975) 143.

EUCLIDEAN QUANTUM FIELD THEORY

FRANCESCO GUERRA*

Istituto di Fisica dell' Università di Salerno, Italy

1. Introduction

In this talk I will deal mainly with two aspects of Euclidean quantum field theory, one related to the problem of mathematical construction, the other related to the problem of physical interpretation.

For the problem of the construction I will emphasize methods based on the deep analogy between Euclidean quantum field theory and classical statistical mechanics, which are the core of a program advocated and developped in joint work [27] with Rosen and Simon in the last two years. In particular I will describe the lattice approximation strategy and its applications.

For the problem of physical interpretation I will sketch the main ideas of a proposal [31], made by Ruggiero and myself, according to which the Euclidean field theory can be interpreted as a stochastic field theory in the physical Minkowski space-time.

During the talk I will try to describe the main physical and structural ideas stating some of the results in detail but keeping all technicalities to a bare minimum. In any case I will give detailed references to the existing literature, where more complete considerations and full proofs can be found.

The organization of the talk is as follows. In Section 2 we introduce the general structure of Euclidean quantum field theory of Boson systems, following ideas advocated by Symanzik [62, 63, 64] and Nelson [38, 39, 40]. We show also the connection of the Euclidean structure with classical statistical mechanics. In Section 3 we describe the lattice approximation strategy. This strategy is based on the idea that Euclidean field theory can be approximated by a lattice of continuous spins interacting through a nearest neighbor ferromagnetic coupling. In Section 4 we give some applications and results especially in relation to the infinite volume limit and the problem of phase transitions. In Section 5 we deal with the new proposed physical interpretation. For the basic concepts of probability theory and stochastic processes we refer to [9] and [47].

2. The Structure of Euclidean Quantum Field Theory and its Connection with Classical Statistical Mechanics

Firstly we consider free fields on the Euclidean space $\boldsymbol{R}^d$, where d is the number of space-time dimensions ($d = 4$ is the physical case). We introduce the free two-point Schwinger function

$$S(x-y) = \frac{1}{(2\pi)^d} \int \frac{e^{ip\cdot(x-y)}}{p^2+m^2} dp, \qquad x, y, p \in \boldsymbol{R}^d,$$

which satisfies the differential equation

$$(-\Delta+m^2)\, S(x-y) = \delta^{(d)}(x-y),$$

where Δ is the Laplacian in d dimensions.

Then we define the Sobolev–Hilbert space N of real tempered distributions f on $\boldsymbol{R}^d$, with symmetric scalar product

$$\langle f, g\rangle_N = \langle f, (-\Delta+m^2)^{-1} g\rangle,$$

where $\langle\ ,\ \rangle$ is the usual Lebesgue scalar product on Fourier transforms. Formally one has

$$\langle f, g\rangle_N = \iint f(x)\, S(x-y)\, g(y)\, dx\, dy.$$

DEFINITION 1. The *free Euclidean–Markov field* is the real Gaussian random field $\varphi(f)$, indexed by N and defined by the expectations

$$E(\varphi(f)) = 0, \qquad E(\varphi(f)\varphi(g)) = \langle f, g\rangle_N.$$

Formally if we write

$$\varphi(f) = \int f(x)\, \varphi(x)\, dx,$$

then

$$E(\varphi(x)\varphi(y)) = S(x-y),$$

therefore the free Schwinger function S is the covariance of the Euclidean field φ.

We call (Q, Σ, μ) the underlying probability space [9]. Then the fields $\varphi(f)$ are represented by $L^p(Q, \Sigma, \mu)$ functions on Q, $1 \leqslant p < \infty$, which we still call $\varphi(f)$, and the expectations are expressed as integrals

$$E(\varphi(f_1) \dots \varphi(f_n)) = \int_Q \varphi(f_1) \dots \varphi(f_n)\, d\mu.$$

One peculiar feature of the field introduced in the Definition 1 is the Markov property isolated by Nelson [38, 39, 40]. Let us associate to each closed region A of $\boldsymbol{R}^d$ the sub-σ-algebra Σ_A of Σ generated by fields $\varphi(f)$ with $\operatorname{supp} f \subseteq A$. We call E_A the *conditional expectation with respect to* Σ_A. Then we have

PROPOSITION 2. Let π be a smooth $(d-1)$-dimensional closed manifold dividing $\boldsymbol{R}^d$ in two closed regions A and B such that $A \cup B = \boldsymbol{R}^d$ and $A \cap B = \pi$. Then

$$E_\pi = E_A E_B .$$

This property is a generalization of the usual Markov property of stochastic processes. Roughly speaking it tells us that all mutual informations transmitted from the region A to B and vice versa are 'felt' by the separating region π.

When we introduce the interaction we must consider firstly the field restricted to some region Λ of $\boldsymbol{R}^d$ and then consider the limit as $\Lambda \to \infty$ (thermodynamic limit). Therefore we are allowed to choose various boundary conditions on the boundary $\partial\Lambda$ of Λ. To specify the boundary conditions we can consider the free Schwinger function $S^X(x, y)$, with $x, y \in \Lambda \subset \boldsymbol{R}^d$ and X specifying the boundary conditions, as solution of the equation

$$(-\Delta_X + m^2) S^X(x, y) = \delta^d(x-y),$$

where now Δ_X is the Laplacian in Λ with appropriate boundary conditions X on $\partial\Lambda$.

For example we can consider Dirichlet or Neumann boundary conditions ($X = D$ or $X = N$) or, if Λ is a hyperrectangle, periodic boundary conditions ($X = P$), etc. We put $X = \emptyset$ to denote the free Schwinger function $S(x-y)$.

Using S^X we can define the Markov field φ^X in the region Λ with boundary conditions X by specifying

$$E(\varphi^X(x)) = 0, \qquad E(\varphi^X(x)\varphi^X(y)) = S^X(x, y).$$

When the interaction is turned on it is expected that the interacting fields $\hat{\varphi}$ are equal to the free fields as functions on Q space, but there is a change in the measure, so that the expectations of the interacting fields are given by

$$E(\hat{\varphi}(f_1)\ldots\hat{\varphi}(f_n)) = \int_Q \varphi(f_1) \ldots \varphi(f_n)\, d\hat{\mu},$$

where $\hat{\mu}$ is a new measure on Q space associated to the given interaction.

From now on we consider the case of a two-dimensional field theory with the interaction given by a real polynomial P bounded below. This is the so called $P(\varphi)_2$ model for which an impressive amount of informations have been obtained in the last years in the context of the successful program of constructive quantum field theory [66, 37, 11, 14, 15, 48, 54, 59, 67, 17, 27, 18], advanced by Glimm and Jaffe and their followers. For results in $(\varphi^4)_3$ see [12, 16, 6].

Given general boundary conditions X, X', to each compact region Λ of $\boldsymbol{R}^2$ we associate the Euclidean action $U_\Lambda^{(X,X')}$ defined by

$$U_\Lambda^{(X,X')} = \int_\Lambda :P\big(\varphi^X(x)\big):_{X'} dx,$$

where the local limit is obtained through the removal of an ultraviolet cut off and the symbol $:\ :_{X'}$ denotes Wick substractions with respect to the covariance $S^{X'}$. For example

$$:\varphi(x)\varphi(y):_{X'} = \varphi(x)\varphi(y) - S^{X'}(x, y), \text{ etc.}$$

This procedure of Wick substractions eliminates all ultraviolet divergences of the theory and can be introduced in a purely stochastic framework as explained for example in [53] and [40].

The volume cut off expectation values of the interacting field in volume Λ are given in [51, 64]

$$S_\Lambda(f_1, \dots, f_n) = Z_\Lambda^{-1} \int_Q \varphi(f_1) \dots \varphi(f_n) e^{-U_\Lambda} d\mu,$$

$$Z_\Lambda = \int_Q e^{-U_\Lambda} d\mu,$$

where $d\mu$ is the free measure and we have suppressed in the notation all reference to the boundary conditions X, X'. Of particular interest are the cases $X = D, X' = \emptyset$ (Half–Dirichlet) and $X = N$, $X' = \emptyset$ (Half–Neumann).

This expression of the Schwinger functions S of the interacting theory is very similar to the expression of the expectation values in Gibbsian ensembles of classical statistical mechanics. This is the starting point of the analogy of Euclidean field theory with classical statistical mechanics which has been exploited in a series of works [26, 27, 55, 56, 28, 29, 30] in order to use the modern techniques of rigorous statistical mechanics [50] for the study of constructive Euclidean quantum field theory.

In particular we can consider a variational principle for the entropy density, as introduced by Ruelle [49, 50] in statistical mechanics, and the equilibrium equations of the type considered by Dobrushin [2, 3, 4] and Lanford and Ruelle [34], for which we refer to [27].

The quantity Z_Λ plays the role of partition function, therefore it is very natural to consider the limit of $|\Lambda|^{-1}\log Z_\Lambda$ as $\Lambda \to \infty$. Extending earlier work [21, 24, 25, 27] it is possible to prove [28, 29] the following

THEOREM 3. The limit

$$\lim_{\Lambda\to\infty}|\Lambda|^{-1}\log Z_\Lambda = \alpha_\infty$$

exists and is independent of the choice of boundary conditions.

The analogy with classical statistical mechanics can be further deepened if we consider the lattice approximation [27, 42], the subject of our next section.

3. The Lattice Approximation

For the sake of simplicity we consider only the two-dimensional case. We approximate the Euclidean space $\boldsymbol{R}^2$ with the lattice $\boldsymbol{L}_\varepsilon^2$ of spacing ε. If n is a point of the two-dimensional unit lattice $\boldsymbol{Z}^2$, $n = \{n_{(1)}, n_{(2)}\}$, $n_{(i)} = 0, \pm 1, \pm 2, \ldots$, then x_n will be the corresponding point in $\boldsymbol{L}_\varepsilon^2$, with $(x_n)_{(i)} = \varepsilon n_{(i)}$, $i = 1, 2$. The lattice approximation of the integral of a smooth function f on $\boldsymbol{R}^2$ is obviously given by

$$\int f(x)\,dx \simeq \sum_{n\in\boldsymbol{Z}^2} \varepsilon^2 f(x_n).$$

We introduce the operator $A = -\Delta + m^2$ and its lattice approximation $A_{n,n'}$ obtained using the finite difference approximation for the Laplacian. Therefore we have

$$A_{n,n'} = \begin{cases} m^2 + 4\varepsilon^{-2} & \text{if } n = n', \\ -\varepsilon^2 & \text{if } |n-n'| = 1, \\ 0 & \text{otherwise.} \end{cases}$$

Then the lattice Schwinger function is defined by

$$\sum_{n'} A_{n,n'} S_{n',n''} = \varepsilon^{-2}\delta_{n,n''},$$

where the normalization is chosen so that

$$\iint f(x)\, S(x-y) g(y)\, dx\, dy \simeq \sum_{n,n'} \varepsilon^4 f(x_n)\, S_{n,n'} g(x_{n'}).$$

DEFINITION 4. The lattice field φ_n is the real Gaussian random field indexed by Z^2 and defined by the expectations

$$E(\varphi_n) = 0, \qquad E(\varphi_n \varphi_{n'}) = S_{n,n'}.$$

The lattice field φ_n provides the lattice approximation to the Euclidean Markov field $\varphi(f)$. This approximation is particularly useful when bounded regions are considered, so that only a finite number of variables φ_n must be taken into account and the standard representation of Gaussian random variables [9, 47] can be used. Taking also into account the boundary conditions it is possible to prove the following proposition, using methods of [27]. Consider a compact region Λ and let Z^2_Λ be the subset of Z^2 such that, for $n \in Z^2_\Lambda$, $x_n \in \Lambda$. Define

$$U^D_\Lambda(q) = \tfrac{1}{2} \sum_j (\varepsilon^2 m^2 + 4) q_j^2 - \sum_{(jj')} q_j q_{j'},$$

$$U^N_\Lambda(q) = \tfrac{1}{2} \sum_j \varepsilon^2 m^2 q_j^2 - \tfrac{1}{2} \sum_{(jj')} (q_j - q_{j'})^2,$$

where q is a point in the Euclidean space R^K with dimensions K equal to the cardinal of Z^2_Λ, $\sum_j$ is the sum over all $j \in Z^2_\Lambda$ and $\sum_{(jj')}$ is the sum over all couples of nearest neighbors $j, j' \in Z^2_\Lambda$.

PROPOSITION 5. For sufficiently regular functions f, g with support on Λ,

$$\int_\Lambda \int_\Lambda f(x)\, S^X(x, y) g(y)\, dx\, dy$$
$$= \lim_{\varepsilon \to 0} \int_{R^K} \hat{f}(q) \hat{g}(q) e^{-U^X_\Lambda(q)}\, dq \cdot \Big[\int_{R^K} e^{-U^X_\Lambda(q)} dq \Big]^{-1}, \qquad X = N, D,$$

where dq is the Lebesque measure on R^K and $\hat{f}(q) = \sum_n \varepsilon^2 f(x_n) q_n$.

This proposition shows that the lattice cutoff theory is given by a set of Gaussian random variables q_n, each associated to a lattice point, interacting through a nearest neighbor coupling which is of ferromagnetic type in the Dirichlet case and of elastic type in the Neumann case. It is also clear that the expressions of U^N and U^D differ only be terms concentrated at the boundary of Z^2_Λ so that in general the coupling between nearest neighbors can be considered always as ferromagnetic. These considerations extend to other types of boundary conditions [27].

Let us suppose now that we consider the interacting theory as described in the previous section. Then the interesting feature of the lattice approximation is that the introduction of the interaction modifies only the distribution of the single variables q_n, which will be no longer Gaussian, but does not affect the ferromagnetic coupling between nearest neighbors. This fact is related to the locality of the interaction.

In fact, let us modify the expressions of U^X given previously by adding the term

$$\sum_j \varepsilon^2 :P(q_j):,$$

where : : denote Wick subtractions with respect to the covariance $S_{n,n'}$. Then, following [27], we can prove

THEOREM 6. Consider the half-Dirichlet and half-Neumann Schwinger functions S^X, $X = D, N$ in the region Λ. Then

$$S_\Lambda^X(f_1, \ldots, f_s) = \lim_{\varepsilon\to 0} \int_{R^K} \hat{f}_1(q) \ldots \hat{f}_s(q) e^{-U_\Lambda^X(q)} dq \cdot \Big[\int_{R^K} e^{-U_\Lambda^X(q)} dq \Big]^{-1}.$$

Theorem 6 extends obviously to more general boundary conditions.

The structure of the lattice approximation makes possible to prove correlation inequalities of Griffiths–Kelly–Sherman type [20, 33, 10] or of Fortuin–Kasteleyn–Ginibre type [7] well known in statistical mechanics. This possibility is based on the ferromagnetic nature of the nearest neighbor coupling and on the fact that the interaction affects only the distribution of the single variables q_n.

Typical results are the following [27].

THEOREM 7. Consider the $P(\varphi)_2$ Euclidean Markov theory with interaction $P(\varphi) = P_e(\varphi) - \lambda\varphi$, where P_e is even and $\lambda \geqslant 0$. Then, for the volume cut off Schwinger functions with general boundary conditions, one has

$$S_\Lambda(x_1, \ldots, x_n) \geqslant 0$$

and

$$S_\Lambda(x_1, \ldots, x_n) \geqslant S_\Lambda(x_1, \ldots, x_s) S_\Lambda(x_{s+1}, \ldots, x_n)$$

(Griffiths inequalities).

THEOREM 8. For an arbitrary interaction P, if F and G are increasing functions of the fields then they are positively correlated in the sense that

$$\langle FG\rangle \geqslant \langle F\rangle \langle G\rangle,$$

where the averages are taken with respect to the measure $d\hat{\mu} = Z_{\Lambda}^{-1}\, e^{-U_{\Lambda}} d\mu$ for various boundary conditions (F.K.G. inequalities).

The inequalities carry to the infinite volume limit (when it exists).

An interesting application of F.K.G. inequalities has been given by Simon [55] who proved that 'the field couples the vacuum to the first excited state', therefore the mass gap can be found looking at the two-point function. For precise statements and results see [55, 57].

4. Applications to the Infinite Volume Limit

The main problem of the $P(\varphi)_2$ theory is to prove the existence of the thermodynamic limit as $\Lambda \to \infty$, and investigate its properties especially in relation to the existence of multiple phases, the occurrence of a mass gap, the existence of particles, etc.

For this problem the most detailed informations have been obtained in the region of small coupling constant by Glimm, Jaffe and Spencer in a remarkable series of papers [17, 18, 19], using techniques of cluster expansions, reminiscent of those used in classical statistical mechanics for dilute systems. For an up to date account see the contributions of J. Glimm and A. Jaffe to these Proceedings.

On the other hand results related to the existence of multiple phases for large coupling have been announced by Dobrushin and Minlos [5]. In this section we will describe some of the typical results which can be obtained using the lattice approximation strategy.

The problem of the existence of the infinite volume limit can be easily solved using an ingenious remark of Nelson [41, 42] who noticed that for an even theory with Dirichlet boundary conditions the Schwinger functions are increasing in the volume as a consequence of the second Griffiths inequality (Theorem 7), since new ferromagnetic bonds are introduced as the volume is enlarged. Obviously some bounds are necessary on the Schwinger functions to be sure that they do not diverge to ∞. These bounds were established in [27] using the Glimm–Jaffe φ-bounds [13], see also [8]. Slightly more general interactions can also be handled [27], adding a term of the type $\lambda\varphi$. Then the following can be proven [41, 42, 27].

THEOREM 9. The volume cut off half-Dirichlet Schwinger functions S_{Λ}^{D} have a unique limit as $\Lambda \to \infty$, for all interactions of the type $P_e(\varphi)+\lambda\varphi$ where P_e is even and bounded below.

Related results for exponential interactions have been obtained by Albeverio and Høegh-Krohn [1]. By the very structure of the proof of Theorem 9 it is reasonable to expect that it could be extendend also in the case of higher space-time dimensions.

The next problem to consider is to see whether the infinite volume theory so obtained describes a single phase or multiple phases. For the case of polynomials of fourth order Griffiths and Simon [58] have been able to extend the Lee–Yang circle theorem [50] to the Euclidean theory. Then Simon [56] has proven the following

THEOREM 10. For all interactions of the type $a\varphi^4+b\varphi^2+\mu\varphi$, $a > 0$, $\mu \neq 0$, the infinite volume Schwinger functions of Theorem 9 describe a unique phase.

These results are in complete agreement with the conventional wisdom of the Goldstone picture of phase transitions [67]. Using results of Spencer [60] and techniques inspired by recent work of Lebowitz and Penrose [35] it is also possible to prove [29, 30] the following

THEOREM 11. In the conditions of Theorem 10 the infinite volume theory has a mass gap above zero in its spectrum.

For other results we refer to [27, 55, 56, 28, 29, 30].

In conclusion it seems that the consistent use of statistical mechanics ideas, in particular the lattice approximation strategy, has been particularly useful for the study of Euclidean field theory, and that more progress is to be expected in the future.

In the next section we turn to the problem of physical interpretation.

5. The Stochastic Interpretation

From an historical point of view [51, 52, 36, 62, 63] Euclidean field theory has been obtained from Minkowski field theory making an analytic continuation of the Wightman functions [61] to imaginary time in order to get the Schwinger functions.

Therefore the most natural way of using Euclidean field theory in physics is to try to continue back to real time the Schwinger functions. The first results in this direction were obtained by Nelson [38, 39] who isolated a set of properties for the Euclidean Markov fields which are sufficient to derive quantum fields satisfying the Wightman axioms. Subsequently, motivated by the analysis of Nelson, Osterwalder and Schrader considered the more general problem of the reconstruction of a Wightman theory starting from the Schwinger functions directly and without using the additional informations provided by the existence of Euclidean fields. Their interesting results are contained in [45, 46].

In the applications it is also useful the method originally used by Glimm and Spencer in [19].

In this section we will show that it is possible to give a physical interpretation of the Euclidean Markov field theory without making the analytic continuation back with respect to the time variables. This interpretation [31] is based on the extension of Nelson's stochastic mechanics [43, 44] to systems with an infinite number of degrees of freedom [32]. Nelson's stochastic mechanics can be considered as a method of quantization based on the theory of Markov processes and stochastic differential equations. For a general Lagrangian system the classical configuration variables $q_i(t)$ are promoted to a Markov process satisfying stochastic differential equations which describe the kinematical and dynamical behavior of the theory. For all details we refer to the original works [43, 44] and to [31, 32, 22, 23]. As a result of the dynamical assumptions the Markov process $q_i(t)$, corresponding to a given state, has the same distribution, at each fixed time, as the corresponding Heisenberg operators, which evolve according to the quantum unitary evolution. For all interactions of physical interest the stochastic scheme is completely equivalent to the usual quantum scheme. In particular the state of a system is completely specified giving the distribution of $q_i(t)$ at some fixed time and the correlations of the Markov process in the infinitesimal interval $[t, t+dt]$.

Our main result is the following [31, 32, 22, 23].

THEOREM 12. The Euclidean Markov field coincides with the lowest energy generalized stochastic process associated with classical field theory through the procedure of Nelson's stochastic mechanics.

The theorem is proven through explicit computation in the free field case and using the Feynman–Kac formula for stochastic processes in the interacting case.

Since the Markov processes of stochastic mechanics evolve in the physical (real) time, we have that, using the interpretation suggested by Theorem 12, the underlying four-dimensional manifold on which the Euclidean Markov field is defined can be considered as the physical Minkowski space-time and not as a ficititious space-imaginary time.

Poincaré trnsformations for the Markov field can be easily realized, as explained in [31, 32], they are not of purely kinematical nature but have a certain amount of dynamical content.

At present some work is in progress in order to describe the scattering processes directly inside the Euclidean Markov framework.

Reference

* Permanent postal address: Via A. Falcone 70, 80127 Napoli, Italy.

Bibliography

[1] S. Albeverio, and R. Høegh-Krohn, 'The Wightman Axioms and the Mass Gap for Strong Interactions of Exponential Type in Two-Dimensional Space-Time', Oslo preprint, 1973.

[2] R. L. Dobrushin, 'Gibbsian Random Fields for Lattice Systems with Pairwise Interactions', *Funct. Anal. Applic.* **2** (1968) 292.

[3] R. L. Dobrushin, 'The Problem of Uniqueness of a Gibbsian Random Field and the Problem of Phase Transitions', *Funct. Anal. Applic.* **2** (1968) 302.

[4] R. L. Dobrushin, 'Gibbsian Random Fields, The General Case', *Funct. Anal. Applic.* **3** (1969) 22.

[5] R. L. Dobrushin, and R. A. Minlos, 'Construction of a One-Dimensional Quantum Field Via a Continuous Markov Field', Moscow preprint, 1973.

[6] J. Feldman, 'The $\lambda\varphi_3^4$ Field Theory in a Finite Volume', Harvard preprint, 1974.

[7] C. Fortuin, P. Kasteleyn, and J. Ginibre, 'Correlation Inequalities on Some Partially Ordered Sets', *Comm. Math. Phys.* **22** (1971) 89.

[8] J. Fröhlich, 'Schwinger Functions and their Generating Functionals', Harvard preprint, 1973.

[9] I. I. Gikhman, and A. V. Skorokhod, *Introduction to the Theory of Random Processes,* W. B. Saunders Co., Philadelphia, 1969.

[10] J. Ginibre, 'General Formulation of Griffith's Inequalities', *Comm. Math. Phys.* **16** (1970) 310.

[11] J. Glimm, 'Boson Fields with Nonlinear Self-Interaction in Two Dimensions', *Comm. Math. Phys.* **8** (1968) 12.

[12] J. Glimm, 'Boson Fields with the :φ^4: Interaction in Three Dimensions', *Comm. Math. Phys.* **10** (1968) 1.

[13] J. Glimm, and A. Jaffe, 'The $\lambda(\varphi^4)_2$ Quantum Field Theory without Cutoffs IV, Perturbations of the Hamiltonian', *J. Math. Phys.* **13** (1972) 1558.

[14] J. Glimm, and A. Jaffe, 'Quantum Field Models', in *Statistical Mechanics and Quantum Field Theory* (ed. by C. de Witt and R. Stora), Gordon and Breach, New York, 1971.

[15] J. Glimm, and A. Jaffe, 'Boson Quantum Field Models', in *Mathematics of Contemporary Physics* (ed. by R. Streater), Academic Press, New York, 1972.

[16] J. Glimm, and A. Jaffe, 'Positivity of the φ_3^4 Hamiltonian', *Fortschritte der Physik* **21** (1973) 327.

[17] J. Glimm, A. Jaffe, and T. Spencer, 'The Wightman Axioms and Particle Structure in the $P(\varphi)_2$ Quantum Field Model', *Ann. Math.*, to appear.

[18] J. Glimm, A. Jaffe, and T. Spencer, 'The Particle Structure of the Weakly Coupled $P(\varphi)_2$ Model and Other Applications of High Temperature Expansions', Part I: 'Physics of Quantum Field Models', Part II: 'The Cluster Expansion', in [65].

[19] J. Glimm, and T. Spencer, 'The Wightman Axioms and the Mass Gap for the $P(\varphi)_2$ Quantum Field Theory', N.Y.U. preprint, 1972.

[20] R. Griffiths, 'Correlations in Ising Ferromagnets', I, II, III, *J. Math. Phys.* **8** (1967) 478, 484, *Comm. Math. Phys.* **6** (1967) 121.

[21] F. Guerra, 'Uniqueness of the Vacuum Energy Density and Van Hove Phenomenon in the Infinite Volume Limit for Two-Dimensional Self-Coupled Bose Fields', *Phys. Rev. Lett.* **28** (1972) 1213.

[22] F. Guerra, 'On the Connection between Euclidean Markov Field Theory and Stochastic Quantization', to appear in the *Proceedings of the "Scuola Internazionale di Fisica 'Enrico Fermii'"*, Varenna 1973.

[23] F. Guerra, 'On Stochastic Field Theory', in *Proceedings of the IInd Aix-en-Provence International Conference on Elementary Particles, 6–12 September 1973*, Supplement *J. de Physique* **34**, Cl–95, 1973.

[24] F. Guerra, L. Rosen, and B. Simon, 'Melson's Symmetry and the Infinite Volume Behavior of the Vacuum in $P(\varphi)_2$', *Comm. Math. Phys.* **27** (1972) 10.

[25] F. Guerra, L. Rosen, and B. Simon, 'The Vacuum Energy for $P(\varphi)_2$: Infinite Volume Limit and Coupling Constant Dependence', *Comm. Math. Pyhs.* **29** (1973) 233.

[26] F. Guerra, L. Rosen, and B. Simon, 'Statistical Mechanics Results in the $P(\varphi)_2$ Quantum Field Theory', *Phys. Lett.* **44B** (1973) 102.

[27] F. Guerra, L. Rosen, and B. Simon, 'The $P(\varphi)_2$ Euclidean Quantum Field Theory as Classical Statistical Mechanics', *Ann. Math.*, to appear.

[28] F. Guerra, L. Rosen, and B. Simon, 'The Pressure is Independent of the Boundary Conditions for $P(\varphi)_2$ Fields Theories', Salerno–Toronto–Princeton preprint, 1974.

[29] F. Guerra, L. Rosen, and B. Simon, 'Boundary Conditions for the $P(\varphi)_2$ Euclidean Quantum Field Theory', preprint in preparation.

[30] F. Guerra, L. Rosen, and B. Simon, 'Correlation Inequalities and the Mass Gap in $P(\varphi)_2$', III, 'Mass Gap for a Class of Strongly Coupled Theories', preprint in preparation.

[31] F. Guerra, and P. Ruggiero, 'New Interpretation of the Euclidean Markov Field in the Framework of Physical Minkowski Space-Time', *Phys. Rev. Lett.* **31** (1972) 1022.

[32] F. Guerra, and P. Ruggiero, in preparation.

[33] D. Kelly, and S. Sherman, 'General Griffiths' Inequalities in Ising Ferromagnets', *J. Math. Phys.* **9** (1968) 466.

[34] O. Lanford, and D. Ruelle, 'Observables at Infinity and States with Short Range Correlations in Statistical Mechanics', *Comm. Math. Phys.* **13** (1969) 194.

[35] J. Lebowitz, and O. Penrose, 'Decay of Correlations', *Phys. Rev. Lett.* **31** (1973) 749.

[36] T. Nakano, 'Quantum Field Theory in Terms of Euclidean Parameters', *Prog. Theor. Phys.* **21** (1959) 241.

[37] E. Nelson, 'A Quartic Interaction in Two Dimensions', in *Mathematical Theory of Elementary Particles* (ed. by Goodman and I. Segal), M.I.T. Press, Cambridge, 1966.

[38] E. Nelson, 'Quantum Fields and Markoff Fields', in *Proceedings of Summer Institute of Partial Differential Equations, Berkeley 1971*, Amer. Math. Soc., Providence, 1973.

[39] E. Nelson, 'Construction of Quantum Fields from Markoff Fields', *J. Funct. Anal.* **12** (1973) 97.

[40] E. Nelson, 'The Free Markoff Field', *J. Funct. Anal.* **12** (1973) 211.

[41] E. Nelson, 'Talk given at the Meeting on Constructive Quantum Field Theory', New York, September 1972.

[42] E. Nelson, 'Probability Theory and Euclidean Field Theory', in [65].

[43] E. Nelson, 'Derivation of the Schrödinger Equation from Newtonian Mechanics', *Phys. Rev.* **150** (1966) 1079.

[44] E. Nelson, *Dynamical Theories of Brownian Motion*, Princeton University Press, Princeton, 1967.

[45] K. Osterwalder, 'Euclidean Green's Functions and Wightman Distributions', in [65].

[46] K. Osterwalder, and R. Schrader, 'Axioms for Euclidean Green's Functions', *Comm. Math. Phys.* **31** (1973) 83.

[47] M. Reed, 'Functional Analysis and Probability Theory', in [65].

[48] L. Rosen, 'A $\lambda\varphi^{2n}$ Field Theory without Cutoffs', *Comm. Math. Phys.* **16** (1970) 157.

[49] D. Ruelle, 'A Variational Formulation of Equilibrium Statistical Mechanics and the Gibbs Phase Rule', *Comm. Math. Phys.* **5** (1967) 324.

[50] D. Ruelle, *Statistical Mechanics*, Benjamin, New York, 1969.

[51] J. Schwinger, 'On the Euclidean Structure of Relativistic Field Theory', *Proc. Mat. Acad. Sc.* **44** (1958) 956.

[52] J. Schwinger, 'Euclidean Quantum Electrodynamics', *Phys. Rev.* **115** (1959) 721.

[53] I. Segal, 'Non Linear Functions of Weak Processes, I', *J. Funct. Anal.* **4** (1969) 404.

[54] I. Segal, 'Construction of Non Linear Local Quantum Processes', *Ann. Math.* **92** (1970) 462.

[55] B. Simon, 'Correlation Inequalities and the Mass Gap in $P(\varphi)_2$', I, 'Domination by the Two Point Function', *Comm. Math. Phys.* **31** (1973) 127.

[56] B. Simon, 'Correlation Inequalities and the Mass Gap in $P(\varphi)_2$', II, 'Uniqueness of the Vacuum for a Class of Strongly Coupled Theories', *Ann. Math.*, to appear.

[57] B. Simon, *The $P(\varphi)_2$ Euclidean (Quantum) Field Theory*, Princeton University Press, to appear.

[58] B. Simon, and R. Griffiths, 'The $(\varphi^4)_2$ Field Theory as a Classical Ising Model', *Comm. Math. Phys.* **33** (1973) 145.

[59] B. Simon, and R. Høegh-Krohn, 'Hypercontractive Semigroups and Two-Dimensional Self-Coupled Bose Fields', *J. Funct. Anal.* **9** (1972) 121.

[60] T. Spencer, 'The Mass Gap for the $P(\varphi)_2$ Quantum Field Model with a Strong External Field', N.Y.U. preprint, 1973.

[61] R. Streater, and A. S. Wightman, *PCT, Spin and Statistics and All That*, Benjamin, New York, 1964.

[62] K. Symanzik, 'A Modified Model of Euclidean Quantum Field Theory', N.Y.U. preprint, 1964.

[63] K. Symanzik, 'Euclidean Quantum Field Theory', I, 'Equations for a Scalar Model', *J. Math. Phys.* **7** (1966) 510.

[64] K. Symanzik, 'Euclidean Quantum Field Theory', in *Local Quantum Theory* (ed. R. by Jost), Academic Press, New York, 1969.

[65] G. Velo, and A. Wightman (eds.), *Constructive Quantum Field Theory*, Springer Verlag, Berlin, 1973.

[66] A. S. Wightman, 'An Introduction to Some Aspects of the Relativistic Dynamics of Quantized Fields', in *1964 Cargèse Summer School Lectures* (ed. by M. Levy), Gordon and Breach, New York, 1967.

[67] A. S. Wightman, 'Constructive Field Theory: Introduction to the Problem', Lecture at the 1972 Coral Gables Conference, Gordon and Breach, to appear.

BOUND STATES, CRITICAL POINTS AND RECENT RESULTS IN QUANTUM FIELD MODELS

ARTHUR JAFFE*

Institut des Hautes Etudes Scientifiques, 91440 Bures-sur-Yvette, France

1. Introduction

DEFINITION. The focus of mathematical physics is problems of substance for both mathematics and physics.

In this respect, constructive quantum field theory provides good examples. For the point of view of physics, field theory has for almost fourty years been the framework for the description of elementary particles. The recent progress of constructive field theory shows that in space-time dimension $d = 2$ relativistic, local field theory models exist. For $d = 3$, strong partial results are established. These models describe non-trivial scattering of particles. Depending on the interaction, some models, e.g. $\lambda(\varphi^6 - \varphi^4)_2$, have two particle bound states, while other models, e.g. $\lambda\varphi_2^4$, do not. The parameter space of the models gives a picture of broken symmetry and critical phenomena. We understand ultraviolet renormalization in the superrenormalizable case.

From the point of view of mathematics, interesting new structures arise in the study of quantum fields. The Hamiltonians of field theory are the most singular linear operators on Hilbert space that are known. Formally they can be written as a limit of a sum, $H = \lim_n (H_{\mathrm{on}} + H_{\mathrm{In}} + H_{\mathrm{Cn}})$, where the individual parts H, n have a physical significance, but no limits over n (even as sesquilinear forms). Other aspects of field theory lead to the notion of local states on C^* algebras, and to the study of elliptic systems with an infinite number of variables. The popular current formulations of Euclidean field theory (see the lecture of Guerra) yield interesting new problems for probability theory and measures on function spaces. Field theories give rise to a new class of non-Gaussian measures on $\mathscr{S}'(R^n)$, whose construction is equivalent to the construction of non-trivial quantum fields.

For a recent survey of results, and a general list of references, see the article 'Physics of quantum field models', in the 1973 Erice lectures [7]. We give here only more recent references.

2. Mass Spectrum for φ_2^4

Let $M = (H^2 - P^2)^{1/2}$ be the positive self-adjoint mass operator which labels the hyperboloids of the energy momentum spectrum. Consider a $\lambda\varphi_2^4$ model with bare mass m_0.

THEOREM 1 [7]. Let $\lambda m_0^{-2} \ll 1$. Then the spectrum of M consists of two eigenvalues 0 and $m > 0$, and the interval $(2m, \infty)$.

Remarks. The vacuum vector Ω spans the one dimensional eigenspace of mass zero.

Let $\mathscr{H}_m$ denote the eigenspace for $M = m$. These are the single particle states, and they span an infinite dimensional manifold. The representation of the Lorentz group, restricted to $\mathscr{H}_m$, is irreducible. In this sense only a single particle exists.

The absence of spectrum in the two particle bound state interval $(m, 2m)$ can be interpreted as the absence of two particle bound state spectrum [7]. It is a consequence of the assumption of small coupling (to reduce the question to the study of the two point and four point Schwinger functions) and the Lebowitz inequality for the four point Green's function,

$$0 \geqslant G(x_1, \ldots, x_4) = S(x_1, \ldots, x_4) - S(x_1, x_2)S(x_3, x_4) - \\ - S(x_1, x_3)S(x_2, x_4) - S(x_1, x_4)S(x_2, x_3).$$

This inequality is special to φ^4, as can be seen in perturbation theory.

CONJECTURE 1. No bound states occur in an even $\lambda\varphi^4$ model.

CONJECTURE 2. In an even $\lambda\varphi^4$ model, the (connected) Green's functions satisfy

$$G_{4n}(x_1, \ldots, x_{4n}) \leqslant 0.$$

THEOREM 2 (Feldman, Spencer [2]). In an even theory the mass operator, restricted to the even subspace, has no spectrum in $(m, 2m)$.

THEOREM 3 [7]. For $0 < \lambda/m_0^2 \ll 1$, the mass operator M for the $\lambda(\varphi^6 - \varphi^4)_2$ model has spectrum in the interval $(m, 2m)$.

I mention that Spencer (work in progress) is studying the bound states and the continuity of the spectrum above the two particle threshold, by means of the Bethe–Salpeter equation.

3. Classical Picture

In the classical (Goldstone) picture of symmetry breaking, one approximates the ground state(s) as localized near the minimum of the interaction polynomial, $\mathscr{P}(\varphi) = \mathscr{P}(\varphi_{cl})$, and the (mass)^2 as the curvature, $\mathscr{P}''(\varphi_{cl})$, at the minimum. In this picture, the model

$$\mathscr{P}(\varphi) = \varphi^4 + \tfrac{1}{2}\sigma\varphi^2$$

has a mass

$$m_{cl}(\sigma) = \begin{cases} (\sigma-\sigma_c)^{1/2}, & \sigma > \sigma_c, \\ \sqrt{2}(\sigma-\sigma_c)^{1/2}, & \sigma < \sigma_c, \end{cases}$$

where we assume $m_{cl}(\sigma_c) = 0$.

DEFINITION 1. Let σ_c be the supremum of σ for which $m = 0$ or $\langle\varphi\rangle \neq 0$. (Thus $\sigma > \sigma_c$ is the 'single phase region'.)

THEOREM 4 [6]. In the single phase region, $m(\sigma)$ is differentiable. If $\sigma_c \neq -\infty$ and $m(\sigma) \to 0$ as $\sigma \to \sigma_c$, then

$$m(\sigma) \leqslant m_{cl}(\sigma) = (\sigma-\sigma_c)^{1/2}.$$

CONJECTURE 3. The Goldstone picture is qualitatively correct in predicting the number of phases. For instance, for exactly one value μ_c of μ, the interaction

$$\varphi^6 + \varphi^5 - \sigma\varphi^2 + \mu\varphi$$

has multiple phases. This would correspond to recent results of Sinai and Piragov for spin system.

CONJECTURE 4. Far from the critical point, and in a pure phase,

$$\langle\varphi\rangle = \varphi_{cl} + 0(\sigma^{-1}).$$

A step in this direction is given by Heep [8].

Near the critical point, we expect that the Goldstone picture is qualitatively incorrect. We define critical exponents ν and γ for the mass (inverse correlation length) in the propagator at zero momentum (susceptibility) by

$$m \approx \text{const}(\sigma - \sigma_c)^\nu,$$

$$\chi = \int \langle \Phi(x)\Phi(y)\rangle dy \approx \text{const}(\sigma - \sigma_c)^{-\gamma}.$$

THEOREM 5 [6]. Assume $\sigma_c \neq -\infty$ in the φ_2^4 model, and $m(\sigma) \to 0$ as $\sigma \to \sigma_c$ in the single phase region. Then

$$\nu \geqslant \nu_{cl} = \tfrac{1}{2}, \qquad \gamma \geqslant \gamma_{cl} = 1.$$

CONJECTURE 5. The field theory exponents ν, γ do not equal their classical values, i.e. the infra-red behavior of φ_2^4 is anomalous.

We note the two and three dimensional Ising exponents are

	ν	γ
I_2	1	$\frac{7}{4}$
I_3	$\frac{5}{8}$	$\frac{5}{4}$
classical	$\frac{1}{2}$	1

Current ideas and calculations of Wilson and others suggest that the I_d exponents are more likely for the φ_d^4 quantum field model than the classical exponents. The proof of a result in this direction is an interesting and challenging research problem.

4. Three Dimensions

Let $V_\varkappa(\Phi_\varkappa)$ denote the Euclidean $\lambda\varphi_3^4$ action, with renormalization counterterms given by second and third order perturbation theory. Let $d\Phi$ denote the Euclidean Gaussian measure for the free field. (See the lecture of Guerra.)

THEOREM 6 [Glimm–Jaffe]. Let the space-time volume Λ be fixed Then

$$\exp[-V_\varkappa(\Phi_\varkappa)] \in L_1(d\Phi),$$

uniformly in the ultraviolet cutoff $\varkappa$.

COROLLARY. The renormalized φ_3^4 Hamiltonian is bounded from below.

Let

$$dq_\varkappa = \frac{\exp[-V_\varkappa(\Phi_\varkappa)]\,d\Phi}{\int \exp[-V_\varkappa(\Phi_\varkappa)]\,d\Phi}.$$

THEOREM 7 (Feldman [1]). For $\lambda m_0^{-1} \ll 1$,

$$dq_\varkappa \to dq$$

in the sense that the moments

$$\int \Phi(f_1) \ldots \Phi(f_n) dq_\varkappa$$

converge to the moments of a unique measure dq on $\mathscr{S}'(R^3)$.

5. Conclusion

The outstanding open problems, aside from cleaning up the super-renormalizable models, are:

1. Investigate the infra-red (critical) behavior in $d \leqslant 4$ dimensions.
2. Investigate the ultraviolet behavior of a renormalizable model, e.g. φ_4^4.

I note in this regard, that these problems are not fully understood on a formal level by physicists. In fact, any real progress on these questions should yield qualitative progress in physics insight. These problems appear on order of magnitude more difficult than past accomplishments in constructive field theory.

Reference

* Supported in part by the National Science Foundation, Grant NSF GP 40354X.

Bibliography

[1] J. Feldman, 'The φ_3^4 Quantum Field Theory in a Finite Volume', *Commun. Math. Phys.* **37** (1974) 93–120.

[2] J. Feldman, 'On the Absence of Bound States in the φ^4 Quantum Field Model without Symmetry Breaking', *Can. J. Phys.* **52** (1974) 1583–1587.
T. Spencer, The Absence of Even Bound State for $\lambda\varphi^4$', *Cammun. Math. Phys.* **39** (1974) 77–79.

[3] J. Fröhlich, 'Verification of Axioms for Euclidean and Relativistic Fields and Haag's Theorem in a Class of $\mathscr{P}(\varphi)_2$ Models', *Adv. Math.*, to appear.

[4] J. Fröhlieh and K. Osterwalder, 'Is there a Euclidean Field Theory for Fermions', *Helv. Phys. Acta* **47** (1974) 781–805.

[5] J. Glimm and A. Jaffe, 'Critical Point Dominance in Quantum Field Models', *Ann. de l'Institut Henri Poincaré* **21**.

[6] J. Glimm and A. Jaffe, 'The φ_2^4 Quantum Field Model in the Single Phase Region: Differentiability of the Mass and Bounds on Critical Exponents', *Phys. Rev.* **D10** (1974) 536–539.

[7] J. Glimm, A. Jaffe and T. Spencer, 'The Particle Structure of the Weakly Coupled $\mathscr{P}(\varphi)_2$ Model and Other Applications of High Temperature Expansions', Parts I, II, in *Constructive Quantum Field Theory* (edited by G. Velo and A. Wightman), Springer Lecture Notes in Physics, 1973.

[8] K. Hepp, 'The Classical Limit for Quantum Mechanical Correlation Functions', *Commun. Math. Phys.* **35** (1974) 265–278.

[9] O. Mc Bryan and Y. Park, 'Lorentz Covariance of the Yukawa$_2$ Quantum Field Theory', *J. Math. Phys.* **16** (1975) 104–110.

[10] R. Schrader, 'On the Euclidean Version of Haag's Theorem in $\mathscr{P}(\varphi)_2$ Theories', *Commun. Math. Phys.* **36** (1974) 133–136; **38** (1974) 81–82.

[11] R. Schrader and D. Uhlenbrock, 'Markov Structure on Clifford Algebras', Berlin preprint.

CHARGE CONSERVATION AND GAUGE INVARIANCE*

I. BIAŁYNICKI-BIRULA

Institute of Theoretical Physics, Warsaw University, Warsaw, Poland

Abstract

Direct relation between charge conservation and gauge invariance of quantum electrodynamics is exhibited. If one insists on the charge conservation at all levels of the theoretical description one is led to consider new types of propagators – charge-conserving propagators. Charge-conserving propagators after renormalization are free of ultraviolet divergencies even in the case of massive neutral vector meson theory without the introduction of indefinite metric. One class of charge-conserving propagators is particularly useful in the description of scattering phenomena because it leads to transition amplitudes which are free of all infrared divergencies.

> "*I remember that when someone had started to teach me about creation and annihilation operators, that this operator creates an electron, I said, "how do you create an electron? It disagrees with the conservation of charge*".
>
> R. P. Feynman, *Nobel Lecture.*

1. Introduction

Charge conservation is perhaps the only conservation law which is universally valid[1] and yet the restrictions which it imposes on all physical processes are completely ignored in the field-theoretic description of electromagnetic processes involving charged particles.

The basic concept employed in this description is that of the propagator or the vacuum matrix element of the product of field operators. Propagators represent quantum probability amplitudes for creation of charges from the vacuum and destruction of charges into the vacuum, so that they automatically carry within themselves the violation of local charge conservation.

Real charges are, of course, indestructible.

Creation and destruction of charges never takes place in experiments and therefore some modifications are necessary to make the propagator description of electromagnetic processes more realistic. As we shall see later, these modifications lead to a theory which is also formally more appealing. Abstraction and simplifications must always be introduced, it is true, in a theoretical description of every real process, but these should not be made at the expense of the fundamental laws of nature.

The compliance with the requirements of local charge conservation leads to a highly welcome property of the formalism, namely, gauge invariance becomes manifest at all levels of the theoretical description.

In order to account for the charge conservation in the description of all processes one must take into consideration the additional currents that flow during every experiment in which beams of charged particles are produced. These additional currents flow between the source and the detector of charged particles and we will call them the *compensating currents* because they compensate for the charge lost by the source and gained by the detector. Taking into account the compensating currents, instead of creation and destruction one simply has the separation and the recombination of charges.

The idea of compensating currents is in full accord with the observation that the natural state of all matter (at least under terrestial conditions) is an electrically neutral state.

The form of the compensating current will, of course, depend on the conditions of the experiment, but we expect this form to have only a negligible influence on the results of the experiment, provided these currents flow far from the region in which the interaction of particles occurs.

Compensating currents can be described at various levels of sophistication, ranging from classical mechanics to quantum field theory. In this paper we shall adopt the simplest approach assuming that all dynamical aspects of the compensating current can be neglected so that the current can be described by a set of four given c-number functions $J^{\mu}(z)$ of space-time coordinates. Such an external current can be considered to be a part of the apparatus used for the initial preparation and the final measurement of the quantum system. Even this greatly simplified model is quite sufficient to produce renormalizability of neutral vector meson theory in the physical Hilbert space and the disappearance of all infrared divergencies.

In what follows we shall restrict ourselves, for definiteness, to the study of electrons interacting with the electromagnetic field (pure quantum electrodynamics), but our method can be applied equally well to the description of processes involving other charged particles. Many of the ideas presented in Secs. 2–4 are not new. They can be found in Brandeis Lectures by Johnson and in the book by Schwinger [11], [15] and also in earlier works of the present author [2], [3], [5]. We give here a unified derivation which is best suited for the discussion of Sec. 5.

We found it more convenient to construct quantum electrodynamics, with charge conservation built in, in two steps. In the first step (Sec. 2) we introduce the quantized electron field operators $\psi(x)$ and $\bar{\psi}(x)$ but we treat the electromagnetic field as an external field, described by four functions $\mathscr{A}_\mu(z)$ of space-time coordinates. In the second step (Sec. 3) we quantize the electromagnetic field with the help of a method first introduced by Feynman. In Sec. 4 we shall prove the existence of a very simple relationship between the new charge-conserving, gauge-invariant propagators and the old charge-nonconserving, gauge-dependent propagators. In Sec. 5 we discuss the role of charge-conserving propagators in the removal of ultraviolet and infrared divergencies.

2. External Electromagnetic Field

Let us consider n electrons negatons and/or positons moving in a prescribed electromagnetic field which vanishes in the remote past and in the distant future.

The vacuum matrix element of the time-ordered product of electron field operators

$$(1)\qquad T[x_1, \ldots, x_n, y_n, \ldots, y_1|\mathscr{A}] = \big(\Omega^{\text{out}}|T\big(\psi(x_1), \ldots, \psi(x_n)\bar{\psi}(y_n), \ldots, \bar{\psi}(y_1)\big)\Omega^{\text{in}}\big)$$

has the interpretation of the probability amplitude of the process in which negatons first are being created from the vacuum at the points $y_1, \ldots, y_n$ or positons created at the points $x_1, \ldots, x_n$ and subsequently annihilated at the points $x_1, \ldots, x_n$ or positons annihilated at the points $y_1, \ldots, y_n$. The state vectors Ω^{in} and Ω^{out} describe respectively the state which is the vacuum state in the past and in the future.

In the simplified version of the theory now under consideration, the vacuum matrix element (1) can be expressed ([8], [13], [14]) as a sum of products of the vacuum to vacuum transition amplitude $C[\mathscr{A}]$ and one-electron Feynman propagators $K_F[x, y|\mathscr{A}]$

$$(2)\qquad T[x_1, \ldots, x_n, y_n, \ldots, y_1|\mathscr{A}] = (-i)^n C[\mathscr{A}] K_F[x_1, \ldots, x_n, y_n, \ldots, y_1|\mathscr{A}],$$

where

$$(3)\qquad K_F[x_1, \ldots, x_n, y_n, \ldots, y_1|\mathscr{A}] \equiv \sum_{\text{perm}\, i} \varepsilon_p K_F[x_1, y_{i_1}|\mathscr{A}], \ldots, K_F[x_n, y_{i_n}|\mathscr{A}].$$

The introduction of an external current changes the probability amplitude $T[x_1, \ldots, x_n, y_n, \ldots, y_1|\mathscr{A}]$ only by the following phase factor

$$(4)\qquad \exp\left\{-i \int d^4z \mathscr{A}_\mu(z) J^\mu(z)\right\}.$$

The compensating current, which we want to use in formula (4), in the case of n electrons depends not only on z but also on the coordinates $x_1, \ldots, x_n$ and $y_1, \ldots, y_n$ at which electrons appear and/or disappear. We shall indicate this dependence explicitly by writing $J^\mu(z;\, x_1, \ldots, x_n, y_1, \ldots, y_n)$. The conservation of charge requires that the compensating current satisfies the following inhomogeneous continuity equation

$$(5)\qquad \partial_\mu J^\mu(z;\, x_1, \ldots, x_n, y_1, \ldots, y_n) = e \sum_{i=1}^{n} \left(\delta^{(4)}(z-x_i) - \delta^{(4)}(z-y_i)\right).$$

The n-electron propagator in the presence of a compensating current will be denoted by $T[x_1, \ldots, x_n, y_n, \ldots, y_1|\mathscr{A}, J]$,

$$(6)\qquad T[x_1, \ldots, x_n, y_n, \ldots, y_1|\mathscr{A}, J] = (-i)^n \exp\left\{-i \int d^4z \mathscr{A}_\mu(z) J^\mu(z;\, x_1, \ldots, x_n, y_1, \ldots, y_n)\right\} \times \times C[\mathscr{A}]\, K_F[x_1, \ldots, x_n, y_n, \ldots, y_1|\mathscr{A}].$$

Even though the phase factor (4) has no effect on the probabilities when external current is not dynamically coupled to electrons, it makes the charge conserving propagator (6) gauge invariant under gauge transformations of the vector potential of the external field

$$(7)\qquad \mathscr{A}_\mu(z) \to \mathscr{A}_\mu(z) + \partial_\mu \Lambda(z).$$

This gauge invariance of $T[x_1, \ldots, x_n, y_n, \ldots, y_1|\mathscr{A}, J]$ follows from the gauge invariance of the vacuum to vacuum transition amplitude

$C[\mathscr{A}]$, from eq. (5) and from the following transformation property of K_F's under the change of gauge

$$K_F[x, y|\mathscr{A}+\partial\Lambda] = \exp(-ie\Lambda(x))K_F[x, y|\mathscr{A}]\exp(ie\Lambda(y)). \tag{8}$$

The gauge invariance is such an appealing theoretical feature, after all, all measurable quantities must be gauge invariant, that we believe that the modified propagators (6) are much more suitable for the description of real processes than ordinary propagators, which are always gauge dependent.

The idea of introducing phase factors to construct various gauge-invariant quantities from gauge-dependent object is an old one. It goes back to Dirac [6], [7] and Heisenberg [10][2] who were the first to realize the importance of line integrals in the construction of physical quantities.

A line integral $\Phi[x, y|\mathscr{A}]$ of the vector-potential evaluated along a certain trajectory in space-time,

$$\Phi[x, y|\mathscr{A}] \equiv -e\int_y^x d\xi^\mu \mathscr{A}_\mu(\xi), \tag{9}$$

represents a special case of our more general expression:

$$\int d^4z \mathscr{A}_\mu(z) J^\mu(z; x, y). \tag{10}$$

In order to see this it suffices to make the following identification

$$J^\mu(z; x, y) = -e\int_y^x d\xi^\mu \delta^{(4)}(z-\xi) \tag{11}$$

and then check that indeed J^μ obeys the right equation:

$$\partial_\mu J^\mu(z; x, y) = e\big(\delta^{(4)}(z-x)-\delta^{(4)}(z-y)\big). \tag{12}$$

For n electrons one must introduce n trajectories. Extensive use of line integrals was made in the gauge-invariant formulation of quantum electrodynamics given by Mandelstam [12].

3. Quantized Electromagnetic Field

The transition from the external field case (first step) to the quantized field case (second step) can be accomplished [9] by expanding the external field propagators into power series in external potentials, selecting pairs of potential vectors in all possible ways and replacing each pair, say $\mathscr{A}_\mu(z)$ and $\mathscr{A}_\nu(z')$, by the Feynman photon propagator $D^F_{\mu\nu}(z-z')$,

$$(13)\qquad D^F_{\mu\nu}(z-z') \equiv -g_{\mu\nu} D_F(z-z'),$$

multiplied by $-i$. The full n-electron propagator, with all the effects of interaction via photon exchanges included, will be denoted by $\underline{T}[x_1, \ldots, x_n, y_n, \ldots, y_1|\mathscr{A}, J]$. It can be written in the following compact form with the use of functional derivatives

$$(14)\qquad \underline{T}[x_1, \ldots, x_n, y_n, \ldots, y_1|\mathscr{A}, J]$$
$$= \exp\left(\frac{1}{2i}\int \frac{\delta}{\delta\mathscr{A}} D^F \frac{\delta}{\delta\mathscr{A}}\right) T[x_1, \ldots, x_n, y_n, \ldots, y_1|\mathscr{A}, J],$$

where we have used the following shortened notation

$$(15)\qquad \int \frac{\delta}{\delta\mathscr{A}} D^F \frac{\delta}{\delta\mathscr{A}} \equiv \int d^4z \int d^4z' \frac{\delta}{\delta\mathscr{A}_\mu(z)} D^F_{\mu\nu}(z-z') \frac{\delta}{\delta\mathscr{A}_\nu(z')}.$$

We will not consider here mixed electron-photon propagators, because the presence of real photons has no bearing on the question of charge conservation.

Owing to our inclusion of the compensating currents, full electron propagators $\underline{T}[x_1, \ldots, x_n, y_n, \ldots, y_1|\mathscr{A}, J]$ are invariant under the gauge transformations (7) of the external potential. This invariance in turn leads to the invariance under gauge transformations of the propagator

$$(16)\qquad D^F_{\mu\nu}(z-z') \to D^F_{\mu\nu}(z-z') + \partial_\mu f_\nu(z-z') + \partial_\nu f_\mu(z'-z),$$

where $f_\mu(z)$ is any vector function of space-time coordinates.[3]

The f-dependent gauge terms in the photon propagator do not contribute to the charge-conserving propagators $\underline{T}[x_1, \ldots, x_n, y_n, \ldots, y_1| \mathscr{A}, J]$ as seen from the equation

$$(17)\qquad \partial_\mu \frac{\delta}{\delta\mathscr{A}_\mu(z)} \underline{T}[x_1, \ldots, x_n, y_n, \ldots, y_1|\mathscr{A}, J] = 0,$$

which follows directly from the invariance under the infinitesimal gauge transformations of the potential

$$(18)\qquad 0 = \underline{T}[\ldots|\mathscr{A} + \partial\delta\Lambda, J] - \underline{T}[\ldots|\mathscr{A}, J]$$
$$= -\int d^4z\, \delta\Lambda(z)\, \partial_\mu \frac{\delta}{\delta\mathscr{A}_\mu(z)} \underline{T}[\ldots|\mathscr{A}, J].$$

It thus follows that the charge-conserving propagators are invariant under two types of gauge transformations: of the external potential and of the free photon propagator. Free photon propagator in any gauge therefore can be used in the exponential operation in formula (14) leading to the same final result. The Feynman gauge, of course, is the simplest and for that reason we have chosen it in formula (13).

It should be stressed, however, that our charge-conserving propagators depend on the form of the compensating current. This compensating current should in principle be appropriately chosen for each experimental situation.

4. Equivalence Theorem

We will show in this section, that under certain natural assumptions on the compensating current, our current-dependent but gauge-invariant propagators become numerically equal to the gauge-dependent but current-independent propagators of the standard theory[4] evaluated in a particular gauge. To this end we shall rewrite the current-dependent propagator (6) in the form

(19) $$\underline{T}[x_1, \ldots, x_n, y_n, \ldots, y_1|\mathscr{A}, J]$$
$$= (-i)^n \exp\left\{-i\int d^4z \mathscr{A}_\mu(z) J^\mu(z; x_1, \ldots, x_n, y_n, \ldots, y_1)\right\} \times$$
$$\times \exp\left\{\frac{1}{2i}\int\left(\frac{\delta}{\delta\mathscr{A}} - iJ\right) D_F\left(\frac{\delta}{\delta\mathscr{A}} - iJ\right)\right\} C[\mathscr{A}] \times$$
$$\times K_F[x_1, \ldots, x_n, y_n, \ldots, y_1|\mathscr{A}].$$

Let us now choose the simplest solution of eq. (5) for the compensating current, namely the one which can be expressed in terms of one vector function $a^\mu(z)$ of space-time coordinates

(20) $$J^\mu(z; x_1, \ldots, x_n, y_n, \ldots, y_1) = e\sum_{i=1}^{n}\left(a^\mu(z-x_i) - a^\mu(z-y_i)\right),$$

where a^μ obeys the equation

(21) $$\partial_\mu a^\mu(z) = \delta^{(4)}(z).$$

The current described by $a^\mu(z)$ has one point-like source of unit strength. The physical meaning of the formula (20) for the compensating current is that we have assumed the same universal mechanism for the production and the capture of all electrons by the apparatus.

In order to transform the r.h.s. of eq. (19) to a more useful form we will use the following property of the Feynman n-electron propagators

$$(22)\qquad \partial_\mu \frac{\delta}{\delta \mathscr{A}_\mu(z)} K_F[x_1, \dots, x_n, y_n, \dots, y_1|\mathscr{A}]$$
$$= ie \sum_{i=1}^{n} \big(\delta^{(4)}(z-x_i) - \delta^{(4)}(z-y_i)\big) K_F[x_1, \dots, x_n, y_n, \dots, y_1|\mathscr{A}],$$

which follows from the transformation law (8) of K_F under the change of gauge (cf. also formula (18)).

From (20), (21) and (22) we obtain

$$(23)\qquad T[x_1, \dots, x_n, y_n, \dots, y_1|\mathscr{A}, J]$$
$$= (-i)^n \exp\Big\{-ie \sum_{i=1}^{n} \int d^4z \mathscr{A}_\mu(z)\big(a^\mu(z-x_i) - a^\mu(z-y_i)\big)\Big\} \times$$
$$\times \exp\Big\{\frac{1}{2i}\int d^4z_1 d^4z_2 \Big(\frac{\delta}{\delta\mathscr{A}_\mu(z_1)} - \int d^4w_1 a^\mu(z_1 - w_1) \times$$
$$\times \partial_\lambda \frac{\delta}{\delta \mathscr{A}_\lambda(w_1)}\Big) D^F_{\mu\nu}(z_1 - z_2)\Big(\frac{\delta}{\delta\mathscr{A}_\nu(z_2)} - \int d^4w_2 a^\nu(z_2 - w_2) \times$$
$$\times \partial_\varrho \frac{\delta}{\delta\mathscr{A}_\varrho(w_2)}\Big)\Big\} C[\mathscr{A}] K_F[x_1, \dots, x_n, y_n, \dots, y_1|\mathscr{A}].$$

Choosing the gauge for the external potential in such a way that it obeys the condition

$$(24)\qquad \int d^4z \mathscr{A}_\mu(z) a^\mu(z-x) = 0$$

we obtain finally

$$(25)\qquad \underline{T}[x_1, \dots, x_n, y_n, \dots, y_1|\mathscr{A}, J]$$
$$= (-i)^n \exp\Big(\frac{1}{2i}\int \frac{\delta}{\delta\mathscr{A}} D^F[a] \frac{\delta}{\delta\mathscr{A}}\Big) C[\mathscr{A}] \times$$
$$\times K_F[x_1, \dots, x_n, y_n, \dots, y_1|\mathscr{A}],$$

where the new photon propagator D^F $[a]$ is defined as follows

$$(26)\quad D^F_{\mu\nu}[z_1-z_2|a] \equiv \int d^4w_1\, d^4w_2 \times \times \left(\delta^\lambda_\mu \delta^{(4)}(z_1-w_1)-\partial_\mu a^\lambda(z_1-w_1)\right)\left(-g_{\lambda\varrho} D_F(w_1-w_2)\right)\times \times \left(\delta^\varrho_\nu \delta^{(4)}(z_2-w_2)-\partial_\nu a^\varrho(z_2-w_2)\right).$$

This propagator differs from the Feynman propagator (13) only by a gauge transformation (16).

This completes the proof of the following equivalence theorem: Under the choice (20) of the compensating current the charge-conserving, gauge-invariant propagator (14) becomes equal to the gauge-dependent propagator of the standard theory evaluated in a special gauge which is defined by eqs. (24), (25) and (26).

5. Ultraviolet and Infrared Divergencies

Our motivation for introducing charge-conserved propagators was the desire to restore local charge conservation and bring with it full gauge invariance of all objects considered in the theory. We will show now that these new propagators have additional attractive features which only indirectly follow from local charge conservation.

The first property which we would like to discuss is the renormalizability of charge conserving propagators in electrodynamics with massive photons.[5] It is known[6] that in the standard formulation massive electrodynamics can be renormalized only at the expense of enlarging the space of state vectors by including unphysical state vectors with zero and negative norms. More precisely the off-mass-shell values of ordinary propagators can not be made finite by renormalization in the Hilbert space of physical state vectors even though all renormalized S-matrix elements between physical states are finite. One usually does not view this fact to be a serious deficiency because off-mass-shell values of the propagators being gauge-dependent are not considered to have any physical significance.

The nonrenormalizability of propagators in massive electrodynamics can be directly linked to the presence of the term $\mu^{-2}\partial_\mu\partial_\nu\Delta_F$ in the massive photon propagator. The presence of such terms increases the degree of ultraviolet divergencies in momentum-space integrals making the number of primitively divergent diagrams growing with the increasing order of perturbation expansion. Charge-conserving propagators are renormalizable because all dangerous terms cancel in every order of perturbation theory. This cancellation follows directly from the property (17) of these propagators. In order to see this we will write the expression for the propagators in the massive theory in the form (14) used before in massless electrodynamics

$$(27)\quad \underline{T}[x_1, \ldots, x_n, y_n, \ldots, y_1|\mathscr{A}, J] = \exp\left(\frac{1}{2i}\int \frac{\delta}{\delta\mathscr{A}}\Delta^F \frac{\delta}{\delta\mathscr{A}}\right) T[x_1, \ldots, x_n, y_n, \ldots, y_1|\mathscr{A}, J],$$

where Δ^F is chosen in the Feynman gauge, i.e.

$$(28)\quad \Delta^F_{\mu\nu}(z-z') = -g_{\mu\nu}\Delta_F(z-z').$$

Thus the ultraviolet divergencies in the perturbation expansion of charge-conserving propagators in the theory with massive photons are not any worse than the divergencies of standard propagators in the massless theory.

The second property of charge conserving propagators is the disappearance of infrared divergencies, provided the compensating current is properly chosen.

For simplicity we shall consider one-electron propagator $\underline{T}[x, y|\mathscr{A}, J]$ in the presence of a compensating current

$$(29)\quad \underline{T}[x, y|\mathscr{A}, J] = -i\exp\left(\frac{1}{2i}\int \frac{\delta}{\delta\mathscr{A}} D^F \frac{\delta}{\delta\mathscr{A}}\right) C[\mathscr{A}]\exp\left(-i\int \mathscr{A}\cdot J\right) K_F[x, y|\mathscr{A}].$$

Infrared divergencies arise from the interaction of charged particles with soft photons. In formula (29) this interaction is described by the soft part of the expression:

$$(30)\quad \int \frac{\delta}{\delta\mathscr{A}} D^F \frac{\delta}{\delta\mathscr{A}}.$$

We shall show now that soft photons are decoupled from the charges if the compensating currents do not flow to infinity. The strength of this coupling of the system to photons with four-momentum k is determined by the functional derivative:

$$(31)\quad \frac{\delta}{\delta\tilde{\mathscr{A}}_\mu(k)} = \int d^4z e^{ik\cdot z}\frac{\delta}{\delta\mathscr{A}_\mu(z)}.$$

At $k = 0$, we obtain

$$\frac{\delta}{\delta \tilde{\mathscr{A}}_\mu(0)} = \int d^4 z \frac{\delta}{\delta \mathscr{A}_\mu(z)}. \tag{32}$$

For one-electron propagator $K_F[x, y|\mathscr{A}]$ we have the relation

$$\frac{\delta}{\delta \tilde{\mathscr{A}}_\mu(0)} K_F[x, y|\mathscr{A}] = -ie(x-y)^\mu K_F[x, y|\mathscr{A}], \tag{33}$$

which is proved in the Appendix. The compensating current contributes the following term to the derivative (31) of (29)

$$\frac{\delta}{\delta \tilde{\mathscr{A}}_\mu(0)} \exp\left(-i\int \mathscr{A}\cdot J\right) \tag{34}$$

$$= -i\int d^4 z J^\mu(z; x, y) \exp\left(-i\int \mathscr{A}\cdot J\right).$$

For the line current (11) we obtain

$$\int d^4 z e \int_y^x d\xi^\mu \delta^{(4)}(z-\xi) = e(x-y)^\mu \tag{35}$$

and the same result will hold for any superposition, with total charge e, of line currents flowing between the points x and y. On the other hand each current flowing in a bounded region of space-time between the points x and y can be regarded to be a superposition (perhaps continuous) of line currents. Therefore, the electron propagator in the presence of such compensating currents decouples from photons with zero four-momentum:

$$\frac{\delta}{\delta \tilde{\mathscr{A}}_\mu(0)} \left(\exp\left(-i\int \mathscr{A}\cdot J\right) K_F[x, y|\mathscr{A}]\right) = 0. \tag{36}$$

This ends the proof of the disappearance of infrared divergencies since the vacuum polarization described by $C[\mathscr{A}]$ in the formula (29) is known not to contribute infrared-infinite terms.

Appendix

Our starting point will be the inhomogeneous Dirac equation obeyed by K_F

$$[m - i\gamma\cdot\partial + e\gamma\cdot\mathscr{A}] K_F[x, y|\mathscr{A}] = \delta^{(4)}(x-y). \tag{A.1}$$

After differentiating functionally both sides with respect to $\mathscr{A}_\mu(z)$ and integrating over z we obtain

$$\text{(A.2)}\quad [m-i\gamma\cdot\partial+e\gamma\cdot\mathscr{A}(x)]\int d^4z\frac{\delta}{\delta\mathscr{A}_\mu(z)}K_F[x,y|\mathscr{A}]$$
$$= -e\gamma^\mu K_F[x,y|\mathscr{A}].$$

The same equation and the same (Feynman) boundary conditions are satisfied by

$$-ie(x-y)^\mu K_F[x,y|\mathscr{A}]$$

and this establishes the equality (33).

References

* Earlier versions of this paper appeared as University of Pittsburgh preprints: Charge-Conserving Propagators (September 1972), Charge-Conserving Propagators and Ultraviolet and Infrared Divergencies (June 1973). This work has been partially supported by the NSF Grant No GF 36217.

[1] Charge can also mean the barionic or leptonic charge. Energy, momentum and angular momentum conservation laws are not so universally valid because these quantities in general can not be unambiguously defined in the presence of gravitation.

[2] The Dirac–Heisenberg line-integral method was rediscovered many years later and was unjustly given the name of later discoverers.

[3] It is often convenient to impose restrictions of the functions $f_\mu(z)$ such that the photon propagators will obey proper boundary conditions (cf. [4]).

[4] An extensive study of gauge-dependent or gauge-covariant propagators was made by B. Zumino [18].

[5] More detailed discussion of renormalizability is contained in the Ph. D. dissertation of A. Bechler [1].

[6] See for example [16], [17].

[7] There is a very lucid discussion of this question in a recent paper by D. Zwanziger [19].

Bibliography

[1] A. Bechler, *Acta Physica Polonica* **B5** (1974) 353, 453, 645.

[2] I. Białynicki-Birula, in: *Proceedings of Seminar on Unified Theories of Elementary Particles*, University of Rochester, Rochester, N.Y. (1963) 274.

[3] I. Białynicki-Birula, *Bull. Acad. Polon. Sci. Cl. III* **11** (1963) 483.

[4] I. Białynicki-Birula, *Phys. Rev.* **D2** (1970) 2877.

[5] I. Białynicki-Birula and Z. Białynicka-Birula, *Quantum Electrodynamics*, PWN and Pergamon Press, Warszawa–Oxford (1975) 352.

[6] P. A. M. Dirac, *Proc. Cambr. Phil. Soc.* **30** (1934) 150.
[7] P. A. M. Dirac, *Canadian Journal of Physics* **33** (1955) 650.
[8] R. P. Feynman, *Phys. Rev.* **76** (1949) 76.
[9] R. P. Feynman, *Phys. Rev.* **80** (1950) 440.
[10] W. Heisenberg, *Z. Phys.* **80** (1934) 209.
[11] K. Johnson, in: *Brandeis Summer Institute in Theoretical Physics 1964*, vol. 2, Prentice Hall Inc., Englewood Cliffs N.J. (1965) 28.
[12] S. Mandelstam, *Ann. of Physics* **19** (1962) 1.
[13] J. Schwinger, *Phys. Rev.* **92** (1953) 1283.
[14] J. Schwinger, *Phys. Rev.* **93** (1954) 615.
[15] J. Schwinger, *Particles, Sources, and Fields*, Addison-Wesley, Reading, Mass. (1970) 254.
[16] W. Zimmermann, *Commun. Math. Phys.* **8** (1968) 66.
[17] W. Zimmermann, in: *1970 Brandeis University Summer Institute in Theoretical Physics*, vol. 1, The M.I.T. Press, Cambridge, Mass. (1970) 552.
[18] B. Zumino, *J. Math. Phys.* **1** (1960) 1.
[19] D. Zwanziger, *Phys. Rev.* **D7** (1973) 1082.

CONFORMAL INVARIANCE OF QUANTUM FIELDS IN TWO-DIMENSIONAL SPACE-TIME

J. KUPSCH, W. RÜHL, B. C. YUNN

Universität Kaiserslautern, Kaiserslautern, Federal Republic of Germany

Free massless fields in Minkowski spaces of even space-time dimensions D exhibit a symmetry under the conformal group $SO(D, 2)$

$$(1) \qquad U_g \Phi(x) U_g^{-1} = \mu(x, g)\Phi(x_g)$$

where U_g is a unitary representation of canonical dimension d and spin s of the conformal group and $\mu(x, g)$ is a multiplier (in general a matrix). The canonical dimension and spin are connected by

$$(2) \qquad d = s+\tfrac{1}{2}(D-2).$$

Free massless fields in odd space-time dimensions show the phenomenon of 'reverberation': The commutator has support inside the light cone and not only on its boundary. It has been argued [1] that this reverberation is responsible for the break-down of symmetry (1) for these fields. This is true, but a generalization of (1) survives in these cases [7]

$$(3) \qquad U_g \Phi^{(\pm)}(x) U_g^{-1} = \mu^{(\pm)}(x, g)\Phi^{(\pm)}(x_g)$$

where $\Phi^{(\pm)}$ are the positive and negative frequency parts of the field and $\mu^{(-)}(x, g)$ ($\mu^{(+)}(x, g)$) is the boundary value of a holomorphic (antiholomorphic) function in the forward tube domain. This definition (3) can be extended to all quasi-free fields with homogeneous Lehmann spectral function, provided the dimension d fulfills the constraint of positivity

$$(4) \qquad d \geqslant s+\tfrac{1}{2}(D-2).$$

The unitary operators U_g form a representation of the universal covering group of $SO(D, 2)$ in general. The multiplier $\mu^{(-)}(x, g)$ is determined by first taking x in the forward tube domain, then continuing it analytically from one at the group unit along a path that represents a group element g of the universal covering group, and finally going to the boundary of the forward tube. In order to distinguish between (1) and (3) we call the former 'strong local conformal symmetry' and the latter 'strong non-local conformal symmetry'. From massless free fields in even dimensional space-time respectively quasi-free fields with homogeneous spectral function these concepts can be carried over to Wick products and integral (non-negative) rank tensor products of these fields, respectively.

We then study the Thirring model. We generalize Klaiber's approach [2] and find for any fixed s and d, restricted only by $0 < s \leqslant d$, an infinite number of explicit operator solutions in the Fock space of the free field $\Psi(x)$. Namely, for any point w in the forward tube domain there is one operator field $\Phi_w(x)$, and all these fields have the same n-point Wightman functions (i.e. Klaiber's [2] n-point functions, his infrared cutoff parameter μ can still be chosen arbitrarily for our solutions). Spin, dimension and coupling constant are related as in Klaiber's solution,

$$(5) \qquad d = \pm s + \frac{g^2}{4\pi^2}.$$

Since any pair w, w' of the forward tube domain defines uniquely an element g of the Weyl group (i.e. the group consisting of inhomogenous Lorentz transformations and dilations) such that

$$(6) \qquad w \underset{g}{\rightarrow} w' = w_g$$

we may write

$$(7) \qquad W_g(w)\,\Phi_w(x)\,W_g(w)^{-1} = \Phi_{w_g}(x)$$

where $W_g(w)$ is unitary. Inspection of the Wightman functions yields that the Weyl group is a symmetry group

$$(8) \qquad U_g(w)\,\Phi_w(x)\,U_g(w)^{-1} = \mu(0, g)\,\Phi_w(x_g).$$

We then prove that the unitary operators

$$(9) \qquad V_g = W_g(w)\,U_g(w)$$

are independent of w and transform the free field locally as a representation of the Weyl group. Whereas the generators of V_g are obtained by the standard canonical quantization procedure of the free field, the generators of $U_g(w)$ can be obtained using Lowenstein's normal product [4] and the energy momentum tensor of Lowenstein and Schroer [5].

Strong local conformal invariance of the Thirring model (that includes the special conformal transformations) occurs only if the characteristic exponents δ_1, δ_2, δ in the Wightman functions [2] are integers. There is an infinite number of non-trivial solutions to the model of this kind with varying spins and dimensions. Among them are all fields with $s = d > \frac{1}{2}$ integral or halfintegral. The latter solutions can be viewed upon as tensor products of the free field. The other solutions may be obtained from the tensor product

$$\Phi(x) = \begin{bmatrix} \prod_{k=1}^{\delta_1-\delta} \Psi_1^{(1,k)}(x_+) \prod_{l=1}^{\delta_2-\delta} \Psi_2^{(2,l)}(x_-) \prod_{m=1}^{\delta} \Psi_1^{(3,m)}(x_+)\Psi_2^{(3,m)}(x_-) \\ \prod_{l=1}^{\delta_2-\delta} \Psi_1^{(2,l)}(x_+) \prod_{k=1}^{\delta_1=\delta} \Psi_2^{(1,k)}(x_-) \prod_{m=1}^{\delta} \Psi_1^{(3,m)}(x_+)\Psi_2^{(3,m)}(x_-) \end{bmatrix} \tag{10}$$

where

$$[\Psi_\sigma^{(\varrho,k)}(x), \Psi_{\sigma'}^{(\varrho',k')}(x')^+]_+ = [\Psi_\sigma^{(\varrho,k)}(x), \Psi_{\sigma'}^{(\varrho',k')}(x')]_+ = 0 \tag{11}$$

for $\sigma \neq \sigma'$ or $\varrho \neq \varrho'$ or $k \neq k'$

and

$$\Psi(x) = \begin{bmatrix} \Psi_1(x_+) \\ \Psi_2(x_-) \end{bmatrix}, \quad x_\pm = x^0 \pm x^3 \tag{12}$$

is the free spin $-\frac{1}{2}$ field, by a continuation in $\delta_1 - \delta$, $\delta_2 - \delta$, δ to arbitrary, eventually negative integers by the method proposed by Migdal [6].

All other Thirring fields can be interpreted as non-integral rank tensor products of the kind (10) in the same sense, and exhibit at least weak conformal invariance in the sense of Hortaçsu *et al.* [1]. This notion includes again the transition to the universal covering group. On the level of the two-point function the concepts of strong non-local and weak conformal invariance are identical. Finally we were able to find an explicit non-local operator conformal symmetry of the Thirring model. For all details we refer to a forthcoming paper [3].

Bibliography

[1] M. Hortaçsu, R. Seiler and B. Schroer, *Phys. Rev.* **D5** (1972) 2519.
[2] B. Klaiber, *Boulder Lectures in Theoretical Physics*, Vol. XA, ed. by A. O. Barut and W. E. Brittin, New York 1968, p. 141.
[3] J. Kupsch, W. Rühl and B. C. Yunn, 'Conformal Invariance of Quantum Fields in Two-Dimensional Space-Time', *Ann. Phys. (N. Y.)* **89** (1975) 115.
[4] J. H. Lowenstein, *Comm. Math. Phys.* **16** (1970) 265.
[5] J. H. Lowenstein and B. Schroer, *Phys. Rev.* **D3** (1971) 1981.
[6] A. A. Migdal, 'Four-Dimensional Soluble Models of Conformal Field Theory', Chernogolovka preprint 1972 and *Phys. Letters* **45B** (1973) 272.
[7] W. Rühl, *Comm. Math. Phys.* **34** (1973) 149.

CONFORMAL EXPANSION FOR EUCLIDEAN GREEN FUNCTIONS*

I. T. TODOROV**

Bulgarian Academy of Sciences, Institute of Physics, Sofia, Bulgaria

Abstract

We review recent work on conformal quantum field theory. The discussion is restricted to a model of a self-interacting scalar field in an arbitrary (even) number of space-time dimensions. Euclidean conformal partial wave expansions are used to diagonalize and solve an infinite set of coupled integral equations for connected Green functions without recourse to perturbation theory. The problem of incorporating crossing symmetry of the solution is left open.

1. Introduction

The beginning of the 1970's was marked by a major revival of interest in conformal quantum field theory and its application to the study of small distance behaviour of Green functions (for a review and bibliography see [28]). We are goind to describe here a recent development in the subject, chiefly due to G. Mack (see [13] and [5]; similar ideas were also expounded in [22] and [17]). It is based on an explicit tensor product decomposition formula for the Euclidean conformal group which allows to diagonalize and solve the set of coupled integral equations for the renormalized Green functions of ref. [25]. The equations correspond to the φ^3 interaction which is strictly renormalizable in six space-time dimensions. In order to incorporate this model along with the physical 4-dimensional case, we consider a $2h$-dimensional space-time throughout these notes.

First, in Chapter 1, we give a summary of graphical notation and write down the set of renormalized integro-differential equations for the connected Green functions, in the form given to them by Symanzik and Mack [25], [15], [13]. We observe that the set of equations (without the boundary conditions) is conformal invariant. Then we replace the boundary conditions by the requirement that the solution is also conformal invariant.

Chapter 2 is devoted to the description of a class of (induced) infinite dimensional representations of the pseudo-orthogonal group $O(2h+1,1)$ (which plays the role of the Euclidean conformal group in $2h$ dimensions) and to the study of the tensor product decomposition of two spin zero representations. The main point here is the evaluation of the normalization of the Clebsch–Gordan kernel, satisfying the Plancherel formula. A helpful device in solving this problem is the use of the homogeneous polynomial techniques for dealing with symmetric traceless tensors (cf. [29], [32], [33]).

Chapter 3 deals with the application of conformal partial wave expansions to the solution of the infinite set of invariant equations of the first chapter. Assuming certain analyticity properties in the continuous label c of the irreducible representations of $O(2h+1,1)$ (which is related to the field dimension) one can derive conformal covariant operator product expansions of the type considered in [6]. The unresolved problem of crossing symmetry is briefly discussed in the last section.

The argument throughout these notes is rather sketchy. The interested reader will find a reasonably self-contained exposition in the set of original papers [25], [16], [15], [13], [5]. Some additional material (contained mainly in Sec. 2D) is based on unpublished results of V. Dobrev, V. Petkova, S. Petrova and the author.

2. Dynamical Equations for Connected Green Functions in a φ^3 Model

A. EUCLIDEAN GREEN FUNCTIONS FOR A SCALAR FIELD. GRAPHICAL NOTATION

As stated in the introduction, we shall deal with the simplest model of a single (self-interacting) neutral scalar field $\varphi(x)$ in $2h$ space-time dimensions. The basic objects of the theory will be the time-ordered Green functions

$$\tau(x_1, \ldots, x_4) = \langle T\varphi(x_1) \ldots \varphi(x_n)\rangle_0, \qquad x = (x^0, \underline{x})$$

and their Euclidean region continuation $s(x_1, \ldots, x_n)$ $(x = (\underline{x}, x^{2h} = -ix^0))$. In the Euclidean region the $2h$-components of the x's are real and so are the s-functions. The present status of axiomatic Euclidean quantum field theory (including the equivalence problem with Wightman's Minkowski space formulation) is reviewed in the lecture notes [19], which also contain a comprehensive guide to the literature on the subject.

We shall be interested in the connected Green functions $G(x_1, \dots, x_n)$, which are defined in a straightforward way in terms of the s-functions:

$$(1.1)\quad \begin{aligned} &G(x_1) = S(x_1) = 0, \quad G(x_1, x_2) = S(x_1, x_2),\\ &S(x_1, x_2, x_3, x_4) = G(x_1, x_2, x_3, x_4) + G(x_1, x_2)G(x_3, x_4) +\\ &+G(x_1, x_3)G(x_2, x_4) + G(x_1, x_4)G(x_2, x_3), \dots \end{aligned}$$

(The first pair of equations expresses the assumption of absence of tadpoles in the theory.)

We shall use the following graphical notation

(1.2) $G(x_1, x_2) =$ 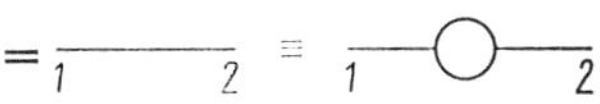 , $G(x_1, \dots, x_n) =$  .

Let $G^{-1}(x_1, x_2)$ be the inverse propagator in the sense of convolution:

$$\int G(x_1, x)G^{-1}(x_1, x_2)dx = \delta(x_1 - x_2).$$

We define the 1-particle irreducible Green function, for the channel $12 \to 3 \dots n$, by

$$(1.3)\quad \begin{aligned} G_{1i}(x_1, x_2; x_3, \dots, x_n) &= G(x_1, \dots, x_n) -\\ &- \int dy_1 \int dy_2 G(x_1, x_2, y_1)G^{-1}(y_1, y_2)G(y_2, x_3, \dots, x_n), \end{aligned}$$

or graphically

(1.3′) 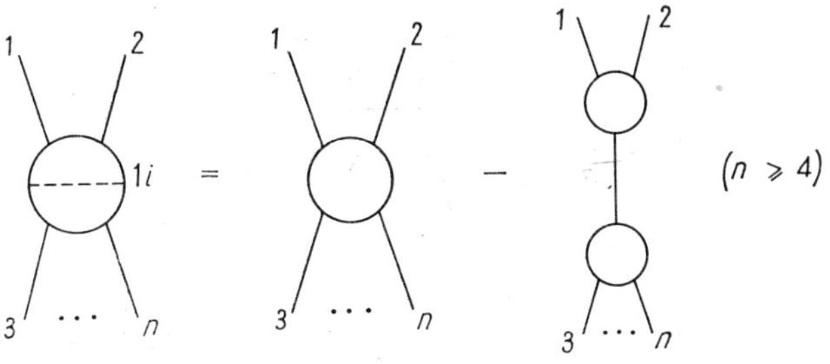

Further, we define the Bethe–Salpeter (BS) kernel

(1.4) $B(x_1, x_2; x_3, x_4) =$ 

as the solution of the integral equation

(1.5) 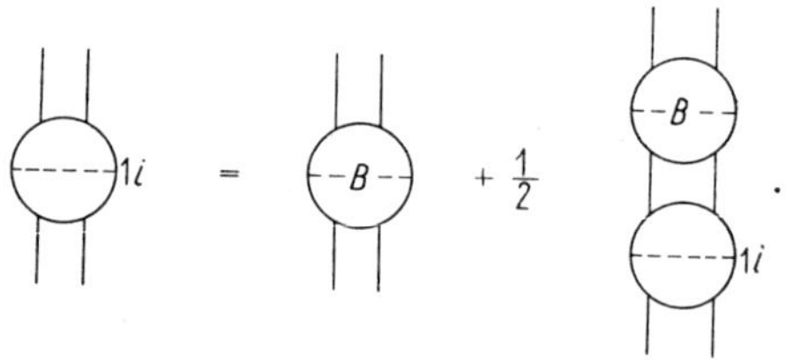

Knowing the BS-kernel we can define 2-particle irreducible Green functions for $n \geqslant 5$:

(1.6) $G_{2i}(x_1, x_2; x_3, \ldots, x_n) \equiv$ 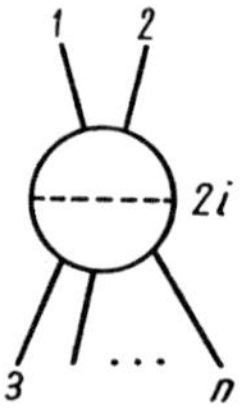

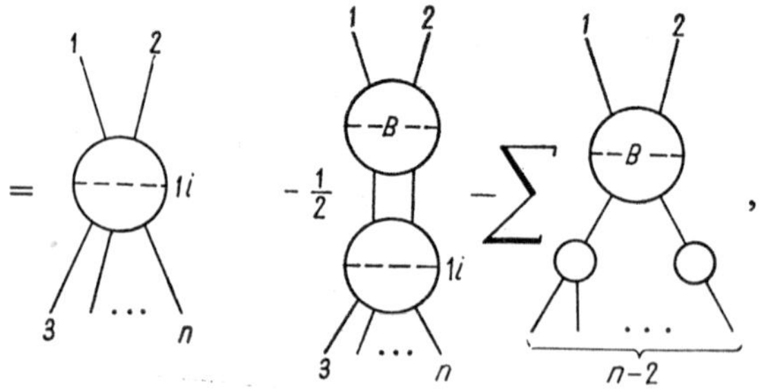

where the sum in the last term is over all $2^{n-3}-1$ partitions of the set of external lines $3 \ldots n$ $(n \geqslant 5)$ into two non-empty subsets.

We shall also use the following notation for the (amputated) vertex function:

$$\Gamma(x_1, x_2, x_3) \equiv$$ 

$$= \iiint G(x_1', x_2', x_3') G^{-1}(x_1', x_1) G^{-1}(x_2', x_2) \times$$
$$\times\, G^{-1}(x_3', x_3)\, dx_1'\, dx_2'\, dx_3' .$$

B. RENORMALIZED EQUATIONS AND BOUNDARY CONDITIONS

We shall write down without derivation the set of integro-differential equations for the G-functions in the theory with interaction Lagrangian $-\frac{g}{3!}:\varphi^3:$ (see [25], [15]). They were originally derived (in [25]) in canonical Lagrangian quantum field theory in 6 space-time dimensions (without reference to perturbation theory). The canonical dimension of the field in that case is $d_\varphi = h-1 = 2$ and the coupling constant g is dimensionless. The equations are expected to be also valid in the super-renormalizable case for $2h < 6$.

For n-point Green functions, $n > 4$, we encounter no (primitive) divergences and obtain the following set of simple integral equations (presented in the graphical notation of the preceding section):

(1.8) 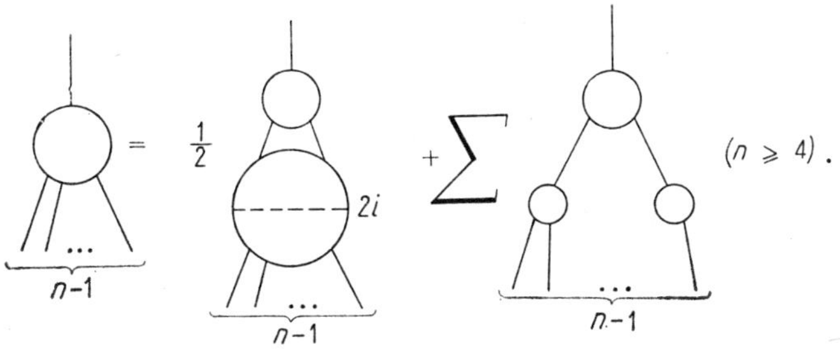

The Schwinger–Dyson equations for the 2- and 3-point functions require explicit renormalization. We can relegate all renormalization constants to the boundary conditions by multiplying the equations for the (amputated) vertex function and for the inverse propagator by appropriate coordinate differences (or equivalently by differentiating the corresponding equations in momentum space with respect to some transverse momenta). As a result, we obtain the following set of equations

(1.9) 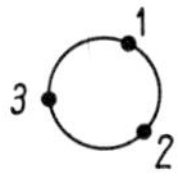 $(x_1-x_2)_\mu = \frac{1}{2}$ 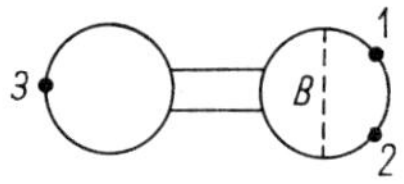 $(x_1-x_2)_\mu,$

(1.10) 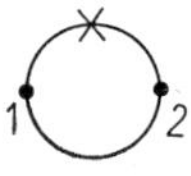 $(x_1-x_2)_\mu(x_1-x_2)_\nu = \frac{1}{2}$ 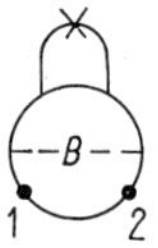 $(x_1-x_2)_\mu(x_1-x_2)_\nu,$

where

(1.11)  $= (x_1-x_2)_k\frac{\partial}{\partial x_{1\lambda}}, \qquad \varkappa \neq \lambda.$

They have to be supplemented by the usual boundary (or 'initial') conditions in Minkowski momentum space:

$$G^{-1}(p) = 0 \quad \text{for } p^2(=p_0^2-\underline{p}^2) = m^2,$$

$$\text{(1.12)} \quad \frac{\partial}{\partial p^2}G^{-1}(p) = -iZ \quad \text{for } p^2 = \mu^2,$$

$$\Gamma(p_2, p_2, p_3) = -igZ^{3/2} \quad \text{for } p_ip_j = \tfrac{1}{2}\mu^2(3\delta_{ij}-1),$$

where

$$G^{-1}(p) = \int G^{-1}(x, 0)e^{ipx}dx,$$

$$\text{(1.13)} \quad (2\pi)^{2h}\delta(p_1+p_2+p_3)\Gamma(p_1, p_2, p_3)$$

$$= \int \Gamma(x_1, x_2, x_3)e^{i(p_1x_1+p_2x_2+p_3x_3)}dx_1dx_2dx_3.$$

The parametr z fixes the normalization of the field and may be chosen according to convenience. (Customarily, one sets $z = 1$, $\mu = m$.)

C. DILATATION AND CONFORMAL INVARIANCE

The form (1.8–10) of the dynamical equations has a remarkable property: it does not depend explicitly on the mass m, the coupling constant g or any other external parameter; the equations only involve the G-functions, while the external parameters enter the boundary conditions (1.12). In particular, there is no dimensional parameter in the whole set of dynamical equations, hence it is natural to expect that they are dilatation invariant. We shall demonstrate that they admit an even wider symmetry – the (Euclidean) conformal invariance.

Indeed, we see that for any choice of the (real) number c and for any $\varrho > 0$ eqs. (1.8–10) are invariant under the dilatation transformation law:

$$(1.14)\quad \begin{aligned} G(x_1, \dots, x_n) &\to \varrho^{n(h+c)} G(\varrho x_1, \dots, \varrho x_n), \\ \Gamma(x_1, \dots, x_n) &\to \varrho^{n(h-c)} \Gamma(\varrho x_1, \dots, \varrho x_n), \quad \varrho > 0 \end{aligned}$$

(in particular, $G^{-1}(x_1, x_2) = \Gamma(x_1, x_2) \to \varrho^{2(h-c)} G^{-1}(\varrho x_1, \varrho x_2)$). For fixed c, the number $d = h+c$ would correspond to the (dynamical) dimension of the field $\varphi(x)$.

Eqs. (1.8–10) (unlike the boundary conditions (1.12)) preserve their form in the Euclidean region. We shall see that they are invariant under the action of the Euclidean conformal group $\mathscr{C}_E$, generated by the Euclidean transformations ($x' = \Lambda x + a$) and the conformal inversion

$$(1.15)\quad x' = Rx = -\frac{x}{x^2}, \qquad x^2 = x_1^2 + \dots + x_{2h}^2,$$

or equivalently by the (proper) special conformal transformations

$$(1.15')\quad N(a)x = RT_{-a}Rx = \frac{x + x^2 a}{1 + 2ax + a^2 x^2}.$$

The group $\mathscr{C}_E$ includes dilatations and is isomorphic to the $(h+1)(2h+1)$-parameter pseudo-orthogonal group $O^{\uparrow}(2h+1,1)$ (the arrow indicates that we exclude reflections of the axis $2h+2$). The connected component of the identity of $\mathscr{C}_E$ will consist of matrices of determinant $1: \mathscr{C}_E^0 = SO^{\uparrow}(2h+1,1)$.

Both the dynamical equations and the boundary conditions are Poincaré invariant and their solutions are required to be Poincaré invariant as well. Therefore, the analytically continued G-functions for pure imaginary times will be invariant under the Euclidean group $E(2h) = T^{2h} \cdot O(2h)$. The Green functions transformation law under conformal inversion is related to the scaling law (1.14):

$$(1.16)\quad G(x_1, \dots, x_n) \to \frac{1}{(x_1^2 \dots x_n^2)^{h+c}} G(Rx_1, \dots, Rx_n)$$

(the parameter c which fixes the representation of $O^{\uparrow}(2h+1,1)$ is the same in both cases).

The R-invariance of eqs. (1.8) follows from the covariance law for the volume element[1] $dRx = (x^2)^{-2h}\det(r_{\mu\nu})dx = (x^2)^{-2h}dx$. The same is true for eq. (1.9) (which is verified first for noncoinciding arguments and then extended to arbitrary x's). In order to make explicit the conformal invariance of eq. (1.10) we shall write it in another form (involving the local stress energy tensor $\theta_{\mu\nu}(x)$), which exhibits its relation to the local dynamics of the problem (see [15]).

Let $\theta_{\mu\nu}(x)$ be a conserved symmetric tensor field which generates the local dynamics (in Minkowski space) in the sense that

$$(1.17a)\quad [\varphi(x), P_\mu] = \int_{x_0=y_0} d^3y[\varphi(x), \theta_{0\mu}(y)] = i\partial_\mu\varphi(x)$$

or

$$(1.17b)\quad [\varphi(x), \theta_{0\mu}(y)]_{x_0=y_0} = i\delta(\underline{x}-\underline{y})\,\partial_\mu\varphi(x) + \text{gradient terms}$$

(see [14, 15]). Then the 3-point function

$$(1.18)\quad G_{\mu\nu}(x_1, x_2; x_3) = \langle 0|T^*\varphi(x_1)\varphi(x_2)\theta_{\mu\nu}(x_3)|0\rangle =$$

(the star on T indicating that we are dealing with covariant (Wick) time-ordered product) satisfies certain Ward–Takahashi identity (see [15] and eq. (1.33) below). The corresponding vertex function $\Gamma_{\mu\nu}(x_1, x_2; x_3)$ (amputated with respect to the scalar legs 1 and 2) satisfies for noncoinciding arguments the equation

$$(1.19)\quad \Gamma_{\mu\nu}(x_1, x_2; x_3) \equiv$$

$$= \tfrac{1}{2}$$

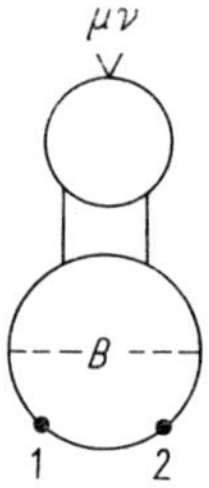

$$(x_1 \neq x_2)$$

which is equivalent to (1.10) (modulo the Ward–Takahashi identity). The transformation law of $\Gamma_{\mu\nu}$ (continued to Euclidean coordinates) with respect to the conformal inversion is:

$$(1.20)\quad \Gamma_{\mu\nu}(x_1, x_2, x_3) \to \frac{1}{(x_1^2 x_2^2)^{h-c}} \frac{1}{(x_3^2)^{2h}} r_{\mu\mu'}(x_3) \times$$
$$\times\, r_{\nu\nu'}(x_3)\Gamma_{\mu'\nu'}(Rx_1, Rx_2; Rx_3),$$

where

$$(1.21)\quad r_{\mu\nu}(x) = 2\frac{x_\mu x_\nu}{x^2} - \delta_{\mu\nu} = r_{\mu\nu}(Rx)$$

$$(r_{\mu\nu}x_\nu = x_\mu, r^2 = 1).$$

Now, eq. (1.19) is readily verified to remain invariant under the substitution (1.16), (1.20), and hence, eq. (1.10) is also conformal invariant.

D. FIXED POINTS, ASYMPTOTIC LIMITS, SKELETON GRAPH EXPANSION

Here we shall further exploit the scale invariance of the dynamical equations in Minkowski space.

Assume that there is a unique Poincaré invariant solution of eqs. (1.8–10), satisfying the boundary conditions (1.12). Since the equations are scale invariant then the transformed (under dilatation) Green functions (1.14) are also solutions of these equations, satisfying new boundary conditions characterized by new parameters m'^2, μ'^2, g', z'. We could exclude the inessential normalization parameter z by replacing the last two boundary conditions (1.12) by

$$(1.22)\quad \left[i\frac{\partial}{\partial p^2}G^{-1}(p)\right]^{-3}[i\Gamma(p_1, p_2, p_3)]^2|_{p^2=p_i^2=\mu^2} = g^2$$

$$(p_1+p_2+p_3 = 0).$$

Setting in (1.14) $\varrho = \lambda^{-1}$ we have the following transformation law under dilatation in momentum space

$$G'(p_1, \ldots, p_n) = \lambda^{(n-2)h-nc}G(\lambda p_1, \ldots, \lambda p_n),$$

$$(1.23)\quad \Gamma'(p_2, \ldots, p_n) = \lambda^{(n-2)h+nc}\Gamma(\lambda p_1, \ldots, \lambda p_n)$$

$$(p_1+ \ldots +p_n = 0,\ \Gamma(p_1-p) \equiv G^{-1}(p)).$$

We see that $G^{-1}(p) = 0$ for $p^2 = m^2$ implies that

$$(1.24)\quad G'^{-1}(p) \equiv \lambda^{2c}G^{-1}(\lambda p) = 0 \quad \text{for } p^2 = m'^2 = \frac{m^2}{\lambda^2}.$$

The transformation law for g is more complicated. We only need to know that $g' = g'(\lambda, g)$ satisfies the composition law

$$(1.25)\quad g'(\lambda_1\lambda_2, g) = g'(\lambda_1, g'(\lambda_2, g)),$$

and the 'initial condition' $g'(1, g) = g$. It implies that $g'(\lambda, g)$ is determined by its derivative for $\lambda = 1$:

$$(1.26) \quad \beta(g) = \frac{\partial}{\partial \lambda^2} g'(\lambda, g)|_{\lambda=1}.$$

Since the equations are scale invariant the question arises whether they also have a dilatation invariant solution (besides the trivial one, corresponding to a free 0-mass field with $c = -1$). If such a solution does exist it leaves practically no freedom in the choice of the boundary conditions. Indeed, assuming that $G' = G$ ($\Gamma' = \Gamma$) we see that (1.24) can only be satisfied for either $m^2 = 0$ or $m^2 = \infty$. Then the invariance of eq. (1.22) implies that it is only possible for some special ('critical') value $g = g_{cr}$ of the coupling constant for which

$$(1.27) \quad g'(\lambda, g_{cr}) = g_{cr}, \quad \text{so that } \beta(g_{cr}) = 0.$$

This solution corresponds to the so called Gell-Mann–Low limit (for a review see [2] or [31]); the invariant Green functions will be temporarily denoted by G_{GML}. According to the analysis of [3], [26], if the derivative $\beta'(g_{cr})$ does not vanish, we have the following alternatives:

$$G(\lambda p_1, \ldots, \lambda p_n) \approx G_{GML}(\lambda p_1, \ldots, \lambda p_n)$$

for

$$(1.28a) \quad \lambda \to \infty \quad \text{if} \quad \beta'(g_{cr}) < 0,$$

$$(1.28b) \quad \lambda \to 0 \quad \text{if} \quad \beta'(g_{cr}) > 0$$

and $m^2 = 0$. In the first case g_{cr} is called an ultraviolet stable point and is usually denoted by g_∞; in the second case it is termed infrared stable. Physically, we are primarily interested in ultraviolet stable fixed points, however formally, we shall only use the conformal invariance of the limit-theory and will not distinguish between the two cases.

It turns out that conformal invariance fixes 2- and 3-point functions uniquely up to a multiplicative constant. Choosing the normalization of the propagator according to convenience, we have

$$(1.29a) \quad G(x_1, x_2) = \Delta_c(x_{12}) = \frac{1}{(2\pi)^h} \frac{\Gamma(h+c)}{\Gamma(-c)} \left(\frac{2}{x_{12}^2}\right)^{h+c}.$$

$$x_{12} = x_1 - x_2,$$

$$(1.29b) \quad G^{-1}(x_1, x_2) = \Delta_{-c}(x_{12});$$

$$(1.30a)\quad G(x_1, x_2, x_3) = gV(x_1, c, x_2, c, x_3, c)$$

$$\equiv g\frac{1}{(2\pi)^h}\frac{\Gamma(\frac{1}{2}h+\frac{1}{2}c)}{\Gamma(\frac{1}{2}h-\frac{1}{2}c)}\left[\frac{\Gamma(\frac{1}{2}h+\frac{1}{2}c)\Gamma(\frac{1}{2}h+\frac{3}{2}c)}{\Gamma(\frac{1}{2}h-\frac{1}{2}c)\Gamma(\frac{1}{2}h-\frac{3}{2}c)}\right]^{3/2}\times$$

$$\times\left(\frac{8}{x_{12}^2 x_{13}^2 x_{23}^2}\right)^{\frac{1}{2}(h+c)},$$

$$(1.30b)\quad \Gamma(x_1, x_2, x_3) = gV(x_1, -c, x_2, -c, x_3, -c).$$

(The meaning of the function V and of the normalization factor in (1.30a) will become clear in Chapter 2 below.) Similarly for the continuation of the 3-point function (1.18) into the Euclidean region, we have

$$(1.31)\quad G_{\mu\nu}(x_1, x_2; x_3) = \frac{h!}{(2\pi)^{2h}}\frac{\Gamma(c+h+1)}{(2h-1)\Gamma(-c)}\left(\frac{2}{x_{12}^2}\right)^{c+1}\times$$

$$\times\left(\frac{4}{x_{13}^2 x_{23}^2}\right)^{h-1}\frac{1}{2}\left(\frac{\lambda_\mu\lambda_\nu}{\lambda^2}-\frac{\delta_{\mu\nu}}{2h}\right),$$

where

$$(1.32)\quad \lambda_\mu = 2\left[\frac{(x_{13})_\mu}{x_{13}^2}-\frac{(x_{23})_\mu}{x_{23}^2}\right] = \partial_{3\mu}\ln\frac{x_{23}^2}{x_{13}^2}.$$

The normalization is chosen in such a way that the (generalized) Ward–Takahashi identity

$$(1.33)\quad \partial_3^\mu G_{\mu\nu}(x_1, x_2; x_3) = \sum_{i=1}^{2}\Big[\delta(x_3-x_i)\,\partial_{i\nu}G(x_1, x_2)-$$

$$-\frac{c+h}{2h}G(x_1, x_2)\,\partial_{i\nu}\,\delta(x_3-x_i)\Big]$$

holds true with $G(x_1, x_2)$ given by (1.29).

With these propagators and vertex functions one could attempt to construct all n-point Green functions (with $n \geqslant 4$) in terms of skeleton diagram expansions. (A skeleton diagram is a Feynmann diagram with dressed propagators and vertices and with no self-energy and vertex function corrections.) This path was followed in [16] and it was shown that the theory is free from ultraviolet and catastrophic infrared divergences in a certain range of field dimensions; in our case the convergence range is

$$(1.34)\quad -\tfrac{1}{3}h < c < \tfrac{1}{3}h, \qquad c \neq 0.$$

In particular, no subtraction (in p-space) or multiplication by $(x_1-x_2)_\mu$ is needed for the equations (1.9), (1.10) (or (1.19)) in that case. Eqs. (1.9), (1.19) become in fact (transcendental) numerical equations for the two parameters g and c of the theory. They are equivalent to the Gell-Mann–Low (or Callan–Symanzik) equations (of type (1.27)). The existence problem for a solution of these equations (with c in the range (1.34)) is still open; the same is true for the convergence problem for the skeleton diagram expansion.

Here we review a different attempt to exploit conformal invariance ([13], [5]). In the following chapters we shall show how (forgetting, for the time being, about permutation symmetry) a conformal covariant solution can be constructed of all dynamical equations by the method of $O^\uparrow(2h+1,1)$ 'partial wave' expansion.

3. Clebsch–Gordan Expansion for $O^\uparrow(2h+1,1)$

A. A FAMILY OF INFINITE DIMENSIONAL REPRESENTATIONS OF $O^\uparrow(2h+1,1)$

The irreducible representations (IR) of the Euclidean conformal group $\mathscr{C}_E = O^\uparrow(2h+1,1)$ can be obtained as irreducible components of induced representations in the following way (see e.g. [30]).

We start with the Iwasawa decomposition

$$\mathscr{C}_E = KAN$$

where $K = O(2h+1)$ is the maximal compact subgroup of $O^\uparrow(2h+1,1)$, A is in our case the one-parameter subgroup of dilatations (or, equivalently, rotations in the $(2h+1,2h+2)$-plane) and N is the $2h$-parameter Abelian subgroup of special conformal transformations. We single out the maximal subgroup $M = O(2h)$ of K which commutes with A. M is commonly called the centralizer of A in K. Then we define the inducing subgroup as MAN and construct our induced representations on the set of vector-valued functions on the homogeneous space

$$S^{2h} = \mathscr{C}_E/MAN$$

(homeomorphic to the unit sphere in $2h+1$ dimensions), which will be identified with the (compactified) $2h$-dimensional Euclidean space (of points x). We shall consider only finite dimensional IR of the inducing subgroup which are necessarily trivial on N. According to a famous theorem of Harish Chandra (see, e.g. [30]) they induce all (topologically completely) IR of G_E. Physically, such representations correspond to the so-called basic (conformal) fields (cf. [28]). The induced representations of th is type are given by

$$D^{\chi}_{\alpha\beta}(uan) = \sigma_c(a) D^{\vec{l}}_{\alpha\beta}(u),$$

where $\sigma_c(a)$ is a 1-dimensional representation of the dilatation subgroup of transformations $x \to \varrho x$ ($\varrho > 0$) specified by the (complex) number c such that $\sigma(\varrho) = \varrho^{-c-h}$; and D^l is an arbitrary IR of $U = O(2h)$ which is specified, in general, by h integers ($\vec{l} = (l_1, \ldots, l_h)$). (A detailed description of the unitary IR of SO(n, 1) may be found in [10], [12], [20].)

Here we shall restrict ourselves to a two-parameter family $\chi = [l, c]$, for which D^l is the (rank l) symmetric traceless tensor representation of $O(2h)$ (l being a nonnegative integer).

Let C_χ be the space of infinitely differentiable (symmetric, traceless) tensor-valued functions

$$f(x) = \{f_{\mu_1 \ldots \mu_l}(x)\}, \quad x = (x_1, \ldots, x_{2h}) \in \boldsymbol{R}^{2h}, \quad \mu_i = 1, \ldots, 2h,$$

whose behaviour at infinity depends on $\chi = [l, c]$ and will be specified below. The representation χ of the Euclidean conformal group $O^{\uparrow}(2h+1,1)$ will be defined through the action of a set of generating transformations used already in Sec. 1.C.

(a) Euclidean transformations $x \to \Lambda x + a$ are represented by

$$(2.1) \quad [U(a, \Lambda)f](x) = \Lambda^{\otimes l} f(\Lambda^{-1}(x-a)).$$

(b) Dilatations $x \to \varrho x$ are implemented by

$$(2.2) \quad [U(\varrho)f](x) = \varrho^{-h-c} f\left(\frac{x}{\varrho}\right).$$

(c) The R-inversion (1.15) is represented by

$$(2.3) \quad [U(R)f](x) = \frac{r(x)^{\otimes l}}{(x^2)^{h+c}} f(Rx)$$

where $r(x)$ ($= r_{\mu\nu}(x)$) is given by (1.21).

(We notice that Euclidean translations and the conformal inversion generate the entire conformal group; thus the transformation laws (2.1–3) are not in fact independent.)

Now the behaviour of $f(x)$ at infinity is fixed in such a way that the space C_χ is left invariant by the conformal inversion: $f \in C_\chi$ iff $U(R)f \in C_\chi$.

Recalling that $r^2 = \mathbf{1}$, we obtain that

$$(x^2)^{h+c} r(x)^{\otimes l}[U(R)f(x)] \underset{x\to\infty}{\sim} f(0) - \left[\frac{x}{x^2}\,\partial f\right](0) + \dots$$

Accordingly we shall postulate that each $f \in C_\chi$ admits an asymptotic expansion for $x \to \infty$ of the form

$$(2.4)\quad (x^2)^{h+c} f(x) \underset{x\to\infty}{\sim} r(x)^{\otimes l} \sum_{k=0}^{\infty} A^{(l,k)} \left(\frac{x}{x^2}\right)^{\otimes k}$$

$$\left(= r(x)_{\mu_1\nu_1} \dots r(x)_{\mu_l\nu_l} \sum_k A^{(l,k)}_{\nu_1\dots\nu_l,\lambda_1\dots\lambda_k} \frac{x_{\lambda_1} \dots x_{\lambda_k}}{(x^2)^k}\right)$$

where $A^{(l,k)}$ are constant tensors (symmetric in each set of indices (ν) and (λ) separately).

The representation $\chi = [l, c]$ so defined is irreducible unless one of the numbers $\pm c - h - l$ is a nonnegative integer or both l and $h+l-1-|c|$ (c real) are positive integers. The exceptional integer points, for which the representation χ is (reducible but) nondecomposable will be dealt with in Sec. 2.D.

One can introduce a natural topology on C_χ (with a denumerable set of semi-norms) which will render it a complete locally convex topological vector space (cf. [8]). The representation χ defined by (2.1–3) will then be a continuous representation on the topological space C_χ.

We close this section with an important technical remark.

Let z be a complex isotropic vector in $2h$ dimensions:

$$(2.5)\quad z \in \mathbf{C}^{2h}, \quad z^2 = z_1^2 + \dots + z_{2h}^2 = 0.$$

Then, there is a one-to-one correspondence between symmetric traceless tensors (or tensor fields) of rank l, $f(x) = \{f_{\mu\dots\mu_l}(x)\}$, and homogeneous polynomial functions of lth degree on the isotropic cone (2.5),

$$(2.6)\quad f(x, z) = f(x)\frac{z^{\otimes l}}{\sqrt{l!}} = \frac{1}{\sqrt{l!}} f_{\mu_1\dots\mu_l}(x) z_{\mu_1} \dots z_{\mu_l}.$$

Indeed, each such polynomial function can be extended in a unique way to a harmonic polynomial in $\zeta \in C^{2h}$ by taking an arbitrary homogeneous polynomial extension f^E of f and setting

$$(2.7)\quad f(x, \zeta) = H_l(\zeta, \partial_z) f^E(x, z) = f^E(x, \partial_\eta) H_l(\zeta, \eta),$$

where $H_l(\zeta, \eta)$ is the $O(2h)$-invariant homogeneous harmonic polynomial

$$(2.7') \quad H_l(\zeta, \eta) = \frac{2^l l!(2h+l-3)!}{(2h+2l-3)!} (\zeta^2\eta^2)^{l/2} P_l^{(h-\frac{3}{2}, h-\frac{3}{2})}\left(\frac{\zeta\eta}{\sqrt{\zeta^2\eta^2}}\right)$$

$$= \binom{l+h-2}{h-2}^{-1} \left(\frac{1}{4}\eta^2\zeta^2\right)^{l/2} C_l^{h-1}\left(\frac{\eta\zeta}{\sqrt{\eta^2\zeta^2}}\right)$$

($\Delta_\zeta H_l = \Delta_\eta H_l = 0$, $H_l(\zeta, \eta) = H_l(\eta, \zeta)$). The right-hand side of (2.7) does not depend on the particular choice f^E of the extension of f since any two extensions differ by a function of the type $f_{l-2}(p, \partial_\eta)\Delta_\eta$ and H_l is harmonic in its second argument.

The Jacobi polynomial $P_l^{(\alpha,\beta)}(t)$ is defined as the solution of the differential equations

$$(2.8) \quad \left\{(1-t^2)\frac{d^2}{dt^2} + [\beta-\alpha-(\alpha+\beta+2)t]\frac{d}{dt} + \right.$$

$$\left. + l(l+\alpha+\beta+1)\right\} P_l^{(\alpha,\beta)}(t) = 0,$$

which satisfies the normalization condition

$$(2.9) \quad \frac{d^l}{dt^l} P_l^{(\alpha,\beta)}(t) = \frac{\Gamma(\alpha+\beta+2l+1)}{2^l\Gamma(\alpha+\beta+l+1)}.$$

(Thus $H_l(\zeta, \eta) = (\zeta\eta)^l + a_1(\zeta\eta)^{l-2}\zeta^2\eta^2 + \ldots$; C_l^{h-1} is the corresponding Gegenbauer polynomial.)

The inverse formula to (2.6) then reads

$$(2.10) \quad f_{\mu_1\ldots\mu_l}(x) = \frac{1}{\sqrt{l!}} \frac{\partial}{\partial\zeta_{\mu_1}} \cdots \frac{\partial}{\partial\zeta_{\mu_l}} f(x, \zeta).$$

The contraction of two tensors f and g can be written in terms of the polynomials (2.7) as

$$(2.11) \quad f_{\mu_1\ldots\mu_l} g_{\mu_1\ldots\mu_l} = f(\partial)g(z) = g(\partial)f(z).$$

We notice that the representations of the Euclidean conformal group corresponding to the exceptional points are related to the so called analytic representations of SO(2h, 2) which may well be of physical importance (see [9]).

B. INVARIANT BILINEAR FORMS, INTERTWINING OPERATORS, AND UNITARY REPRESENTATIONS

We shall determine the most general conformal invariant two-point function

$$(2.12)\quad G_\chi(x_1, x_2; z_1, z_2) = \frac{1}{l!} z_1^{\otimes l} G_\chi(x_1, x_2) z_2^{\otimes l}, \qquad \chi = [l, c].$$

According to the definitions and the rules of the preceding section we require

$$(2.13a)\ G_\chi(\Lambda x_1 + a, \Lambda x_2 + a; \Lambda z_1, \Lambda z_2) = G_\chi(x_1, x_2 : z_1, z_2),$$

$$(2.13b)\ \varrho^{2(h+c)} G_\chi(\varrho x_1, \varrho x_2; z_1, z_2) = G_\chi(x_1, x_2; z_1, z_2) \qquad (\varrho > 0),$$

$$(2.13c)\ \frac{1}{(x_1^2 x_2^2)^{h+c}} G_\chi(Rx_1, Rx_2; r(x_1) z_1, r(x_2) z_2) = G_\chi(x_1, x_2; z_1, z_2),$$

$$(2.14)\quad G_\chi(x_1, x_2; a z_1, z_2) = G_\chi(x_1, x_2; z_1, a z_2) = a^l G_\chi(x_1, x_2; z_1, z_2)$$

The general form of a function G, satisfying the above conditions, is

$$(2.15)\quad G_\chi(x_1, x_2; z_1, z_2) = \frac{n(\chi)}{(2\pi)^h} \left(\frac{2}{x_{12}^2} \right)^{h+c} \frac{(-z_1 r(x_{12}) z_2)^l}{l!}$$

$$\left(x_{12} = x_1 - x_2, \qquad -z_1 r(x_{12}) z_2 = z_1 z_2 - 2\, \frac{(z_1 x_{12})(z_2 x_{12})}{x_{12}^2} \right)$$

where $n(\chi)$ is a normalization constant. The Euclidean and dilatation invariance of (2.15) are obvious. The verification of R-invariance (2.13c) relies on the identity

$$(2.16)\quad r(x_1) r(Rx_1 - Rx_2) r(x_2) = r(x_1 - x_2),$$

which is a consequence of (1.21).

The homogeneity property of G_χ with respect to x_{12} is a consequence of dilatation invariance alone. The tensor structure of G_χ is fixed by R-invariance (see [28]).

We see that (apart from the exceptional case, in which $h+c+l$ is a non-positive integer) G_χ is not an element of $C_\chi \otimes C_\chi$ but could rather be considered as a bilinear form on a suitable space of test functions.

Define the dual representation $\tilde{\chi}$ to $\chi = [l, c]$ by

(2.17) $\tilde{\chi} = [l, -c]$.

Then, $G_{\tilde{\chi}}$ defines an invariant bilinear form on $C_\chi \otimes C_\chi$

$$(2.18)\quad (f, g)_\chi = \iint f(x_1) G_{\tilde{\chi}}(x_1, x_2) g(x_2) dx_1 dx_2$$

$$= \iint G_{\tilde{\chi}}\left(x_1, x_2; \frac{\partial}{\partial z_1}, \frac{\partial}{\partial z_2}\right) f(x_1, z_1) g(x_2, z_2) dx_1 dx_2$$

where the explicit form $G_\chi(x_1, x_2; \zeta_1, \zeta_2)$ of the harmonic extension of (2.15) is

$$(2.19)\quad G_{\tilde{\chi}}(x_1, x_2; \zeta_1, \zeta_2) = \frac{n(\tilde{\chi})}{(2\pi)^h}\left(\frac{2}{x_{12}^2}\right)^{h-c} \times$$

$$\times \frac{(-2)^l (\zeta_1^2 \zeta_2^2)^{l/2} \Gamma(2h+l-2)}{\Gamma(2h+2l-2)} P_l^{\left(h-\frac{3}{2}, h-\frac{3}{2}\right)}\left(\frac{\zeta_1 r(x_{12}) \zeta_2}{\sqrt{\zeta_1^2 \zeta_2^2}}\right)$$

$$\left(= \frac{n(\tilde{\chi})}{(2\pi)^h}\left(\frac{2}{x_{12}^2}\right)^{h-c} \frac{H_l(\zeta_1, -r(x_{12})\zeta_2)}{l!}\right)$$

(cf. (2.7)).

The question arises for which values of χ does the bilinear form (2.18) define a (real) positive scalar product on C_χ. That would allow us to single out the supplementary series of unitary representations (among the two-parameter family under consideration). It is clear from the outset that a necessary condition for the positivity of (2.18) is the reality of c. For real values of c we shall assume that C_χ consists of real (tensor) valued functions $f(x) = \{f_{\mu_1 \ldots \mu_l}(x)\}$.

A standard way to study the positivity condition is to expand the Fourier transform of G_χ in projection operators $\Pi^{ls}(p)$ on the irreducible subspaces of the stability subgroup $O(2h-1)_p \subset O(2h)$ of the vector p (cf. [29]). The calculation is rather intricate, and this is one place where it is important to have the compact functional form (2.15) of the Green function in order to be able to use the theory of Jacobi polynomials in studying the Fourier transform of G. We shall just write down the results which (once written) are not difficult to check (see [5] for a derivation):

$$(2.20)\quad G_\chi(p; z_1, z_2) = \int G_\chi(x, 0; z_1, z_2) e^{-ipx} dx$$

$$= \frac{n(\chi)\Gamma(-c)}{\Gamma(c+h+l)} \left[\frac{(pz_1)(pz_2)}{-\frac{1}{2}p^2} \right]^l \left(\frac{p^2}{2} \right)^c P^{(c-l,h-z)}_{l(\omega)}$$

$$= \frac{n(\chi)\Gamma(-c)\,(\frac{1}{2}p^2)^c}{\Gamma(c+h+l)\Gamma(c+h-1)\Gamma(c-h-l+2)} \sum_{s=0}^{l} (-1)^{l-s} \times$$

$$\times \Gamma(c+h+s-1)\Gamma(c-h-s+2)\Pi^{ls}(p; z_1, z_2)$$

where

$$(2.21) \quad \omega = 1 - \frac{p^2 z_1 z_2}{(pz_1)(pz_2)}$$

and the projectors Π^{ls} are given by

$$(2.22) \quad \Pi^{ls}(p; z_1, z_2) = \frac{1}{l!} z_1^{\otimes l} \Pi^{ls}(p) z_2^{\otimes l}$$

$$= \frac{(2h+2s-3)\,(h+l-2)!(2h+s-4)!}{(l-s)!(h+s-2)!(2h+l+s-3)!} (-1)^s$$

$$\times \left[\frac{(pz_1)(pz_2)}{\frac{1}{2}p^2} \right]^l P^{(h-2,h-2)}_{s(\omega)}, \qquad s = 0, \ldots, l.$$

The easiest way to obtain (or verify) the numerical factor is to use the completeness relation

$$(2.23) \quad \sum_{s=0}^{l} \Pi^{ls}(p) = \mathbf{1} \quad \text{or} \quad \sum_{s=0}^{l} \Pi^{ls}(p; z_1, z_2) = \frac{(z_1 z_2)^l}{l!}.$$

Having the functional form (2.22) of the projection operators it is not difficult to write them down also in a tensor form. We have

$$\zeta_1^{\otimes l} \Pi^{l0}(p) \zeta_2^{\otimes l} = \frac{(2h-3)!(h+l-2)!}{(h-2)!(2h+l-2)!} \left(\frac{2}{p^2} \right)^l H_l(p, \zeta_1) H_l(p_1 \zeta_2),$$

$$H_l(p, \zeta) = \frac{l!(h-2)!}{(l+h-2)!} \left(\frac{p^2 \zeta^2}{4} \right)^{l/2} C_l^{h-1} \left(\frac{p\zeta}{\sqrt{\zeta^2 p^2}} \right),$$

so that

$$\Pi^{10}_{\mu\nu}(p) = \frac{1}{p^2} p_\mu p_\nu, \quad \Pi^{20}_{\mu_1\mu_2\nu_1\nu_2}(p) = \frac{2h}{2h-1} \left(\frac{1}{p^2} p_{\mu_1} p_{\mu_2} - \right.$$

$$\left. - \frac{1}{2h} \delta_{\mu_1\mu_2} \right) \left(\frac{1}{p^2} p_{\nu_1} p_{\nu_2} - \frac{1}{2h} \delta_{\nu_1\nu_2} \right), \quad \text{etc.};$$

on the other hand,

$$\Pi^{11}_{\mu\nu}(p) = \delta_{\mu\nu} - \frac{p_\mu p_\nu}{p^2} \equiv \Pi_{\mu\nu}(p),$$

so that

$$\Pi^{22}_{\mu_1\mu_2\nu_1\nu_2}(p) = \frac{1}{2}\left(\Pi_{\mu_1\nu_1}\Pi_{\mu_2\nu_2} + \Pi_{\mu_1\nu_2}\Pi_{\mu_2\nu_1} - \frac{2}{2h-1}\Pi_{\mu_1\mu_2}\Pi_{\nu_1\nu_2}\right)$$

$$\Pi^{21}_{\mu_1\mu_2\nu_1\nu_2}(p) = \frac{1}{2p^2}(\Pi_{\mu_1\nu_1}p_{\mu_2}p_{\nu_2} + \Pi_{\mu_1\nu_2}p_{\mu_2}p_{\nu_1} + \Pi_{\mu_2\nu_1}p_{\mu_1}p_{\nu_2} + \\ + \Pi_{\mu_2\nu_2}p_{\mu_1}p_{\nu_1}).$$

There is no unique normalization which would be equally convenient for all purposes.

The simple choice

(2.24a) $$n(\chi) = n_u(\chi) \equiv \frac{1}{\Gamma(-c)}$$

has the virtue to correspond to a universal 2-point function $G^{(u)}_\chi$ defined for all values of $\chi = [l, c]$. Indeed, using that

$$(-1)^{l-s}\frac{\Gamma(c-h-s+2)}{\Gamma(c-h-l+2)} = \frac{\Gamma(l+h-c-1)}{\Gamma(s+h-c-1)},$$

we find

(2.25a) $$G^{(u)}_\chi(p) = \frac{\Gamma(l+h-c-1)}{\Gamma(l+h+c)\Gamma(h+c-1)}\left(\frac{1}{2}p^2\right)^c \sum_{s=0}^{l} \frac{\Gamma(h+s+c-1)}{\Gamma(h+s-c-1)}\Pi^{ls}(p).$$

The only points c for which the distribution $(\frac{1}{2}p^2)^c$ is not well defined are $c = -h, -h-1, \ldots$ But it is easily seen that the x-space expression $G^{(u)}_\chi(x)$ is well defined for such c and therefore the same is true for the sum (2.25a). We shall make use of this 'universal' normalization in Sec. 2D below, where we study the properties of representations at integer points. Note that for $h = 1$, $l \geqslant 1$ the right-hand side should be multiplied by c^{-1} in order to have a non-zero limit for $c \to 0$.

Away from the integer points the function G_χ has an inverse proportional to $G_{\tilde{\chi}}$ and it is convenient in the study of the problems of tensor-product expansion, we are concerned with, to normalize the proportionality constant to 1. We shall use a particular choice with this property for which the coefficient to $\left[\frac{p^2}{2}\right]^c \Pi^{l0}(p)$ in the expansion (2.20) is 1. The result is

$$(2.24\text{b})\; n(\chi) = \frac{\Gamma(c+h+l)\Gamma(h-c-1)}{\Gamma(-c)\Gamma(l+h-c-1)} = \frac{\Gamma(c+h+l)\Gamma(h-c-1)}{\Gamma(l+h-c-1)} n_u(\chi),$$

$$(2.25\text{b})\; G_\chi(p) = G_{\tilde{\chi}}^{-1}(p) = [\Pi^{l0}(p) + \alpha_1(c)\Pi^{l1}(p) + \cdots + \alpha_l(c)\Pi^{ll}(p)]\left(\frac{p^2}{2}\right)^c,$$

where

$$(2.26)\quad \alpha_s(c) = \frac{\Gamma(c+h+s-1)\Gamma(h-1-c)}{\Gamma(h+s-1-c)\Gamma(h-1+c)} = \frac{1}{\alpha_s(-c)}.$$

The scalar distribution $[\frac{1}{2}p^2]^c$ defines a positive functional over the space of fast decreasing test functions of p for all $c > -h$. For $l > 0$ the coefficients α_s are all positive in the smaller domain $-(h-1) < c < h-1$. However, the scalar product (2.18) (with χ replaced by $\tilde{\chi}$) can be written in momentum space without recourse to (say, analytic) regularization only for $c < 0$. Thus, the problem of positivity of the form (2.18) for $c > 0$ requires an additional argument. We shall use here the equivalence of the representations χ and $\tilde{\chi}$ for non-exceptional c.

Whenever $G_\chi(p)$ and $G_{\tilde{\chi}}(p)$ are both finite (which means $\pm c \notin \{h-1, h, \ldots, h+l-2\}$ (for $l \geqslant 1$) and $\pm c - h$ is not a non-negative integer) the 2-point functions G_χ and $G_{\tilde{\chi}}$ are kernels of the intertwining operators which display the equivalence between the dual representations χ and $\tilde{\chi}$:

$$(2.27)\quad U_\chi G_\chi = G_\chi U_{\tilde{\chi}}, \qquad G_{\tilde{\chi}} U_\chi = U_{\tilde{\chi}} G_{\tilde{\chi}}.$$

According to (2.25b), G_χ and $G_{\tilde{\chi}}$ are inverse to each other:

$$(2.28)\quad \int G_\chi(x_1, y; z_1, \partial_z) G_{\tilde{\chi}}(y, x_2; z, z_2)\, dy = \delta(x_1 - x_2)\frac{(z_1 z_2)^l}{l!}.$$

Note that G_χ in (2.27) is a short-hand for the integral operator with kernel given by (2.12) (2.15); it is a simple exercise to verify (using (2.4)), that if $f(x) \in C_\chi$, then

$$\int G_{\tilde{\chi}}(x-y) f(y)\, dy \in C_{\tilde{\chi}}.$$

From the equivalence of the representations χ and $\tilde{\chi}$ and from the established above positivity of G_χ for a range of negative c it follows, that χ can be extended to a unitary IR in a (complete) real Hilbert space $\mathscr{H}_\chi \supset C_\chi$ for

$$\begin{array}{lll} l = 0, & -h < c < h, & \\ l = 1, 2, \ldots, & -h+1 < c < h-1, & c \neq 0. \end{array} \tag{2.29}$$

We should add to our list of unitary IR the representations $\chi = [l, c]$ of the principal series for which $c = i\sigma$ is pure imaginary and the scalar product is given by

$$(f, g) = \int \bar{f}(x) g(x) dx .$$

The reader interested in more general mathematical treatment of intertwining operators for semi-simple Lie groups of real rank 1 is referred to [24].

C. TENSOR PRODUCT EXPANSION AND NORMALIZED CLEBSCH-GORDAN KERNELS

Consider the tensor product space $C_{\chi_{01}} \otimes C_{\chi_{02}}$ $(\chi_{0a} = [0, c_a], a = 1, 2)$ of elements $f(x_1, x_2)$ such that the functions

$$\varphi(x_1, x_2) = \left(\frac{1+x_1^2}{2}\right)^{h+c_1} \left(\frac{1+x_2^2}{2}\right)^{h+c_2} f(x_1, x_2),$$

$$\varphi(Rx_1, x_2), \quad \varphi(x_1, Rx_2), \quad \varphi(Rx_1, Rx_2)$$

are infinitely smooth. We notice that the underlying space of pairs of arguments (x_1, x_2) (the points at infinity included) is a homogeneous space of $\mathscr{C}_E$ isomorphic to

$$\mathscr{C}_E^0/MA = \mathrm{SO}^\uparrow(2h+1, 1)/\mathrm{SO}(2h)\mathrm{SO}(1, 1).$$

It is instructive to exhibit this isomorphism, parametrizing the factor space $\mathscr{C}_E^0/MA$ by the two $2h$-dimensional Abelian subgroups of $\mathscr{C}_E$ – the subgroup of translations $T(a)$ and the subgroup $N(b)$ of special conformal transformations (1.15′). That is achieved by noticing that every point (x_1, x_2) can be obtained from a special point, say $(0, \infty)$, by a superposition of transformations N and T:

$$(x_1, x_2) = T(x_1) N\left(\frac{x_2 - x_1}{x_{12}^2}\right)(0, \infty). \tag{2.30}$$

It should be kept in mind that coinciding arguments ($x_{12} \to 0$) is a limit point for a (non compact) family of transformations.

Let c_a be real numbers satisfying

$$(2.31) \quad |c_1|+|c_2| < h.$$

Then the direct product representation $\chi_{01} \otimes \chi_{02}$ can be expanded in unitary irreducible representations of the principal series. In other words, for each $f(x_1, x_2) \in C_{\chi_{01}} \otimes C_{\chi_{02}}$ we can write

$$(2.32) \quad f(x_1, x_2) = \oint d\chi \int V(x_1, c_1, x_2, c_2; x\tilde{\chi}) F_\chi(x) dx,$$

where

$$(2.33) \quad \chi = [l, i\sigma], \qquad \oint d\chi = \sum_{l=0}^{\infty} \int_{-\infty}^{\infty} \frac{d\sigma}{2\pi} \varrho_l(\sigma) \ldots$$

and the 3-point function V remains invariant under the action of the representation $\chi_{01} \otimes \chi_{02} \otimes \tilde{\chi}$ of $O^\uparrow(2h+1, 1)$:

$$\frac{1}{(x_1^2)^{h+c_1}} \frac{1}{(x_2^2)^{h+c_2}} \frac{1}{(x^2)^{h-i\sigma}} r(x)^{\otimes l} V(Rx_1, c_1, Rx_2, c_2; Rx, \tilde{\chi})$$
$$= V(x_1, c_1, x_2, c_2; x, \tilde{\chi}) \quad \text{etc.}$$

Here $\varrho_l(\sigma)$ defines the Plancherel measure on the principal series which according to [11], can be witten as

$$(2.34) \quad \varrho_l(\sigma) = \frac{(l+h-1)!}{2(2\pi)^h l!} \left| \frac{\Gamma(h-1+i\sigma)}{\Gamma(i\sigma)} \right|^2 [\sigma^2+(h+l-1)^2]$$
$$= \frac{(l+h-1)!}{2(2\pi)^h l!} n(\chi) n(\tilde{\chi})$$

($n(\chi)$ is given by (2.24)).

The Clebsch–Gordan kernel V is determined up to a constant factor from conformal invariance. Its normalization will be fixed by the following requirements. The inverse of the map $C_\chi \to C_{\chi_{01}} \otimes C_{\chi_{02}}$ (2.32) should have the form

$$(2.35) \quad F_\chi(x) = \int dx_1 \int dx_2 V(x_1, -c_1, x_2, -c_2; x, \chi) f(x_1, x_2)$$

where $V(x_1, -c_1, x_2, -c_2; x, \chi)$ is obtained from $V(x_1, c_1, x_2, c_2; x, \chi)$ by 'amputation of external legs'. This covariance under amputation is incorporated in the following symmetry properties:

$$\text{(2.36)}\quad V(x_1, \chi_1; x_2, \chi_2; x_3, \tilde{\chi}_3) = \int V(x_1, \chi_1; x_2, \chi_2, x, \chi_3) G_{\tilde{\chi}_3}(x, x_3)\, dx,$$

$$\text{(2.37)}\quad V(x_1, \chi_1; x_2, \chi_2; x_3, \chi_3) = V(x_1, \chi_1, x_3, \chi_3; x_2, \chi_2) = V(x_2, \chi_2; x_1, \chi_1; x_3, \chi_3).$$

They also imply a symmetry relation for the conformal 'Fourier transform':

$$\text{(2.38)}\quad F_\chi(x) = \int G_\chi(x, y) F_{\tilde{\chi}}(y)\, dy.$$

Conformal invariance leads to the following general form for the Clebsch–Gordan kernel (see [17], [28]):

$$\text{(2.39)}\quad V(x_1, c_1, x_2, c_2; x_3, \chi_l, z) = \frac{N_l(c_+, c_-, \delta)}{(2\pi)^h} \times$$
$$\times \frac{(\lambda z)^l}{\sqrt{l!}} \left(\frac{2}{x_{12}^2}\right)^{h+c_+-\delta+\frac{1}{2}l} \left(\frac{2}{x_{13}^2}\right)^{\delta+c_--\frac{l}{2}} \left(\frac{2}{x_{23}^2}\right)^{\delta-c_--\frac{l}{2}},$$

where λ is defined by (1.32) and

$$\text{(2.40)}\quad c_\pm = \tfrac{1}{2}(c_1 \pm c_2), \qquad \delta = \tfrac{1}{2}(h + c_l) \quad \text{for } \chi_l = [l, c_l].$$

We proceed to the determination of the factor N_l from the symmetry properties (2.36), (2.37) and the completeness relation

$$\text{(2.41)}\quad \int dx_1 \int dx_2\, V(x_1, -c_1, x_2, -c_2; x, \chi) \otimes V(x_1, c_1, x_2, c_2; x', \tilde{\chi}') = \mathbf{1}\delta(\chi, \chi') + G_\chi \delta(\chi, \tilde{\chi}'),$$

where

$$\text{(2.42)}\quad \delta(\chi, \chi') = \frac{2\pi\delta(\sigma - \sigma')}{\varrho_l(\sigma)} \delta_{ll'}; \qquad z_1^{\otimes l} \mathbf{1} z_2^{\otimes l} = \delta(x - x')\,(z_1 z_2)^l$$

((2.41) is a consequence of the inversion formula (2.35) and the symmetry properties (2.36–38)). We shall only outline the results refferring to [5] for the rather intricate calculations.

The symmetry properties (2.36), (3.38) applied to the scalar legs, together with the identity (see [4], [27])

$$(2.43)\quad \frac{1}{(2\pi)^h}\int \frac{\Gamma(\delta_1)}{(\frac{1}{2}x_{14}^2)^{\delta_1}}\frac{\Gamma(\delta_2)}{(\frac{1}{2}x_{24}^2)^{\delta_2}}\frac{\Gamma(\delta_3)}{(\frac{1}{2}x_{34}^2)^{\delta_3}}\,dx_4 = \frac{\Gamma(h-\delta_1)}{(\frac{1}{2}x_{23}^2)^{h-\delta_1}}\frac{\Gamma(h-\delta_2)}{(\frac{1}{2}x_{13}^2)^{h-\delta_2}}\frac{\Gamma(h-\delta_3)}{(\frac{1}{2}x_{12}^2)^{h-\delta_3}}$$

$$\text{for } \delta_1+\delta_2+\delta_3 = 2h,$$

give

$$(2.44)\quad \begin{aligned} \frac{N_l(c_+, c_-, \delta)}{N_l(-c_+, -c_-, \delta)} &= \frac{\Gamma\left(h+c_+-\delta+\frac{l}{2}\right)\Gamma\left(\delta+c_++\frac{l}{2}\right)}{\Gamma\left(h-c_+-\delta+\frac{l}{2}\right)\Gamma\left(\delta+c_++\frac{l}{2}\right)},\\ \frac{N_l(c_+, c_-, \delta)}{N_l(-c_+, c_-, \delta)} &= \frac{\Gamma\left(h+c_+-\delta+\frac{l}{2}\right)\Gamma\left(\delta+c_++\frac{l}{2}\right)}{\Gamma\left(h-c_+-\delta+\frac{l}{2}\right)\Gamma\left(\delta-c_++\frac{l}{2}\right)}. \end{aligned}$$

In order to exploit (2.41) we first regularize the integrand replacing c_1 in the second factor by $c_1-2\varepsilon$, and go to the limit $\varepsilon \to +0$ after smearing with a holomorphic function $f(\sigma')$ analytic in a strip along the real axis. Equating the coefficient of $G_\chi\delta(\chi, \tilde{\chi})$ in both sides of (2.41) we obtain

$$(2.45)\quad N_l(c_+, c_-, \delta)N_l(-c_+, -c_-, \delta) = \frac{\Gamma\left(\delta-c_-+\frac{l}{2}\right)\Gamma\left(\delta+c_-+\frac{l}{2}\right)}{\Gamma\left(h-\delta-c_-+\frac{l}{2}\right)\Gamma\left(h-\delta+c_-+\frac{l}{2}\right)}.$$

Multiplying (2.44) and (2.45) we obtain

$$(2.46)\quad N_l(c_+, c_-, \delta) = \left\{\frac{\Gamma\left(h+c_+-\delta+\frac{l}{2}\right)\Gamma\left(c_++\delta+\frac{l}{2}\right)}{\Gamma\left(h-c_+-\delta+\frac{l}{2}\right)\Gamma\left(\delta-c_++\frac{l}{2}\right)}\times \frac{\Gamma\left(\delta-c_-+\frac{l}{2}\right)\Gamma\left(\delta+c_-+\frac{l}{2}\right)}{\Gamma\left(h-\delta-c_-+\frac{l}{2}\right)\Gamma\left(h-\delta+c_-+\frac{l}{2}\right)}\right\}^{1/2}.$$

The sign of the square root is fixed by requiring that for real dimensions, satisfying (2.31) and

$$(2.47)\quad |c_-| < \delta < h-|c_+|$$

the factors N_l are positive.

If we restiict ourselves to the subspace $C^0_{1,2}$ of $C_{\chi_{10}} \otimes C_{\chi_{20}}$ of test functions vanishing together with all their derivatives for $x_1 = x_2$, then the analytic continuation of the conformal Fourier transform $F_{[l,c]}(x)$ with respect to c is expected to satisfy some additional symmetry properties [besides (2.38)], in the exceptional points

$$(2.48)\quad \chi^{\pm}_{ln} = [l, \pm(h+l+n-1)], \qquad \chi_{ln} = [l+n, 1-h-l],$$
$$\tilde{\chi}_{ln} = [l+n, h+l-1], \qquad n = 1, 2, \ldots$$

(cf. the Paley–Wiener type of theorem for the Lorentz group, proven in [8] as well as its generalization to arbitrary semisimple complex Lie groups reviewed in [34]). A complete characterization of the functions $F_\chi(x)$ which are conformal Fourier transforms of smooth functions of $C^0_{1,2}$ is not yet worked out. Some partial results in this direction are presented in the next section.

D. REPRESENTATIONS, INTERTWINING OPERATORS AND CLEBSCH–GORDAN KERNELS AT EXCEPTIONAL (INTEGER) POINTS

Let us have a closer look at the properties of representations at the exceptional points of type (2.48).

First of all we shall demonstrate that they are reducible non-decomposable representations of $O^\uparrow(2h+1, 1)$.

For

$$(2.49)\quad \chi = \chi^-_{ln} = [l, c^-_{ln}], \qquad c^-_{ln} = 1-h-l-n, \qquad n = 1, 2, \ldots$$

the space C_χ contains a finite dimensional invariant subspace E_{ln} which can be described in the following way in the (x, z)-picture. The space E_{ln} consists of all polynomial functions $P(x, z)$ $(x \in R^{2h},\ z^2 = 0)$ which are homogeneous of degree l with respect to z and satisfy the equation

$$(2.50)\quad \left(z\frac{\partial}{\partial x}\right)^n P(x, z) = 0.$$

The general (polynomial) solution of this equation may be written down in the form

$$(2.51)\quad P(x, z) = \sum_{\nu=0}^{n-1} P_\nu(x)\,(x^2)^{n-\nu-1} H_{l\nu}(xz;\, x_\Lambda z;\, x^2 z - 2(xz)x;\, z),$$

$$x_\Lambda z = (x_\mu z_\nu - x_\nu z_\mu)$$

where $P_\nu(x)$ is a polynomial of degree $\leqslant \nu$, and $H_{l\nu}$ is a homogeneous polynomial of degree l (of its $(h+1)(2h+1)$ arguments). Thus the total degree of $P(x, z)$ in x does not exceed $2(l+n-1)$. We leave it to the reader to verify that the representation $\chi_{\overline{ln}}$ defined by (2.1)–(2.3) with $c_{\overline{ln}}$ given by (2.49) acts as a (finite dimensional) IR on E_{ln}.

The factor space $C_{\chi_{\overline{ln}}/E_{ln}}$ is isomorphic to an infinite dimensional invariant subspace F_{ln} of $C_{\chi^+_{ln}}$, where

$$(2.52)\quad \chi^+_{ln} = \tilde{\chi}_{\overline{ln}} = [l,\, h+l+n-1].$$

The space F_{ln} can be characterized by the condition

$$f(x, \zeta) \in F_{ln} \quad \text{iff} \quad f(x, \zeta) \in C_{\chi^+_{ln}} \quad (\Delta_\zeta f(x, \zeta) = 0)$$

and

$$(2.53)\quad \int P(x, \partial_\zeta) f(x, \zeta)\, dx = 0, \qquad \forall P(x, z) \in E_{ln}.$$

The space $C_{\chi_{ln}}$ contains an infinite dimensional invariant subspace D^-_{ln} of functions of the form

$$(2.54)\quad f(x, z) = (z\partial)^n g(x, z) \qquad \left(\partial \equiv \frac{\partial}{\partial x}\right), \qquad g \in C_{\chi_{\overline{ln}}}.$$

The factor space $C_{\chi_{ln}}/D^-_{ln}$ is isomorphic to an infinite dimensional invariant subspace D^+_{ln} of $C_{\tilde{\chi}_{ln}}$ of functions $f(x, \zeta)$ satisfying the differential equation

$$(2.55)\quad (\partial_\zeta \partial)^n f(x, \zeta) = 0 \qquad (\Delta_\zeta f(x, \zeta) = 0).$$

The situation is summarized on the following diagram (cp. with a similar diagram for the case of the connected Lorentz group, presented in [8], Ch. III, Sec. 3.3, p. 155):

Here every directed sequence is exact and the upper and lower triangle diagrams are commutative.

The $G^{(u)}$'s are integral operators with densities defined by (2.15) (or (2.19)) and (2.24a). In particular, $G^{(u)}_{\chi_{\overline{ln}}}(x_{12})$ is a polynomial in x_{12}

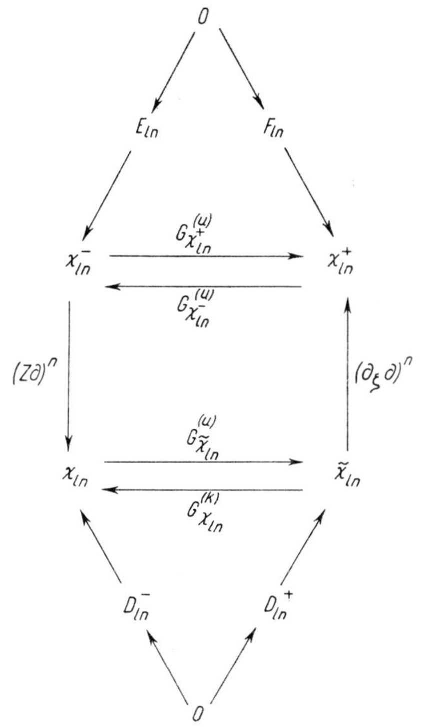

Fig. 1. Diagram of intertwining operators at exceptional points

and annihilates the invariant subspace F_{ln} of $C_{\chi^+_{ln}}$. The Fourier transforms of $G^{(u)}_{\tilde{\chi}_{ln}}$ and $G^{(u)}_{\chi_{ln}}$ are orthogonal to each other (according to (2.25a)):

(2.56a) $G^{(u)}_{\tilde{\chi}_{ln}}(p)$

$$= \frac{(\frac{1}{2}p^2)^{h+l-1}}{(2h+2l+n-2)!} \sum_{k=1}^{n} \frac{(n-1)!(2l+2h+k-3)!}{(k-1)!(l+2h-3)!} \Pi^{l+n} \overset{l+k}{(p)},$$

(2.56b) $G^{(u)}_{\chi_{ln}}(p)$

$$= \frac{l!}{n!} \left(\frac{2}{p^2}\right)^{h+l-1} \sum_{s=0}^{l} \frac{(-1)^s}{(l-s)!} \frac{(2l+2h+n-3)!}{(l+s+2h-3)!} \Pi^{l+n}\overset{s}{(p)}.$$

The normalized Green functions $G_{\tilde{\chi}_{ln}}$ and $G_{\chi_{ln}}$ can also be defined on the invariant subspaces D^-_{ln} and D^+_{ln}, respectively. For instance, the mapping

$$G_{\tilde{\chi}_{ln}} : D_{\bar{l}n}^{-} \to C_{\chi_{ln}}/D_{\bar{l}n}^{+}$$

is given (in momentum space) by

$$G_{\tilde{\chi}_{ln}}(p) = \left(\frac{p^2}{2}\right)^{h+l-1} \sum_{s=0}^{l} (-1)^s \frac{(l-s)!(2h+l+s-3)!}{l!\,(2h+l-3)!} \overset{s}{\Pi}{}^{l+n}(p).$$

We note that the bilinear form

$$(2.57)\quad \int dx_1 \int dx_2 f(x_1)\, G_\chi(x_1, x_2) g(x_2)$$

where G_χ is normalized according to (2.24b), (2.25b), also has a limit for $\chi \to \chi_{\bar{l}n}$ provided that the test functions f and g belong to the invariant subspace F_{ln} of $C\chi_{ln}^{+}$ (i.e. satisfy (2.53)). To see that, we observe that according to (2.15), (2.24b)

$$G_{[l,\,1-h-l-n+\varepsilon]}(x, 0; z_1, z_2)$$
$$\underset{\varepsilon\downarrow 0}{\approx} \frac{1}{(2\pi)^h} \frac{(2h+l+n-3)!}{(h+l+n-2)!\,(2h+2l+n-3)!} \frac{(-1)^{n-1}}{(n-1)!} \times$$
$$\times \left(\frac{x^2}{2}\right)^{n+l-1} \left(\frac{1}{\varepsilon} - \ln\frac{x^2}{2} + O(\varepsilon)\right) \frac{(-z_1 \tau(x) z_2)^l}{l!}$$

(cf. [1]). The first term $\left(\text{of order } \frac{1}{\varepsilon}\right)$ gives zero contribution to (2.18) for f, g satisfying (2.53). Moreover, for such f and g the bilinear form

$$(2.58)\quad (f, G_{\chi_{\bar{l}n}} g)$$
$$\equiv \frac{1}{(2\pi)^h} \frac{(-1)^n}{(n-1)!} \frac{(2h+l+n-3)!}{(h+l+n-2)!\,(2h+2l+n-3)!} \times$$
$$\times \int dx_1 \int dx_2\, f(x_1, \zeta_1) \left(\frac{x_{12}^2}{2}\right)^{n+l-1} \times$$
$$\times \ln\left(\frac{x_{12}^2}{2}\right) \frac{(\overleftarrow{\partial}_1 r(x_{12}) \vec{\partial}_2)^l}{l!} g(x_2, \zeta_2)$$

is conformal invariant.

In addition to the symmetry property (2.36) at nonexceptional points the Clebsch–Gordan kernels V satisfy the following type of relations at the points $\chi_{ln}^{(\pm)}$ of Fig. 1:

$$(2.59)\quad (z\partial_3)^n V(x_1, c_1, x_2, c_2; x_3, \chi_{\bar{l}n}^{-}, z) = a_{ln} V(x_1, c_1, x_2, c_2; x_3, \chi_{ln}, z),$$

$$(2.60)\quad (\partial_\zeta \partial_3)^n V(x_1, c_1, x_2, c_2; x_3, \tilde{\chi}_{ln}, \zeta)$$
$$= b_{ln} V(x_1, c_1, x_2, c_2; x_3, \chi_{ln}^{+}, \zeta).$$

The constant in (2.59) is found to be

$$(2.61) \quad a_{ln} = \sqrt{\frac{(l+n)!}{l!}}\,.$$

Eqs. (2.59), (2.60) imply similar relations for the conformal Fourier components (2.35) whenever they are well defined for exceptional c's.

4. Conformal Partial Wave Expansion

A. EXPANSION OF THE 4-POINT FUNCTION AND SOLUTION OF THE BS EQUATION

The direct product decomposition, studied in the previous section, is readily extended to elements of the (real) Hilbert space $H_{1,2} = H_{\chi_{01}} \otimes H_{\chi_{02}}$ equipped with the scalar product

$$(3.1) \quad (f, g) = \int \dots \int f(x_1, x_2) G_{\tilde{\chi}_{01}}(x_1, y_1) G\chi_{02}(x_2, y_2) \times \\ \times g(y_1, y_2) dx_1 dx_2 dy_1 dy_2 .$$

The analysis of the ultraviolet behaviour of conformal quantum field theory performed in [16] indicates that (at least in any finite order of the skeleton perturbation theory) the 1-particle irreducible Green functions (1.3), regarded as functions of x_1, x_2 (for fixed $x_3, \dots, x_n$), satisfy the integrability conditions (3.1) for c in the domain (1.34) (while the 1-particle reducible part does not). The conformal expansion formula (2.32) for G_{1i} is written as

$$(3.2) \quad G_{1i}(x_1, x_2; x_3, \dots, x_n) = \sum\!\!\!\!\!\!\int d\chi \int dx\, V(x_1, c, x_2, c; x\tilde{\chi}) \times \\ \times G_\chi(x; x_3, \dots, x_n) .$$

The image G_χ of G_{1i} is a covariant $(n-1)$-point function transforming according to the representations χ with respect to the first argument and according to the representations $[0, c]$ with respect to the remaining $n-2$ arguments. Hence, for $n = 4$ it should be proportional to the Clebsch–Gordan kernel (2.39):

$$(3.3) \quad G_\chi(x; x_3, x_4) = g(\chi) V(x_3, c, x_4, c; x, \chi).$$

The symmetry properties (2.36), (2.38) and (2.59), (2.60) imply that the conformal partial waves $g(\chi)$ obey the simple symmetry relations

(3.4a) $g(\tilde{\chi}) = g(\chi)$,

(3.4b) $g(\chi_{l\bar{n}}) = g(\chi_{ln})$, $\quad g(\tilde{\chi}_{ln}) = g(\chi^{+}_{l\bar{n}})$.

Using (2.41), (2.36) we can express $g(\chi)$ in terms of G_{1i}:

$$g(\chi)V(x_1, c, x_2, c; x\chi z)$$
$$= \tfrac{1}{2}\int dx_3 \int dx_4\, G_{1i}(x_1, x_2; x_3, x_4)V(x_3, -c, x_4, -c; x\chi z)$$

or, after integration in x and application of the operator $\left[x_{12}\dfrac{\partial}{\partial z}\right]^l$ to both sides,

(3.5) $$g(\chi) = \frac{l!}{2}\,\frac{\Gamma\left(h-c-\delta+\frac{l}{2}\right)\Gamma\left(\delta-c+\frac{l}{2}\right)\Gamma\left(h-\frac{1}{2}\right)}{\Gamma\left(h+c-\delta+\frac{l}{2}\right)\Gamma\left(\delta+c+\frac{l}{2}\right)\Gamma\left(l+h-\frac{1}{2}\right)}\times$$
$$\times\left(\frac{x_{12}^2}{2}\right)^{c+\delta}\int dx_3\int dx_4\, G_{1i}(x_1, x_2; x_3, x_4)\left(\frac{2}{x_{34}^2}\right)^{\delta-c}\times$$
$$\times P_l^{\left(h-\frac{3}{2}, h-\frac{3}{2}\right)}\left(\frac{x_{12}x_{34}}{\sqrt{x_{12}^2 x_{34}^2}}\right).$$

In deriving (3.5) we have used that $\left[\xi\dfrac{\partial}{\partial z}\right]^l H_l(z, x) = l!\, H_l(\xi, x)$, where H_l is defined by (2.7′). A similar formula can be obtained by exchanging the places of (x_1, x_2) and (x_3, x_4); the two formulas are consistent between each other because of the symmetry of G_{1i} with respect to the substitution $(x_1, x_2) \rightleftarrows (x_3, x_4)$.

Conformal expansion of the type (3.2), (3.3) is applicable for any 4-point function, satisfying the integrability condition (3.1) in each pair of arguments. In particular, it can be written for the BS kernel (1.4):

(3.6) $$B(x_1, x_2; x_3, x_4)$$
$$= \displaystyle\sum\!\!\!\!\!\!\int d\chi b(\chi)\int dx V(x_1, c, x_2, c; x, \tilde{\chi})\; V(x_3, c, x_4, c; x, \chi).$$

The conformal partial wave expansion allows us to diagonalize the BS equation. Indeed, inserting (3.2), (3.3) and (3.6) in eq. (1.5) and using the completeness relation (2.41) and the symmetry property (2.36), we obtain the simple algebraic equation

$$g(\chi) = b(\chi)+b(\chi)g(\chi). \tag{3.7}$$

Solving (3.7) with respect to $g(\chi)$,

$$g(\chi) = \frac{b(\chi)}{1-b(\chi)}, \tag{3.8}$$

we see that $g(\chi)$ has a pole in each (analyticity) point $\chi = \chi_a$ of $b(\chi)$ for which

$$b(\chi_a) = 1. \tag{3.9}$$

It turns out that the conformal transforms G_χ of the n-point functions also have poles for such χ's. This is a simple consequence of eq. (1.6) and of the expansion formulas (3.2), (3.6).

B. SOLUTION OF THE DYNAMICAL EQUATIONS

Now we proceed to the solution of the dynamical equations by first exploring the consequences of the so-called 'boot-strap' equations (1.9) and (1.19). According to the results of [16] the equation (1.9) for the vertex function can be rewritten as an equation for the full 3-point function without subtractions:

$$\begin{aligned} G(x_1, x_2, x_3) &= \tfrac{1}{2}\int \dots \int G(x_1', x_2', x_3)G^{-1}(x_1', x_1'')G^{-1}(x_2', x_2'') \times \\ &\times B(x_1'', x_2''; x_1, x_2)dx_1'\, dx_2' dx_1'' dx_2''. \end{aligned} \tag{3.10}$$

Expressing $G(x_1, x_2, x_3)$ in terms of V according to (1.30a), inserting the conformal partial wave expansion (3.6) for the BS kernel and using the symmetry relations (2.36), (2.37), we obtain

$$\begin{aligned} V(x_1, c, x_2, c, x_3, c) &= \sum\!\!\!\!\!\!\int d\chi b(\chi) \times \\ &\times \tfrac{1}{2}\int dx_1' \int dx_2' \int dx V(x_1', -c, x_2', -c, x_3, c) \times \\ &\times V(x_1', c, x_2', c; x, \tilde{\chi})V(x_1, c, x_2, c; x, \chi). \end{aligned}$$

Finally, continuing analytically the completeness relation (2.41) from the points $\chi = [0, i\sigma]$ of the principal series to $\chi = \chi_0 = [0, c]$, we find

$$(3.11) \quad b(\chi_0) \equiv b(c) = 1 \qquad (\chi_0 = [0, c]).$$

The analytic continuation in the dimension, associated with an external line of a Green function, can be justified in the framework of the skeleton diagram expansion by the argument contained in [16].

The results of the preceding section together with eq. (3.11) imply that the conformal partial waves $g(\chi)$ and G_χ have a pole in χ for $\chi = \chi_0$. We shall assume (in accord with the indications from the skeleton expansion) that it is a simple pole, so that

$$(3.12) \quad b_0 = \frac{db}{dc} = -[\operatorname*{Res}_{\chi=\chi_0} g(\chi)]^{-1} \neq 0.$$

We assume here that $-\frac{1}{3}h < c < 0$. This ensures the positivity of the residue at the physical pole $\chi = \chi_0$ and the cancellation of the 'shadow pole' at $\chi = \tilde{\chi}_0$ (see eq. (3.26) and Sec. 3.C below).

Similarly, starting with eq. (1.19), we find that

$$(3.13) \quad b(\chi_2) = 1 \quad \text{for} \quad \chi_2 = [2, h].$$

We shall again assume that

$$(3.14) \quad b_2 = \left[\frac{db([2, c])}{dc}\right]_{c=h} \neq 0.$$

It follows that $\chi = \chi_2$ is also a (simple) pole of the conformal partial waves.

Consider next eq. (1.8), inserting in the right-hand side the partial wave expansion for the $2i$-kernel and the expression (1.30) for the 3-point function. Using again the analytically continued completeness relation (2.41), we obtain

$$(3.15) \quad G(x_1, \ldots, x_n) = g\Big[G_{2i}^{\chi_0}(x_1; x_2, \ldots, x_n) +$$
$$+\sum \int dy_1 \int dy_2 \, V(x_1, c, y_1, -c, y_2, -c) \times$$
$$\times \, G(y_1, x_{i_2}, \ldots, x_{i_k}) G(y_2, x_{i_{k+1}}, \ldots, x_{in}),$$

where the sum is over all $2^{n-2}-1$ partitions of the external lines $2, \ldots, n$ into two non-empty sets. Applying on the other hand the conformal expansion to both sides of eq. (1.6), which defines the $2i$-kernel and going to the limit $\chi \to \chi_0$:

$$(3.16)\quad \lim_{\chi\to\chi_0}[1-b(\chi)]G_\chi(x; x_2, \ldots, x_n) = -b_0 \operatorname*{Res}_{\chi=\chi_0} G_\chi(x; x_2, \ldots, x_n), \qquad \chi_0 = [0, c],$$

we can express the right-hand side of (3.15) in terms of the partial wave of the $(n+1)$-point function. The result is

$$(3.17)\quad G(x_1, \ldots, x_n) = -gb_0 \operatorname*{Res}_{\chi=\chi_0} G_\chi(x_1; x_2, \ldots, x_n),$$

or, equivalently,

$$(3.18)\quad \operatorname*{Res}_{\chi=\chi_0} \int V(x_1, c, x_2, c; x, \tilde{\chi}) G_\chi(x; x_3, \ldots, x_n)dx = \frac{-1}{g^2 b_0}\int dx \int dy\, G(x_1, x_2, x) G^{-1}(x, y) G(y, x_3, \ldots, x_n).$$

Thus, we see that the dynamical equations imply some factorization properties for the residues of the conformal partial waves.

Similar relations are valid for Green functions involving (symmetric) tensor fields.

$$O_l(x, z) = \frac{1}{\sqrt{l}} O^{\mu_1\ldots\mu_l}(x) z_{\mu_1} \ldots z_{\mu_l}.$$

The connected Green functions

$$(3.19)\quad G_{\chi_l}(x, z; x_1, \ldots, x_n) = \langle 0|TO_l(x, z)\varphi(x_1) \ldots \varphi(x_n)|0\rangle_{\mathrm{con}}$$

(and their Euclidean counterparts) satisfy an infinite set of integral equations analogous to (1.8) (1.19):

(3.20) 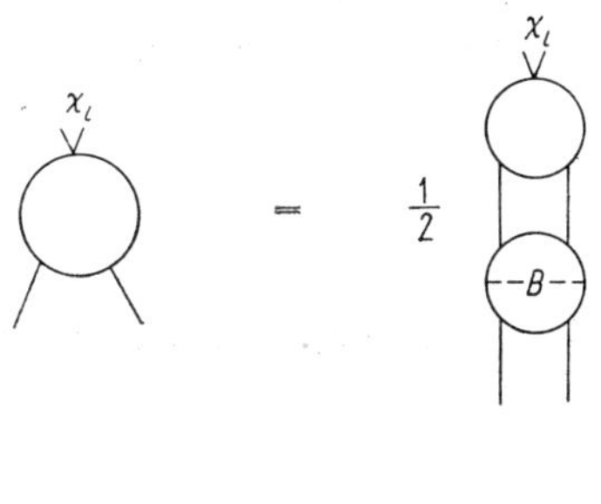

, $\chi_l = [l, c_l]$,

(3.21)

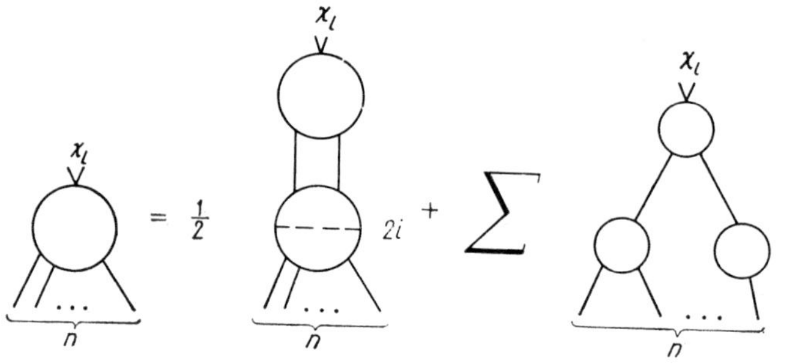

$, \quad n \geqslant 3.$

(In particular, for $l = 2$, $c_l = h$, $O_{\mu\nu} = \theta_{\mu\nu}$ eq. (3.20) is equivalent to eq. (1.19).)

Conformal invariance implies that

$$(3.22)\quad G_{\chi_l}(x; x_1, x_2) = a_l V(x_1, c, x_2, c; x, \chi_l);$$

$a_0 = g$ (according to (1.30a)), a_2 is fixed by the Ward identity

$$\left(a_2 = h!\,\Gamma(c+1+h)\left[2(2h-1)\,(2\pi)^h \Gamma(-c) N_2(c, 0, h)\right]^{-1}\right).$$

If $a_l \neq 0$, then

$$(3.23)\quad b(\chi_l) = 1$$

and all conformal partial waves $G_\chi(x; x_1, \ldots, x_n)$ have poles for $\chi = \chi_l$. Moreover, the following generalization of (3.17) holds:

$$(3.24)\quad G_{\chi_l}(x; x_1, \ldots, x_n) = -a_l b_l \operatorname*{Res}_{\chi=\chi_l} G_\chi(x; x_1, \ldots, x_n)$$

where

$$(3.25)\quad b_l = \left.\frac{db([l, c])}{dc}\right|_{c=c_l} = -\left[\operatorname*{Res}_{\chi=\chi_l} g(\chi)\right]^{-1}.$$

Further exploration of the Schwinger–Dyson equation for the 2-point function in the lines of [21] allows to relate the residue b_0^{-1} with the effective coupling constant g:

$$(3.26)\quad \operatorname*{Res}_{\chi=\chi_0} g(\chi) = -\frac{1}{b_0} = \frac{g^2}{2\varrho_0(-ic)};$$

here ϱ_l is the Plancherel weight (2.34) (cf. [13] where similar relations are also derived for the residues b_l^{-1}).

C. RELATION TO CONFORMAL COVARIANT OPERATOR PRODUCT EXPANSION

The conformal expansion is related to the operator product decomposition of ref. [6] in a similar way as the Sommerfeld–Watson representation (which is a $SO^{\uparrow}(2, 1)$ expansion of the scattering amplitude) is related to the asymptotic Regge pole decomposition.

First of all, we shall find the analog of the splitting of the Legendre function P_l into two second kind functions Q_l and Q_{-l-1}. The role of P_l is played in our case by the Clebsch–Gordan kernel V.

Noting that the Fourier transform of V in x_3 can be written as

$$(3.27)\quad V^{\chi}(x_1, x_2; p) = \int V(x_1, c, x_2, c; x_3, \chi)e^{-ipx_3}dx_3$$

$$= \frac{N_l(c, 0, \delta)}{(2\pi)^h}\left(\frac{2}{x_{12}^2}\right)^{h+c-\delta+\frac{l}{2}} \vec{D}_l \int e^{-ipx_3}\left(\frac{4}{x_{13}^2 x_{23}^2}\right)^{\delta-\frac{l}{2}} dx_3$$

$$= 2\frac{N_l(c, 0, \delta)}{\Gamma^2\left(\delta-\frac{l}{2}\right)}\left(\frac{2}{x_{12}^2}\right)^{h+c-\delta+\frac{l}{2}} \vec{D}_l \left(\frac{x_{12}^2}{p^2}\right)^{\frac{1}{2}(h+l)-\delta} \int_0^1 du \times$$

$$\times e^{-ip[ux_1+(1-u)x_2]}\,[u(1-u)]^{\frac{1}{2}h-1}\, K_{h-2\delta+l}\left(\sqrt{u(1-u)p^2x_{12}^2}\right),$$

where

$$K_\nu(z) = \frac{1}{2}\left(\frac{z}{2}\right)^{\nu} \int_0^\infty e^{-t-\frac{z^2}{4t}} \frac{dt}{t^{\nu+1}},$$

$$(3.28)\quad \vec{D}_l(z) = \frac{1}{\sqrt{l!}}\sum_{k=0}^{l}\binom{l}{k}\frac{\Gamma\left(\frac{l}{2}-\delta-k+1\right)\Gamma\left(k-\frac{l}{2}-\delta+1\right)}{\Gamma^2\left(\frac{l}{2}-\delta+1\right)} \times$$

$$\times (z\vec{\partial}_1)^{l-k}(-z\vec{\partial}_2)^k \qquad (z^2 = 0),$$

and that

$$K_\nu(t) = \frac{\pi}{2}(\sin\pi\nu)^{-1}[I_{-\nu}(t) - I_\nu(t)]$$

we shall set

$$(3.29)\quad V^{\chi}(x_1, x_2; p)$$

$$= \frac{\pi(-1)^l}{\sin\pi(2\delta-h)}\,[Q^{\chi}(x_1, x_2; p) - G^{\chi}(p)Q^{\tilde{\chi}}(x_1, x_2; p)],$$

with

$$(3.30)\quad Q^{\varkappa}(x_1, x_2; p) = -\frac{N_l(c, 0, \delta)}{\Gamma^2\left(\delta - \frac{l}{2}\right)} \left(\frac{2}{x_{12}^2}\right)^{h+c-\delta+\frac{l}{2}} \times$$

$$\times \vec{D}_l \left\{ \left(\frac{x_{12}^2}{2p^2}\right)^{\frac{1}{2}(h+l)-\delta} \int_0^1 du e^{-ip[ux_1+(1-u)x_2]} [u(1-u)]^{\frac{1}{2}h-1} \times\right.$$

$$\left. \times I_{h-2\delta+l}\left(\sqrt{u(1-u)p^2 x_{12}^2}\right)\right\}.$$

The decomposition (3.29), (3.30) has the following characteristic properties (reminiscent to known properties of the splitting of Legendre functions

$$P_{-\frac{1}{2}+\lambda}(t) = \frac{\cos \pi\lambda}{\pi} [Q_{-\frac{1}{2}+\lambda}(t) + Q_{-\frac{1}{2}-\lambda}(t)]$$

in the Regge theory).

(i) It makes obvious the symmetry property (2.36):

$$(3.31)\quad V^{\varkappa}(x_1, x_2; p) = G^{\varkappa}(p) V^{\tilde{\varkappa}}(x_1, x_2; p).$$

(ii) The small x_{12} behaviour of $Q^{\varkappa}$ is given by

$$(3.32)\quad Q^{\varkappa}(x_1, x_2; p) \underset{x_{1,2}\to 0}{\approx} -N_l \frac{B\left(h-\delta+\frac{l}{2}, h-\delta+\frac{l}{2}\right)}{\Gamma^2\left(\delta-\frac{l}{2}\right)} \times$$

$$\times \left(\frac{2}{x_{12}^2}\right)^{h+c-\delta+\frac{l}{2}} \vec{D}_l \left(\frac{x_{12}^2}{2}\right)^{h+l-2\delta}$$

$$= -N_l \left[\frac{\Gamma\left(h-\delta+\frac{l}{2}\right)\Gamma\left(1-\delta-\frac{l}{2}\right)}{\Gamma\left(\delta-\frac{l}{2}\right)\Gamma\left(1+\frac{l}{2}-\delta\right)}\right]^2 \times$$

$$\times \frac{\Gamma(2-2\delta)}{\Gamma(2h-2\delta+l)\Gamma(2-2\delta-l)} \left(\frac{2}{x_{12}^2}\right)^{c+\delta} \frac{(x_{12}z)^l}{\sqrt{l!}}.$$

It has a power-type singularity in x_{12}^2 whose degree is an increasing function of $\operatorname{Re}\delta$.

(iii) Q_l is an entire analytic function of p.

The relation between the conformal partial wave expansion (3.2) and the operator product expansion is established by the following procedure. First, we express V in terms of Q, writing eq. (3.2) in the form

$$(3.33)\quad G_{1i}(x_1, x_2; x_3, \dots, x_n) = 2\pi \sum\!\!\!\!\!\!\int \frac{d\chi}{\sin\pi(2\delta-h)} [Q^{\tilde{\chi}}(x_1, x_2; i\partial_x) \times \times G_\chi(x; x_3, \dots, x_n)]_{x=0}.$$

Second, we expand the 1-particle reducible and the disconnected Green function in terms of Q^χ. This can be done by using (3.18) and shifting the integration path in the complex δ-plane. It turns out that due to the relation (3.26) between b_0 and g, the expansion of the 1-particle reducible part of G cancels the so-called shadow singularity of G_χ for $\chi = \tilde{\chi}_0 = [0, -c]$ of the basic field $\varphi(x)$. Finally, we shift the contour of δ-integration in (3.33) to the right and pick up the contribution of the encountered pole terms. In order to perform this last step, we have to assume that G_χ is a meromorphic function of δ with poles corresponding to the dimensions of the physical tensor fields $O_l(x)$, considered in the preceding section. Here we encounter a technical problem: are all poles coming from the factor $[\sin\pi(2\delta-h)]^{-1}$ in (3.33) cancelled and if so which is the precise mechanism of such a cancellation? Modulo this unresolved problem we obtain an expansion of the complete s-functions of the following type:

$$(3.34)\quad S(x_1, \dots, x_n) = S(x_1, x_2) S(x_3, \dots, x_n) + +2 \sum_l c_l [Q^{\tilde{\chi}_l}(x_1, x_2; i\partial_x) S_l(x, x_3, \dots, x_n)]$$

where $S_l(x; x_3, \dots, x_n)$ is the Euclidean region continuation of the function

$$\langle 0 | T O_l(x) \varphi(x_3) \dots \varphi(x_n) | 0 \rangle,$$

and $O_0(x)$ is identified with $\varphi(x)$. Eq. (3.34) corresponds to the operator product expansion

$$(3.35)\quad T\varphi(x_1)\varphi(x_2) = \langle 0 | T\varphi(x_1)\varphi(x_2) | 0 \rangle \mathbf{1} + +2 \sum_l c_l [Q^{\tilde{\chi}_l}(x_1, x_2; i\partial_x) O_l(x)]_{x=0}.$$

We have written all conformal expansions for the pair of variables (x_1, x_2), i.e. for a selected channel. It is therefore not surprising that the problem of expressing the permutation symmetry of $s(x_1, ..., x_n)$ (or $G(x_1, ..., x_n)$) in terms of the conformal partial waves is a non-trivial one. It was studied in [22] and [13], but is far from being completely resolved yet.

5. Acknowledgments

I would like to thank Professor G. Mack for a helpful correspondence and Professors F. Gürsey and D. Želobenko for enlightening discussions. I benefited greatly from the stimulating atmosphere at the seminars in Scuola Normale Superiore and I wish to thank Professors L. A. Radicati and A. Di Giacomo for organizing the present lectures. I am very grateful to Dr. D. R. Dumitru for his help in editing these notes.

It is a pleasure to thank Professors K. Maurin and R. Rączka for their hospitality at the Warsaw Symposium.

References

* Lectures presented in Scuola Normale Superiore, Pisa (June, 1974). Extended version of a talk given at the International Symposium on Recent Progress in Mathematical Physics, Warsaw (25–30 March, 1974).

Supported by Consiglio Nazionale delle Ricerche.

** On leave of absence from the Institute of Nuclear Research and Nuclear Energy, Bulgarian Academy of Sciences, Sofia.

[1] Here $r_{\mu\nu}(x) = x^2\partial_\mu(Rx)_\nu = \dfrac{2x_\mu x_\nu}{x^2} - \delta_{\mu\nu}$ (see (1.21) below).

Bibliography

[1] I. Bars and F. Gürsey, *J. Math. Phys.* **13** (1972) 131.

[2] N. N. Bogolubov and D. V. Shirkov, *Introduction to the Theory of Quantized Fields*, Interscience, N. Y. 1959.

[3] C. G. Callan Jr., *Phys. Rev.* **D2** (1970) 1541.

[4] M. D'Eramo, L. Peliti and G. Parisi, *Nuovo Cimento Letters* **2** (1971) 878.

[5] V. Dobrev, G. Mack, V. Petkova, S. Petrova and I. Todorov, 'On Clebsch–Gordan expansions for $O(2h+1,1)$', Preprint JINR, E2-7977 Dubna (1974) (to be submitted to *Reports on Math. Phys.*).

[6] S. Ferrara, A. Grillo and R. Gatto, *Ann. Phys.* (*N. Y.*) **67** (1973) 1.

[7] I. M. Gel'fand and G. E. Shilov, *Generalized Functions*, Vol. 1, Academic Press, N. Y. 1964.

[8] I. M. Gel'fand, M. I. Graev and N. Ya. Vilenkin, *Generalized Functions*, Vol. 5, Academic Press, N. Y. 1966.
[9] F. Gürsey and S. Orfanidis, *Phys. Rev.* **D7** (1973) 2415.
[10] T. Hirai, *Proc. Japan Acad.* **38** (1962) 258.
[11] T. Hirai, *Proc. Japan Acad.* **42**, (1966) 323.
[12] G. Kurian, N. Mukunda and E. C. G. Sudarshan, *Commun. Math. Phys.* **8** (1968) 204.
[13] G. Mack, in: *Renormalization and Invariance in Quantum Field Theory* (edited by F. R. Caianello, Plenum Press, N. Y., 1974, pp. 123–157, and *J. de Physique* **34**, C1 Suppl. No 10 (1973) 99–106.
[14] G. Mack and Abdus Salam, *Ann. Phys. (N. Y.)* **53** (1969) 174.
[15] G. Mack and K. Symanzik, *Commun. Math. Phys.* **27** (1972) 247.
[16] G. Mack and I. T. Todorov, *Phys. Rev.* **D8** (1973) 1764.
[17] A. A. Migdal, '4-Dimensional Soluble Models of Conformal Field Theory', Preprint, Landau Institute for Theoretical Physics, Chernogolovka (1972).
[18] M. A. Naimark, *Trudy Mosc. Mat. Obsc.* **10** (1961) 181 (English transl.: *Ann. Math. Soc. Transl.* **36** (1964) 189); *DAN SSSR* **130** (1960) 261.
[19] K. Osterwalder, in: *Constructive Quantum Field Theory* (edited by G. Velo and A. Wightman), *Lecture Notes in Physics* **25**, Springer, Berlin 1973, 71–93.
[20] U. Ottoson, *Commun. Math. Phys.* **8** (1968) 228.
[21] G. Parisi, *Nuovo Cimento Letters* **4** (1972) 777.
[22] A. M. Polyakov, 'Non-Hamiltonian Approach to the Quantum Field Theory at Small Distances', Preprint, Landau Institute for Theoretical Physics, Chernogolovka (1973).
[23] W. Rühl, *Commun. Math. Phys.* **30** (1973) 287.
[24] G. Schiffmann, *Bull. Soc. Math. France* **99** (1971) 3.
[25] K. Symanzik, in: *Lectures in High Energy Physics*, edited by B. Jaksič, Zagreb 1961, Gordon and Breach, N. Y. 1965, 485–517.
[26] K. Symanzik, *Commun. Math. Phys.* **18** (1970) 227; *ibid.* **23** (1971) 49.
[27] K. Symanzik, *Nuovo Cimento Letters* **3** (1972) 734.
[28] I. T. Todorov, *Acta Phys. Austr., Suppl.* **11** (1973) 241; I. T. Todorov, 'Conformal Invariant Quantum Field Theory with Anomalous Dimensions', Cargèse lecture notes, Ref. TH. 1697-CERN (1973).
[29] I. T. Todorov and R. P. Zaikov, *J. Math. Phys.* **10** (1969) 2014.
[30] G. Warner, *Harmonic Analysis on Semi-Simple Lie Groups*, Springer, Berlin 1972 (see, in particular, Sec. 5.5).
[31] K. Wilson, *Phys. Rev.* **D3** (1971) 1818.
[32] R. P. Zaikov, 'Spectral Representation of Conformal Invariant Two-Point Function for Fields of Arbitrary Spin', *Bulgarian Journal of Physics* (to be published).
[33] R. P. Zaikov, 'Conformal Invariant Euclidean Two-Point Function for Tensor Fields', Preprint, Higher Pedag. Institute, Shumen (1974).
[34] D. P. Želobenko, in: *Mathematical Analysis*, Moscow, VINITI (1973) 51–90 (in Russian).

PART TWO

NONLINEAR PROBLEMS IN FIELD THEORY

SOME AD-HOC TECHNIQUES FOR NONLINEAR PARTIAL DIFFERENTIAL EQUATIONS*

W. F. AMES

Georgia Institute of Technology, Atlanta, Ga. 30332, U.S.A.

1.0. Introduction

Much of the progress in the solution methodology in the linear theory of partial differential equations has resulted from various assumptions such as separability, invariance and progressing wave forms. The ad hoc techniques of the linear theory may prove useful in specific nonlinear problems. However, their great utility rests upon the principle of (linear) superposition. In accordance with the principle, elementary solutions of the equations are combined to yield more flexible ones, namely, solutions which can satisfy the auxiliary (initial and boundary) conditions arising from the modeled phenomena. In nonlinear problems this general principle no longer applies. While its loss is a severe one there exist classes of nonlinear (and even linear!) equations which possess nonlinear superposition principles. Of course there is no universal nonlinear superposition – such a situation would be beyond our fondest dreams!

The analytic processes in this field, like Bunyan's road, have a quagmire on one side and a deep ditch on the other, and branch off into innumerable bypaths that end in a wilderness. In spite of all the obstacles and lack of effective machinery, considerable information has been extracted from nonlinear partial differential equations by utilizing special procedures, which, herein are called ad-hoc methods. The originators[1] of some of these techniques may be compared to a designer of garments, who is utterly oblivious of the creatures whom his garments may fit. To be sure his art originated in the necessity for clothing such creatures, but this was long ago. To this day a shape will occasionally appear which will fit into the garment as if the garment had been made for it. Then there is no end of surprise and of delight! The exact ad hoc procedures detailed here are not intended to be exhaustive but are typical of those which have been found useful in a variety of problems in science and technology. Some of these have a long history while others have an original flavor. Some are easily generalized while others seem to be dead ends. Additional material and references are available in Ames [6, 12]. The procedures as arbitrarily divided into several categories which form the titles of the various sections.

2.0. Separation and Generalizations

2.1. SIMPLE SEPARATION

Perhaps the first idea to come to mind is that of simple separation of variables. The technique is indeed useful and has been employed by Smith [115] in one-dimensional anisentropic gas flow while Tomotika and Tamada [121], Tamada [119], Sedov [113], Lidov [76] and Keller [65] obtain exact solutions for other gas flows by this method. Irmay [59] discusses simple separation of nonlinear parabolic equations. Other detailed examples can be found in Ames [6]. Since this method is so well known only one example will be considered mainly to illustrate limitations.

The transverse oscillations of a flexible string of length L are governed by the equation

$$(2.1) \quad (T\sin\theta)_x = my_{tt}.$$

Using the classical small vibration assumption, $\sin\theta \sim \tan\theta = y_x$ but replacing the constant tension assumption, $T = T_0$, by $T = T(t)$ and using the linear stress-strain (ε) relation

$$T - T_0 = E\varepsilon = Ea(S-L)/L,$$

where

$$(2.2) \quad S = \int_0^L [1+y_x^2]^{1/2}\,dx = \int_0^L [1+y_x^2+O(y_x^4)]\,dx \sim L+\tfrac{1}{2}\int_0^L y_x^2\,dx,$$

the dynamic nonlinear equation for the displacement $y(x, t)$ is

$$(2.3) \quad y_{tt} - c^2\left[1+\lambda^2\int_0^L y_x^2\,dx\right]y_{xx} = 0,$$

with $c^2 = T_0/m$, $\lambda^2 = Ea/2LT_0$. Assuming

$$(2.4) \quad y(x, t) = F(x)G(t),$$

there follows $\int_0^L y_x^2\,dx = IG^2(t)$, and the separation equations become

$$(2.5)\quad \begin{aligned} &F''+\nu^2 F = 0, \\ &G''+\nu^2 c^2[1+\lambda^2 IG^2]G = 0. \end{aligned}$$

The first equation is that of the harmonic oscillator and the second is solvable in terms of elliptic functions. Since (linear) superposition no longer applies the form of possible boundary and initial conditions is very restricted. Indeed the class of possibilities is limited to those provided by the separation solutions. One class found useful by Oplinger [94, 95] is

$$(2.6)\quad y(0,t) = 0, \quad y(L,t) = y(L,t+\omega)$$

whereupon, the solution becomes

$$(2.7)\quad y(x,t) = \alpha \sin\nu x\, \mathrm{cn}(ft, K)/\sin\nu L$$

with f, K, ν and α constants.

2.2. IMPLICIT SEPARATION

As might be expected the classical method of simple separation can be generalized. In fact, it is a special case of implicit separation. Irmay [59] discusses some special cases of this idea while the terminology and additonal examples are due to Ames [13]. In illustration of the idea consider the semilinear wave equation

$$(2.8)\quad u_{\xi\xi}-u_{\theta\theta} = H(u)$$

which is employed as a model equation for a variety of wave phenomena by Scott [112] and Lamb [70] with particular interest in $\exp u$, $\sin u$ and $\sinh u$. Equation (2.8) is generally not simply separable although special cases (e.g. $H(u) = u$) exist for which it is. To simplify (2.8) set $x = \xi+\theta$, $y = \xi-\theta$ (the characteristics of the wave operator) whereupon (2.8) becomes

$$(2.9)\quad u_{xy} = H(u)/4.$$

While u is not generally obtainable as the product of two functions, $f(x)g(y)$, *it may be that some function of u is*. Thus assume $F(u) = f(x)g(y)$ or equivalently,

$$(2.10)\quad u = G(\eta), \quad \eta = f(x)g(y).$$

Under this assumption (2.9) becomes

$$(2.11)\quad 4f'(x)g'(y) = H(G)/[G'+\eta G'']$$

which is simply separable if (2.11) equals $k(\eta) = k_1(f)k_2(g)$, simultaneously. This occurs, nontrivially, if $k(\eta) = c_1\eta^p$ for any p (taken as real, herein), in which case the separation equations become

$$(2.12)\quad \begin{aligned} f' = c_2 f^p, \quad c_2 g' &= c_1 g^p, \\ c_1\eta^p[\eta G']' &= H(G)/4. \end{aligned}$$

The infinity of solutions constructed from (2.12) satisfy (2.9).

Similar analyses can be applied to the semilinear equation

$$u_{xx}+u_{yy} = F(u)$$

which arises in heat conduction in systems undergoing chemical reaction and steady vortex motion of an incompressible fluid (Bateman [20]).

2.3. INVARIANT SOLUTIONS BY SEPARATION OF VARIABLES

The last remarks of this section concern the development of invariant (similarity) solutions by a process which, because of its motivation, is called "similarity by separation of variables" by Abbott and Kline [1] (see also the more accessible works of Hansen [49] and Ames [6]); we shall return to the general topic of invariant (similarity) solutions in a later section. For a general equation in two variables, (x, y) say, we seek new variables $\xi = \xi(x, y)$, $\eta = \eta(x, y)$ such that the transformed equation is separable in the new variables. In illustration let us examine a problem due to Pattle [101] and later analyzed by Boyer [27]. Consider the nonlinear diffusion equation

$$(2.13)\quad r^{1-q}[r^{q-1}C^nC_r]_r = C_t \qquad (n > 0)$$

with q-dimensional spherical symmetry. Functions $U(\xi)$, $Y(\eta)$, and $R(t)$, with $\xi = t$, $\eta = r/R(t)$, are sought so that, in these new variables (ξ, η), C is separable. Thus a trial solution in the form

$$(2.14)\quad C(r, t) = U(t)Y\big(r/R(t)\big)$$

is sought. If $R(t) = \big[U(t)\big]^{-A}$ then the equation separates into

$$\frac{\eta^{1-q}(d/d\eta)[\eta^{q-1}Y^nY']}{Y+A\eta Y'} = \frac{U^{-2A}U'}{U^{n+1}} = \text{const}$$

and exact solutions are constructable for specific, physically meaningful values of A.

This idea seems to have been introduced by Boltzmann in 1894 [25].

3.0. Progressing Waves and Generalizations

Since $u = f(x+ct)$ and $g(x-ct)$ satisfy the (linear) wave equation $u_{tt} - c^2 u_{xx} = 0$, an obvious question is whether nonlinear equations can support waves of permanent profile.

3.1. EXPLICIT PROGRESSING WAVES

It is well known (Lamb [69]) that nonlinear dispersive media can support steady progressive wave forms such as the cnoidal wave and the solitary wave (soliton). Waveforms which satisfy the conditions of uniform propagation without change of type can be sought in the form

$$(3.1)\quad f(x+\lambda t) \quad \text{or} \quad f(ax+bt)$$

or their generalizations to higher dimension, e.g.

$$f(\alpha x+\beta y+\gamma t), \qquad g(r)h(\alpha\theta+ct).$$

Zabusky [134] shows that the generalized Korteweg–de Vries equation

$$(3.2)\quad u_t \pm u^p u_x + \delta^2 u_{xxx} = 0$$

has a solitary-wave pulse solution ('soliton') constructable by assuming $u = U(x-ct)$. In a series of associated papers from the Princeton school (e.g., Zabusky and Kruskal [133] and Gardner *et al.* [44]) and in van Wijngaarder [129], further work for equations of Korteweg–de Vries form is developed. See also Jeffrey and Kakutani [60].

The nonlinear diffusion equation

$$(3.3)\quad c_t = (c^n)_{xx}$$

is known (Ames [8]) to admit progressing wave solutions of uniform propagation. Wave solutions of the form $kw = f(\varphi+\lambda\psi)$ are used by Tomotika and Tamada [121] and Tamada [119] to explore

$$(3.4)\quad (kw)_{\psi\psi} = [(kw)^2]_{\phi\phi},$$

together with other ad-hoc forms.

Some very complicated systems will support explicit progressing waves. Advani [3, 4], while examining the nonlinear transverse vibrations of a spinning membrane, whose equations (von Karman) are

$$
\begin{aligned}
&\{r^{-1}\varphi_{rr}+r^{-2}\varphi_{\theta\theta}\}w_{rr}+\{r^{-1}w_r+r^{-2}w_{\theta\theta}\}\varphi_{rr}-\\
&-2\{(r^{-1}w_\theta)_r(r^{-1}\varphi_\theta)_r\}-\\
(3.5)\quad &-\tfrac{1}{2}\varrho\omega^2r^2h\nabla^2w-\varrho\omega^2rhw_r = \varrho hw_{tt},\\
&\nabla^4\varphi = Eh\{-[r^{-1}w_r+r^{-2}w_{\theta\theta}]w_{rr}+r^{-2}(w_{r\theta})^2-\\
&-2r^{-3}(w_r\ )w\ +r^{-4}(w_\theta)^2\}+2(1-\nu)\varrho\omega^2h,
\end{aligned}
$$

found exact solutions

$$
\begin{aligned}
&w = Ar^2\sin 2(\theta\pm ct),\\
(3.6)\quad &\varphi = Br^4\cos 4(\theta\pm ct)+[2hEA^2+(1-\nu)\varrho\omega^2h]\frac{r^4}{32}-Cr^2.
\end{aligned}
$$

Here the wave velocity c is related to the wave ‘amplitude’ by

$$c^2-\frac{5-\nu}{8}\omega^2 = \frac{E}{4\varrho}A^2-\frac{12B}{\varrho h}$$

thus admitting the possibility of standing waves. The linear theory of a spinning membrane with a free edge excludes the possibility of standing waves. While (3.6) does not precisely fit the earlier forms, the motivation for the ad-hoc form $w\ = H(r)\sin n(\theta\pm ct)$ is apparent.

3.2. GENERALIZED EXPLICIT PROGRESSING WAVES

Several possible generalizations of the classical progressing wave come to mind. Perhaps the most general explicit form is

$$(3.7)\quad u = F(\omega),\qquad \omega = a(x,t)x+b(x,t)t$$

where F, a, and b are to be determined. Some linear equations of the form

$$u_{tt} = \gamma(x,t)u_{xx}+\alpha(x,t)u_x+\beta(x,t)u_t$$

have solutions given by Eq. (3.7) where a and b are functions determined by α, β, and γ (Ames [13]).

The transformed Burgers’ equation (Ames [6])

$$(3.8)\quad v_t+\tfrac{1}{2}v_x^2 = \nu v_{xx}$$

has a solution $v = F(t+g(x))$ where F and g satisfy nonlinear differential equations

$$\begin{aligned} \nu f''-\tfrac{1}{2}(f')^2-cf' &= 0, \\ \nu g''+c(g')^2 &= 1. \end{aligned} \tag{3.9}$$

These are first order equations in the first derivatives and can be integrated without difficulty. Similarly the two-dimensional momentum equations of fluid mechanics

$$\begin{aligned} u_t+uu_x+vu_y &= \nu(u_{xx}+u_{yy}), \\ v_t+uv_x+vv_y &= \nu(v_{xx}+v_{yy}), \end{aligned} \tag{3.10}$$

after substitution of $u = \psi_x$, $v = \psi_y$, both reduce to

$$\psi_t+\tfrac{1}{2}(\psi_x^2+\psi_y^2) = \nu(\psi_{xx}+\psi_{yy}) \tag{3.11}$$

upon integration of the first with respect to x and the second with respect to y. With

$$\psi = F(\omega), \qquad \omega = t+g(x)+h(y)$$

the solution of

$$\begin{aligned} \nu F''-\tfrac{1}{2}(F')^2 &= cF', \\ \nu g''+c(g')^2 &= 1-c_1, \\ \nu h''+c(h')^2 &= c_1 \end{aligned} \tag{3.12}$$

is

$$\begin{aligned} F &= B\nu\{1-\tfrac{1}{2}\exp(2\omega A/\nu)\}, \\ g &= D\ln[\cos(\nu cD)(x+D_1)]+D_2, \\ h &= E\ln[\cosh(cE/\nu)\,(y+E_1)]+E_2. \end{aligned}$$

These constitute a solution of the momentum equations but the continuity equation $u_x+v_y = 0$ is not satisfied, in general.

3.3. IMPLICIT PROGRESSING WAVES

Generalization of the preceding ideas to implicit traveling waves, whose velocity and shape vary as a function of the dependent variable, is amply motivated by noting that any function u which satisfies

$$u(x,t) = f[x+t\varphi(u)] \tag{3.13}$$

is a solution of the equation

$$u_t-\varphi(u)u_x = 0. \tag{3.14}$$

Indeed (3.13) is also a solution of the nth order equation (Ames [13])

$$\partial^n u/\partial t^n - (\partial^{n-1}/\partial x^{n-1})\,[\varphi^n(u)\,(\partial u/\partial x)] = 0, \qquad n = 1, 2, \ldots \tag{3.15}$$

for any appropriately differentiable f. The case $n = 2$ has been employed to study the evolution of discontinuities (wave breakdown) in a variety of problems (Ames [10] and references therein). Applications occur in gas dynamics, shallow water wave theory, electromagnetic transmission lines and wave propagation in solids.

In two papers Noh and Protter [93] and Protter [105], while introducing the idea of 'soft' solutions, showed that the system

$$\begin{aligned} &u_t + \alpha(u)u_x + \beta(v)u_y + \gamma(w)u_z = 0,\\ &v_t + \alpha(u)v_x + \beta(v)v_y + \gamma(w)v_z = 0,\\ &w_t + \alpha(u)w_x + \beta(v)w_y + \gamma(w)w_z = 0 \end{aligned} \tag{3.16}$$

has implicit progressing wave solutions

$$u = f(\omega, \lambda, \mu), \qquad v = g(\omega, \lambda, \mu), \qquad w = h(\omega, \lambda, \mu),$$
$$\omega = x - t\alpha(u), \qquad \lambda = y - t\beta(v), \qquad \mu = z - t\gamma(w).$$

Noh and Protter [93] derive results for one-, two-, and three-dimensional gas flows.

The possibility that a nonlinear diffusion equation can support explicit and implicit progressing wave has been explored by Ames [13]. For example, the equation

$$u_{xx} - \frac{\partial}{\partial u}\,[B^2/(G' - B'x)]\,u_t = 0$$

has the implicit progressing wave solution

$$u = F[B(u)x + t], \qquad F = G^{-1}.$$

4.0. Miscellaneous Results

4.1. ASSUMED SOLUTION STRUCTURE

As a means of exploring an equation, a variety of ad hoc forms may pay off. For example, the equation

$$uu_x + \frac{dp}{dx} = u(uu_\varphi)_\varphi \tag{4.1}$$

is obtained when the pressure boundary-layer equations are transformed by means of the von Mises transformation $(x, y) \to (x, \psi)$, ψ the stream fuction. Ames and Adams [14] assume

$$(4.2) \quad u = f(x) + g(x)\psi^a + h(x)\psi^b$$

and show that a solution of (4.1) is possible for $a = 2$, $b = 0$ and $a = 2$, $b = 1$. A family of exact solutions for the boundary layer equations is constructed. Typical of these is the form $(A < 0)$

$$u = \frac{4(-A)\alpha(x)Q(x, y)}{P(x)\,[1+\alpha(x)Q(x, y)]^2},$$

where $P(x) = (6x)^{1/3}$, $Q(x, y) = \exp[-2(-A)^{1/2}y/P^2(x)]$, with $u(0, y) = 0$, $\lim_{y\to\infty} u = 0$ and

$$u(x, 0) = 4(-A)\alpha(x)/P(x)[1+\alpha(x)]^2$$

where $\alpha(x)$ and A are arbitrary. Such solutions are useful in studying the boundary layer during blowing or suction at the wall.

Tomotika and Tamada [121] found that (3.4) possessed solutions of the form

$$\text{(a)} \quad kw = f(\varphi) + g(\psi),$$

$$\text{(b)} \quad kw = f_0(\psi) + f_1(\psi)\varphi^2,$$

$$\text{(c)} \quad kw = F(\varphi + \psi^2) + 2\psi^2$$

as well a progressing wave and separated form. Similarly the equation

$$w_{\eta\eta} + \eta^{-1} w_\eta = (w^2)_{\xi\xi}$$

possesses solutions with the forms (a), (b), (c) and $w = f(\eta)\,g(\xi)$.

The untransformed Burgers equation

$$(4.3) \quad u_t + uu_x = R^{-1}u_{xx}, \qquad R = U/\nu k$$

has been examined by Benton [21] who discovered the exact solutions

$$u = (R\alpha + t)^{-1}\{x + \beta\tan[\beta Rx/2(R\gamma + t)]\},$$

$$u = (R\alpha + t)^{-1}\{x - \beta\tanh[\beta Rx/2(R\alpha + t)]\}.$$

We shall return to (4.3) in the section where equations equivalent to integrable forms are discussed.

4.2. DEPENDENT VARIABLE RELATIONSHIPS

On occasion insight into a problem may be obtained by assuming an unspecified relation between two or more of the dependent variables. In 1858 Earnshaw [36] made such an assumption (hence Earnshaw's method) and drew conclusions concerning finite amplitude sound waves in gases. Nowinski [94] made an Earnshaw-like assumption in studying the propagation of finite disturbances in one-dimensional bars of rubberlike materials. Ames [8] introduces a functional relationship between the velocity components in boundary-layer flows and finds some exact solutions. A typical application is given by Ames *et al.* [15] for longitudinal wave propagation on a traveling threadline. Banta [18] introduces an Earnshaw-like assumption in his study of finite amplitude sound waves. In illustration of the idea Pai's [100] example of a magneto-gasdynamic problem will be employed.

The governing equations for an inviscid, nonheat conducting, infinite electrical conductivity, ideal plasma are

$$(4.4)\qquad \varrho_t + (\varrho u)_x = 0,$$

$$(4.5)\qquad \varrho u_t + \varrho u u_x + p_x + \mu_e H H_x = 0,$$

$$(4.6)\qquad H_t + (uH)_x = 0,$$

$$(4.7)\qquad \varrho h_t + \varrho u h_x - p_t + \mu_e u H H_x = 0.$$

Here ϱ, u, p, H, and h are density, velocity, pressure, magnetic field, and stagnation enthalpy ($h = c_p T + u^2/2$), respectively. Since (4.4) and (4.6) are identical their solutions have the same functional form – that is

$$(4.8)\qquad H = A\varrho = \frac{H_0}{\varrho_0}\varrho.$$

For anisentropic flow

$$p\varrho^{-\gamma} = b\exp[S/c_v] = b\theta,$$

where $b = (p_0/\varrho_0^\gamma)\exp[-S_0/c_v]$ with S the entropy and $(\ )_0$ refers to values at the stagnation point. Some critical speeds are useful: First, the local speed of sound is defined as

$$c^2 = \left(\frac{\partial p}{\partial \varrho}\right)_\theta = \frac{\gamma p}{\varrho} = b\gamma\theta\varrho^{\gamma-1}.$$

Second, the local speed of Alfven's wave

$$V_H^2 = \mu_e H^2/\varrho = \mu_e A^2 \varrho = \mu_e (H_0^2/\varrho_0^2)\varrho = V_{H_0}^2 \frac{\varrho}{\varrho_0}.$$

Third, the effective speed of sound

$$c_e^2 = c^2 + V_H^2.$$

Generally, c_e is a function of both θ and ϱ, except that for isentropic flow (θ = constant everywhere) c_e depends only upon ϱ. For *isentropic flow* assume

(4.9) $\quad \varrho = f(u)$

where f is to be determined. Then (4.4) becomes

(4.10) $\quad u_t + uu_x = -fu_x/f'.$

Utilizing (4.8) and (4.9) Eq. (4.5) becomes

(4.11) $\quad fu_t + fuu_x + (\gamma b\theta\varrho^{\gamma-1})f'u_x + \mu_e A^2 \varrho f' u_x = 0.$

But since the effective speed is defined as

$$c_e^2 = \gamma b\theta\varrho^{\gamma-1} + \mu_e A^2 \varrho$$

(4.11) becomes

(4.12) $$u_t + uu_x = -\frac{c_e^2 f'}{f} u_x.$$

Of course (4.10) and (4.12) must agree so that

$$c_e^2 \frac{f'}{f} = \frac{f}{f'}$$

or

(4.13) $$\frac{f'}{f} = \frac{1}{\varrho}\frac{d\varrho}{du} = \pm\frac{1}{c_e},$$

which when integrated yields

(4.14) $$u = \pm \int_{\varrho_0}^{\varrho} \frac{c_e(\varrho)}{\varrho}\, d\varrho.$$

Actually, (4.13) when substituted into (4.10) gives us the proper information to proceed without the use of (4.14). Thus (4.10) becomes

$$u_t + (u \pm c_e) u_x = 0$$

and (4.6) becomes

$$\varrho_t + (u \pm c_e)\varrho_x = 0.$$

Direct integration by means of the Lagrange subsidiary equations gives

$$u = F[x-(u \pm c_e)t], \quad \varrho = G[x-(u \pm c_e)t],$$

where

$$u(x, 0) = F(x), \quad \varrho(x, 0) = G(x).$$

5.0. Transformations

Ad hoc transformations can involve only the independent variables, only the dependent ones, or, more usually, a combination. A number of useful transformations will be given together with selected examples for some of them.

5.1. KIRCHHOFF'S TRANSFORMATION

One of the older transformations is due to Kirchhoff [67], who introduced the relation $\psi = \int_{C_0}^{C} K(t)\, dt$ into the 'pseudo-Laplacian'

$$\text{(5.1)} \quad \operatorname{div}[K(C)\operatorname{grad} C] = 0$$

thereby reducing (5.1) to Laplace's equation $\nabla^2\psi = 0$. The Kirchhoff transformation may also be applied to the nonlinear diffusion equation $C_t = \operatorname{div}[K(C)\operatorname{grad} C]$ thereby generating an alternative nonlinear equation having some advantages according to Carslaw and Jaeger [29].

5.2. VON MISES TRANSFORMATION

The von Mises transformation [82] (see also Ames [6]), originally introduced to convert the two-dimensional boundary-layer equations into a nonlinear diffusion equation, has been utilized by many authors. Kalman [62], in his discussion of the oscillations of a zero temperature plasma, was able to construct exact solutions by means of this transformation. Since his application is typical, we describe it here. The first of the governing equations

$$\text{(5.2)} \quad \varrho_t + (\varrho v)_x = 0, \quad v_t + v v_x = \alpha E, \quad \alpha E_x = \omega^2(\varrho - 1)$$

is satisfied by introducing an auxiliary function ψ by means of $\psi_t = \varrho v$, $\psi_x = -\varrho$. A change of variables, from (x, t) to (ψ, t), is then accomplished. The second equation becomes $v_t(t, \psi) = \alpha E$ and the third ultimately leads to $x_{tt} + \omega^2 x = -\omega^2(\psi + g(t))$. Applications of the von Mises idea in boundary-layer flow are contained in Pai [99] while Ames [10] and Ames *et al.* [15] employ it to study wave propagation. The reduction of several nonlinear hyperbolic equations to standard forms has been accomplished by means of this and related transformations by Ames [10]. The construction of exact solutions is thereby facilitated.

5.3. QUASI-LINEAR HYPERBOLIC EQUATIONS

A number of engineering and applied physics problems are modeled with quasi-linear (linear in the highest order derivative) equations. In the general case of n coupled equations in n unknowns

$$\text{(5.3)}\qquad \sum_{i=1}^{n} [a_{ji} u_x^i + b_{ji} u_y^i] + d_j = 0, \qquad j = 1, 2, \ldots, n$$

where the coefficients are functions of x, y, and the u^i, there exists the quasi-linear theory (see Courant and Hilbert [34] or Ames [6]). Via the characteristics, this constructive theory permits the development of a canonical form whose individual equations contain differentiation in only one direction. If the F_i and G_i, $i = 1, 2, 3, 4$ are functions of u and v only, then the equations

$$\text{(5.4)}\qquad \begin{aligned} F_1 u_x + F_2 u_y + F_3 v_x + F_4 v_y &= 0, \\ G_1 u_x + G_2 u_y + G_3 v_x + G_4 v_y &= 0 \end{aligned}$$

are called reducible because the interchange of dependent and independent variables transforms (5.4) into a linear system. This is the essence of the so-called hodograph transformation. A number of physical examples modeled by reducible equations are given in Ames [6], Courant and Hilbert [34], von Mises [83], and Courant and Friedrichs [33]. Examples from plasticity are found in Hill [52]. A typical vibration problem concerns the equation $y_{tt} = [F(y_x)]^2 y_{xx}$ which is given by Zabusky [132]. With $u = y_x$ and $v = y_t$ the equation becomes

$$\text{(5.5)}\qquad u_t - v_x = 0, \qquad v_t - F^2(u) u_x = 0.$$

While this system can be treated by direct interchange, Zabusky found it more convenient to introduce the Riemann invariants as the new variables. The Riemann invariants, resulting from the quasi-linear theory, are functions which are invariant in the characteristic directions. Mei [80] separated the problem of collapse of a homogeneous fluid mass in a stratified fluid into two parts. During the initial stages the quasi-linear theory was employed to generate an exact solution. Ames and Vicario [16] found a similar analysis generated exact solutions for a longitudinal wave propagation problem.

5.4. CONTACT TRANSFORMATIONS

First employed by Lagrange (see e.g. Forsyth [39, p. 131]) and originally (and sometimes still) called a tangential transformation, the contact transformation establishes a correspondence between surface elements having the property that two hypersurfaces tangent at a point P are mapped into two hypersurfaces tangent at P^I, the map of P. Perhaps the best known of these is the Lagrange–Legendre transformation, often erroneously called the Legendre transformation (see Courant and Hilbert [34], Madelung [78], Ames [6, 12]).

(a) *Lagrange–Legendre Transformation*

With $\partial u/\partial x_i = \xi_i$, $\partial \omega/\partial \xi_i = x_i$, $i = 1, 2, \dots, n$ the n-dimensional form of the Lagrange–Legendre transformation is

$$u(x_1, x_2, \dots, x_n) + \omega(\xi_1, \xi_2, \dots, \xi_n) = \sum_{i=1}^{n} x_i \xi_i. \tag{5.6}$$

In two variables (x, y) and (ξ, η) Eq. (5.6) becomes

$$\begin{aligned} &u(x, y) + \omega(\xi, \eta) = x\xi + y\eta, \\ &\omega_\xi = x, \quad u_x = \xi, \\ &\omega_\eta = y, \quad u_y = \eta \end{aligned} \tag{5.7}$$

and in second derivatives

$$\begin{aligned} &u_{xx} = j\omega_{\eta\eta}, \quad u_{xy} = -j\omega_{\xi\eta}, \quad u_{yy} = j\omega_{\xi\xi}, \\ &j = u_{xx} u_{yy} - u_{xy}^2. \end{aligned}$$

Many applications have been made in compressible fluid mechanics (see e.g. von Mises [83, p. 261 *et seq.*]) and in wave propagation. For example, the quasi-linear wave equation (see Ames [12, p. 23])

$$(5.9)\quad u_{yy} = f(u_x)u_{xx}$$

transforms into the linear equation

$$(5.10)\quad \omega_{\xi\xi} = f(\xi)\,\omega_{\eta\eta}, \qquad j \neq 0$$

under the Lagrange–Legendre transformation.

Since (5.9) is quasi-linear, it may appear that contact transformations are not useful on other nonlinear problems. That this is not the case has been demonstrated by several authors (see e.g. Varley [122], Ames [12]) by employing a parametrization in combination with several Lagrange–Legendre transformations. Consider two simultaneous nonlinear equations

$$(5.11)\quad F[u_x, u_y, v_x, v_y] = 0, \qquad G[u_x, u_y, v_x, v_y] = 0$$

which can be parametrized (non-uniquely), with parameters (α, β) as

$$(5.12)\quad u_x = p(\alpha, \beta), \qquad u_y = q(\alpha, \beta), \qquad v_x = r(\alpha, \beta), \qquad v_y = s(\alpha, \beta)$$

so that (5.11) are satisfied.

Variables $\bar{u}$ and $\bar{v}$ are introduced by means of the Legendre transformation

$$(5.13)\quad \begin{aligned} u &= xp+yq+\bar{u}, \\ v &= xr+ys+v. \end{aligned}$$

If $p, q, r, s, \bar{u}$, and $\bar{v}$ depend upon (x, y) in such a way that

$$(5.14)\quad xdp+ydq+d\bar{u} = xdr+yds+d\bar{v} = 0$$

then (5.13) satisfy (5.12). (A trivial way of satisfying (5.14) is to take $\alpha, \beta, \bar{u}, \bar{v}$ constant.)

Generally we satisfy (5.14) by finding $x, y, \bar{u}$, and $\bar{v}$ as functions of (α, β) such that

$$(5.15)\quad \begin{aligned} [xp_\alpha+yq_\alpha+\bar{u}_\alpha]d\alpha+[xp_\beta+yq_\beta+\bar{u}_\beta]d\beta &= 0, \\ [xr_\alpha+ys_\alpha+\bar{v}_\alpha]d\alpha+[xr_\beta+ys_\beta+\bar{v}_\beta]d\beta &= 0 \end{aligned}$$

for all α and β. We now assume that solutions of (5.12) are desired for which β is not a function of α; solutions with $\beta = \beta(\alpha)$ can readily be derived from (5.14). Consequently, each of the bracketed terms in (5.15) must vanish. Therefore (5.15) imply that

$$(5.16)\qquad \begin{aligned} x &= \frac{\partial_{\alpha\beta}(q, \bar{u})}{\partial_{\alpha\beta}(p, q)} = \frac{\partial_{\alpha\beta}(s, \bar{v})}{\partial_{\alpha\beta}(r, s)}, \\ -y &= \frac{\partial_{\alpha\beta}(p, \bar{u})}{\partial_{\alpha\beta}(p, q)} = \frac{\partial_{\alpha\beta}(r, \bar{v})}{\partial_{\alpha\beta}(r, s)} \end{aligned}$$

where

$$\partial_{\alpha\beta}(f, g) = \begin{vmatrix} f_\alpha & g_\alpha \\ f_\beta & g_\beta \end{vmatrix}.$$

Equations (5.16) are two linear first order equations for $\bar{u}$ and $\bar{v}$ as functions of α and β and they also provide the mapping from the (α, β) to the (x, y) plane.

Varley [122] applied this method to the study of two-dimensional flow of a dilatant fluid whose equations are

$$(5.17)\qquad F = u_x + v_y = 0,$$

$$(5.18)\qquad G = \tfrac{1}{2}[u_x^2 + v_y^2] + \tfrac{1}{4}[u_y + v_x]^2 - 1 = 0.$$

These are to be solved in the neighborhood of a curve

$$\Gamma\colon\ x = x_0(l), \qquad y = y_0(l)$$

subject to the conditions that on

$$\Gamma\colon\ u = v = 0.$$

Equations (5.17) and (5.18) can be written parametrically as

$$\begin{aligned} u_x &= -\sin 2\alpha, & u_y &= \beta + \cos 2\alpha, \\ v_x &= -\beta + \cos 2\alpha, & v_y &= \sin 2\alpha \end{aligned}$$

whereupon (5.16) imply that

$$x = \bar{v}_\beta, \qquad y = \bar{u}_\beta$$

and

$$\begin{aligned} \bar{u}_\alpha + (2\sin 2\alpha)\bar{u}_\beta - (2\cos 2\alpha)\bar{v}_\beta &= 0, \\ \bar{v}_\alpha - (2\sin 2\alpha)\bar{v}_\beta - (2\cos 2\alpha)\bar{u}_\beta &= 0. \end{aligned}$$

This linear system has the two real characteristics

$$\alpha \pm \tfrac{1}{2}(\beta - 1) = \text{const}$$

and is therefore hyperbolic.

This parametrization procedure also applies to a suitably reduced Monge–Ampère equation

(5.19) $u_{xy} = f(u_{xx}, u_{yy})$.

An application to wave propagation is given in Ames [12].

(b) *Bäcklund Transformations*

In this section, our discussion relates to a transformation that had its origin in some investigations by Bäcklund [17] (see also Forsyth [39]) concerning the transformation of surfaces in differential geometry. The goal of this transformation, like others, is to provide a constructive method for obtaining various classes of 'equivalent' equations, thereby leading to the integrals of the original problem. A *Bäcklund transformation* for a partial differential equation of second order in two independent variables is a pair of first-order partial differential equations that relate the dependent variable satisfying the given equation to another dependent variable that satisfies either the same or another second-order partial differential equation. A method for constructing these transformations, due to Clairin [31], can be found in Forsyth [39] and Ames [12]. Many of the early results are summarized in Goursat [46], which volume contains an extensive bibliography of early papers.

The details of Clairin's method are rather lengthy and are omitted here. Instead some transformations will be presented and recent literature cited. The notation $p = u_x$, $q = u_y$, $r = u_{xx}$, $s = u_{xy}$, $t = u_{yy}$ and $\alpha = u_{xxx}$ will be employed throughout.

(i) *Liouville's equation* (Forsyth [39]) $s = e^u$ has associated with it the Bäcklund transformation

$$p+p' = \sqrt{2}\exp[\tfrac{1}{2}(u-u')],$$
$$q-q' = \sqrt{2}\exp[\tfrac{1}{2}(u+u')]$$

where u satisfies $s = e^u$ and u' satisfies $s' = 0$.

(ii) *The sine-Gordon equation* (Ames [12]), which in characteristic coordinates may be written $s = \sin u$ has associated with it the Bäcklund transformation

$$\tfrac{1}{2}(p+p') = a\sin[\tfrac{1}{2}(u-u')],$$
$$\frac{1}{2}(q-q') = \frac{1}{a}\sin\left[\frac{1}{2}(u+u')\right] \tag{5.20}$$

where both u and u' satisfy the sine-Gordon equation and a is an arbitrary constant. (The sinh-Gordon equation, $s = \sinh u$ has the same transformation as (5.20) with sinh replacing sin in both equations of (5.20)). Although these do not lead to a simpler equation for which the general solution is known they are useful. They lead to a so-called 'theorem of permutability' by means of which four solutions of $s = \sin u$ are interrelated. Thus an infinite sequence of solutions to the sine-Gordon equation can be constructed without additional quadrature.

(iii) *The Korteweg–de Vries equation* (Lamb [72], [71]) in the form

$$z_y + 6zz_x + z_{xxx} = 0 \tag{5.21}$$

with $z = \partial u/\partial x$ one finds that u satisfies

$$q + 3p^2 + \alpha = 0$$

and a Bäcklund transformation is

$$\begin{aligned} p &= p' + c(u, u'), \\ q &= q' - 2[ap + 2(u')^2 c(u, u') + \tfrac{1}{2}p'(u-u')^2 - r'(u-u')] \end{aligned} \tag{5.22}$$

where a is an arbitrary constant and $c(u, u') = a - \frac{1}{2}(u-u')(u+3u')$. In obtaining this result the coefficients of p' and q', which are identical functions of u and u', have been set equal to unity. If they are set equal to -1 the result of Wahlquist and Estabrook [127]

$$\begin{aligned} p + p' &= a - \tfrac{1}{2}(u-u')^2, \\ q + q' &= -2(p^2 + pp' + p'^2) + (u-u')(r-r') \end{aligned} \tag{5.23}$$

is obtained. From (5.23) a theorem of permutability is obtained which interrelates four solutions of (5.21). Notice will also be taken of these theorems on permutability in our section on nonlinear superposition.

One may also obtain Bäcklund transformations in which u and u' satisfy equations of different form. The recently discovered (Miura [84]) relation between the Korteweg–de Vries (KdV) equation (5.21) and the modified KdV equation

$$q + 6u^2 p + \alpha = 0, \tag{5.24}$$

as well as the Hopf–Cole [51, 32] transformation relating the Burgers' equation

$$q' + u'p' + r' = 0 \tag{5.25}$$

to the linear diffusion equation

(5.26) $q+r = 0$

are examples of Bäcklund transformations. The transformation relating (5.25) and (5.26) is

(5.27) $p = \frac{1}{2}uu'$,

(5.28) $q = -\frac{1}{2}(up'+pu')$.

This 'linearizing' transformation has been known since 1906 (Forsyth [39, Vol. 6, pp. 97–101]). Equation (5.27) is the Hopf–Cole transformation.

Lastly, Loewner [77], Power *et al.* [104], Rogers [107] study the application of Bäcklund transformations and generalized ideas to wave propagation problems in gases, nonlinear elastic materials, etc.

5.5. TRANSFORMATIONS OF THE BOUNDARY-LAYER EQUATIONS

A great many special transformations have been devised and applied to the dimensionless laminar boundary layer equations

$$\begin{aligned} u_x+v_y &= 0, \\ uu_x+vu_y &= UU'(x)+u_{yy} \end{aligned} \tag{5.29}$$

and their generalizations. Some of these are summarized in Ames [11] while a more detailed description is available in Sun [118] and Schlichting [110].

5.6. MISCELLANEOUS

Classes of coupled nonlinear elliptic equations are examined by Dasarathy [35]. By judicious pretransformations he finds classes of equations which can be linearized by inversion of the dependent and independent variables. Vein [123, 124], employing Appell and inverse functions, establishes exact solutions for a number of nonlinear parabolic and hyperbolic equations. Typical examples include $\varphi_{xx} = \varphi_x^2\varphi_y$, $\varphi^2\varphi_{xx} = \varphi_y$, and $\varphi\varphi_{xxx} = \varphi_{xy}$.

Exact solutions are constructed for diffusion-conduction problems by Fujita [40, 41, 42], Irmay [58, see also 59] and Philip [103]. Philip's procedure is based upon the reduction of $\theta_t = [D(\theta)\theta_x]_x$, by means of the similarity variable $\psi = xt^{-1/2}$ to

$$D = -\tfrac{1}{2}(d\psi/d\theta) \int_0^\theta \psi \, d\theta .$$

Solutions of the original equation exist in exact form, so long as $D(\theta)$ is expressible in the form

$$D = -\tfrac{1}{2}(dF/d\theta) \int_0^\theta F d\theta$$

where F is any single valued function of θ.

6.0. Equations with 'Built in' Solutions

Let us consider the linear wave equation $u_{tt} - k^2 u_{xx} = 0$ and make a simple dependent variable transformation $u = f(v)$. Then v satisfies the equation

$$(6.1) \qquad v_{tt} - k^2 v_{xx} = (f''/f')(k^2 v_x^2 - v_t^2).$$

Consequently, large classes of nonlinear equations exist which are 'equivalent' to a linear equation in the sense that if u satisfies the linear equation, then $v = f^{-1}(u)$ satisfies Eq. (6.1). This equivalence has been employed during studies of longitudinal wave propagation in moving media by Ames and Vicario [16].

Hopf [51] and Cole [32] found that Burgers' equation

$$(6.2) \qquad v_t + v v_x = \nu v_{xx}, \qquad \nu > 0$$

is solved by

$$(6.3) \qquad v = -2\nu \frac{\partial}{\partial x}[\ln \theta]$$

where θ is any positive solution of $\theta_t = \nu\theta_{xx}$. This transformation can be generated in the same manner as that applied to (6.1). In fact $\theta_t = \nu\theta_{xx}$ under $\theta = f(v)$ becomes

$$v_t = \nu v_{xx} + f'' v_x^2 / f'.$$

It has already been observed that this is the Bäcklund transformation (5.27). An easy generalization to higher dimension occurs. If θ satisfies the diffusion equation $\theta_t = k\nabla^2\theta$, then $\bar{u} = -2\nu\nabla(\ln\theta)$ satisfies the pressure free momentum equations of fluid mechanics. However, the continuity equation is generally not satisfied, thereby limiting the physical utility of the result.

Motivated by the techniques applied to Burgers' equation, Chu [30] shows that the class of equations

$$\frac{\partial u_i}{\partial t} + F_j(u_i)\frac{\partial u_i}{\partial x_j} = G_i(u_i)\frac{\partial u_i}{\partial x_j}\frac{\partial u_i}{\partial x_j} + k\frac{\partial^2 u_i}{\partial x_j \partial x_j} + H_i R_i, \tag{6.4}$$

$$i, j = 1, 2, \ldots, n$$

is reducible to a single linear diffusion equation when certain restrictions are placed on the coefficient functions. The incompressible momentum equations with a pressure term and the inhomogeneous Burgers' equation are included as special cases. Rodin also studies the inhomogeneous Burgers equation [106]

$$u_t + uu_x = \nu u_{xx} + f(x, t)$$

by means of Eq. (6.3) and a similarity variable. The inhomogeneous Burgers' equation for burning gas in a rocket was shown in Ames [6, p. 25] to be reducible to a linear diffusion equation without the use of similarity.

Montroll [85] and Goel *et al.* [45] discuss a number of equations appearing in kinetics, population growth, diffusion, separation cascades, combat, and fluid mechanics which possess quadratic nonlinearities as a common feature. Various transformations are employed to obtain equations possessing built in solutions. For example, the linear diffusion equation $\theta_t = \nu\theta_{xx}$ transforms into

$$u_t = \nu\left\{u_{xx} + \frac{F''}{F'}u_x^2\right\} + 2\nu F(u)u_x$$

with $\theta = \exp\left[\int^x F(u)\,dx\right]$, and into

$$u_t = \nu\left\{u_{xx} + \frac{F''}{F'}u_x^2\right\} - f'F/fF'$$

with $\theta = f(t)F(u)$. Montroll also demonstrates, with proper boundary conditions, that equations with built in solutions are asymptotic solutions of other nonlinear problems not possessing built-in solutions. Much of this material is summarized in Ames [6, 12].

7.0. Splitting

There exist at least two basic concepts which bear the title of this section. Perhaps the term 'principle' should be used here instead of method for both concepts are guides rather than techniques. The first of these was introduced by Ames [8] and the second by Yanenko [131]. While Yanenko's principle is primarily aimed at numerical solutions it holds promise for the development of analytic solutions.

7.1. SPLITTING PRINCIPLE À LA AMES

This splitting principle is fundamentally simple – although not always simple to employ. Disregarding the inviolate nature of the equation(s) to be solved it is decomposed into parts, which are equated to a common term, in such a manner that a general solution, containing an appropriate number of arbitrary functions, can be constructed for at least one part. The form of the arbitrary function(s) is then specified, although some may remain free, by means of the requirement that the other part be satisfied.

In illustration of this concept consider the problem (Ames [8]) specified by the Navier–Stokes equations in two dimensions. In terms of the stream function ψ defined by means of $u = \psi_x$ and $v = -\psi_y$ the dimensionless equations take the form

$$\psi_y\psi_{xy} - \psi_x\psi_{yy} = -p_x + \psi_{yyy} + \mathrm{Re}^{-1}\psi_{yxx}, \tag{7.1}$$

$$p_y + \mathrm{Re}^{-1}[\psi_x\psi_{xy} - \psi_y\psi_{xx} + \psi_{xyy}] = -\mathrm{Re}^{-2}\psi_{xxx}. \tag{7.2}$$

Here

$$u = \frac{\bar{u}}{U}, \quad v = \mathrm{Re}^{1/2}\frac{\bar{v}}{U}, \quad p = \frac{\bar{p}-\bar{p}_\infty}{\varrho U^2}, \quad x = \bar{x}/L,$$

$$y = \mathrm{Re}^{1/2}\bar{y}/L, \quad \mathrm{Re} = UL/\nu$$

are the dimensionless relations for (7.1) and (7.2).

Suppose that $p = p(y)$ and split (7.1) into the two parts

$$\psi_y \psi_{xy} - \psi_x \psi_{yy} = F(x, y, \psi, \psi_x, \psi_y, \psi_{yy}, \psi_{xy}, \psi_{xx}),$$

$$\psi_{yyy} + \mathrm{Re}^{-1} \psi_{yxx} = F(x, y, \psi, \psi_x, \psi_y, \psi_{yy}, \psi_{xy}, \psi_{xx})$$

in such a way that the *general* solution of at least one of these can be developed. The form of the arbitrary functions is then determined by requiring that the other equation is also satisfied. There will remain certain arbitrary constants which are selected in such a way that the y-momentum equation (7.2) is satisfied – that is so that $p = p(y)$.

It is clear that the choice made for F strongly influences the general solution obtained, the labor involved and the final result. For simplicity it is herein chosen as zero, other forms have been used. With this choice the system becomes

(7.3) $\quad p_x = 0,$

(7.4) $\quad \psi_y \psi_{xy} - \psi_x \psi_{yy} = 0,$

(7.5) $\quad \psi_{yyy} + \mathrm{Re}^{-1} \psi_{yxx} = 0$

and p_y is defined by

(7.6) $\quad p_y = -\mathrm{Re}^{-2} \psi_{xxx} - \mathrm{Re}^{-1} [\psi_x \psi_{yx} - \psi_y \psi_{xx} + \psi_{xyy}].$

The general solution of (7.4) has the explicit form, with arbitrary φ and η,

(7.7) $\quad \psi = \varphi[y + \eta(x)]$

although it is the usual situation that the general solution is implicit. In such a case the details of the computation are more complicated. Upon substituting (7.7) into (7.5), with $w = y + \eta(x)$, there results

$$\varphi'''(w)[1 + \alpha^2 (\eta')^2] + \alpha^2 \varphi''(w) \eta''(x) = 0,$$

where $\alpha^2 = \mathrm{Re}^{-1}$. To eliminate the dependence on x, and thus determine η, set

(7.8) $\quad 1 + \alpha^2 (\eta')^2 = A \alpha^2 \eta'',$

where A is an arbitrary constant. Equation (7.8) has the solution

(7.9) $\quad \eta(x) = -A \ln \cos[(\mathrm{Re}^{1/2} x/A) + C] + c_1.$

Then φ satisfies

$$\varphi''' + A^{-1} \varphi'' = 0$$

or

$$\varphi(w) = \bar{\Gamma} + \gamma A w + \varepsilon \exp[-w/A].$$

Finally,

$$\psi(x, y) = \Gamma + \gamma A y - \gamma A^2 \ln\cos[(\mathrm{Re}^{1/2} x/A) + C]$$
$$+ \varepsilon e^{-y/A} \cos[(\mathrm{Re}^{1/2} x/A) + C],$$

where Γ, γ, A, C, ε are arbitrary constants.

Lastly, are there any values of these constants for which $p = p(y)$? From (7.6) this will be the case when $\gamma \equiv 0$. Hence,

$$\psi = \Gamma + \varepsilon \exp[-y/A] \cos[(\mathrm{Re}^{1/2} x)/A + C],$$
$$u = -\varepsilon A^{-1} \exp[-y/A] \cos[(\mathrm{Re}^{1/2} x)/A + C],$$
$$v = \varepsilon A^{-1} \mathrm{Re}^{1/2} \exp[-y/A] \sin[(\mathrm{Re}^{1/2} x)/A + C],$$
$$p(y) = -\tfrac{1}{2}(\varepsilon A)^{-2} \exp[-2y/A] + D.$$

With $C = 0$ this might be interpreted as the flow through a porous flat plate on the y axis. Note especially the appearance of the $\mathrm{Re}^{1/2}$.

This splitting principle has been successfully employed in constructing solutions for nonlinear dynamics and wave propagation in rate-type materials by Suliciu, Lee and Ames [117] and Lee and Ames [74].

7.2. SPLITTING PRINCIPLE À LA YANENKO [131, p. 21 *et seq.*]

Originally introduced as a concept for reducing computation in two and three dimensions into a sequence of one dimensional problems this idea offers possibilities for the construction of analytic solutions. Some of these are hinted at by Yanenko [131, p. 108, 109] but considerable research remains to be done. Here only one example is presented.

The unsteady flow of a weakly compressible fluid is governed by the equations

$$\begin{aligned} & u_t + u u_x + v u_y + p_x = \nu \nabla^2 u, \\ (7.10)\quad & v_t + u v_x + v v_y + p_y = \nu \nabla^2 v, \\ & \varepsilon(p_t + u p_x + v p_y) + p(u_x + v_y) = 0 \end{aligned}$$

(here $\varrho = 1$, $\varepsilon = 1/k$, $p = a^2 \varrho^k$). A splitting scheme in coordinates x, y can be applied to (7.10). At the first half step (with $t \in [n\tau, (n+\frac{1}{2})\tau]$)

solve

$$\tfrac{1}{2}u_t + uu_x + p_x = \nu u_{xx},$$
$$\tfrac{1}{2}v_t + uv_x = \nu v_{xx},$$
$$\tfrac{1}{2}\varepsilon p_t + \varepsilon u p_x + p u_x = 0$$

and at the second half step (with $t \in [(n+\frac{1}{2})\tau, (n+1)\tau]$) solve

$$\tfrac{1}{2}u_t + vu_y = \nu u_{yy},$$
$$\tfrac{1}{2}v_t + vv_y + p_y = \nu v_{yy},$$
$$\tfrac{1}{2}\varepsilon p_t + \varepsilon v p_y + p v_y = 0.$$

These are usually discretized with appropriate discrete labels. The suggestion is made herein that this may be a fruitful way to proceed analytically.

8.0. Invariant (Similar) Solutions via Groups

Probably commencing in 1894 with Boltzmann [26], who treated the diffusion equation $c_t = [D(c)c_x]_x$ the application of invariant (similar) solutions has been widespread in diffusion, heat transfer, and fluid mechanics. Applications in solid mechanics, plasmas, and wave propagation are appearing with increasing frequency in the modern literature. Because of this, it is impossible to be complete. Nevertheless the highlights will be included together with a sample of the applications. In 1908 Blasius [23] carried out a new classical analysis on the boundary-layer equations. Both of these early works would be classified as similarity by separation of variables, a procedure already discussed. Additional literature, pertaining to the method, is available in Hansen [49] and Ames [6, 12].

Construction of invariant solutions by means of continuous transformation groups was probably initiated by Birkhoff [22] and placed upon a rigorous foundation, for simple groups, by Michal [81] and Morgan [89]. In the Michal–Morgan method a group is selected, the forms of the group invariants (similarity variables) are fixed by seeking invariance of the equations, and the results are examined to ascertain if the auxiliary conditions can be written in the group invariants without inconsistencies. If so, these group invariants can be employed as similarity variables for the composite system of equations and auxiliary conditions. The transformed system, expressed in terms of the similarity variables, is then a similarity representation for the composite. Typical applications of the Michal–Morgan method in fluid mechanics are presented by Manohar [79] in laminar boundary layers, Lee and Ames [73] for two-dimensional power-law non-Newtonian fluids, and in three dimensions by Na and Hansen [91]. A general discussion is presented by Na *et al.* [90]. Additional examples and theory are presented in Hansen [49] and Ames [6, 12].

Studies of nonlinear diffusion by Michal–Morgan method have been carried out by Ames [7]. Similarity studies in gases have occupied the interests of Rosen [108, 109] and Ames [9] in unsteady flow in one dimension. Kuchemann [68] employs similarity variables in his examination of some three-dimensional transonic flows. Several other original self-similar solutions are known in gas dynamics (Guderley, Landau, and Staniukovich-shock wave problem, Zel'dovich and Weizsacker-short shock). These are given in Zel'dovich and Raizer [135] and Brushlinskii and Kazhdan [28]. Problems of filtration are treated in self-similar form by Barenblatt and Sivashinskii [19] and Kerchman [66] using the method of groups as presented in Sedov's work [114].

Self-similar solutions in plasmas are due to Gurevich *et al.* [48] and Alexeff *et al.* [5]. Wave propagation studies in solid mechanics have been published by von Karman and Duwez [63], Schultz [111], Vicario [126], and Ames *et al.* [15].

The shortcomings of the Michal–Morgan process are rather obvious. Applications have been based on particular assumed transformation groups – for example, the linear, $\bar{x}_i = a^{\alpha_i} x_i$, and spiral groups. After assuming such a specific form, it is established whether or not the differential equations could be transformed invariantly. If so, a similarity representation for the differential equations is known to exist. However, even if such a similarity representation were found, there may be other independent representations corresponding to different groups. The question of auxiliary conditions also arises. Secondly, upon obtaining a transformation group which transforms the differential equations invariantly, there is no systematic method for establishing the required set of functionally independent absolute invariants. Since the groups employed have been simple, the invariants have been determined by inspection or trial. Consequently, an undue amount of trial – of assumed groups, of absolute invariants, of suitability for expressing auxiliary conditions in fewer independent variables – has been involved in applications of the Michal–Morgan method.

Several attempts (Hellums and Churchill [50], Morgan and Gaggioli (discussed below)) have been made to ease or remove these limitations. The formal deductive theory of Moran-Gaggioli [43, 86–88] consists of the following steps: (a) an appropriate group or class of finite transformation groups is deduced, (b) the procedure explicitly considers the auxiliary conditions as well as the equations, and (c) sets of absolute invariants are derived systematically. Applications to fluid mechanics are given in the cited literature and a number are developed in Woodard and Ames [130] together with those from infinitesimal groups.

Infinitesimal transformations have been successfully applied in analysis for many years since their introduction by Lie in 1895. Ovsjannikov [98] describes their application to differential equations and similarity analyses. Bluman and Cole [24] consider applications in nonlinear diffusion while Nariboli [92] examines the diffusion equation $u_{xx}+x^{-1}ku_x = f(u)u_t$ and the transonic equation $u_{xx} = u_y u_{yy}$. All this work, together with additional references, is summarized in Chapter 2 of Ames [12].

9.0. Nonlinear Superposition

There is no reason to believe that linear operators have the 'monopoly' of the superposition principle, nor is it essential that the superposition be additive in order to obtain solutions of an equation by composing known solutions. All that is required is to know how to compose solutions to arrive at other solutions. If the composition is not linear we call it a nonlinear superposition principle.

The Riccati equation $y'+Q(x)y+R(x)y^2 = P(x)$ forms a link between the nonlinear and linear realms because by means of the transformation $y = u'/Ru$ the linear second order equation $Ru''-(R'-QR)u'-$ $-PR^2u = 0$ is obtained. This equation has the important cross-ratio theorem: *The cross-ratio of any four linearly independent solutions of the Riccati equation is a constant.* Thus from any three solutions y_1, y_2, y_3

we can calculate a fourth, y_4, from the cross-ratio

$$(9.1) \quad R = \frac{(y_1-y_3)(y_2-y_4)}{(y_1-y_4)(y_2-y_3)},$$

without quadrature. This is one of the earliest known nonlinear superpositions in the sense that it provides a rule for constructing solutions from other solutions.

Temple [120] ascribes the idea of superposition for nonlinear operators to Vessiot [125] in 1893 although Abel [2] preceded him by many years. Vessiot's theory for first order equations was immediately generalized by Guldberg [47] to systems of first order differential equations. (Lie pointed out that Vessiot's theory was a special case of his own but the generality makes Lie's work of less direct use.) Temple [120] in his reexamination of the problem arrived at the conclusion that there is no general analogue of the principle of (linear) superposition for nonlinear equations. There are, however, various types of differential equations soluble by composition. *The superposition principle as treated by the foregoing authors is understood to be a way of expressing the general solution of a nonlinear equation as a function of a certain finite number of particular solutions, the function having the same form whatever the particular solutions.*

At approximately the same time Inselberg [53] and Jones and Ames [61] removed the general solution restriction and inquired how two or more solutions could be combined in order to form another solution. This notion of superposition is an extension of the idea considered by Vessiot and Temple. After these papers there followed a number of published results consolidating and applying the theory laid down. In particular there are papers by Oppenheim [97], Foerster *et al.* [38], Inselberg [54, 55, 56, 57], Weston [128], Peterson [102], Levin [75], Spijker [116] and Keckic [64]. For some of the ensuing results applications have been found (see Oppenheim [97], Foerster *et al.* [38], Inselberg [54, 55], Weston [128] and Peterson [102]).

While most of the basic theory seems applicable to nonlinear partial differential operators the usual application has been to ordinary differential operators. Exceptions include the papers by Jones and Ames [61], Levin [75] and Keckic [64]. In particular it is known that a linear equation may possess a noncountable infinity of nonlinear superposition principles. An example (Jones and Ames [61]) is the equation

$$u_x + u_y = u \tag{9.2}$$

which has the superposition

$$u = (u_1^n + u_2^n)^{1/n} \tag{9.3}$$

for any real n, $n \neq 0$. As $n \to 0$ the superposition becomes $u = (u_1 u_2)^{1/2}$. Superpositions are presented in Jones and Ames [61] for Burgers' equation and for some additional first order quasi-linear equations. For example, the superposition for

$$a(x, y)v_x + b(x, y)v_y = c(x, y)v^n, \quad n \neq 1 \tag{9.4}$$

is

$$v = \left\{\log\left[\exp\left(\frac{1}{1-n}v_1^{1-n}\right) + \exp\left(\frac{1}{1-n}v_2^{1-n}\right)\right]^{1-n}\right\}^{1/(1-n)}. \tag{9.5}$$

A superposition is also given for Burgers' equation and for generalizations of Burgers' equation by Levin [75].

Lastly it should be remarked that the Bäcklund transformation sometimes permits development of an extension of the Riccati cross-ratio. For example, Eisenhart [37] has shown that if three solutions of the sine-Gordon equation $u_{xy} = \sin u$ are known, a fourth, u_4, and hence infinitely many, can be calculated from

$$\tan\left(\frac{u_4 - u_1}{4}\right) = \frac{a_1 + a_2}{a_1 - a_2}\tan\left(\frac{u_2 - u_3}{4}\right), \tag{9.6}$$

without additional quadrature. This relation can also be obtained from the Bäcklund transformation as demonstrated by Lamb [71].

Reference

[1] Paraphrased from an article by T. Dantzig, 'Number – The Language of Science', Circa, 1930.

Bibliography

[1] D. E. Abbott and S. J. Kline, 'Simple Methods for Classification and Construction of Similarity Solutions of Partial Differential Equations', Rept. MD-6, Dept. Mech. Eng., Stanford University, Stanford, Calif. AFOSR-TN-60-1163 (1960).

[2] N. H. Abel, 'Recherche de fonctions de deux quantités variable independantes x et y, telles que $f(x, y)$, qui ont la propriété que $f(z, f(x, y))$ est une fonction

symétrique de z, x et y', *J. Reine Angew. Math.* **1** (11), (1826) (*Oeuvres completes I*, 61–65, Christiania 1881).

[3] S. H. Advani, 'Large Amplitude Asymmetric Wave in a Spinning Membrane', *J. Appl. Mech.* **34** (1967) 1044.

[4] S. H. Advani, 'Stationary Waves in a Thin Spinning Disk', *Int. J. Mech. Sci.* **9** (1967) 307.

[5] I. Alexeff, K. Estabrook and M. Widner, 'Spreading of a Pseudowave Front', *Phys. Fluids* **14** (1971) 2355.

[6] W. F. Ames, *Nonlinear Partial Differential Equations in Engineering*, Volume I, Academic Press, New York 1965.

[7] W. F. Ames, 'Similarity for the Nonlinear Diffusion Equation', *Ind. Eng. Chem. Fund.* **4** (1965) 72.

[8] W. F. Ames, 'Ad Hoc Exact Techniques for Nonlinear Partial Differential Equations', *Nonlinear Partial Differential Equations*, Academic Press, New York 1967, p. 55.

[9] W. F. Ames, *Nonlinear Ordinary Differential Equations in Transport Processes*, Academic Press, New York 1968, p. 114.

[10] W. F. Ames, 'Discontinuity formation in solutions of homogeneous nonlinear hyperbolic equations possessing smooth initial data', *Int. J. Nonlinear Mech.* **5** (1970) 605.

[11] W. F. Ames, 'Ad Hoc Methods for Nonlinear Partial Differential Equations', *Appl. Mech. Rev.* **25** (1972) 1021.

[12] W. F. Ames, *Nonlinear Partial Differential Equations in Engineering*, Volume II, Academic Press, New York 1972.

[13] W. F. Ames, 'Implicit Ad Hoc Methods for Nonlinear Partial Differential Equations', Abstract in *Proceedings of Third Canadian Congress of Applied Mechanics*, Calgary, Alberta, May (1971) 749; *Journal Math. Anal. App.* **42** (1973) 20.

[14] W. F. Ames and E. Adams, 'A Family of Exact Solutions for Laminar Boundary Layer Equations', to appear in *Zeit. Angew. Math. u. Mech.*, 1974.

[15] W. F. Ames, S. Y. Lee, and A. A. Vicario, Jr., 'Longitudinal Wave Propagation on a Traveling Threadline II', *Int. Jl. Nonlinear Mech.* **5** (1970) 413.

[16] W. F. Ames and A. A. Vicario, Jr., 'On the Longitudinal Wave Propagation on a Traveling Threadline', in: *Developments in Mechanics V, Proceedings of the 11th Midwestern Mechanics Conference*, Ames, Iowa (1969) 733.

[17] A. V. Bäcklund, 'Om ytor med konstant negativ krökning', *Lunds Universitets Arrskrift* **19** (1883).

[18] E. D. Banta, 'Lossless Propagation of One-Dimensional Finite Amplitude Sound Waves', *J. Math. Anal. Appl.* **10** (1965) 166.

[19] G. I. Barenblatt and G. I. Sivashinskii, 'Self-Similar Solutions of the Second Kind in Nonlinear Filtration', *Prikl. Math. Mech.* **33** (1969) 861.

[20] H. Bateman, *Partial Differential Equations of Mathematical Physics*, Cambridge U. Press, London and New York 1959, p. 166 *et seq.*

[21] E. R. Benton, 'Some New Exact, Viscous, Nonsteady Solutions of Burgers' Equation', *Phys. Fluids* **9** (1966) 1247.

[22] G. Birkhoff, *Hydrodynamics*, Princeton University Press, Princeton, N. J., Ch. V, 1st ed. (1950), 2nd ed. (1960).

[23] H. Blasius, 'Grenzschichten in Flüssigkeiten mit kleiner Reibung', *Zeit. Ang. Math. Phys.* **56** (1908) 1; also NACA TM No. 1256 (1950).

[24] G. W. Bluman and J. D. Cole, 'The General Similarity Solution of the Heat Equation', *J. Math. Mech.* **18** (1969) 1025.

[25] L. Boltzmann, *Ann. Physik* (3) **53** (1894) 959.

[26] L. Boltzmann, 'Zur Integration der Diffusionsgleichung bei Variabeln Diffusions-Coefficient', Ann. Physik (N.F.) **53** (1894) 959.

[27] R. H. Boyer, 'On Some Solutions of a Nonlinear Diffusion Equation', *J. Math. and Phys.* **41** (1962) 41.

[28] K. B. Brushlinskii and Ia. M. Kazhdah, 'On the Self-Similar Solutions of Certain Problems of Gas Dynamics', *Usp. Matem. Nauk.* **18** (1963).

[29] H. S. Carslaw and J. C. Jaeger, *Conduction of Heat in Solids*, 2nd ed., Oxford Univ. Press, London (1960) 11.

[30] C. W. Chu, 'A class of Reducible Systems of Quasi-Linear Partial Differential Equations', *Quart. Appl. Math.* **23** (1965) 275.

[31] M. J. Clairin, *Annales de Toulouse*, 2e Ser., **5** (1903) 437.

[32] J. D. Cole, 'On a Quasi-Linear Parabolic Equation Occurring in Aerodynamics', *Quart. Appl. Math.* **9** (1951) 225.

[33] R. Courant and K. O. Friedrichs, *Supersonic Flow and Shock Waves*, Ch. II, Wiley (Interscience), New York 1948.

[34] R. Courant and D. Hilbert, *Methods of Mathematical Physics*, Vol. 2, Ch. 2, Wiley (Interscience), New York 1962.

[35] B. V. Dasarathy, 'Classes of Coupled Non-Linear Systems Described by Inverse Laplace Equations', Report of Computer Science Corporation, Huntsville, Ala. (1970).

[36] S. Earnshaw, 'On the Mathematical Theory of Sound', *Phil. Trans. Roy. Soc.* **150** (1858) 133.

[37] L. P. Eisenhart, *A Treatise on the Differential Geometry of Curves and Surfaces*, Ch. 8, p. 286, Dover, New York 1960.

[38] H. von Foerster, A. Inselberg and P. Weston, 'Memory and Inductive Inference', in: *Bionics Symposium* (ed. by H. Oestreicher and D. Moore), Gordon and Breach, New York 1968, pp. 31–68.

[39] A. R. Forsyth, *Theory of Differential Equations*, Vols. 5 and 6, Dover, New York 1959.

[40] H. Fujita, 'The Exact Pattern of a Concentration-Dependent Diffusion on a Semi-Infinite Medium, Part I', *Tex. Res. J.* **22** (1952) 757.

[41] H. Fujita, *ibid.*, Part II, *Tex. Res. J.* **22** (1952) 823.

[42] H. Fujita, *ibid.*, Part III, *Tex. Res. J.* **24** (1954) 234.

[43] R. A. Gaggioli and M. J. Moran, 'Group Theoretic Techniques for the Similarity Solutions of Systems of Partial Differential Equations with Auxiliary

Conditions', MRC Tech. Summary Rept. No. **693** (1966); 'Similarity Analyses of Compressible Boundary Flows via Group Theory', MRC Tech. Summary Rept. No. **838** (1967), Univ. of Wis., Madison, Wis.

[44] C. S. Gardner *et al.*, 'A Method for Solving the Korteweg–de Vries Equation', *Phys. Rev. Letters* (1967) 1095.

[45] N. S. Goel, S. C. Maitre and E. W. Montroll, *Nonlinear Models of Interacting Populations*, Academic Press, New York 1971.

[46] E. Goursat, 'Le problème de Bäcklund', *Memorial Sci. Math.* **6**, Gauthier-Villars, Paris 1925.

[47] A. Guldberg, 'Sur les equations différentielles ordinaire qui possident un système fundamental d'integrales', *Compt. Rend. Acad. Sci.* **116** (1893) 964.

[48] A. V. Gurevich., L. V. Pariiskaya and L. P. Pitaevskii, 'Self Similar Motion of Rarefied Plasma', *Sov. Phys. JETP* **22** (1966) 449.

[49] A. G. Hansen, *Similarity Analyses of Boundary Value Problems in Engineering*, Prentice-Hall, Englewood Cliffs, N. J. 1964.

[50] J. D. Hellums and S. W. Churchill, 'Generalized Similarity Analysis', *Am. Inst. Chem. Eng. J.* **10** (1964) 110.

[51] E. Hopf, 'The Partial Differential Equation $u_t+uu_x = u_{xx}$', *Commun. Pure Appl. Math.* **3** (1950) 201.

[52] R. Hill, *The Mathematical Theory of Plasticity*, Ch. VI, Oxford Univ. Press, London and New York 1950.

[53] A. Inselberg, 'On Classification and Superposition Principles for Nonlinear Operators', AF Grant 7–64, Tech. Rept. **4**, Elect. Eng. Res. Lab., Univ. of Illinois, Urbana, Ill., 1965.

[54] A. Inselberg, 'Linear Solvability and the Riccati Operator', *Jour. Math. Anal. Appl.* **22** (1968) 577.

[55] A. Inselberg, 'Phase Plane Solutions of Langmuir's Equation', *Jour. Math. Anal. Appl.* **26** (1969) 438.

[56] A. Inselberg, 'Noncommutative Superpositions for Nonlinear Operators', *Jour. Math. Anal. Appl.* **29** (1970) 294.

[57] A. Inselberg, 'Superpositions for Nonlinear Operators', *Journ. Math. Anal. Appl.* **40** (1972) 494.

[58] S. Irmay, 'Extension of Darcy's Law to Unsteady Unsaturated Flow Through Porous Media', *Proc. Symp. Darcy, U.G.G.I., IASH*, **41** (1956) 57.

[59] S. Irmay, 'Solutions of the Nonlinear Diffusion Equation with a Gravity Term in Hydrology', *Int. Assoc. Scient. Hydrology, Proc. Wageningen Symposium* (1966) 478.

[60] A. Jeffrey and T. Kakutani, 'Weak Nonlinear Dispersive Waves: A discussion centered around the Korteweg–deVries equation', *SIAM Review* **14** (1972) 582.

[61] S. E. Jones and W. F. Ames, 'Nonlinear Superposition', *Jour. Math. Ann. Appl.* **17** (1967) 484.

[62] G. Kalman, 'Nonlinear Oscillations and Nonstationary Flow in a Zero Temperature Plasma', *Ann. Phys.* **10** (1960) 1.

[63] T. von Karman and P. Duwez, 'The Propagation of Plastic Deformation in Solids', *J. Appl. Phys.* **21** (1950) 987.

[64] J. D. Keckic, 'On Nonlinear Superposition', personal communication, Dept. Math., Fac. Electrotechnique, Univ. Belgrade, Yugoslavia.

[65] J. B. Keller, 'Spherical, Cylindrical and One-Dimensional Gas Flows', *Quart. Appl. Math.* **14** (1956) 171.

[66] V. I. Kerchman, 'On Self-Similar Solutions of the Second Kind in the Theory of Unsteady Filtration', *Prikl. Math. Mech.* **35** (1971) 189.

[67] G. Kirchhoff, *Vorlesungen ueber die Theorie der Waerme*, Springer (1894).

[68] D. Kuchemann, 'On Some Three-Dimensional Flow Phenomena of the Transsonic Type', *IUTAM Symposium Transsonicum, Aachen 1962* (ed. by K. Oswatitsch), Springer-Verlag, Berlin 1964, p. 218.

[69] H. Lamb, *Hydrodynamics*, 6th ed., Ch. 9, Section 250, Dover, New York 1962.

[70] G. L. Lamb, Jr., 'Propagation of Ultrashort Optical Pulses', United Aircraft Research Laboratories; See also *Phys. Lett.* **25A** (1967) 181.

[71] G. L. Lamb, Jr., in: *In Honor of Philip M. Morse* (ed. by H. Feshbach and K. U. Ingard), MIT Press (1969) 88; See also 'Analytical Descriptions of Ultra-Short Optical Pulse Propagation in a Resonant Medium', *Rev. Mod. Phys.* **43** (1971) 99.

[72] G. L. Lamb, Jr., 'Bäcklund Transformations for Certain Nonlinear Evolution Equations', in press *Journ. Math. Phys.*, 1975.

[73] S. Y. Lee and W. F. Ames, 'Similarity Solutions for Non-Newtonian Fluids', *Amer. Inst. Chem. Eng. J.* **12** (1966) 700.

[74] S. Y. Lee, and W. F. Ames, 'A Class of General Solutions to the Nonlinear Dynamic Equations of Elastic Strings', *Jour. Appl. Mech.* **40** (1973) 1035.

[75] S. A. Levin, 'Principles of Nonlinear Superposition', *Jour. Math. Anal. App.* **30** (1970) 197.

[76] M. L. Lidov, 'Exact Solution of the Equations of One-Dimensional Unsteady Gas Motion Taking into Account Newtonian Gravitational Forces', *Dokl. Akad. Nauk SSSR* **97** (1954) 409.

[77] C. Loewner, 'A Transformation Theory of Partial Differential Equations of Gasdynamics', *NACA Tech. Note* **2065** (1950) 1–56.

[78] E. Madelung, Die mathematischen Hilfsmittel des Physikers, Springer, Berlin 1936, p. 105.

[79] R. Manohar, 'Some Similarity Solutions of Partial Differential Equations of Boundary Layer Theory', University of Wisconsin, Madison, Wis., MRC Tech. Summary Rept. (1963) 375.

[80] C. C. Mei, 'Collapse of a Homogeneous Fluid Mass in a Stratified Fluid', *12th Int. Cong. Applied Mech.*, Stanford University, Stanford, Calif., Aug. 26–31 (1968) 321.

[81] A. D. Michal, 'Differential Invariants and Invariant Partial Differential Equations Under Continuous Transformation Groups in Normed Linear Spaces', *Proc. Natl. Acad. Sci. U.S.A.* **37** (1952) 623.

[82] R. von Mises, 'Bemerkungen zur Hydrodynamik', *Z. Angew. Math. Mech.* **7** (1927) 425.

[83] R. von Mises, *Mathematical Theory of Compressible Fluid Flow*, Art. 12, Academic Press, New York 1958.

[84] R. M. Miura, 'Korteweg–de Vries Equation and Generalizations I: A Remarkable Explicit Nonlinear Transformation', *J. Math. Phys.* **9** (1968) 1202.

[85] E. W. Montroll, 'Lectures on Nonlinear Rate Equations, Especially Those with Quadratic Nonlinearities', Presented at Univ. of Colorado, Theoretical Physics Institute (1967); Copies from Dept. of Physics, Univ. of Rochester, Rochester, N.Y.

[86] M. J. Moran, 'A Unification of Dimensional Analysis and Similarity Analysis via Group Theory', Ph.D. Dissertation, University of Wisconsin, Madison, Wis. (1967).

[87] M. J. Moran and R. A. Gaggioli, 'Similarity for a Real Gas Boundary Layer Flow', MRC Tech. Summary Rept. No. 919 (1968); 'A Generalization of Dimensional Analysis', MRC Tech. Summary Rept. No. 927 (1968); 'On the Reduction of Differential Equations to Algebraic Equations', MRC Tech. Summary Rept. No. 925 (1968); Univ. of Wis., Madison, Wis.

[88] M. J. Moran and R. A. Gaggioli, 'Reduction of the Number of Variables in Systems of Partial Differential Equations with Auxiliary Conditions', *SIAM J. Appl. Math.* **16** (1968) 202; 'Similarity Analyses via Group Theory', *Amer. Inst. Aeron. Astro.* **6** (1968) 2014; 'A New Systematic Formalism for Similarity Analyses', *J. Eng. Math.* **3** (1969) 151.

[89] A. J. A. Morgan, 'The Reduction by One of the Number of Independent Variables in Some Systems of Partial Differential Equations', *Quart. Jour. of Math.* (*Oxford*) **2** (1952) 250.

[90] T. Y. Na, D. E. Abbott and A. G. Hansen, 'Similarity Analysis of Partial Differential Equations', Tech. Rept. NASA Contract 8-20065, University of Michigan, Ann Arbor, Mich. (Apr. 1967).

[91] T. Y. Na and A. G. Hansen, 'Similarity Solutions of a Class of Laminar Three-Dimensional Boundary Layer Equations of Power Law Fluids', *Int. J. Non-Linear Mech.* **2** (1967) 373.

[92] G. A. Nariboli, 'Self-Similar Solutions of Some Nonlinear Equations', *Appl. Sci. Res.* **22** (1970) 449.

[93] W. F. Noh and M. H. Protter, 'Difference Methods and the Equations of Hydrodynamics', *J. Math. Mech.* **12** (1963) 149.

[94] J. L. Nowinski, 'On the Propagation of Finite Disturbances in Bars of Rubber-Like Materials', *Jl. Engng. Ind. Trans. ASME* **87B** (1965) 523.

[95] D. W. Oplinger, 'Frequency Response of a Nonlinear Stretched String', *J. Acoust. Soc. Am.* **32** (1960) 1529.

[96] D. W. Oplinger, 'Nonlinear Vibration of a Viscoelastic String', in: *Proc. 4th Intern. Congr. Rheol.*, Providence, Part 2 (1965) 231.

[97] A. V. Oppenheim, 'Superposition in a Class of Nonlinear Systems', Rept. 432, Res. Lab. Elect. MIT, Cambridge, Mass., 1965.

[98] L. V. Ovsjannikov, *Gruppovye svoystva differentsialny uravneni*, Novosibirsk 1962; *Dokl. Akad. Nauk SSSR* **125** (1959) 492.

[99] S. I. Pai, *Viscous Flow Theory*, Vol. I: *Laminar Flow*, Van Nostrand, Princeton, N. J. 1956, p. 152.

[100] S. I. Pai, 'One Dimensional Flow of Magneto-Gasdynamics,' *Proc. 5th Midwest. Conf. Fluid Mech.*, Ann Arbor 1957, p. 251.

[101] R. E. Pattle, 'Diffusion from an Instantaneous Point Source with a Concentration-Dependent Coefficient', *Quart. J. Mech. Appl. Math.* **12** (1959) 407.

[102] L. Peterson, 'Cascades of Transformations', Rept. 5.9, Biological Computer Lab., Univ. of Illinois, Urbana, Ill., 1969.

[103] J. R. Philip, 'General Method of Exact Solution of the Concentration Dependent Diffusion Equation', *Aust. Jl. Phys.* **13** (1960) 1.

[104] G. Power, C. Rogers, and R. A. Osburn, Bäcklund and Generalized Legendre Transformations in Gas Dynamics, *Zeit. Angew. Math. Mech.* **49** (1969) 333.

[105] M. H. Protter, 'Difference Methods and Soft Solutions', in: *Nonlinear partial differential equations* (ed. by W. F. Ames), Academic Press, New York 1967, p. 161.

[106] E. Y. Rodin, 'A Riccati Solution for Burgers' Equation', *Quart. Appl. Math.* **27** (1970) 541.

[107] C. Rogers, 'Iterated Bäcklund-Type Transformations and the Propagation of Disturbances in Nonlinear Elastic Materials', to appear in *Jour. Math. Anal. Appl.* 1974.

[108] G. Rosen., 'Integration Theory for One-Dimensional Viscous Flows', *Phys. Fluids* **2** (1959) 517.

[109] G. Rosen, 'Exact Solutions for the One-Dimensional Viscous Flow of a Perfect Gas', *Phys. Fluids* **3** (1960) 191.

[110] H. Schlichting, *Boundary Layer Theory*, McGraw-Hill, New York 1960.

[111] A. B. Schultz, 'Large Dynamic Deformations Caused by a Force Traveling on an Extensible String', *Int. J. Solids Structures* **4** (1968) 799.

[112] A. Scott., *Active and Nonlinear Wave Propagation in Electronics*, Wiley–Interscience, New York (1970) 247.

[113] L. I. Sedov., 'On the Integration of the Equations of One-Dimensional Gas Motion', *Dokl. Akad. Nauk SSSR* **90** (1953) 735.

[114] L. I. Sedow, *Similarity and Dimensional Methods in Mechanics*, Ch. IV, Academic Press, New York 1959.

[115] P. Smith, 'Anisentropic Rectilinear Gas Flows', *Appl. Sci. Res.* **A12** (1963) 66; see also R. M. Gundersen, 'Nonisentropic Rectilinear Hydromagnetic Flow', *Ann. Soc. Sci. Bruxelles* **T82** (1968) 225.

[116] M. N. Spijker, 'Superposition in Linear and Nonlinear Ordinary Differential Equations', *Jour. Math. Anal. Appl.* **30** (1970) 206.

[117] I. Suliciu, S. Y. Lee, and W. F. Ames, 'Nonlinear Traveling Waves for a Class of Rate-Type Materials', *Jour. Math. Anal. Appl.* **42** (1973) 313.

[118] E. Y. Ch. Sun, 'A Compilation of Coordinate Transformations Applied to the Boundary Layer Equations for Laminar Flows', Deutsche Versuchsanstalt für Luftfahrt E. V., Nr. **121**, 1960.

[119] K. Tamada, 'Studies on the Two-Dimensional Flow of a Gas with Special Reference to the Flow through Various Nozzles', Ph.D. Thesis, Univ. Kyoto, Japan (1950).

[120] G. Temple, 'A Superposition Principle for Ordinary Nonlinear Differential Equations', in: *Lectures on Topics in Nonlinear Differential Equations*, Report 1415, David Taylor Model Basin, Carderock, Md. (1960) 1–15.

[121] S. Tomotika and K. Tamada, 'Studies on Two-Dimensional Transonic Flows of Compressible Fluid, Part I', *Quart. Appl. Math.* **7** (1950) 381.

[122] E. Varley, A Class of Nonlinear Partial Differential Equations, *Comm. Pure Appl. Math.* **15** (1962) 91.

[123] P. R. Vein, 'Functions which Satisfy Abel's Differential Equation', *SIAM Jl. Appl. Math.* **15** (1967) 618.

[124] P. R. Vein, 'Nonlinear Ordinary and Partial Differential Equations Associated with Appell Functions', *Journal of Differential Equations* **11** (1972) 221.

[125] M. E. Vessiot, 'Sur une classe d'equations différentielles', *Ann. Sci. Ecole Norm. Sup.* **10** (1893) 53.

[126] A. A. Vicario, Jr., 'Longitudinal Wave Propagation Along a Moving Threadline', Ph.D. Dissertation, Univ. Delaware, Newark, Del. (1968).

[127] H. D. Wahlquist, and F. B. Estabrook, *Phys. Rev. Letters* **31** (1973) 1386.

[128] P. Weston, 'Counting Binary Rooted Trees Arboreal Numbers of the First and Second Kind', in: *Rooted Trees*, Rept. 2.0 Biological Computer Lab., Univ. of Illinois, Urbana, Ill., 1966.

[129] L. van Wijngaarden, 'On the Equations of Motion for Mixtures of Liquid and Gas Bubbles', *J. Fluid Mech.* **33** (1968) 465.

[130] H. S. Woodard, and W. F. Ames, 'Similarity Solutions for Partial Differential Equations Generated by Finite and Infinitesimal Groups', Iowa Institute of Hydraulic Research Rept. No. 132, Univ. of Iowa, Iowa City 1971.

[131] N. N. Yanenko, *The Method of Fractional Steps*, Springer-Verlag, New York, Heidelberg 1971.

[132] N. J. Zabusky, 'Exact Solution for the Vibrations of a Nonlinear Continuous Model String', *J. Math. Phys.* **3** (1962) 1028.

[133] N. J. Zabusky, and M. D. Kruskal, 'Interaction of Solitons in a Collisionless Plasma and the Recurrence of Initial States', *Phys. Rev. Letters* **15** (1965) 240.

[134] N. J. Zabusky, 'A Synergetic Approach to Problems of Nonlinear Dispersive Wave Propagation and Interaction', in: *Nonlinear Partial Differentia Equations* (ed. by W. F. Ames), Academic Press, New York 1967, p. 223; See also P. D. Lax, 'Integrals of Nonlinear Equations of Evolution and Solitary Waves', *Comm. Pure App. Math.* **21** (1968) 467.

[135] Ia. B. Zel'dovich, and Iu. P. Raizer, *Physics of Shock Waves and High Temperature Hydrodynamic Phenomena*, 2nd ed., Nauka 1966.

RECENT DEVELOPMENTS IN NONLINEAR SPINOR THEORIES

H. P. DÜRR

Max-Planck-Institut für Physik und Astrophysik, Munich, Federal Republic of Germany

Abstract

Nonlinear spinor theories based on spinor fields with anomalous dimension 1/2 (subcanonical fields) are shortly reviewed. Such theories appear very interesting because they can be scale invariant and, therefore, formally renormalizable, and they do allow the incorporation of local gauge invariance without the addition of independent vector fields. The main part of the lecture is devoted to a discussion of a canonical imbedding of such subcanonical theories. In this extended framework a more conventional approach is again possible and proves highly effective. In particular, the skeleton theory in this framework proves to be conformally invariant under the same condition as required by gauge invariance. The lowest approximation for gauge field type bound states is nondivergent and automatically yields massless solutions.

1. Introduction

Nonlinear field theories are conceptionally the simplest theories which are capable of describing interaction. In fact, the φ^4-theory (in the Lagrangian formulation), i.e. a field theory with a selfinteracting scalar field $\varphi(x)$ based on a field equation

(1.1) $$(\partial^2+m^2)\varphi(x)+\lambda:\varphi^3:(x) = 0$$

is one of the simplest interacting theories, and therefore is commonly used as a model theory. Such a theory cannot, of course, be seriously considered to be of fundamental importance in describing the spectrum of elementary particles and their interaction.

A realistic theory of elementary particles certainly has to contain, at least, a spinor field. Since, in principle, all fields interpolating the asymptotic fields of elementary particles can be constructed from spinor fields and their products, it appears most attractive for a unified description of elementary particles to consider on a fundamental level a nonlinear spinor theory, i.e. a theory containing only a fundamental spinor field $\psi(x)$ which is locally coupled to itself. Investigations of such theories have been started about 20 years ago by Heisenberg.[1] Recently theories of this type have drawn appreciable attention in connection with the quarks and the supposed quark-structure of hadrons. Since we do not necessarily want to attach a SU_3-property to our fundamental spinor field – in fact, we do *not* want this for various reasons which I will not try to spell out here – we don't want to call our spinor field 'quark field' but prefer to call it 'the constituent field' to emphasize its character as a building block for elementary particles.

Since the internal degrees of freedom of the spinor field will only be of minor importance in the considerations I intend to present, I wish to concentrate in my lecture on the simplest possible nonlinear spinor theory, namely a theory for a self-coupled 2-component Weyl spinor field $\psi(x)$ which obeys the nonlinear field equation ($\sigma^\mu = (I, \vec{\sigma})$)

$$(1.2) \qquad i\sigma \cdot \partial\psi(x) + g\sigma^\mu : \psi(\psi^* \sigma_\mu \psi) : (x) = 0 .$$

This is essentially the Heisenberg nonlinear spinor theory (without SU_2-degrees of freedom) in the form as given by Dürr [2]. This theory has some beautiful properties. E.g. it automatically admits only a 'current-current' type interaction, and higher non-derivative interaction terms seem to be impossible due to the antisymmetry of the spinor fields ('Pauli principle'), at least if we disregard possible modifications due to the 'finite part' prescription of the product indicated by the double dots.

This theory, however, has some very bad properties. In particular, it is *non-renormalizable*, in contrast to the nonlinear scalar theory (1.1) which is renormalizable. In a perturbation expansion of the theory in which $\psi(x)$ is assumed to be a perturbed canonical field $\psi_c(x)$ with the equal-time anticommutator

$$(1.3) \qquad \left\{\psi_c\left(\frac{x}{2}\right), \psi_c^*\left(-\frac{x}{2}\right)\right\}_{t=0} = \delta(\vec{x})$$

the coupling constant g has the dimension of a length square or an inverse mass square like in local weak interaction theory. Theories with coupling constants of inverse mass dimension in higher order of a perturbation expansion lead to worse singularities which exhibits their non-renormalizable character. This is the main reason why nonlinear theories of the type (1.2) are usually discarded.

To provide a chance for a nonlinear theory of type (1.2) to be physically acceptable is to require it to be *renormalizable* or to have a *dimensionless* coupling constant g. In this case the theory would become *scale invariant*, at least at small distances where possible mass terms should not contribute essentially (skeleton theory) [22]. Scale invariance of the field equation requires the spinor field to transform under dilation like

$$(1.4)\qquad \psi(x) \xrightarrow{D} \psi'(x') = \eta^{-\frac{1}{2}}\psi(\eta^{-1}x') \qquad \eta \geqslant 0 \text{ real}$$

which indicates that the field must have the *anomalous dimension* (mass dimension)

$$(1.5)\qquad \dim\psi(x) = \tfrac{1}{2}$$

which is *smaller* than the dimension of the massless canonical field

$$\dim\psi_c(x) = 3/2.$$

We call the dimension of $\psi(x)$ a '*subcanonical*' *dimension*, and the field a 'subcanonical' field, or a 'spinor potential' [15]. As a consequence of (1.4) the invariant commutation rules cannot be the canonical rules but the ψ-field has to obey the less singular anticommutation relations ($\bar{\sigma}^\mu = (I, -\vec{\sigma})$)

$$(1.6)\qquad \begin{aligned} &\lim_{t\to 0}\left\{\psi\left(\frac{x}{2}\right), \psi^*\left(-\frac{x}{2}\right)\right\} = \lim_{t\to 0}\left[-\frac{1}{2}(\bar{\sigma}\cdot x)\Delta(x;0)\right],\\ &\Delta(x;0) = \frac{1}{2\pi}\varepsilon(x^0)\,\delta(x^2)\end{aligned}$$

where we have arbitrarily chosen a normalization such that the 'canonical field' $\psi_c(x) = i\sigma\cdot\partial\psi(x)$, the 'Weyl-derivative' of $\psi(x)$, has again canonical anticommutation relations. In momentum space this corresponds to a propagator for large p of the form

$$(1.7)\qquad \psi\psi^* \sim \frac{\bar{\sigma}\cdot p}{(p^2)^2}$$

in contrast to the canonical propagator $\bar{\sigma}\cdot p/p^2$.

The additional power of (p^2) in the propagator (1.7) has the very serious consequence that the field operators have to be assumed to be linear operators in a state space with *indefinite metric*. The appearance of an indefinite metric in the quantum mechanical state space does, however, jeopardize the probability interpretation of the theory and, in particular, the *unitarity* of the S-matrix. I do not have the time to discuss this important problem [23], [29]–[31]. I just want to emphasize that the occurrence of an indefinite metric is *by no means fatal* to the theory as many physicists believe, but only requires to consider the physical states as a certain subset of the complete set of states of the larger space, more precisely: a subset of positive norm states. This situation, in fact, we have already encountered in Gupta–Bleuler quantum electrodynamics. Physically speaking it is a consequence of the fact that the concept of an 'interaction' is more general than the concept of 'virtual particle exchange' if we interpret 'particle' in the physical sense of possible asymptotic states. The virtually exchanged 'ghosts' describe a 'genuine' interaction similar to the Coulomb interaction in quantum electrodynamics [4]. Hence we have to be careful to conceive $\psi(x)$ as an ordinary 'particle-like' spinor field which we encounter in usual perturbation theory. We will shortly come back to this point later on.

The 'subcanonical' nature of our spinor field has the important consequence that the 'finite part' or the 'Wick product' of the local bilinear spinor form

$$(1.8)\qquad :\psi^*\sigma_\mu\psi:(x) = \overline{\lim_{\zeta\to 0}}\left\{\psi^*\left(x-\frac{\zeta}{2}\right)\sigma_\mu\psi\left(x+\frac{\zeta}{2}\right) - \frac{1}{4\pi^2 i}\frac{\zeta_\mu}{\zeta^2 - i\varepsilon}\right\}$$

under local gauge transformations (phase gauge transformations)

$$(1.9)\qquad \psi(x) \stackrel{G}{\to} \psi'(x') = e^{-i\alpha(x)}\psi(x)$$

transforms *inhomogeneously*, namely as

$$(1.10)\qquad :\psi^*\sigma_\mu\psi:(x) \stackrel{G}{\to} :\psi^*\sigma_\mu\psi:(x) - \frac{1}{16\pi^2}\partial_\mu\alpha(x),$$

i.e. exactly like a gauge vector field [15]. Hence, if we write the field equation (1.2) in the suggestive form

$$(1.11)\qquad \begin{aligned} &i\sigma\cdot\partial\psi(x) - g\sigma^\mu : R_\mu\psi:(x) = 0,\\ &R_\mu(x) = -:\psi^*\sigma_\mu\psi:(x) \end{aligned}$$

we realize that the equation is gauge invariant if $R_\mu(x)$ transforms as

$$(1.12)\qquad R_\mu(x) \stackrel{G}{\to} R_\mu(x) + \frac{1}{g}\partial_\mu\alpha(x).$$

This can be arranged by choosing for the dimensionless coupling constant the special value [15]

(1.13) $g = 16\pi^2$

and by assuming that the finite part prescription $:R_\mu\psi:$ in (1.11) does not induce any additional gauge dependence, which can be secured [16]. Therefore, local gauge invariance can be established in such a subcanonical spinor theory *without* additional vector fields. The gauge field arises as a local 'compound' of the constituent field.

The nonlinear field equation (1.11) can be imagined to arise from a 2-component canonical (neutrino) field equation

(1.14) $i\sigma \cdot \partial\psi = 0$

in which a self-interaction is switched on. This self-interaction forces the canonical dimension 3/2 of the field down to the lower value 1/2 in such a way that the interaction term gets less singular and allows to be incorporated into a 'covariant' derivative

(1.15) $\partial_\mu \rightarrow \nabla_\mu = \partial_\mu + igR_\mu(x)$

which establishes the local gauge invariance and as a consequence the renormalizability of the theory.

From this point of view the *change of the dimension* of the spinor field is a *consequence* of the interaction. To exhibit more clearly the structure of such a theory we will consider in the following the somewhat artificial *intermediate theory* where we take full account of the dimensional shift $3/2 \rightarrow 1/2$ of the spinor field but still neglect the nonlinear interaction term, i.e. we switch off the interaction without letting the field 'slip off' its 'interaction dimension'. In this way we obtain a linear theory for a subcanonical field which may serve as the corresponding 'free' theory for our interacting theory in conventional terms. The linear theory for a subcanonical field provides a canonical framework, a 'canonical imbedding' for the subcanonical theory.

2. Linear Theory for a Subcanonical Field

The free 'classical' theory related to a $\dim\psi = 1/2$ spinor field is the *third order derivative theory* [1]

(2.1) $-i(\sigma \cdot \partial)\partial^2\psi(x) = 0.$

It derives from the Lagrangian density

$$(2.2)\qquad \mathscr{L} = \frac{i^3}{2}\{\psi^*(\sigma\cdot\partial)\,\partial^2\psi - ((\sigma\cdot\partial)\,\partial^2\psi)^*\psi\}$$

which obviously is invariant under the dilatation transformation (1.4) of the spinor field. In a quantized version of the theory the anticommutator of the spinor field is, as usual, connected with the causal invariant solution of the homogeneous field equation (2.1) with a normalization given by the Lagrangian density [7]. We obtain

$$(2.3)\qquad \left\{\psi\left(\frac{x}{2}\right), \psi^*\left(-\frac{x}{2}\right)\right\} = -\frac{1}{2}(\bar{\sigma}\cdot x)\Delta(x;0)$$

$$= \frac{i}{(2\pi)^4}\oint d^4p \frac{\bar{\sigma}\cdot p}{(p^2)^2} e^{-ipx}.$$

Actually we could proceed discussing in detail the general properties of this third order derivative theory. This, in fact, leads to a very compact formalism [1], [7]. To emphasize, however, the relationship to conventional theories we prefer to cast the third order derivative theory into a *canonical* form with only first order derivatives [1]. This we achieve by introducing two additional spinor fields $\hat{\psi}(x)$ and $\hat{\hat{\psi}}(x)$ which are essentially connected with the first and second time derivative of $\psi(x)$. Such a relationship is established by the following conjecture for the Lagrangian density

$$(2.4)\qquad \mathscr{L} = \frac{i}{2}\{\hat{\hat{\psi}}^*\sigma\cdot\overleftrightarrow{\partial}\psi + \hat{\psi}^*\bar{\sigma}\cdot\overleftrightarrow{\partial}\hat{\psi} + \psi^*\sigma\cdot\overleftrightarrow{\partial}\hat{\hat{\psi}}\} - \{\hat{\hat{\psi}}^*\hat{\psi} + \hat{\psi}^*\hat{\hat{\psi}}\}$$

which yields the field equations

$$(2.5)\qquad \begin{aligned} i\sigma\cdot\partial\psi &= \hat{\psi},\\ i\bar{\sigma}\cdot\partial\hat{\psi} &= \hat{\hat{\psi}},\\ i\sigma\cdot\partial\hat{\hat{\psi}} &= 0. \end{aligned}$$

The first two equations may be considered as defining equations for $\hat{\psi}$ and $\hat{\hat{\psi}}$. If we insert them into the last one we recover the old third order equation (2.1). Hence (2.2) and (2.4) are equivalent Lagrangian densities.

We realize that, if $\psi(x)$ is a right-chirality field of $\dim\psi = 1/2$, the $\hat{\psi}(x)$ is a *left*-chirality field of dimension

$$\dim\hat{\psi} = \tfrac{3}{2}, \tag{2.6}$$

i.e. a field of canonical dimension, and the

$$\hat{\hat{\psi}} = -\partial^2\psi \tag{2.7}$$

is a right-chirality field of dimension

$$\dim\hat{\hat{\psi}} = \frac{5}{2}. \tag{2.8}$$

From the Lagrangian density (2.4) we further deduce that $(\psi, \hat{\hat{\psi}}{}^*)$, $(\hat{\psi}, \hat{\psi}^*)$ and $(\hat{\hat{\psi}}, \psi^*)$ form canonical conjugate pairs, i.e. we have the nonvanishing equal-time anticommutators

$$\begin{aligned}
&\left\{\psi\left(\frac{x}{2}\right), \hat{\hat{\psi}}{}^*\left(-\frac{x}{2}\right)\right\}_{t=0} = \delta(\vec{x}),\\
&\left\{\hat{\psi}\left(\frac{x}{2}\right), \hat{\psi}^*\left(-\frac{x}{2}\right)\right\}_{t=0} = \delta(\vec{x}),\\
&\left\{\hat{\hat{\psi}}\left(\frac{x}{2}\right), \psi^*\left(-\frac{x}{2}\right)\right\}_{t=0} = \delta(\vec{x}).
\end{aligned} \tag{2.9}$$

For the general 2-point function we actually find [8], [28]

$$\begin{aligned}
\left\langle 0 \left| \begin{bmatrix} \psi\psi^* & \psi\hat{\psi}^* & \psi\hat{\hat{\psi}}{}^* \\ \hat{\psi}\psi^* & \hat{\psi}\hat{\psi}^* & \hat{\psi}\hat{\hat{\psi}}{}^* \\ \hat{\hat{\psi}}\psi^* & \hat{\hat{\psi}}\hat{\psi}^* & \hat{\hat{\psi}}\hat{\hat{\psi}}{}^* \end{bmatrix} \right| 0 \right\rangle &= \begin{bmatrix} -\frac{1}{2}\bar{\sigma}\cdot x & -i & \bar{\sigma}\cdot\partial \\ -i & \sigma\cdot\partial & 0 \\ \bar{\sigma}\cdot\partial & 0 & 0 \end{bmatrix} \Delta(x;0) \\
&= \frac{i}{(2\pi)^4}\int d^4p e^{-ipx} \begin{bmatrix} \dfrac{\bar{\sigma}\cdot p}{(p^2)^2} & \dfrac{1}{p^2} & \dfrac{\bar{\sigma}\cdot p}{p^2} \\ \dfrac{1}{p^2} & \dfrac{\sigma\cdot p}{p^2} & 0 \\ \dfrac{\bar{\sigma}\cdot p}{p^2} & 0 & 0 \end{bmatrix}
\end{aligned} \tag{2.10}$$

which can be derived from the $\psi\psi^*$-2-point function (2.3) by appropriate differentiation.

To make these relationships more transparent we expand $\psi(x)$ in terms of creation and annihilation operators [8]. From the form of the 2-point function in momentum space

$$(2.11)\quad \frac{\bar{\sigma}\cdot p}{(p^2)^2} = \frac{1}{(\sigma\cdot p)(\bar{\sigma}\cdot p)(\sigma\cdot p)} = \frac{1}{(p^0-\vec{\sigma}\cdot\vec{p})^2(p^0+\vec{\sigma}\cdot\vec{p})}$$

we immediately deduce that $\psi(x)$ creates ordinary massless particless (n-mode) of negative helicity

$$(2.12)\quad p^0 = -\vec{\sigma}\cdot\vec{p}$$

but 'null-ghost couples' [13], [14] of the Heisenberg dipole type [19] with positive helicity

$$(2.13)\quad p^0 = \vec{\sigma}\cdot\vec{p}$$

giving rise to a massless double pole. A 'null-ghost-couple' consists of two non-orthogonal norm zero states, a 'good ghost' and a 'bad ghost' [13], [14]. In our case, the 'good ghost' is an eigenstate of energy with mass zero, the 'bad ghost', the Heisenberg 'dipole ghost', has a pathological time dependence $\sim te^{-iEt}$ and therefore is *not* an energy eigenstate. The $\psi(x)$ has the expansion

$$(2.14)\quad \psi(x) = \frac{1}{(2\pi)^{3/2}}\int d^3p\,\frac{1}{2|\vec{p}|}\,[a_g(+,\vec{p})-i|\vec{p}|\,ta_b(+,\vec{p})]h_+(\vec{p})+$$

$$+a_n(-,\vec{p})h_-(\vec{p})\}e^{-i(|\vec{p}|t-\vec{p}\cdot\vec{x})}+\text{creation operators.}$$

The $(\pm)$ refer to the helicity of the states and

$$(2.15)\quad h_\pm(\vec{p}) = \frac{1}{2|\vec{p}|}(|\vec{p}|\pm\vec{\sigma}\cdot\vec{p})$$

are the projection operators on the helicity states. The a_b and a_g operators obey the anticommutation relations typical for a 'null-ghost-couple'

$$\{a_g(\vec{p}),\, a_b^*(\vec{p}')\} = \delta(\vec{p}-\vec{p}'),$$

$$(2.16)\quad \{a_b(\vec{p}),\, a_g^*(\vec{p}')\} = \delta(\vec{p}-\vec{p}'),$$

all other zero,

whereas a_n has ordinary anticommutation relations

$$(2.17)\quad \{a_n(\vec{p}),\, a_n^*(\vec{p}')\} = \delta(\vec{p}-\vec{p}').$$

The $*$-operation refers to the pseudo-hermitian conjugation. The 1-particle states (g, b, n) have the metrical tensor

$$(2.18)\quad \eta = \begin{bmatrix} 0 & 1 & 0 \\ 1 & 0 & 0 \\ 0 & 0 & 1 \end{bmatrix}$$

where the diagonal element refers to the n-state.

The expansion (2.14) of the Weyl field shows that it contains physical as well as unphysical degrees of freedom. In fact, the situation bears some similarity to Gupta–Bleuler quantum electrodynamics. There the vector potential $A_\mu(x)$ contains the transverse modes as physical degrees of freedom and the longitudinal and timelike modes as unphysical degrees of freedom. These latter modes in their plus and minus combinations also form a 'null-ghost-couple' with the 'good ghost' being connected with the gauge degrees of freedom [13], [14]. The 'bad ghosts' on the other hand do not fulfil the Lorentz condition, and are suppressed by the Gupta condition as a condition for the physical states

$$(2.19)\quad \partial^\mu A_\mu^{(+)}|\text{phys}\rangle = 0 \quad \text{or} \quad a_b(\vec{p})|\text{phys}\rangle = 0\,.$$

This condition provides for unitarity if an interaction is considered.

Similarly, to arrive at a physically interpretable theory in our case we have to forbid the 'bad ghosts', the dipole ghosts, to occur asymptotically, i.e. we have to require for the physical subspace

$$(2.20)\quad a_b(+, \vec{p})|\text{phys}\rangle = 0$$

which also secures that the physical states are all eigenstates of energy. Without the 'bad ghosts' the 'good ghosts' are harmless because of their vanishing norm. To secure the subsidiary condition (2.20) not only for the incoming but also the outgoing states the admixture of 'good ghosts' in the incoming states has to be fixed in a certain way contrary to the simpler situation in quantum electrodynamics [19].[2] In the local anticommutator not only the physical operators but *all* operators equally contribute, and, in fact, this is precisely the reason why the anticommutator and as a consequence the selfinteraction is less singular than in the canonical case. In contrast to quantum electrodynamics the *ghost states* in our case *do change the short distance behavior* of the interaction and in such a way as to essentially cancel the ultraviolet divergencies.

It is instructive to look also at the expansions of the $\hat{\psi}(x)$ and $\hat{\hat{\psi}}(x)$ field operators which can be derived from the expansion of $\psi(x)$ by an appropriate differentiation. We obtain

$$(2.21)\quad \hat{\psi}(x) = \frac{1}{(2\pi)^{3/2}}\int d^3p\left\{\frac{1}{2}a_b(+,\vec{p})h_+(\vec{p})+ + a_n(-,\vec{p})h_-(\vec{p})\right\}e^{-i(|\vec{p}|t-\vec{p}\cdot\vec{x})}+\text{creation part.}$$

I.e. $\hat{\psi}(x)$ contains effectively the negative helicity part since the positive helicity contributions connected with the 'bad ghosts' have zero norm. Finally

$$(2.22)\quad \hat{\hat{\psi}}(x) = \frac{1}{(2\pi)^{3/2}}\int d^3p|\vec{p}|\, a_b(+,\vec{p})h_+(\vec{p})e^{-i(|\vec{p}|t-\vec{p}\cdot\vec{x})}+ +\text{creation part}$$

only contains the zero-norm positive helicity 'bad ghost' contributions. This explains why $\hat{\hat{\psi}}(x)$ has vanishing anticommutator with $\hat{\hat{\psi}}^*(x)$ and $\hat{\psi}^*(x)$ but not with $\psi^*(x)$ which contain the 'good ghosts'. The subsidiary condition on the physical states (2.20) hence requires

$$(2.23)\quad \langle\text{phys}\,|\hat{\hat{\psi}}(x)\,|\text{phys}\rangle = 0$$

or in short

$$\hat{\hat{\psi}}(x) \triangleq 0$$

if $\triangleq$ is meant to express an equality only for physical matrix elements. It is important to notice that in a quantum mechanical interpretation of the equations (2.5) we cannot set $\hat{\hat{\psi}} = 0$ as an operator condition without destroying the algebraic structure of the theory which according to (2.9) states, in particular, that $\hat{\hat{\psi}}^*$ is the canonical conjugate field to $\psi(x)$.

It is interesting to observe that the situation in the state space is less pathological if we include mass terms in the theory. Adding to (2.4) the Lagrangian density

$$(2.24)\quad \mathscr{L}_m = -\tfrac{1}{2}m^2(\hat{\psi}^*\psi+\psi^*\hat{\psi})$$

leads to the modified field equations

$$(2.25)\quad \begin{aligned} i\sigma\cdot\partial\psi &= \hat{\psi},\\ i\bar{\sigma}\cdot\partial\hat{\psi} &= \hat{\hat{\psi}}+\tfrac{1}{2}m^2\psi,\\ i\sigma\cdot\partial\hat{\hat{\psi}} &= \quad +\tfrac{1}{2}m^2\hat{\psi}\end{aligned}$$

or to the single equation

$$(2.26)\quad -i\sigma\cdot\partial(\partial^2+m^2)\psi = 0$$

in the third order formalism. In this case we have the following 2-point functions [8], [28]

$$(2.27)\quad \left\langle 0\left\|\begin{bmatrix}\psi\psi^* & \psi\hat{\psi}^* & \psi\hat{\hat{\psi}}^*\\ \hat{\psi}\psi^* & \hat{\psi}\hat{\psi}^* & \hat{\psi}\hat{\hat{\psi}}^*\\ \hat{\hat{\psi}}\psi^* & \hat{\hat{\psi}}\hat{\psi}^* & \hat{\hat{\psi}}\hat{\hat{\psi}}^*\end{bmatrix}\right\|0\right\rangle$$

$$= \frac{i}{(2\pi)^4}\int d^4p e^{-ipx}\times$$

$$\times\begin{bmatrix}\dfrac{\bar{\sigma}\cdot p}{p^2(p^2-m^2)} & \dfrac{1}{p^2-m^2} & \dfrac{\bar{\sigma}\cdot p}{p^2-m^2}\left(1-\dfrac{m^2}{2p^2}\right)\\ \dfrac{1}{p^2-m^2} & \dfrac{\sigma\cdot p}{p^2-m^2} & \dfrac{m^2}{2}\,\dfrac{1}{p^2-m^2}\\ \dfrac{\bar{\sigma}\cdot p}{p^2-m^2}\left(1-\dfrac{m^2}{2p^2}\right) & \dfrac{m^2}{2}\,\dfrac{1}{p^2-m^2} & \left(\dfrac{m^2}{2}\right)^2\dfrac{\bar{\sigma}\cdot p}{p^2(p^2-m^2)}\end{bmatrix}$$

$=$ 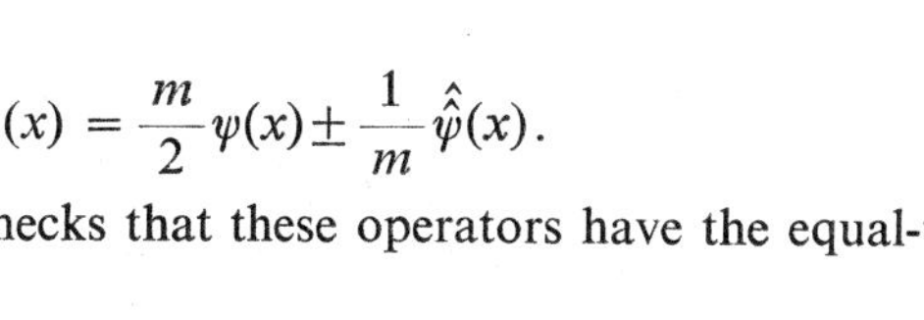

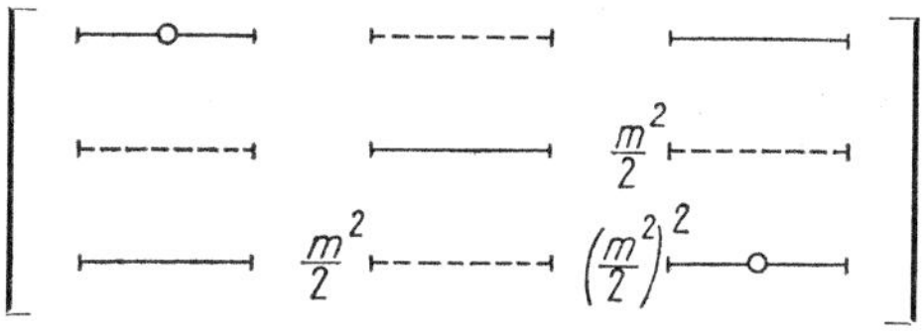

This, however, can be written in a simpler way if we introduce the new operators [1]

$$(2.28)\quad \psi_\pm(x) = \frac{m}{2}\psi(x)\pm\frac{1}{m}\hat{\hat{\psi}}(x).$$

One easily checks that these operators have the equal-time anticommutators

$$(2.29)\quad \begin{aligned}\left\{\psi_+\left(\frac{x}{2}\right),\ \psi_+^*\left(-\frac{x}{2}\right)\right\}_{t=0} &= \delta(\vec{x}),\\ \left\{\psi_-\left(\frac{x}{2}\right),\ \psi_-^*\left(-\frac{x}{2}\right)\right\}_{t=0} &= -\delta(\vec{x})\end{aligned}$$

which shows that ψ_+ is a *canonical* field connected with *positive norm states* whereas ψ_- is a *canonical* field connected with *negative* norm states. This means that in the presence of a mass term the off-diagonal metric tensor of the 'null-ghost-couple' can be diagonalized by forming the linear combinations (2.36), i.e.

$$(2.30)\quad \eta = \begin{bmatrix} 0 & 1 \\ 1 & 0 \end{bmatrix} \overset{\psi, \hat{\psi} \to \psi_\pm}{\rightarrow} \eta = \begin{bmatrix} 1 & 0 \\ 0 & -1 \end{bmatrix}.$$

The field equations (2.25) with the field operators (2.28) can now be written in the very simple form

$$(2.31)\quad \begin{aligned} i\sigma \cdot \partial \psi_+ &= m\hat{\psi}, \\ i\bar{\sigma} \cdot \partial \hat{\psi} &= m\psi_+, \\ i\sigma \cdot \partial \psi_- &= 0 \end{aligned}$$

and the 2-point functions have the diagonal structure

$$(2.32)\quad \left\langle 0 \left| \begin{bmatrix} \psi_+\psi_+^* & \psi_+\hat{\psi}^* & \psi_+\psi_-^* \\ \hat{\psi}\psi_+^* & \hat{\psi}\hat{\psi}^* & \hat{\psi}\psi_-^* \\ \psi_-\psi_+^* & \psi_-\hat{\psi}^* & \psi_-\psi_-^* \end{bmatrix} \right| 0 \right\rangle$$

$$= \frac{i}{(2\pi)^4} \int d^4p e^{-ipx} \begin{bmatrix} \dfrac{\bar{\sigma} \cdot p}{p^2 - m^2} & \dfrac{m}{p^2 - m^2} & 0 \\ \dfrac{m}{p^2 - m^2} & \dfrac{\sigma \cdot p}{p^2 - m^2} & 0 \\ 0 & 0 & -\dfrac{\bar{\sigma} \cdot p}{p^2} \end{bmatrix}.$$

This shows that $\begin{bmatrix} \psi_+ \\ \hat{\psi} \end{bmatrix}$ now form an *ordinary canonical 4-component Dirac spinor operator* with mass m, and the ψ_- is a 2-component canonical neutrino operator relating to negative norm states.

It is interesting to realize that with the introduction of the mass term we generate two new symmetries which will be of decisive physical importance, namely a *parity symmetry* of the massive modes

$$(2.33)\quad \begin{aligned} &\psi_+(\vec{r}, t) \overset{P}{\rightarrow} \hat{\psi}(-\vec{r}, t), \\ &\hat{\psi}(\vec{r}, t) \overset{P}{\rightarrow} \psi_+(-\vec{r}, t) \end{aligned}$$

and a *phase transformation* which transforms the positive and negative norm parts in a different way, e.g.

$$(2.34)\qquad \begin{bmatrix}\psi_+\\ \hat{\psi}\end{bmatrix} \to \begin{bmatrix}\psi_+\\ \hat{\psi}\end{bmatrix},$$

$$\psi_- \to e^{-i\alpha}\psi_- .$$

The transformations (2.33) and (2.34) have no correspondence in the limit $m \to 0$, because only by virtue of the mass terms in (2.25) the ψ_- (which still may be formally defined like (2.28) with an arbitrary mass parameter) decouples from the ψ_+ and the $\hat{\psi}$.

If eventually an interaction is turned on the unitarization problem will consist in preventing the ψ_--field to ever couple singly with a ψ_+- or a $\hat{\psi}$-field because this would lead to transitions in which a single particle transmutes into a ghost state and vice versa. There would be no problem with unitarity if only pairs of particles turn into pairs of ghosts. The situation would be even better if the new phase transformation symmetry (2.34) could also be effectively maintained in the *interacting theory* in which case the particles and ghosts would be separated by an effective superselection rule and could only transmute as particle-antiparticle pairs into each other. Such a possibility will be discussed later on.

3. Symmetries and Currents in the Linear Theory

I wish shortly to comment on the symmetries and the corresponding 'currents' in the linear massless theory (2.2) or (2.4) as discussed in the former section. It is obvious that the theory is invariant under the Weyl group (Poincaré transformations and dilatations) and a phase transformation [15]. The 'currents', i.e. the current proper, the energy-momentum tensor, the spin tensor and the dilatation current can be constructed in a straightforward way [1], [7]. In the first order formalism (2.4) we may in particular, proceed along conventional lines. For the current proper we find [1], e.g.

$$(3.1)\qquad j_\mu = \hat{\hat{\psi}}^* \sigma_\mu \psi + \hat{\psi}^* \bar{\sigma}_\mu \hat{\psi} + \psi^* \sigma_\mu \hat{\hat{\psi}} .$$

In the third order formalism we derive [1] from (2.2) the current

$$(3.2)\qquad j_\mu = -\psi^* \sigma^\varrho [\tfrac{1}{4}(\overline{\partial}^2 + \overleftrightarrow{\partial}^2) g_{\varrho\mu} + \tfrac{1}{2} \overleftrightarrow{\partial}_\varrho \overleftrightarrow{\partial}_\mu + \tfrac{1}{6}(\overline{\partial}^2 g_{\varrho\mu} - \overline{\partial}_\varrho \overline{\partial}_\mu)]\psi$$

with

$$A\overline{\partial}B = \partial(AB),$$
$$A\overleftrightarrow{\partial}B = A(\partial B)-(\partial A)B$$

which agrees with (3.1) up to 'ineffective' terms, i.e. terms the 4-divergence of which vanishes identically and which do not contribute to the charge integral.

If mass terms are included in the Lagrangian density the current can be written in the form

$$(3.3) \qquad j_\mu = \psi_+^* \sigma_\mu \psi_+ + \hat{\psi}^* \bar{\sigma}_\mu \hat{\psi} - \psi_-^* \sigma_\mu \psi_-$$

with the $\psi_\pm$-operators as defined earlier. I.e. the current obtains a 'diagonal' form consisting of the Dirac current (first two terms) and a (negative norm) neutrino current (last term). The higher symmetry indicated by (2.34) is reflected by the fact that the Dirac current and the 'neutrino' current are *separately* conserved. This particular structure of the current will be of importance later on in the formulation of the interacting theory.

There are some pecularities in this theory regarding the dilatation transformation [1]. E.g. it turns out that the conservation of the dilatation current does not imply the vanishing of the trace of the canonical energy-momentum tensor. I have no time to go into this.

I wish to remark, however, on some other interesting features of this theory. A closer investigation reveals that the massless theory is not only invariant under the 11-parameter Weyl group but, in fact, is invariant under the full 15-parameter conformal group [12]. For people working in this field this may not appear very surprising because in canonical theories such a generalization of the symmetry group can be generally demonstrated.[3] The theory described here, however, because of her non-canonical character (anomalous dimensions) does not fulfil the general criteria which are commonly used in this context. Nevertheless the same result holds. I will shortly discuss some of the peculiar features of our theory with regard to conformal transformation [12].

We require the field $\psi(x)$ to transform under special conformal transformation according to an irreducible representation characterized by 'spin 1/2' and 'dimension 1/2', i.e. explicitly to transform under infinitesimal transformations with parameters c_μ as

$$(3.4)\quad \begin{aligned} &\psi(x) \xrightarrow{c} \psi'(x') + \delta\psi(x),\\ &\delta\psi(x) = (c\cdot x + c_\nu \sigma^{\nu\lambda} x_\lambda)\psi(x) = (\bar\sigma\cdot c)(\sigma\cdot x)\psi(x). \end{aligned}$$

To keep the first and second field equation in (2.5) conformal covariant the $\hat\psi(x)$ and $\hat{\hat\psi}(x)$ can*not* transform according to *decomposable*, representations, but must have the transformation property

$$(3.5)\quad \delta\hat\psi(x) = (3c\cdot x + c_\nu\sigma^{\nu\lambda}x_\lambda)\hat\psi(x) - 2i(\sigma\cdot c)\psi(x),$$

$$(3.6)\quad \delta\hat{\hat\psi}(x) = (5c\cdot x + c_\nu\sigma^{\nu\lambda}x_\lambda)\hat{\hat\psi}(x) + 2(\bar\sigma\cdot\partial)(\sigma\cdot c)\psi(x)$$

where the last term indicates the non-decomposable part. The factor 2 appearing there is actually (3-2d) with $d = 1/2$ the subcanonical dimension of our ψ-field. For canonical fields with $d = 3/2$ this factor does not occur. The transformation property (3.6) of $\hat{\hat\psi}$, however, is such that with the validity of the first two equations in (2.5), the Weyl derivative of $\hat{\hat\psi}$, i.e. $i\sigma\cdot\partial\hat{\hat\psi}$ again transforms according to an irreducible representation, i.e.

$$(3.7)\quad \delta(i\sigma\cdot\partial\hat{\hat\psi}) = (7c\cdot x + c_\nu\sigma^{\nu\lambda}x_\lambda)(i\sigma\cdot\partial\hat{\hat\psi}).$$

This establishes the conformal invariance of the last equation in (2.5) and hence the conformal invariance of the theory. This result can be generalized to theories with even higher derivatives [9].

In the explicit construction of the conformal current it turns out [12] that the non-decomposable parts in the $\hat\psi$ and $\hat{\hat\psi}$ transformation law (3.5, 3.6) just cancel an uncommon term arising in connection with the non-canonical dimension of ψ and $\hat{\hat\psi}$.

There is one important consequence of this peculiar feature of the theory, namely that the current j_μ as given by (3.1) does *not* transform according to a decomposable representation. Besides the usual term characteristic for a vector field of dimension 3 we obtain [12] a non-decomposable part of the form

$$(3.8)\quad \delta_n j_\mu = 8\partial^\lambda c_{[\lambda}(\psi^*\sigma_{\mu]}\psi).$$

The non-decomposable part is identically conserved and does not contribute to the charge integral. Hence the charge is still a time-independent conformal scalar. The unusual behavior of j_μ, however, has serious consequences for the conformal invariance of interacting theories of the gauge type in which j_μ appears in the interaction [12]. We will come back to this point in the next section.

4. The Nonlinear Theory

After the discussion of the intermediate theory where we took into account the subcanonical dimension of the interacting spinor field but disregarded the nonlinear terms we now want to return to the nonlinear theory again. The interaction is essentially introduced by the replacement

$$(4.1)\qquad \partial_\mu \to \nabla_\mu = \partial_\mu + igR_\mu(x)$$

where $R_\mu(x)$ is considered to be a bilinear form of the ψ-field. Formally we may start from the Lagrangian density [8].

$$(4.2)\qquad \mathscr{L} = \mathscr{L}_0 + \mathscr{L}_I + \mathscr{L}_R + \mathscr{L}_\lambda$$

where $\mathscr{L}_0$ is the 'free' Lagrangian density (2.4) as discussed before. The $\mathscr{L}_I$ is the interaction term arising from the replacement (4.1) in (2.4), i.e.

$$(4.3)\qquad \mathscr{L}_I = -gj_\mu R^\mu$$

with j_μ the current (3.1). The $\mathscr{L}_R$ is a 'nonminimal' term which we may add with an arbitrary positive constant k simulating the 'free' Lagrangian density of the vector field in a conventional approach

$$(4.4)\qquad \mathscr{L}_R = -\tfrac{1}{4}kgf_{\mu\nu}f^{\mu\nu}$$

where $f_{\mu\nu}$ is the field strength tensor constructed from the vector field

$$(4.5)\qquad f_{\mu\nu} = R_{\mu,\nu} - R_{\nu,\mu}.$$

Finally $\mathscr{L}_\lambda$ is a term which enforces the identification of R_μ with the bilinear form of the ψ-field

$$(4.6)\qquad \mathscr{L}_\lambda = -g\lambda^\mu(R_\mu + :\psi^*\sigma_\mu\psi:)$$

with λ^μ a Lagrangian multiplier.

From the Lagrangian density (4.2) we formally derive the field equations

$$
\begin{aligned}
& i\sigma\cdot\nabla\psi = \hat{\psi}, \\
& i\bar{\sigma}\cdot\nabla\hat{\psi} = \hat{\hat{\psi}}, \\
(4.7)\quad & i\sigma\cdot\nabla\hat{\hat{\psi}} = g\sigma\cdot\lambda\psi, \\
& \lambda_\mu = R\partial^\nu f_{\mu\nu} - j_\mu, \\
& R_\mu = -:\psi^*\sigma_\mu\psi:
\end{aligned}
$$

which may be combined in the single nonlinear equation for

$$(4.8) \qquad -i(\sigma\cdot\nabla)(\bar{\sigma}\cdot\nabla)(\sigma\cdot\nabla)\psi = g(\sigma\cdot\lambda)\psi.$$

The above formulas should be augmented by appropriate definitions of the operator products [8].

The nonlinear theory above has some remarkable properties:

1. The theory is invariant under local gauge transformations (phase gauge transformations) if we choose for the dimensionless coupling constant g the particular value (1.13), provided we take some care in defining the current in a gauge invariant fashion

$$(4.9) \qquad j_\mu(x) = \mathbin{\circ\!\circ} (\hat{\hat{\psi}}^*\sigma_\mu\psi + \hat{\psi}^*\bar{\sigma}_\mu\hat{\psi} + \psi^*\sigma_\mu\hat{\hat{\psi}}) \mathbin{\circ\!\circ}.$$

This can be achieved by using a gauge invariant finite part prescription [16] which, as usual, employs in the limiting process $\zeta\to 0$ an exponential factor

$$(4.10) \qquad \exp\left\{-ig\int_{x-\frac{\zeta}{2}}^{x+\frac{\zeta}{2}} R_\mu(l)\,dl^\mu\right\}.$$

We further have to assume that the finite part prescriptions in the operator products $\mathbin{\circ\!\circ} R_\mu\psi \mathbin{\circ\!\circ}$ and $\mathbin{\circ\!\circ} \lambda_\mu\psi \mathbin{\circ\!\circ}$, necessary in (4.7), do not change their gauge covariance properties.

2. The theory in invariant under the full 15-parameter conformal group. To demonstrate this one essentially has to show that the interaction term $\mathscr{L}_I$ under special conformal transformations transforms according to a Lorentz scalar of dimension 4. The conformal covariance of the other terms in the Lagrangian density (4.2) is obvious from our former considerations in Section 3 and the known properties of quantum electrodynamics. The identification of R_μ with $-:\psi^*\sigma_\mu\psi:$ requires that the

bilinear form transforms homogeneously like R_μ according to an irreducible representation of dimension 1. For the ordinary local product this follows immediately from the transformation peroperty (3.4) of $\psi(x)$. But it can also be shown [8] to hold for the finite part of this product. To demonstrate the convariance of $\mathscr{L}_I$ it, therefore, remains to show that $j_\mu(x)$ transforms homogeneously according to an irreducible representation of dimension 3. This seems to *fail* on the basis of our earlier findings (3.8) exhibiting a nondecomposable part involving $\psi^*\sigma_\mu\psi$ in the transformation law of $j_\mu(x)$. With the proper gauge invariant definition (4.9) of the current j_μ this term, however, should be interpreted as the *gauge invariant* finite bilinear product $\mathring{:}\psi^*\sigma_\mu\psi\mathring{:}$ rather than the gauge variant finite part $:\psi^*\sigma_\mu\psi:$ which we used for defining the vector field R_μ. To establish the correct behavior of the current under conformal transformations this non-decomposable term has to be suppressed and we, therefore, have to require

$$(4.11)\qquad \mathring{:}\psi^*\sigma_\mu\psi\mathring{:}(x) = 0\,.$$

One can show [8] that this condition is *equivalent* to the *identification* of the gauge field R_μ (appearing in (4.11) through (4.10)) with the bilinear spinor form $-:\psi^*\sigma_\mu\psi:$. The theory cannot be made conformal invariant with the gauge field as an independent field [12]. From this point of view one may state that conformal invariance necessarily requires this indentification.

It is interesting to note that the gauge invariance of the theory again implies that the current $j_\mu(x)$ occurs explicitly in the field equations and, therefore, gets a dynamical meaning. With $k = 0$ the equation (4.8) can be considered in some sense to be the 3-dimensional counterpart of the field equation of the Thirring model if we disregard the gauge covariance features.

The equations (4.7) may be interpreted to be an appropriate description of the original nonlinear field equation (1.2) for the subcanonical 2-component Weyl spinor field. The apparent difference between the original and the new formulation arises from a *more elaborate description* of the unspecified nonlinear term in (1.2). According to the new formulation the nonlinear term in (1.2) should be interpreted as

$$(4.12)\qquad \sigma^\mu:\psi(\psi^*\sigma_\mu\psi):(x) = -\sigma^\mu:\psi R_\mu:(x) - \frac{1}{g}\hat{\psi}(x)$$

to coincide with the first equation in (4.7). The new operator $\hat{\psi}(x)$ appearing here should be considered a 'genuine' part of the 3-ψ-operator product which cannot be simply expressed in terms of a local ψ-operator product but only to a certain extent can be related to certain but complicated nonlocal expressions of the ψ-operators which are obtained by solving the other equations in (4.7) for $\hat{\psi}(x)$. In the 'spread-out' form (4.7) the anomalous dimension of the ψ-operator is manifest and the usual relation between Green's functions and 2-point functions is reestablished.

The field equations (4.7) still have the bad feature that in the third equation $\hat{\hat{\psi}}(x)$ couples back to $\psi(x)$. This means that we cannot set $\hat{\hat{\psi}} \triangleq 0$, as required by unitarity, without reducing the theory to the trivial theory $\psi \triangleq 0$, *except* if $g = 0$ or $(\sigma \cdot \lambda) = 0$. The $g = 0$ is the linear case we have discussed before and which is not admissible in the gauge invariant case. The $\sigma \cdot \lambda = 0$ condition may be induced by the stronger condition [8]

$$\begin{aligned} &\lambda_\mu(x) \triangleq 0, \\ \text{(4.13)}\quad &\text{i.e.}\quad \langle \text{phys}|k\partial^\nu f_{\mu\nu} - j_\mu|\text{phys}\rangle = 0 \end{aligned}$$

which means that for physical states the R_μ-field has to fulfil a Maxwell-type equation. This condition, however, seems to be a consequence of the requirement that the physical subspace only contains gauge invariant states, or

$$\text{(4.14)}\quad \langle \text{phys}|Q(\alpha)|\text{phys}\rangle = 0$$

with $Q(\alpha)$ the generator of the local gauge transformations as a functional of $\alpha(x)$. Because, as usual, $Q(\alpha)$ has the form (with df^μ a space-like surface element)

$$\begin{aligned} \text{(4.15)}\quad Q(\alpha) &= \int df^\mu \big(\alpha(x)\, j_\mu(x) + \big(\partial^\nu \alpha(x)\big) R f_{\mu\nu}(x)\big) \\ &= -\int df^\mu \alpha(x)\, \lambda_\mu(x) \end{aligned}$$

for any $\alpha(x)$ which vanishes sufficiently fast in space-like directions.

A closer investigation of a theory of the type (4.7) reveals that there are severe problems in the infrared region which are similar to the infrared difficulties in quantum electrodynamics if one also lets the electron mass go to zero [24], [25]. To remove these infrared difficulties the dilatation symmetry (and conformal symmetry) has to be broken one way or the other, i.e. either by putting genuine but soft mass terms into the Lagrangian (to keep the conformal symmetry still in the skeleton limit), or, as we will prefer, by spontaneous symmetry breaking through a dilatation asymmetrical vacuum. Effective mass terms may be produced in the following way. One realizes that the term $(\hat{\psi}^*\bar{\sigma}_\mu\hat{\psi})(\psi^*\sigma^\mu\psi)$, contained in $\mathscr{L}_I$, by Fierz reordering can be rewritten in the form

$$(4.16) \quad (\hat{\psi}^*\bar{\sigma}_\mu\hat{\psi})(\psi^*\sigma^\mu\psi) = -\tfrac{1}{2}(\varphi_1^2+\varphi_2^2)$$

with the scalar fields

$$(4.17) \quad \begin{aligned} \varphi_1 &= \psi^*\hat{\psi}+\hat{\psi}^*\psi, \\ \varphi_2 &= i(\psi^*\hat{\psi}-\hat{\psi}^*\psi). \end{aligned}$$

Mass terms of the type (2.24) then arise effectively if we require

$$(4.18) \quad \langle 0|\varphi_1|0\rangle = \tfrac{1}{2}m^2 \neq 0$$

which exhibits an asymmetry of the vacuum under dilatation and establishes the mass scale in the theory. The field equations then effectively can be cast into the form (with $\psi_\pm$ as defined in (2.28)):

$$(4.19) \quad \begin{aligned} i\sigma\cdot\nabla\psi_+ &\mathrel{\hat{=}} m\hat{\psi}, \\ i\sigma\cdot\nabla\hat{\psi} &= m\psi_+, \\ i\sigma\cdot\nabla\psi_- &= 0, \\ k\partial^\nu f_{\mu\nu} &\mathrel{\hat{=}} j_\mu, \\ R_\mu &= -:\psi^*\sigma_\mu\psi: \end{aligned}$$

which, with the exception of the last equation (and perhaps the third equation), looks like an ordinary gauge theory.

In particular ψ_+ is linked to ψ_- only via the gauge field R_μ. Hence in this somewhat oversimplified version the theory appears to be invariant under the new phase transformation (2.34) due to the absence of $\psi_-^*\sigma_\mu\psi_+$ terms in j_μ, provided we refrain of ever looking at the ψ-content of R_μ as expressed by the last equation which off-hand does not appear to be an admissible condition. Anyway, if the new phase transformation (2.34) could, indeed, be effectively established for physical matrix elements there will be no problem with unitarity. Saller [28] in a recent investigation has made an interesting proposal how all these requirements can, indeed, be incorporated.

Let me close this section by quoting some interesting results Saller [28] recently obtained by calculating the simplest boson solutions in the lowest Bethe–Salpeter approximation. The Bethe–Salpeter equation for the 4-point correlation function in lowest approximation has the form

(4.20) 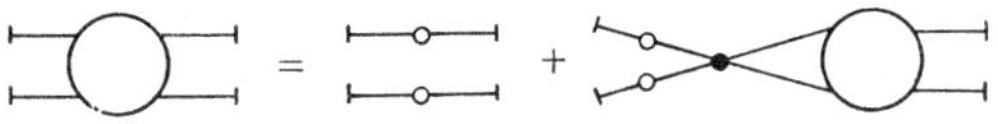

which is formally solved to give

(4.21) 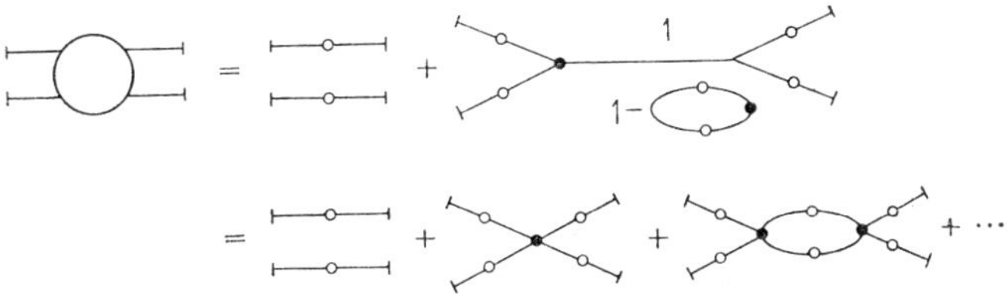

i.e. corresponds to a summation of the bubble graphs. The boson eigenvalue solutions are given as solutions of the algebraic equation for the Bethe–Salpeter amplitude  obtained by factorizing the 4-point function at the pole

(4.22) 

This simple graphical language can only be used in our theory if we identify the Green's function 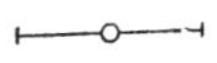 as functions of the 'superspinor' $\begin{bmatrix} \psi \\ \hat{\psi} \\ \hat{\hat{\psi}} \end{bmatrix}$. This is not really a practical language due to the very different behavior of ψ, $\hat{\psi}$ and $\hat{\hat{\psi}}$ in the interaction. A translation of (4.22) to our case yields essentially an equation of the form [28]

(4.23) 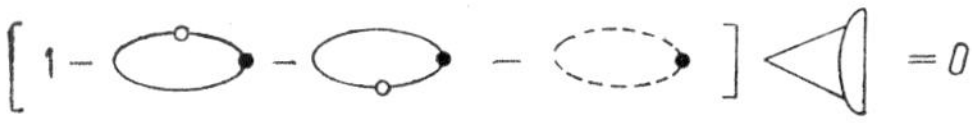

with the propagators as given in (2.27). Except for the  term this is the old boson equation we obtained in the past by employing the lowest New–Tamm–Dancoff approximation[1] and agrees completely with an equation derived earlier by Saller [27]. As Saller noted, in this particular sum of bubble graphs the remaining *logarithmic divergencies* cancel exactly if one chooses the particular value $g = 16\pi^2$ as required by gauge invariance. The kernel function, therefore, yields a *finite* expression. Furthermore Saller showed that the eigenvalue equation (4.23) has always a solution $p^2 = 0$ for the gauge field Bethe–Salpeter amplitude  $= \langle 0|R_\mu(x)|p\rangle$ *provided* a mass $m \neq 0$ is introduced for the spinor field. Finally the coupling constant of this massless gauge field to the spinor field is essentially given by the residue of the pole at $p^2 = 0$ in (4.21) which is finite in non-singular cases.

As a concrete example Saller conjectures [28] for the ψ-propagator the special form

$$(4.24) \quad \psi\psi^* \sim \frac{\bar{\sigma}\cdot p}{(p^2-\mu^2)(p^2-m^2)}$$

in which μ, m are considered arbitrary mass parameters. If the Bethe–Salpeter amplitude is subdivided into a spin $= 0$ part $\varphi^{(0)}(p)$ and a spin $= 1$ part $\varphi_\mu^{(1)}(p)$

$$(4.25) \quad \begin{aligned} \langle 0|R_\mu(x)|p\rangle &= p_\mu\varphi^{(0)}(p)+\varphi_\mu^{(1)}(p), \\ p^\mu\varphi_\mu^{(1)}(p) &= 0 \end{aligned}$$

he derives from (4.23) the eigenvalue equations

$$(4.26) \quad [1-\pi^{(0)}(p^2, m^2, \mu^2)]\varphi^{(0)}(p) = 0,$$

$$(4.27) \quad [1-\pi^{(1)}(p^2, m^2, \mu^2)]\varphi_\mu^{(1)}(p) = 0.$$

On the basis of gauge invariance the first equation turns out to be identically fulfilled

$$(4.28) \quad \pi^{(0)}(p^2, m^2, \mu^2) \equiv 1.$$

If one of the masses m, μ are different from zero (we choose $m \neq 0$) the $\pi^{(1)}(p^2, m^2, \mu^2)$ can be written in the form

$$(4.29) \quad \pi^{(1)}(p^2, m^2, \mu^2) = \frac{m^2}{m^2-\mu^2}\,\pi\left(\frac{p^2}{m^2}\right) - \frac{\mu^2}{m^2-\mu^2}\,\pi\left(\frac{p^2}{\mu^2}\right)$$

with

$$(4.30) \quad \pi(\lambda) = 1-\frac{2}{3}\lambda\log(-\lambda)+\frac{4}{3}\lambda+\frac{2}{3}(2+\lambda)K\left(\frac{\lambda}{4}\right),$$

$$K\left(\frac{\lambda}{4}\right) = 2\left[\frac{\tan^{-1}\sqrt{\dfrac{\lambda}{4-\lambda}}}{\sqrt{\dfrac{\lambda}{4-\lambda}}} - 1\right].$$

For small values of λ (4.28) can be written as

$$(4.31)\quad \pi^{(1)}(p^2, m^2, \mu^2) = 1 - \frac{2}{3}\frac{p^2}{m^2-\mu^2}\log\frac{\mu^2}{m^2} + \ldots$$

which clearly exhibits a solution of (4.27) at $p^2 = 0$ independent of μ and $m \neq 0$. For the coupling constant (defined analog to the fine structure constant) one derives

$$(4.32)\quad \alpha = 4\pi \cdot \frac{3}{2} \cdot \frac{1}{\log \frac{\mu^2}{m^2}},$$

i.e. this constant goes logarithmically to zero if one lets $\mu \to 0$. It goes logarithmically to ∞ if $\mu \to m$. This limiting case, however, is pathological, because (4.24) will have a double pole which needs special handling as we discussed before. There are, of course, more complicated parametrizations of the ψ-propagator, as e.g.

$$(4.33)\quad \psi\psi^* \sim \frac{\bar{\sigma} \cdot p(p^2+m^2)}{(p^2-\mu^2)^2(p^2-m^2)}$$

which have the same ultraviolet behavior as our former propagators. They all will lead to different values of the coupling constant (4.32). Hence, in this context, it appears only of interest to note that this coupling constant, i.e. the mass zero content of the gauge field, depends sensitively on the mass structure of the spinor propagator.

5. Concluding Remarks

In closing let me remark on some particular aspects of the theory described in the previous sections which appear interesting because they have no counterpart in the more conventional theories.

Theories employing fields of subcanonical dimension necessarily have to be formulated in the framework of a quantum mechanical state space with indefinite metric. To allow a quantum statistical interpretation of such theories the physical states can only belong to a certain subspace of this state space which must have a positive definite metric. In a more physical language this means that more states can be reached by virtual processes than by real processes, or that the virtual state space is larger than the asymptotic state space. Virtual processes describe forces which act between physical states. As a consequence the inequivalence between the virtual and the asymptotic state space can be physically interpreted as the existence of a 'genuine interaction', i.e. of an interaction which cannot be traced back to the virtual exchange of physical particles. The indefinite metric state space from this point of view is only a mathematical artifice to allow the luxury of forcing an Einstein causal theory for observables into the narrow framework of a manifestly local and manifestly covariant theory for an unobservable, gauge invariant spinor field. Quantum electrodynamics represents a good example how the formal convenience of a local theory may favor an extension of the physical state space. On the other hand it also indicates that the indefinite metric formulation is by no means unavoidable.

In our former discussion the constituent field in the massless case had physical as well as unphysical components similar to the vector potential in Gupta–Bleuler quantum electrodynamics. It is quite simple, however, to imagine a situation in which the constituent spinor field contains *no* particle-like components at all. If, e.g. the 2-point function has the form of a double pole at a finite mass

$$(5.1)\qquad \psi\psi^* \sim \frac{\bar{\sigma}\cdot p}{(p^2-m^2)^2}$$

there would be no normalizable states connected with the spinor field; we would have the situation of a 'null-ghost-couple'. Nevertheless such a theory may lead to 'compounds' which lie in the physical subspace, e.g. $:\psi\psi^*:(x)$ may create a mass zero boson state, as indicated in the last section, etc. The indefinite metric theories, therefore, may offer an interesting way to formulate a quark theory, in which the quark only plays the role of the constituent field without showing up as a physical particle asymptotically. Such a program, however, will not be without hazards. E.g. one still would have to explain why no quark-like compounds, e.g. $:qq:$ or $:q^*qq:$ etc. occur. Of course, this may have purely dynamical reasons (only repulsive interactions in these modes). We will not pursue this possibility here any further.

Instead we will shortly turn our attention to another interesting possibility, proposed several years ago [11], to give double poles of the type (5.1) a physical meaning, namely to connect them with leptons. Let us imagine, at first, that the 2-point function contains, among other things, an ordinary pole with an exceedingly small residue

$$\psi\psi^* \sim Z_2 \frac{\bar{\sigma} \cdot p}{p^2 - m^2} + \ldots, \tag{5.2}$$
$$0 < Z_2 \ll 1.$$

As a consequence the particle connected with this pole will only be very rarely produced in collisions, or, as one would say, its effective coupling constant to other systems will be $g_{\mathrm{eff}} \sim Z_2 g$, i.e. exceedingly small. However, there will exist an important exception to this general rule, namely in case of an interaction with a gauge field. Due to gauge invariance and the resulting Ward–Takahashi identity the vertex corrections will produce a strong enhancement factor $1/Z_1$, which because of

$$Z_1 = Z_2 \tag{5.3}$$

will just compensate the residue dependence. This leads to a universality of the coupling of particles to a gauge field, i.e. to the well-known universality of the 'charge'. This situation suggests to identify leptons with poles in the 2-point function with extremely small residue or, in fact, with singularities where $Z_2 \to 0$. Since the 'good ghost' buried in a double pole (5.1), which is an eigenstate of energy, acts like a single pole with residue $Z_2 = 0$ we may, indeed, consider *to relate double poles to leptons*. If no single poles occur in the 2-point function the constituent field could then be essentially interpreted as a lepton field. Hadrons one could imagine to result as 'compounds' of the spinor field, or mathematically as poles in higher point Green's functions, and strong interaction as a secondary interaction connected with the virtual exchange of these 'compounds'. These questions, of course, have to be studied in much more detail. Saller in his recent work [28] has made a very promising start in this direction.

It is quite clear that in order to successfully employ a theory of the discussed type for a realistic description of elementary particles it will be necessary to *enrich* its *internal symmetry structure*. For a unified theory of elementary particles it may suffice to start with a 4-component constituent spinor field, i.e. with a nonlinear spinor theory as proposed by Heisenberg. This allows to incorporate a $U_1 \otimes SU_2$ as an internal symmetry group (phase group $\otimes$ isospin group) which may be generalized to a $U_1 \otimes SU_2$ *local* gauge group in the described manner.

In this case an additional isovector-vector field

$$(5.4) \qquad \vec{R}_\mu(x) = -:\psi^* \sigma_\mu \vec{\tau} \psi:(x)$$

will occur and play the role of a Yang–Mills gauge field. A theory in this form has close similarities with a Weinberg-type model in which, however, all fields (gauge fields and scalar fields) appear as combination of the constituent spinor field [26], [28].

Finally we wish to mention that – similarly as the local phase transformation symmetry or the local isospin symmetry lead to gauge fields $R_\mu(x)$ or $\vec{R}_\mu(x)$ constructed from the spinor field – one may also attempt to establish *local Poincaré invariance* in the same manner, i.e. by trying to express the four translation gauge fields $g_{\mu m}(x)$ and the six Lorentz gauge fields $A_{[\mu\nu]m}(x)$ completely in terms of the spinor field. Since in a geometrical interpretation $g_{\mu m}$ have the meaning of 'vierbein fields' and $A_{[\mu\nu]m}$ of 'affine connections' in a Riemannian space (generally with torsion) there is an immediate connection of this theory to General Relativity. First attempts in this direction were undertaken and appear quite promising [6].

References

[1] See W. Heissenberg [18], H. P. Dürr, W. Heisenberg, H. Mitter, S. Schlieder and K. Yamazaki [10]. For a review of the older work see e.g.: W. Heisenberg [20], H. P. Dürr [3].

[2] See M. Karowski [21]. For a general account of the present situation see also: H. P. Dürr [5].

[3] See e.g. S. Ferrara, R. Gatto and A. F. Grillo [17].

Bibliography

[1] I. I. Bigi, H. P. Dürr and N. J. Winter, Preprint MPI-PEA-Th 1, Jan. 1974, Max-Planck-Institut für Physik, München (to be published in *Nouvo Cimento*).

[2] H. P. Dürr, *Zeits. Naturforsch.* **16a** (1961) 327.

[3] H. P. Dürr, *Acta Phys. Austriaca, Suppl III* **3** (1966).

[4] H. P. Dürr, in: *Proccedings of the Symposium on Basic Questions in Elementary Particle Physics*, June 1971, Max-Planck-Institut für Physik, München, 118.

[5] H. P. Dürr, CPT-179, ORO-3992-130, Univ. of Texas Report (1973).

[6] H. P. Dürr, GRG 4 (1973) 29.

[7] H. P. Dürr, *Nuovo Cimento* **22A** (1974) 386.

[8] H. P. Dürr, *Nuovo Cimento* **27A** (1975) 305.

[9] H. P. Dürr and C. C. Chiang, *Nuovo Cimento* **28A** (1975) 89.

[10] H. P. Dürr, W. Heisenberg, H. Mitter, S. Schlieder and K. Yamazaki, *Zeits. Naturforsch.* **14a** (1959) 441.
[11] H. P. Dürr, W. Heisenberg, H. Yamamoto and K. Yamazaki, *Nuovo Cimento* **38** (1965) 1220.
[12] H. P. Dürr and P. du T. van der Merve, *Nuovo Cimento* **23A** (1974) 1.
[13] H. P. Dürr and E. Rudolph, *Nuovo Cimento* **62A** (1969) 411.
[14] H. P. Dürr and E. Rudolph, *Nuovo Cimento* **65A** (1970) 423.
[15] H. P. Dürr and N. J. Winter, *Nuovo Cimento* **70A** (1970) 467.
[16] H. P. Dürr and N. J. Winter, *Nuovo Cimento* **7A** (1972) 461.
[17] S. Ferrara, R. Gatto and A. F. Grillo, *Springer Tracts in Mod. Phys.* **67** (1973).
[18] W. Heisenberg, *Nachr. d. Göttinger Akad. d. Wiss.* **111** (1953).
[19] W. Heisenberg, *Nucl. Phys.* **4** (1957) 532.
[20] W. Heisenberg, *Introduction to the Unified Field Theory of Elementary Particles*, Interscience Publ., London 1966.
[21] M. Karowski, *Zeits. Naturforsch.* **24a** (1969) 510; *Nuovo Cimento* **23A** (1974) 126.
[22] H. Mitter, *Nuovo Cimento* **32** (1964) 1789.
[23] K. L. Nagy, *State Vector Spaces with Indefinite Metric in Quantum Field Theory* Akademiai Kiadó, Budapest 1965.
[24] H. Saller, *Nuovo Cimento* **4A** (1971) 404.
[25] H. Saller, *Nuovo Cimento* **7A** (1972) 779.
[26] H. Saller, *Nuovo Cimento* **12A** (1972) 349.
[27] H. Saller, *Nuovo Cimento* **21A** (1974) 661.
[28] H. Saller, *Nuovo Cimento* **24A** (1974) 391.
[29] S. Schlieder, *Zeits. Naturforsch.* **15a** (1960) 448.
[30] S. Schlieder, *Zeits. Naturforsch.* **15a** (1960) 460.
[31] S. Schlieder, *Zeits. Naturforsch.* **15a** (1960) 555.
[32] E. C. G. Sudarshan, *Phys. Rev.* **123** (1961) 2183.

GRAVITATIONAL WAVES

ANDRÉ LICHNEROWICZ

Physique Mathematique, Collège de France, Paris, France

1. Introduction

One of the most important features of the relativistic theory of gravitation is the existence of gravitational waves. This fact is strictly connected with the hyperbolic character of the Einstein system and of the different systems appearing in the problems of mathematical physics on a curved background. In recent years, an important group of experimental teams have searched to detect gravitational waves, although it appears that all this research is, for the moment, completely disappointing.

I have chosen to speak on the recent progress of the theory of gravitational waves. This theory is very old and the first steps were made by Einstein, Levi-Civita, Fock and myself between the two wars. After 1955 Petrov and Pirani, Trautman, Papapetrou and Treder, Taub, Y. Choquet-Bruhat, Penrose and myself acchieved substantial results, the last of which are two years old.

Here I have chosen to study mainly *gravitational shock waves*, because the corresponding theory is the most interesting from the mathematical and perhaps physical point of view, and because all the algebraic and differential features are already present in this situation, in the richest form; Papapetrou, Taub, Y. Choquet-Bruhat and, more recently, Penrose and myself have studied these waves by means of different methods and under different assumptions. I will give here a method sketched in 1967 and developed in 1971, founded on the technique of distributions. A synthesis of the results of the various authors is then possible, and some new facts appear, in particular the connection between electromagnetic and gravitational shock waves.

2. Gravitational Radiations and Waves

I will first give some general considerations. An interesting general report on waves was given by Trautman during the London conference on gravitation and relativity.

a) It is well known that the 2-forms F onto R^4, equivalent under the Lorentz group, give two algebraic classes, the regular class and the singular or null-class. Petrov has proved that, similarly, the double 2-forms, that is the 4-tensors admitting the same algebraic type as the curvature tensor, equivalent under the Lorentz group, give five algebraic classes. We usually denote these classes by:

$$I, II, III, IV, S$$

where S is the singular class. Pirani has studied these classes and shown the existence for some classes of a vector similar to the Poynting vector (radiational class).

Now, on the space-time, if the curvature tensor $\boldsymbol{R}$ is singular, there exists an isotropic vector l, such that the Ricci tensor has the form

$$R_{\alpha\beta} = \sigma l_\alpha l_\beta.$$

If $\sigma \neq 0$, we have total radiation, if $\sigma = 0$ we have purely gravitational radiation. If l is a gradient, the curvature tensor $\boldsymbol{R}$ is propagating along the gravitational rays defined by l, according to

$$2l^\varrho \nabla_\varrho \boldsymbol{R} + \nabla_\varrho l^\varrho \cdot \boldsymbol{R} = 0$$

for $\sigma = 0$.

Sachs has proved the so-called 'peeling theorem'. Let $(V_4, \boldsymbol{g})$ be an asymptotically flat space-time; we choose a system of timelike paths and we denote by r the corresponding spatial distance of an origin-point to an arbitrary point. If the corresponding gravitational field admits sources in a spatially finite domain, we have for the corresponding curvature tensor:

$$\boldsymbol{R} = \frac{R^{(S)}}{r} + \frac{R^{(IV)}}{r^2} + \frac{R^{(III)}}{r^3} + \frac{R^{(II)}}{r^4} + \frac{R^{(I)}}{r^5},$$

where the index denotes the Petrov type of the corresponding tensor. We see that the distance filters the different algebraic types, the singular case, which is the most degenerated, being the type which remains the furthest.

b) If the energy tensor is continuous, an *ordinary gravitational wave* corresponds to a solution of the Einstein equations for which some significant second derivatives of the gravitational potentials (or equivalently the curvature tensor) are discontinuous across a hypersurface

Σ ($\varphi = 0$, $l = d\varphi$). It easily follows from the theory of the characteristics of a hyperbolic system that Σ is tangent to the elementary cones.

If $\boldsymbol{T}$ is of class C^1, Trautman has proved that $[\boldsymbol{R}]$ is propagating along the gravitational rays defined by l according to the differential system

$$2l^\varrho \nabla_\varrho [\boldsymbol{R}] + \nabla_\varrho l^\varrho [\boldsymbol{R}] = 0.$$

The case of the *gravitational shock waves* corresponds to the case where some significant *first* derivatives of the gravitational potentials are discontinuous across a hypersurface Σ.

c) Y. Choquet-Bruhat and Taub have given an extension of the W. K. B. method to the non-linear case and thus introduced another formalism for the representation of waves, which are here approximate waves defined by limited expansions of the type

$$g(x, \omega\varphi) = \overset{0}{g}(x) + \frac{1}{\omega}\overset{1}{g}(x, \omega\varphi) + \frac{1}{\omega^2}\overset{2}{g}(x, \omega\varphi).$$

φ is a given function, the phase function, ω is a large parameter which can be interpreted as frequency. If we attempt to satisfy approximatively, for example, the Einstein equations of the vacuum, we obtain interesting results: in particular φ is necessarily characteristic and the main metric $\overset{0}{g}$ must satisfy the radiational formula $\overset{0}{R}_{\alpha\beta} = \sigma l_\alpha l_\beta$ ($l = d\varphi$).

Ordinary waves and gravitational shock waves can be considered as convenient limits of high frequency approximate waves, in the Taub sense.

3. Covariant Differentiation and Tensor Distributions

a) Let $(V_4, \boldsymbol{g})$ be a space-time which is a differentiable manifold of class (C^1, piecewise C^3) with a Lorentz metric $\boldsymbol{g}$ of class (C^0, piecewise C^2) and signature $+ - - -$. On a domain Ω, we have in local coordinates:

$$\boldsymbol{g}|_\Omega = g_{\alpha\beta}\, dx^\alpha \otimes dx^\beta \qquad (\alpha, \beta = 0, 1, 2, 3).$$

We consider in Ω a regular hypersurface, with the local equation $\varphi = 0$, with divides Ω into two domains corresponding to $\varphi > 0$ and $\varphi < 0$; we set $l = d\varphi$. Coordinates (y^α) such that $y^0 = \varphi$ are said to be *adapted to* Σ; for these coordinates $l_0 = 1$, $l_i = 0$ $(i, j, \ldots = 1, 2, 3)$ the function φ and the metric tensor g satisfy, in the neighbourhood of Σ, the following assumptions:

1°) φ is of class C^1 onto Ω and of class C^3 onto Ω^+ and Ω^-. The second and third derivatives of φ are regularly discontinuous across Σ.

2°) g is continuous onto Ω, of class C^2 onto Ω^+ and Ω^-. The first and second derivatives of g are regularly discontinuous across Ω.

Let Y^+ (resp. Y^-) the function equal to 1 (resp. 0) onto Ω^+ and equal to 0 (resp. 1) onto Ω^-; these functions define on Ω distributions which are denoted by the same symbols. If π is a 3-form such that $l \wedge \pi$ is equal to the volume element corresponding to g, δ is the Dirac distribution defined by

$$\langle \delta, f \rangle = \int_{\partial\Omega^-} f\pi = -\int_{\partial\Omega^+} f\pi,$$

where $\partial\Omega^+$, $\partial\Omega^-$ are the oriented boundaries on Σ of Ω^+ and Ω^- and where f is a continuous function. We have in terms of distributions

$$(3.1) \qquad dY^+ = l\delta, \qquad dY^- = -l\delta.$$

b) Let ∇ be the operator of covariant differentiation. For a system of coordinates (x^ϱ) of domain Ω, the connection 1-form ω can be written on Ω:

$$\omega = (\omega^\alpha_\beta) = (\Gamma^\alpha_{\beta\gamma} dx^\gamma).$$

Under our assumptions, ω is continuous on Ω^+ and Ω^- and is regularly discontinuous across Σ. Let T be a tensor of Ω, of class C^1 on Ω^+ and Ω^-, such that T and ∇T are regularly discontinuous across Σ; T and ∇T define on Ω tensor distributions $T^D (= Y^+T + Y^-T)$ and $(\nabla T)^D$. If, for example, T is a vector, we can define the covariant derivative of T^D by the tensor distribution

$$(3.2) \qquad \nabla T^D = dT^D + Y^+(\omega T) + Y^-(\omega T)$$

where dT^D corresponds to the distribution components; (3.2) can be extended, in a natural way, to tensors of arbitrary order and it is easy to obtain the relations giving the components of ∇T^D. If ω is continuous across Σ, this definition of ∇T^D coincides with the usual definition of the covariant derivative of a tensor-distribution. We can deduce from

(3.1) and from (3.2)

(3.3) $\nabla T^D = \delta l[T] + (\nabla T)^D$.

The formula (3.3) and similar formulas for the second covariant derivatives are the main tool for the theory of shock waves.

4. Singular 2-Forms and Double 2-Forms

a) Let us consider, at the point x of V_4, a 2-form $F \neq 0$ such that, for a vector $l \neq 0$,

(4.1) $S l^\alpha F_{\beta\gamma} = 0, \quad l^\alpha F_{\alpha\beta} = 0,$

where S denotes summation on the circular permutations of (α, β, γ). If such is the case, F is called *singular* at x and the fundamental vector l is necessarily *isotropic*. There exists a vector b, defined up to a change $b \to b + \sigma l$, such that

(4.2) $F_{\alpha\beta} = l_\alpha b_\beta - l_\beta b_\alpha \quad (\text{with } l^\alpha b_\alpha = 0).$

Let $v_{(1)}$, $v_{(2)}$ be two orthogonal normalized vectors ($v_{(1)}^2 = v_{(2)}^2 = -1$) of the 3-plane tangent along l to the elementary cone at x. We have

(4.3) $F_{\alpha\beta} = \sum_u a_u (l_\alpha v_{(u)\beta} - l_\beta v_{(u)\alpha}) \quad (u = 1, 2).$

The positive scalar

(4.4) $e_{(1)} = -b^\alpha b_\alpha = \sum_u (a_u)^2$

depends only on F and l. Singular 2-forms are the main tool for the representation of the electromagnetic waves and radiations.

b) A *double 2-form* is, by definition, a 4-tensor H with the following symmetry properties:

$$H_{\alpha\beta,\lambda\mu} = -H_{\beta\alpha,\lambda\mu} = -H_{\alpha\beta,\mu\lambda}, \quad H_{\alpha\beta,\lambda\mu} = H_{\lambda\mu,\alpha\beta}.$$

Let us consider at x a double 2-form $H \neq 0$ such that, for a vector $l \neq 0$,

(4.5) (a) $S l_\alpha H_{\beta\gamma,\lambda\mu} = 0$, (b) $l^\alpha H_{\alpha\beta,\lambda\mu} = 0$.

If such is the case, H is called *singular* at x and the fundamental vector l is again necessarily isotropic. There exists a symmetrical 2-tensor $c_{\alpha\beta}$, defined up to a change $c_{\alpha\beta} \to c_{\alpha\beta} + l_\alpha t_\beta + l_\beta t_\alpha$ where $l^\alpha t_\alpha = 0$, such that

(4.6) $H_{\alpha\beta,\lambda\mu} = c_{\alpha\lambda} l_\beta l_\mu + c_{\beta\mu} l_\alpha l_\lambda - c_{\alpha\mu} l_\beta l_\lambda - c_{\beta\lambda} l_\alpha l_\mu \quad (c_{\alpha\beta} l^\beta = 0).$

In terms of the vector $v_{(1)}, v_{(2)}$ we have

$$(4.7) \qquad H_{\alpha\beta,\lambda\mu} = \sum_{u,v} a_{uv}(l_\alpha v_{(u)\beta} - l_\beta v_{(u)\alpha})(l_\lambda v_{(v)\mu} - l_\mu v_{(v)\lambda}),$$

where $a_{uv} = a_{vu}$ $(u, v = 1, 2)$. The system (4.5) implies

$$(4.8) \qquad H_{\alpha\beta} \equiv g^{\varrho\sigma} H_{\alpha\varrho,\beta\sigma} = \sigma l_\alpha l_\beta$$

and (4.8) can be substituted for $(4.5)_b$. We have $\sigma = -(a_{11}+a_{22})$. The positive scalar

$$(4.9) \qquad e_{(2)} = c^{\alpha\beta} c_{\alpha\beta} = \sum_{u,v} (a_{uv})^2$$

depends only on H and l. Singular double 2-forms are the main tool for the representation of gravitational waves.

5. Gravitational Shock Waves

a) If we use coordinates adapted to the hypersurface Σ, then the potentials g_{ij} are called *significant* for Σ, and the potentials $g_{0\lambda}$ *non-significant*. There exist changes of coordinates, tangent to the identity change along Σ, preserving the coordinates of each point x of Σ, the values of the potentials and the values of the first derivatives of the significant potentials, but such that we can create or destroy discontinuities of the derivatives of the non-significant potentials. Such discontinuities are physically meaningless.

A gravitational shock wave is defined by a metric **g** *satisfying in* Ω *the assumptions of the* § 3, *solution of the Einstein equations in a convenient weak sense and such that the first derivatives of the significant potentials are effectively discontinuous across the hypersurface* Σ.

According to the Hadamard formulas, there exist on Σ quantities $b_{\alpha\beta}$ such that

$$[\partial_\gamma g_{\alpha\beta}] = l_\gamma b_{\alpha\beta}.$$

For an admissible change of coordinates tangent to the identity change we have

$$(5.1) \qquad b_{\alpha\beta} \to l_{\alpha\beta} + l_\alpha t_\beta + l_\beta t_\alpha,$$

where t_α is an arbitrary vector; (5.1) defines the gravitational gauge-change for the $b_{\alpha\beta}$. For a C^2-change of coordinates, we obtain for the $b_{\alpha\beta}$ the tensor law.

b) Let us choose on Ω a determined coordinates system (x^α). For each pair (β, μ) as indexes, we consider the corresponding coefficients of the connection as defining on Ω a *local vector* $\Gamma^\alpha_{(\beta\mu)}$. For this system of coordinates and on each of the domains Ω^+ and Ω^-, the components of the curvature tensor can be written as

$$(5.2) \qquad R^\alpha{}_{\beta,\lambda\mu} = \nabla_\lambda \Gamma^\alpha_{(\beta\mu)} - \nabla_\mu \Gamma^\alpha_{(\beta\lambda)}.$$

The vector $\Gamma^\alpha_{(\beta\mu)}$ defines on Ω a local vector distribution $(\Gamma^\alpha_{(\beta\mu)})^D$. We now introduce, systematically, the curvature tensor-distribution Q defined by

$$Q^\alpha{}_{\beta,\lambda\mu} = \nabla_\lambda (\Gamma^\alpha_{(\beta\mu)})^D - \nabla_\mu (\Gamma^\alpha_{(\beta\lambda)})^D,$$

where the covariant differentiation has been defined in § 3. We deduce from (3.3) that Q can be written as

$$(5.3) \qquad Q_{\alpha\beta,\lambda\mu} = \delta H_{\alpha\beta,\lambda\mu} + (R_{\alpha\beta,\lambda\mu})^D,$$

where H is the tensor defined on Σ by

$$(5.4) \qquad H_{\alpha\beta,\lambda\mu} = -\tfrac{1}{2}(b_{\alpha\lambda} l_\beta l_\mu + b_{\beta\mu} l_\alpha l_\lambda - b_{\alpha\mu} l_\beta l_\lambda - b_{\beta\lambda} l_\alpha l_\mu).$$

δH is invariant under the gravitational gauge changes and depends only on the wave. The wave is effective if and only if H is $\neq 0$. The Ricci tensor-distribution corresponding to Q is given by

$$(5.5) \qquad Q_{\alpha\beta} = \delta H_{\alpha\beta} + (R_{\alpha\beta})^D,$$

where

$$(5.6) \qquad H_{\alpha\beta} = \tfrac{1}{2}(b_{\alpha\varrho} l^\varrho l_\beta + b_{\beta\varrho} l^\varrho l_\alpha - b l_\alpha l_\beta - b_{\alpha\beta} l^\varrho l_\varrho) \qquad (b = g^{\alpha\beta} b_{\alpha\beta}).$$

c) We suppose in the following part that *the energy tensor* $\boldsymbol{T}$ *is continuous on* Ω^+ *and* Ω^- *and regularly discontinuous across* Σ. We set

$$T'_{\alpha\beta} = T_{\alpha\beta} - \tfrac{1}{2} g_{\alpha\beta} T.$$

On Ω^+ and Ω^-, the Einstein equations can be written as

$$(5.7) \qquad R_{\alpha\beta} = \chi T'_{\alpha\beta}.$$

We adopt as Einstein equations on Ω, in a weak sense, the following system:

$$(5.8) \qquad Q_{\alpha\beta} = \chi (T'_{\alpha\beta})^D$$

which is compatible with (5.7), on Ω^+ and Ω^-. It follows from (5.7):

$$(5.9) \qquad (R_{\alpha\beta})^D = \chi (T'_{\alpha\beta})^D.$$

According to (5.5), we deduce from (5.8) and (5.9) the shock conditions:

(5.10) $H_{\alpha\beta} = 0.$

It follows from (5.4) and (5.10) that the double 2-form $H \neq 0$ is *singular* at each point of Σ. Therefore l is isotropic and *the wave front Σ is necessarily characteristic* for the Einstein system. The situation is different from that corresponding to the hydrodynamical shock wave. Intuitively, we see that the speed of the spatial wave should be $\geqslant c$ (supersonic) before the shock and therefore, in relativity, is necessarily equal to c. The vector l being isotropic, (5.10) is equivalent to the four conditions

$$(5.11)\quad b_{\alpha\varrho} l^\varrho = \frac{b}{2} l_\alpha .$$

We can introduce from H the *shock energy tensor* defined on Σ by

$$\tau^{(2)}_{\alpha\beta} = -\tfrac{1}{2} b^{\lambda\mu} H_{\alpha\lambda,\beta\mu} .$$

According to (5.11), we have

$$(5.12)\quad \tau^{(2)}_{\alpha\beta} = e_{(2)} l_\alpha l_\beta , \qquad e_{(2)} = \frac{1}{4}\left(b^{\lambda\mu} b_{\lambda\mu} - \frac{b^2}{2}\right).$$

For a gauge such that $b = 0$, we have with the notation of § 4

$$e_{(2)} = \tfrac{1}{4} b^{\alpha\beta} b_{\alpha\beta} = c^{\alpha\beta} c_{\alpha\beta} = \sum (a_{uv})^2 = 2\{(a_{12})^2 - a_{11} a_{22}\} > 0 .$$

The conditions (5.11) express the continuity across Σ of the four harmonic quantities $F^\varrho = \Delta x^\varrho$ corresponding to an arbitrary coordinates system (Y. Choquet-Bruhat).

6. Propagation of the Discontinuities

a) R. Penrose [7] and myself have proved independently the following proposition.

PROPOSITION. The covariant differentiation operator $l^\varrho \nabla_\varrho$ along l is well defined onto Σ. The scalar $\nabla^\varrho l_\varrho$ is continuous across Σ.

In fact, the shock conditions imply the continuity across Σ of the quantities $l^\varrho \Gamma^\alpha_{\varrho\beta}$. Moreover, we have $[\nabla_\lambda l_\mu] = m l_\lambda l_\mu$; l being isotropic, we have in particular on Σ

(6.1) $l^\alpha \nabla_\alpha l^\beta = 0$

and Σ is generated by *rays* satisfying (6.1)

b) We will study the propagation of H along these rays. From the continuity of the harmonic quantities F_β, it follows that there exist on Σ quantities Φ_β such that

$$[\partial_\alpha F_\beta] = l_\alpha \Phi_\beta .$$

From the technique of distributions and from the expression of the components of the Ricci tensor in terms of harmonic quantities, it is possible to deduce that, along the rays, we have on Σ

$$(6.2) \qquad 2l^\varrho \nabla_\varrho b_{\alpha\beta} + \nabla_\varrho l^\varrho b_{\alpha\beta} - l_\alpha \Phi_\beta - l_\beta \Phi_\alpha = -2[T'_{\alpha\beta}],$$

where we have chosen unities such that $\chi = 1$. The formula (6.2) is equivalent to the differential system concerning H:

$$(6.3) \qquad 2l^\varrho \nabla_\varrho H_{\alpha\beta,\lambda\mu} + \nabla_\varrho l^\varrho H_{\alpha\beta,\lambda\mu} = [T'_{\alpha\lambda}] l_\beta l_\mu + [T'_{\beta\mu}] l_\alpha l_\lambda - [T'_{\alpha\mu}] l_\beta l_\lambda - [T'_{\beta\lambda}] l_\alpha l_\mu .$$

If $b'_{\alpha\beta} = b_{\alpha\beta} - (b/2) g_{\alpha\beta}$, the formula (6.2) can be written as

$$(6.4) \qquad 2l^\varrho \nabla_\varrho b'_{\alpha\beta} + \nabla_\varrho l^\varrho b'_{\alpha\beta} - l_\alpha \Phi_\beta - l_\beta \Phi_\alpha + g_{\alpha\beta} l^\varrho \Phi_\varrho = -2[T_{\alpha\beta}] .$$

The shock conditions (5.11) being satisfied, it follows from (6.4) that necessarily

$$(6.5) \qquad [T_{\alpha\beta}] l^\beta = 0,$$

which is equivalent to $\nabla_\alpha (T^{\alpha\beta})^D = 0$.

THEOREM. If the energy tensor is regularly discontinuous across the front Σ of a gravitational shock wave, the tensor H is propagating along the rays which generate Σ according to (6.3); $[T_{\alpha\beta}]$ satisfies (6.5).

It follows from (6.2) that $\tau^{(2)}$ satisfies

$$(6.6) \qquad \nabla_\alpha \tau_\beta^{(2)\alpha} = -\tfrac{1}{2} b^{\varrho\sigma} [T_{\varrho\sigma}] \cdot l_\beta .$$

c) We consider the following situation: T being supposed to be regularly discontinuous (with effective discontinuity) across a characteristic hypersurface Σ, let $\boldsymbol{g}$ be a solution of the Einstein equations (possibly in the weak sense) in the neighbourhood of Σ.

If such is the case, we can consider Σ as the front of a gravitational shock wave, effective or not. *In general, the shock wave is effective*: if that is not the case, $b'_{\alpha\beta} = 0$ and it follows from (6.4) that there exist on Σ quantities ψ_α such that

$$(6.7) \qquad [T_{\alpha\beta}] = l_\alpha \psi_\beta + l_\beta \psi_\alpha - g_{\alpha\beta} l^\varrho \psi_\varrho .$$

If $[T_{\alpha\beta}]$ *does not have the form* (6.7), *we obtain an effective gravitational shock wave.* If $[T_{\alpha\beta}]$ has the form (6.7) and if Σ does not correspond to an effective shock wave, it corresponds necessarily to an effective ordinary wave (discontinuity of the curvature tensor).

Taub has recently [10] studied the extreme relativistic perfect fluids admitting sonic speed equal to c. We note that, for the shock waves of such a fluid, $[T_{\alpha\beta}]$ has precisely the form (6.7)

7. Electromagnetic and Gravitational Shock Waves

a) The preceding approach is valid for an electromagnetic shock wave, the front of which can also be the front of a gravitational shock wave.

We suppose that the vector potential, the 1-form α, is of class C^2 on Ω^+ and Ω^- and that the $\partial_\lambda \alpha_\mu$ are regularly discontinuous across Σ. There exists on Σ a vector $\boldsymbol{b}$ such that

$$[\partial_\lambda \alpha_\mu] = \boldsymbol{l} \wedge \boldsymbol{b},$$

where $\boldsymbol{b}$ is defined up to a gauge change $\boldsymbol{b} \to \boldsymbol{b} + \sigma l$. If F is the corresponding electromagnetic field,

$$(7.1) \qquad [F] = [d\alpha] = \boldsymbol{l} \wedge \boldsymbol{b} \neq 0.$$

We suppose the electric current J to be continuous on Ω^+ and Ω^- and regularly discontinuous across Σ. We have on Ω the Maxwell equations:

$$(7.2) \qquad \delta F^D = \delta(d\alpha)^D = J^D.$$

An argument similar to that of the gravitational case gives the shock conditions:

$$l^\lambda [F]_{\lambda\mu} = 0.$$

$[F]$ is singular and l is isotropic. We now have the unique condition

$$(7.3) \qquad l^\lambda b_\lambda = 0,$$

which expresses the continuity of $g^{\lambda\mu}\partial_\lambda \alpha_\mu$. The energy tensor of the electromagnetic shock wave is

$$\tau^{(1)}_{\alpha\beta} = e_{(1)} l_\alpha l_\beta, \qquad e_{(1)} = -b^\lambda b_\lambda > 0.$$

We can deduce from (7.3) that there exists on Σ a convenient quantity Φ such that along the rays

$$(7.4) \qquad 2l^\varrho \nabla_\varrho b_\lambda + \nabla_\varrho l^\varrho b_\lambda - l_\lambda \Phi = F^{\varrho\sigma} l_\varrho b_{\lambda\sigma} - [J_\lambda].$$

The formula (7.4) is equivalent to the differential system concerning $[F]$:

(7.5) $\quad 2l^{\varrho}\nabla_{\varrho}[F_{\lambda\mu}]+\nabla_{\varrho}l^{\varrho}[F_{\lambda\mu}] = -F^{\varrho\sigma}H_{\lambda\varrho,\mu\sigma}-[l_{\lambda}J_{\mu}-l_{\mu}J_{\lambda}].$

It follows from (7.3) and (7.4) that we have

(7.6) $\quad [J_{\lambda}]l^{\lambda} = 0,$

which is equivalent to $\nabla_{\lambda}(J^{\lambda})^{D} = 0$. For $\tau^{(1)}$, we have

(7.7) $\quad \nabla_{\alpha}\tau^{(1)\alpha}_{\beta} = -F^{\lambda\sigma}l_{\lambda}\cdot b^{\varrho}b_{\varrho\sigma}\cdot l_{\beta}+b^{\lambda}[J_{\lambda}]\cdot l_{\beta}.$

b) We consider the following situation:

We suppose that (g, F) *are a metric and an electromagnetic field solution (possibly in the weak sense) of the Einstein–Maxwell system*

(7.8) $\quad S_{\alpha\beta} = \tau_{\alpha\beta}, \qquad \nabla_{\alpha}F^{\alpha\beta} = -J^{\beta},$

where $S_{\alpha\beta}$ *is the Einstein tensor of* g *and* $\tau_{\alpha\beta}$ *the Maxwell tensor of* F.

Let Σ be the front of a shock wave both electromagnetic and gravitational, satisfying (7.8). We suppose *the electromagnetic shock wave to be effective*. The relation $[\tau_{\alpha\beta}]l^{\beta} = 0$ is identically satisfied. We deduce from (6.6) and (7.7) that

(7.9) $\quad \nabla_{\alpha}\{\tau^{(1)\alpha}{}_{\beta}+\tau^{(2)\alpha}{}_{\beta}\} = b^{\lambda}[J_{\lambda}]\cdot l_{\beta}.$

In particular, if $[J]$ is colinear to l, $(\tau^{(1)}+\tau^{(2)})$ is conservative.

It follows from the previous results that *if* $[\tau_{\alpha\beta}]$ *does not have the form*

(7.10) $\quad [\tau_{\alpha\beta}] = l_{\alpha}\psi_{\beta}+l_{\beta}\psi_{\alpha},$

Σ *is also necessarily the front of an effective gravitational shock wave.*

We have (7.10) if and only if there exists on Σ a scalar k such that on Σ

(7.11) $\quad F_{\alpha\beta}l^{\alpha} = kl_{\beta},$

which is an exceptional case.

We note that *an effective electromagnetic shock wave creates necessarily (except a very rare case) an effective gravitational shock wave and not an ordinary gravitational wave*. The situation is completely different for a hydrodynamic shock wave corresponding to an extreme relativistic fluid.

Bibliography

[1] Y. Choquet-Bruhat, *C. R. Acad. Sc. Paris* **248** (1959) 181.

[2] Y. Choquet-Bruhat, *Ann. Inst. Henri Poincaré* **8** (1968) 327–338; *Comm. in Math. Phys.* **12** (1969) 16–35.

[3] A. Lichnerowicz, *Ann. di Matematica* **50** (1960) 2–95.
[4] A. Lichnerowicz, *C. R. Acad Sc. Paris* **273** (1971) 528; **276** (1973) 1385; *Symposia Mathematica Ist. Naz. Alta Mat. Roma* **12** (1972) 93–110.
[5] M. A. H. Mac Callern et A. H. Taub, *Comm. in Math. Phys.* **30** (1973) 153–169.
[6] A. Papapetrou, H. Treder, *Math. Nachricht. Berlin* **20** (1959) 53; **23** (1961) 371.
[7] R. Penrose, *Selecta in honor of J. L. Synge*, 1972.
[8] F. A. Pirani, Chap. 6 in: *Gravitation: an Introduction to Current Research* (edited by L. Witten), John Wiley, New York 1962.
[9] A. H. Taub, *Illinois J. of Math.* **3** (1957) 85.
[10] A. H. Taub, *Comm. in Math. Phys.* **29** (1973) 79–88.
[11] A. Trautman, *London Conference on Gravitation and Relativity*, 1965.

THE CONSTRUCTIVE APPROACH TO NONLINEAR QUANTUM FIELD THEORY

I. SEGAL

Massachusetts Institute of Technology, Cambridge, Ma. 02139, U.S.A.

1. The General Situation

Following the clarifying but inconclusive axiomatic approach to the murky territory of nonlinear quantum field theory, two definite constructive directions emerged. First, beginning in [3], that through consideration of the associated nonlinear classical equation, in terms of analysis on its solution manifold. This had the virtue of immediate resolution of the fundamental ambiguity of the meaning of the nonlinear terms, but presented at the same time the problem of correlation with the intuitive ideas of the Dirac–Heisenberg–Pauli formalism. At the same time, there were formidable purely mathematical problems. These included the cogent global solution of nonlinear wave equations, the development of suitable varieties of functional integration, and the existence of relevant differential-geometric structures in the solution manifold.

Beginning with the 1966 Conference on Functional Integration and Constructive Quantum Field Theory at M.I.T. [6] (but foreshadowed by a slightly earlier such Conference on Elementary Particles [5]), a different approach emerged. In this the meaning of the nonlinear terms was based on a notion of local nonlinear function of operator-values distributions. By early 1967, the Wick powers of free fields in 2 space-time dimensions were fully understood as uniquely characterizable local powers in a suitably generalized sense, which could alternatively be cogently founded on nonlinear variants of the Weyl relations. At the same time, the irrelevance of spatial cutoffs for the C^*-propagation was shown, and the only remaining technical problem in the establishment of the C*-temporal dynamics was that of perturbing the free hamiltonian by a well-defined, if relatively singular, self-adjoint operator. A simple general procedure for effecting such perturbations was developed by 1969, and the apparent rapidity of progress may have encouraged considerable optimism for the future.

Now however, 5 years (and many facts) later, we still don't know whether there exists a quantized solution to a simple given equation such as $\Box\varphi = m^2\varphi + \varphi^3$ $(m > 0)$ in two-space-time dimensions, even if the space component of space-time is taken as compact (i.e., with periodic boundary conditions). Nor if we had a solution (including of course the vacuum state as well as the operator-valued distribution φ itself) would we know whether it was unique – even with physically reasonable regularity assumptions. In fact, the evidence suggests that non-unicity may well be a problem (cf. [1]). We do know the existence of solutions of e.g., a certain class of equations of the form $\Box\varphi = m^2\varphi + p'(\varphi)$ where p is a polynomial of degree $2r$ (although we would be hard put to determine specific members of this class); but we don't know whether they are strongly asymptotic to free fields at times $t = \pm\infty$ in accordance with physical ideas, or what kinds of constraints there may be on the coefficients of p.

What has happened is, on the one hand, an intensive effort in what might be called experimental mathematical quantum field theory, which deals with well-posed problems in functional analysis rather than conceptually new ground from the mathematical-physical standpoint. This work of Glimm, Jaffe, Guerra, Høegh-Krohn, Rosen, Simon, and their many collaborators as well as others provides a basis for testing some theoretical ideas of a more general nature. However it may well be too close to free-field considerations, and in part too dependent on the scalar character of the field, to provide a really suggestive basis for the treatment of a physically relevant Heisenberg field. As just indicated, a direct way to treat the Heisenberg fields for a given equation in 2 space-time dimensions is still lacking.

Physically, the only sensible place for the free field to come in is at times $t = \pm\infty$, and not at finite times as it does in the model, in which the interaction hamiltonian, – properly given, naturally, in terms of the interacting field (as in the formalism of Heisenberg–Pauli), – is modified by inserting in it the free field. While this is mathematically unexceptionable as an approach to field-theoretic model building, it is physically incorrect as a means of obtaining the Heisenberg field corresponding to the original hamiltonian. (It should be recalled that the substituted free field can not be unitarily equivalent to the interacting field.)

Progress in quantum field theory in the past five years is best understood in the light of the developments of the earlier two decades. Twenty-five years ago it was almost universally believed that the free-field commutation relations, together with irreducibility, uniquely determined the field, within unitary equivalence. It was necessary to develop the true facts in this regard, leading eventually to the use of C^*-algebras as a means of attaining a representation-independent formalism. Another important development was the simultaneous diagonalization of the free-field operators at a sharp time; this work has been technically essential in constructive quantum field theory. This in turn depended on the development of functional integration theory, in the generalized form applicable to fermion as well as boson fields, involving the so-called Clifford distribution as an analogue to the Gaussian. The representation-independent C^*-formalism also involved in turn a re-examination of the concept of vacuum; it could no longer be defined as the lowest eigenstate of the hamiltonian; rather it was a stationary state of a C^*-algebra with specified analyticity (or positivity) properties in relation to its Heisenberg dynamics as a group of automorphisms. The validation of this definition through the treatment of linear fields made it directly available for the two-dimensional relativistic nonlinear case. Finally, it was necessary to understand the concept of a power or other nonlinear expression in a local quantum field. The Wick notion applies only to free fields, and was put forth as a simple means of standardization of otherwise ambiguous objects, rather than as a physically or mathematically fundamental notion. The development of a simple, general notion on the basis of pure locality considerations, which in the case of the free field could be formally identified with Wick's prescription, both rationalized in part the use of the Wick notion of power on a preliminary basis, and led to a local interpretation of the renormalized field, relative to the *physical* rather than *free* vacuum.

Once these fundamental conceptual matters had been effectively settled, rapid technical development of the theory of two-dimensional scalar relativistic fields was possible, and is still continuing. But looking ahead in physical needs and back on the conceptual progress of the past five years, it appears that we are once again in a software crisis. The outstanding mathematical developments of the period in constructive field theory due to Leonard Gross and Edward Nelson are elegant, relatively definitive, and likely to be of enduring value in quantum field theory. But they are not especially in the direction of getting away from free fields as a first approximation to the interacting field, nor of dealing with singularities representative of the very strong ones arising in four-dimensional space-time.

Several new directions of software development might be proposed. I shall discuss here only that of going back once again to the canonical quantization formalism, and developing it to the maximal level attainable on the basis of advances in the past decade in nonlinear scattering theory and functional integration.

2. Geometry of the Solution Manifold of a Nonlinear Wave Equation

The usefulness of the symplectic structure in the solution manifold of a nonlinear wave equation has been increasingly recognized in the past decade. Briefly speaking, a symplectic structure is something analogous to the differential form $\Omega = \sum_{i=1}^{N} dp_i dq_i$ which plays a fundamental role in the mechanics of systems of a finite number $2N$ of degrees of freedom. It determines the field commutators and considerably more for the second quantization of the given equation. Since the lecture of Rączka treats such matters, they need not be elaborated here.

Rather, the general picture will be completed, by treating the generally more familiar Riemannian structure, in the less familiar setting of the infinite-dimensional manifold defined by a wave equation. The basic situation may be briefly summarized as follows:

Let M denote the solution manifold of the nonlinear wave equation

$$(\#) \qquad \Box\varphi = m^2\varphi + p'(\varphi)$$

where p is a smooth function which vanishes to sufficiently high order near zero, is bounded from below, and grows no more rapidly than a polynomial, Then there exists a natural Lorentz-invariant Riemannian structure on M, at least in a neighborhood of the solution $\varphi \equiv 0$.

EXAMPLE. If $p \equiv 0$, the equation is the free one: $\Box\varphi = m^2\varphi$. The usual Lorentz-invariant inner product

$$\langle \varphi_1, \varphi_2 \rangle = \int_{k^2=m^2} f_1(k)\overline{f_2(k)} \frac{d_3 k}{|k_0|}$$

if

$$\varphi_j(x) = \int_{k^2=m^2} e^{ik.x} f_j(k) \frac{d_3k}{|k_0|} \qquad (j = 1, 2)$$

is then Lorentz-invariant; its real part gives the Riemannian metric in question here; its imaginary part, essentially the symplectic structure earlier mentioned.

Note that due to the irreducibility of the action of the Lorentz group in this case, the Lorentz-invariant metric is unique (within a constant factor), assuming it has minimal regularity properties.

Note also that $\varphi \equiv 0$ is an invariant point of the solution manifold M in the nonlinear case, under the action of the Lorentz group. As a consequence, by general principles there is a corresponding linear representation of the Lorentz group in the tangent space to the manifold M at the point $\varphi \equiv 0$. The metric in the tangent space at this point, given by the Riemannian structure, must consequently be invariant under this linear representation. This tangent space is in fact identifiable with the solutions of the free equation, and its metric with the one just defined. For, as will be seen shortly, the tangent space to M at a given solution φ can be naturally identified with the set of all solutions of the corresponding first-order variational equation

$$(*) \qquad \Box\eta = m^2\eta + p''(\varphi)\eta$$

which at the point $\varphi \equiv 0$ is the free equation.

Thus the Riemannian structure may be regarded as an extension of the familiar Lorentz-invariant inner product for solutions of the free equation, to an inner product for two solutions of the equation $(*)$ – this latter inner product being, of course, φ-dependent. This symmetric inner product differs from the anti-symmetric bilinear form corresponding to the symplectic structure, i.e.,

$$A(\varphi, \psi) = \int \big(\varphi(t)\dot{\psi}(t) - \dot{\varphi}(t)\psi(t)\big) d\vec{x},$$

in the same way that the 2-point function for the quantized field in the conventional sense differs from the commutator function. The latter is an entirely local object for a given differential equation, while the former depends on the global situation. It is consequently a great deal more difficult to set up the Riemannian structure, but it is no less relevant than the symplectic structure.

The Riemannian structure is also interesting as a meeting ground for the ideas of quantum field theory and of general relativity. One of the perenni questions of quantum theory has been how to express non-trivial interaction; essentially the only practical possibility has been the addition of nonlinear term(s) to linear ones representing the free system. The present geometrical approach to the phase space suggests that the essential difference is the non-vanishing curvature of the phase space representing the interacting system. This is an extension to a function space of Einstein's idea about geometrical space-time, which is parallel to Dirac's extension of the Heisenberg quantization program to function space from ordinary space.

Before describing in more detail the Riemannian structure in the solution manifold M, it may be helpful to outline a partially analogous structure in the case of systems of finitely many degrees of freedom. If a system of this type has configuration space S, its phase space P is (normally) the space of twice the dimension known as the cotangent bundle. If $q_1, \ldots, q_n$ are local coordinates near a point in S, and $p_1, \ldots, p_n$ are dual coordinates in the space of covectors at the point, the differential form $\Omega = \sum_{i=1}^{n} dp_i \wedge dq_i$ is determined and turns out to have the property of being independent of the choice of the coordinates $q_1, \ldots, q_n$. It is the "fundamental bilinear covariant" of analytical dynamics.

Now it commonly happens that S is given not merely as a manifold, which is all that is required for the foregoing, but also as a Riemannian space. The metric given in S then determines a corresponding metric in P; specifically, e.g., one may choose the q's to be normal coordinates at a point, the p's to be corresponding biorthogonal coordinates, and define $ds^2 = \sum_i (dp_i^2 + dq_i^2)$. Thus one ends up with two bilinear forms on the tangent space to any given point of the phase space; one is symmetric, corresponding to the Riemannian metric in P, the other is anti-symmetric, corresponding to the symplectic structure.

Turing now to the solution manifold M of a nonlinear wave equation, this is physically the phase space (of course, M can always be represented by the Cauchy data of the solution at some particular time t_0, giving an alternative but essentially entirely equivalent form to the phase space). However, there is no unique or distinguished configuration space from which M arises as the contangent bundle. One could so regard it by choosing any maximal space-like surface Σ, and taking S to be all data for fields on this surface; but there is no unique Σ, and there is no natural Riemannian metric in this space S, nor any metric which appears independent of Σ.

Indeed, the possibility of obtaining the desired metric in M depends on nonlinear scattering theory, which is a relatively sophisticated subject. Fortunately, the main presently relevant results are not to difficult to summarize.

These results, obtained in the past 12 years by Morawetz and Strauss, Strauss, von Wahl, and myself, assert the existence of a dense set of solutions φ_{in} of the free equation, at least in a sufficiently small neighborhood of the zero solution, such that

a) there exists a unique regular solution φ in M such that $\varphi \sim \varphi_{\text{in}}$ near $t = -\infty$, in the sense that the difference between their Cauchy data at time t goes to zero as $t \to -\infty$ in the energy norm, and moreover $\varphi = 0(|t|^{-(n/2)})$ uniformly throughout space as $|t| \to \infty$ (n = number of space dimensions);

b) there exists a unique free solution φ_{out} such that the same is true near $t = +\infty$.

The order to which $p(\varphi)$ is required to vanish near $\varphi = 0$ is fairly modest, e.g., in 2 space-time dimensions all the equations $\Box\varphi = m^2\varphi + g\varphi^p$ ($m > 0$, $g > 0$, $p \geqslant 5$) are covered; in at least 4 space-time dimensions, any odd value of $p \geqslant 3$ is satisfactory. Non-polynomial functions p are equally covered.

Remark 1. It is probable that the conclusion is valid even when φ_{in} is not necessarily close to 0, but as yet this is proved only in certain cases in 4 space-time dimensions, including $p = 3$.

Remark 2. The transformation S: $\varphi_{\text{in}} \to \varphi_{\text{out}}$, i.e., the nonlinear scattering transformation, determines p' according to a result of Morawetz and Strauss. It is symplectic, i.e., leaves invariant the fundamental bilinear covariant. S conserves the energy, but is not an isometry of the in-field into the out-field; if it were, it would be essentially linear, which it is not.

Remark 3. The induced action S': $F(\varphi_{\text{in}}) \to F(S^{-1}\varphi_{\text{in}})$ of S on suitable functions F on the free solution manifold M_0 is linear, and gives a representation for the quantum-field theoretic S-matrix in soluble cases, the functions F being suitably restricted. A possible extension of this to nonlinear cases (i.e., hamiltonians of higher than 2nd degree) is one of the goals of the present theory.

To make use of the cited results on nonlinear scattering, it is necessary to identify more precisely the tangent spaces to the solution manifold. Intuitively it is plausible that it is essentially the solution manifold of the equation (*) above, but a more explicit connection is needed in any case. This is perhaps most readily accomplished by rewriting the equation (#) as a first-order abstract evolutionary differential equation, say $u' = Au + K(u)$, where A is linear and K nonlinear. If M is parametrized by the Cauchy data at a time t_0, its tangent vectors are correspondingly parametrized by vectors v in the Cauchy data space, and it can be shown that their dependence on t is governed by the first-order variational equation $v' = Av + (\partial_u K)v$.

It follows that the problem of defining thc indicated Riemannian metric in M is that of defining for two solutions of equation (*) a real inner product, in a way which is natural, Lorentz-invariant, and goes over when $p \equiv 0$ into the usual free Lorentz-invariant inner product. To this end let W_t denote the mapping of the tangent space T_φ at φ into the free solution manifold M_0 which sends η into that element of M_0 which has the same Cauchy data at time t. If we form the inner product of the two elements of M_0 thus corresponding to two elements η_1, η_2 of T_φ, an asymptotically Lorentz-invariant expression is obtained, and temporal invariance may naturally be sought by formation of

$$\lim_{A \to \infty} \int_{-A}^{A} \langle W_t \eta_1, W_t \eta_2 \rangle dt.$$

The existence of this limit is indeed a consequence of nonlinear scattering theory; Lorentz-invariance is similarly a consequence of the known Lorentz-invariance of the nonlinear wave and scattering operators for M_0 in relation to M. It is only necessary to state in precise terms what such invariance means from a general differential-geometric point of view; to use the identification given of the tangent space T; and to observed the form of the induced action (i.e., so-called 'differential') of the transformation effected on M by a given Lorentz transformation on space-time.

There are many possible uses of this canonical Riemannian structure in M (or in M_0, which is related to M by the Lorentz-invariant wave operator, permitting their identification for this restricted purpose). One important use, and the only one I shall attempt to sketch here, is in connection with the introduction of a suitable (generalized) measure in M. The original free space M_0 has defined on it an isotropic normal distribution, which is Lorentz-invariant, and closely related to the quantized free field; the problem is that of setting up a measure which is correspondingly related to the interacting field. One can expect the Riemannian structure to help for a variety of reasons, – generalization from the case of a free system, relation to the heat equation, etc. In the finite-dimensional case a symplectic structure is quite sufficient to give a measure, but in the case, e.g., of a complex Hilbert space with coordinates $z_1, z_2, \ldots$ the formal measure that results (i.e., the analog of Ω^N as $N \to \infty$) is the mathematically hopelessly nebulous object $\prod_{i=1}^{\infty} dx_i dy_i$. The symmetric part of the complex Hilbert structure, i.e., the Riemannian structure, is essential for a kind of renormalization of the measure into the isonormal distribution, whose definition clearly depends on the symmetric part.

It is known that a linear transformation T on a real Hilbert space is absolutely continuous (i.e., has bona fide Jacobian) if and only if $T^*T - I$ is a Hilbert–Schmidt (i.e., square integrable trace) operator. It can be verified (cf. [4] that the Riemannian metric can be expressed in terms of the Lorentz-invariant free metric in M_0, at a particular point u_-, with an operator kernel which is $(1/2)(I + D^*D)$, D being the differential of the scattering transformation S, evaluated at u_-. It follows that it suffices to show that $D^*D - I$ is Hilbert–Schmidt, in order to have an initial basis for the formulation of the interacting measure in terms of the free measure together with a local density. This property of S is indeed valid in 2 space-time dimensions.

In an n-dimensional space time, it is likely that $D^*D - I$ is compact and has a finite pth power trace, for sufficiently large p, dependent on n. This fact might be useful in establishing a Lorentz-invariant weak probability measure in M if there exists such a measure in M_0, for which such transformations D are absolutely continuous. There is no special reason to doubt the existence of a wide class of specifically

Lorentz-invariant weak measures in M_0, but the only presently known ones are those invariant under the full unitary group, i.e., mixtures of isotropic Gaussians, and these do not serve as a basis for the construction of the 'vacuum measure' indicated except when $n = 2$.

Bibliography

[1] C. N. Friedman, 'Renormalized oscillator equations', *Jour. Math. Phys.* **14** (1973) 1378–80.

[2] C. Morawetz and W. Strauss, 'The inverse scattering problem for nonlinear wave equations', Preprint, 1973, to appear.

[3] I. E. Segal, 'Quantization of nonlinear systems', *J. Math. Phys.* **1** (1960) 468–488.

[4] I. Segal, 'Symplectic structures and the quantization problem for wave equations', To appear in *Proceedings of the Conference on Symplectic Geometry and Mathematical Physics*, Rome, January, 1973.

[5] *Conference on the Mathematical Theory of Elementary Particles, September, 1965, Proceedings*, M.I.T. Press, 1966.

[6] *Conference on Functional Integration and Constructive Quantum Field Theory April, 1966*, M.I.T., Proceedings Offset Notes, July, 1966.

SOME RECENT RESULTS IN BIFURCATION THEORY

EBERHARD ZEIDLER

Karl-Marx-Universität, Sektion Mathematik, Leipzig, German Democratic Republic

The purpose of my paper is to give a short account of some recent results on bifurcation theory. Bifurcation theory plays an important role in modern nonlinear functional analysis. We shall study a number of problems connected with eigenfunctions of nonlinear operators. The problem is generally stated thus. Let X be a real Banach space. The equation to be studied is

$$(1) \qquad x \in X\colon\ \mu(Lx+Nx) = x, \qquad x \neq 0,$$

where L is a linear continuous operator, $N\colon X \to X$ a nonlinear continuous operator, $||Nx||/||x|| \to 0$ as $x \to 0$ and μ a real parameter. This equation has the trivial solution $x = 0$. We pose the problem of finding a solution x different from this trivial solution.

An illustrative plot of $||x||$ versus μ, called the *response diagram*, is shown in Fig. 1.

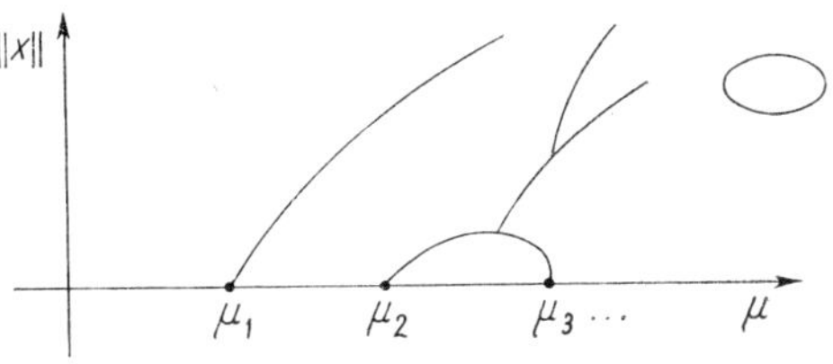

Fig. 1

Bifurcation theory studies the local and global properties of eigenfunction branches.

Nonlinear eigenvalue problems occur in their own right in physics. We mention some important examples:

1) periodic motions of bodies in celestial mechanics (systems of ordinary differential equations),

2) forms of equilibrium of rotating liquids
(nonlinear integral equations),
3) the problem of Taylor vortices in a rotating liquid
(elliptic differential equations),
4) the Bénard problem
(elliptic differential equations),
5) waves on the surface of a heavy liquid
(nonlinear boundary value problems for first order linear elliptic and quasilinear elliptic systems),
6) the motion of a gas pocket in an ideal liquid
(nonlinear boundary value problems in complex function theory),
7) buckled beams
(ordinary differential equations, variational problems),
8) buckled plates
(elliptic differential equations, variational problems),
9) buckled beams and plates with restrictions for the displacements
(variational inequalities),
10) nonlinear heat generation
(parabolic differential equations),
11) spectrum of the Helium atom
(singular ordinary differential equations, Hartree equation),
12) the problem of the heavy rotating chain and of the rotating rod
(ordinary differential equations),
13) neutron transport
(first order partial differential equations),
14) ambi-polar diffusion of ions and electrons in a plasma,
15) bifurcation problems in superconductivity
(Ginzburg–Landau equation).

(For general references see, for example, [19, 37, 62].)

Bifurcation theory is closely related to the theory of stability.

In recent papers, the following questions have been investigated:

a) non-compact operators,
b) global structure of the branches,
c) bifurcation from characteristic values of even algebraic multiplicity,
d) estimations for the number of branches,
e) constructive methods.

1. Linearization

The number μ_n is called a *bifurcation point* if there exists a sequence of solutions $(x^{(k)}, \mu^{(k)})$ of equation (1) such that

$$x^{(k)} \neq 0, \quad (x^{(k)}, \mu^{(k)}) \to (0, \mu_n) \quad \text{as} \quad k \to \infty.$$

The first problem arising in the theory of eigenfunctions is the problem of finding the bifuraction points.

1.1. $L+N$ IS COMPLETELY CONTINUOUS

Let $L+N$ be a completely continuous operator. An operator is called *completely continuous* if it is continuous and if it transforms every bounded set into a relatively compact set. The following well-known theorem is due to Krasnoselskiĭ [24]: Suppose that μ_n is a bifurcation point of (1). Then μ_n is a characteristic value of the operator L,

$$\text{(2)} \qquad \mu_n \text{ bifurcation point } \Rightarrow \exists x : \mu_n Lx = x, \quad x \neq 0.$$

(The inverse of an eigenvalue is called a *characteristic value*.) However, some of these characteristic values may not be bifurcation points. But it is always true that every characteristic value of odd algebraic multiplicity $m(\mu_n)$ is a bifurcation point;

$$\text{(3)} \qquad \mu_n \text{ bifurcation point} \Leftarrow \exists x\colon \mu_n Lx = x,\ x \neq 0,\ m(\mu_n) \text{ odd}.$$

The algebraic multiplicity is defined as

$$m(\mu_n) := \dim \bigcup_{k=1}^{\infty} N[(\mu_n L - I)^k].$$

($N[(\mu_n L-I)^k]$ is the null space of the operator $(\mu_n L-I)^k$.) Let L be a self-adjoint operator in a Hilbert space; then $m(\mu_n) = \dim N(\mu_n L-I)$.

This theorem shows that linearization is not always permissible in the case of a real space X. The proof of this theorem is based on the Leray–Schauder degree theory.

1.2. $L+N$ IS A k-SET CONTRACTION

The theory of k-set contractions plays an important role in modern nonlinear operator theory (cf. [42]). As an important example of a k-set contraction we mention the sum $A+B$, where A is a completely continuous operator and B a k-contraction, i.e. $\|Bx-By\| \leqslant k\|x-y\|$.

Let us give some definitions. The measure of non-compactness of a bounded set $\Omega \subset X$, denoted by $\alpha(\Omega)$, is defined as the infimum of the set of positive numbers δ such that Ω can be covered by a finite number of sets in X of diameter not greater than δ. An operator A: $X \to X$ is said to be a *k-set contraction* if it is continuous and if $\alpha(A(\Omega)) \leqslant k\alpha(\Omega)$ for all bounded sets $\Omega \subset X$. It is useful to associate with a bounded linear operator $L: X \to X$ the non-negative number

$$\gamma(L) := \inf\{k: L \text{ is a } k\text{-set contraction}\}.$$

If L is a bounded self-adjoint operator in a complex Hilbert space, then $\gamma(L) = r_e(L)$, where $r_e(L)$ is the radius of the essential spectrum. The essential spectrum of a closed, densely defined linear operator L is defined as the set of points $\lambda \in \sigma(L)$ ($\sigma(L)$ spectrum of L) such that at least one of the following conditions holds:

(a) the range of $L-\lambda I$ is not closed,

(b) λ is a limit point of $\sigma(L)$,

(c) $m(\lambda)$ is infinite.

The following theorem, due to Stuart [49], Thomas [54], generalizes (2), (3).

Suppose that $L+N$ is a k-set contraction with $\gamma(L) \leqslant k$. Then (2), (3) is also true if we assume that $|\mu_n| < \dfrac{1}{k}$.

This theorem has a local structure. Moreover, there is also a global result. Let S denote the subset of $X \times R^1$ consisting of all non-trivial solutions (x, μ) of (1). Let

$$S' := S \cup \{(0, \mu): \mu \text{ is a characteristic value of } L\}.$$

Let C_{μ_n} denote the component of S' containing $(0, \mu_n)$, where μ_n is a characteristic value of odd algebraic multiplicity.

Then C_{μ_n} satisfies at least one of the following conditions:

(a) C_{μ_n} is unbounded in $X \times R^1$,

(b) C_{μ_n} meets $(0, \mu_m) \in S'$, where $\mu_n \neq \mu_m$,

(c) $\inf\left\{\left|\mu \pm \dfrac{1}{k}\right|: (x, \mu) \in C_{\mu_n}\right\} = 0$.

This theorem, due to Stuart [49], is a generalization of the Rabinowitz alternative [39] concerning the global behaviour of the maximal connected set of nontrivial solutions emanating from a characteristic value of odd algebraic multiplicity of a completely continuous operator.

The proofs are based on a generalized degree theory for k-set contractions due to Nussbaum [35].

In a recent paper Dancer [12] has shown that the following result holds: Let K be a cone in the real Banach space X such that the closed linear hull of K is X and $(L+N)(K) \subseteq K$, $L+N$ being completely continuous. Under this assumption and the assumption that the spectral radius $r(L)$ is positive, the component C_{μ_1} of $S' \cap (K \times R^1)$ containing $(0, r^{-1}(L))$ is unbounded in $X \times R^1$.

To apply these abstract results we consider the equation

$$Dx = \mu(x+Mx), \tag{4}$$

where D is an elliptic differential operator with boundary conditions in a domain $G \subseteq R^n$.

For example, if $G \subseteq R^n$ is a bounded domain and D is a uniformly strongly elliptic second-order operator, then, provided $D^{-1} = L$ exists, it is completely continuous; hence

$$x = \mu(Lx+Nx). \tag{5}$$

This is the case $k = 0$.

Suppose, however, that $G \subseteq R^n$ is an unbounded domain. Thus, in general, L is not completely continuous. If we consider $X = L^2(G)$, and suppose that the differential operator together with the corresponding boundary conditions defines a self-adjoint operator, then the theory of k-set contractions does in fact apply.

From this it follows that the bifurcation theory of k-set contractions plays an important role in the theory of elliptic differential equations in unbounded domains. The equations of quantum theory are of this type. Applications to the Hartree equation are given by Stuart in [48].

2. The Bifurcation Equation

The topological arguments which we have just expounded have an important defect; they do not allow us to find the number of eigenvectors corresponding to a given characteristic value μ of (1). In this section we will describe a method which enables us to construct the solutions of (1) in an explicit manner. This method of the so-called bifurcation equation goes back to Lyapunov and Erhard Schmidt (cf. [56]).

We consider the non-homogeneous linear equation $\mu_n Lx - x = f$, and suppose that there are projection operators P, Q such that

$$\begin{aligned} &\mu_n Lx - x = f \Leftrightarrow Pf = 0, \quad P^2 = P, \\ (6) \qquad &\mu_n Lx - x = 0 \Leftrightarrow Qx = x, \quad Q^2 = Q, \\ &\dim QX, \ \dim PX < \infty. \end{aligned}$$

Here Q is the projection operator from X onto the eigenspace of L. (6) is a generalization of the Fredholm alternative concerning regular integral equations and of the Noether alternative concerning singular integral equations. The index $i(L)$ of L is defined as

$$i(L) := \dim QX - \dim PX.$$

We seek solutions of equation (1) in the form

$$x = y + z, \quad \mu = \mu_n + \varepsilon, \quad y := Qx, \quad z := (I-Q)x$$

$$y = \sum_{i=1}^{n} s_i x_i, \quad x_i \text{ eigenvectors of } L.$$

Multiplying the equation

$$(1) \qquad \mu_n Lx - x = -\varepsilon Lx - (\mu_n + \varepsilon)Nx$$

by $P, I-P$, we obtain the equivalent system

$$(7) \qquad 0 = P\big(-\varepsilon L(y+z) - (\mu_n + \varepsilon)N(y+z)\big),$$

$$(8) \qquad \mu_n Lz - z = (I-P)\big(-\varepsilon L(y+z) - (\mu_n + \varepsilon)N(y+z)\big).$$

From the implicit function theorem it follows that (8) permits a unique solution $z = z(y, \varepsilon)$. Substituting $z(y, \varepsilon)$ into (7), we get the so-called bifurcation equation

$$(7') \qquad f_i(s_1, \ldots, s_n, \varepsilon) = 0, \quad i = 1, \ldots, m.$$

Equation (7′) represents m equations to be solved for n scalar unknown $s_1, \ldots, s_n$.

The basic idea of this method is to reduce the infinite-dimensional problem (1) to the finite-dimensional problem (7′). The solution of (7′) presents considerable difficulty in the cases $n, m > 1$. Although several different approaches using methods of complex function theory, algebra, topology and variational calculus have been developed, none of the methods have so far provide a completely satisfactory answer to the basic problem of finding the number of nontrivial branches.

2.1. ANALYTIC METHODS

Let X be a complex Banach space and suppose that N is an analytic operator. Then (7′) has also an analytic structure. Using the Weierstrass preparation theorem, we get from (7′) the equivalent system (for small solutions)

$$\sum_{j=1}^{k_j} s_1^j a_{ij}(s_2, \ldots, s_n, \varepsilon) = 0$$

(a_{ij} power series of $s_2, \ldots, s_n, \varepsilon$). Applying a purely algebraic reduction theory due to Kronecker, we can eliminate s_1. Of course, it is possible to continue this procedure. In the end we get one equation for two variables. Such analytic equations are solvable with the aid of the so-called Newton diagram. (For more details see the books of Vainberg, Trenogin [55] and Krasnoselskiĭ [25].) The solutions of (1) have in many cases the following form:

$$(8') \qquad x = \sum_{l=1}^{\infty} \varepsilon^{l/k} x_l .$$

However, there are also solutions which do not have such a structure. Precisely, we have the following statements:

a) $n := \dim N(\mu_n L - I) = 1,\ i(L) = 0.$

If the bifurcation equation is not identical to zero, then every small solution of (1) ($x(\varepsilon) \to 0$ as $\varepsilon \to 0$) has the structure (8′).

Every formal solution of the structure (8′) converges in a neighbourhood of ε.

b) $n := \dim N(\mu_n L - I) > 1,\ i(L) = 0.$

If, for fixed ε, (1) permits a finite number of small solutions, then all those solutions are of the form (8′).

If, for fixed ε, (1) permits an infinite number of small solutions, then there is a solution which is not of the form (8′).

If (1) permits only a finite number of formal solutions which are of the structure (8′), then all those formal expansions converge in a neighbourhood of $\varepsilon = 0$.

If (1) permits an infinite number of formal solutions (8′), then there exists a formal solution which diverges.

This theory was developed by Aisengendler, Melamed, Trenogin and Vainberg [cf. 55, 25].

It was shown by Melamed [32] that in the analytic case (X a complex Banach space, $L+N$ an analytic operator) every characteristic value is a (complex) bifurcation point.

Let X be a real Banach space. In the case $m(\mu_n) = 1$ (μ_n has the algebraic multiplicity one) it is possible to obtain the solutions of (1) in a direct manner from (1) (without using the bifurcation equation). Such methods are described in the author's paper [65]. Many applications are related to this case (e.g. the Bénard problem, the Taylor vortices, surface waves). Using these methods, the author was able to develop a unified theory for a wide class of permanent surface waves [61–67]. In a recent paper [6], the existence of tidal waves with an arbitrary distribution of vortices was proved. For the case of simple eigenvalues we refer the reader also to the work of Crandall, Rabinowitz [9] and Keller, Langford [20].

2.2. TOPOLOGICAL AND VARIATIONAL METHODS

In order to solve the bifurcation equation, we may also apply topological and variational methods. Topological methods are based on the Brouwer degree theory. We have to compute the degree of the mapping $(s_1, \ldots, s_n) \to (f_1, \ldots, f_n)$ (see Cronin [11]). If the bifurcation equation (7′) is the Euler equation of a variational problem, then any solution of the variational problem is also a solution of (7′).

Let X be a real Banach space and let μ_n be a characteristic value of even algebraic multiplicity. Then it is not always true that μ_n is a bifurcation point. Moreover, if there exists a decomposition of the operator N:

$$N = H+M$$

(H homogeneous of degree $k \geqq 2$, M higher order terms), then it is possible to formulate conditions with respect to H guaranteeing that μ_n is a bifurcation point. For example, such conditions have been given by Berger [4], Greenlee [15], Keener, Keller [18], Keller, Langford [20], Kirchgässner [21], Kosel [23], V. M. Krasnoselskiĭ [26], Sather [43, 44, 45] and Vogel [57].

In Keller, Langford [20] the idea is to solve first the bifurcation equations, which are but a set of compatibility conditions, and then to solve the resulting operator equation by iteration.

The connection between bifurcation problems and abstract differential equations was investigated by Langenbach [27, 28] (see also [60]).

3. The Modified Bifurcation Equation

In this section, we shall report on some important results due to Böhme [7].

Let H be a Hilbert space. We consider the bifurcation problem

$$(9)\qquad x \in H\colon\ (\mu_n+\varepsilon)(Lx+Nx) = x+Mx, \qquad \mu_n \neq 0.$$

We assume that N, M are continuously differentiable with

$$N(0) = M(0) = 0, \qquad M'(0) = N'(0) = 0.$$

Furthermore, we assume that the projection operators P, Q in (6) have the properties

$$Q = P, \qquad PL = LP.$$

Multiplying (9) by P, we get

$$0 = P\big(-\varepsilon Lx-(\mu_n+\varepsilon)Nx+Mx\big)$$

and therefore

$$(10)\qquad \varepsilon = \bar{\varepsilon}(x) = \frac{\langle PMx-\mu_n PNx,\ Px\rangle}{\langle P(Lx+Nx),\ Px\rangle}.$$

From (9) we obtain the following equivalent system

$$(11)\qquad 0 = P\big(-\bar{\varepsilon}(y+z)L(y+z)-\big(\mu_n+\bar{\varepsilon}(y+z)\big)N(y+z)+M(v+z)\big),$$

$$(12)\qquad \mu_n Lz-z = (I-P)\big(-\bar{\varepsilon}(y+z)L(y+z)- -(\mu_n+\bar{\varepsilon}(y+z))\,N(y+z)+M(y+z)\big),$$

$$x = y+z, \qquad y = Px, \qquad z = (I-P)x.$$

Using a modified version of the implicit function theorem due to Böhme, we can show that (12) has a solution $z = z(y)$. Then the modified bifurcation equation follows from (11)

$$(13)\qquad 0 = P\big(-\bar{\varepsilon}(y+z)\,L(y+z)- -\big(\mu_n+\bar{\varepsilon}(y+z)\big)N(y+z)+M(y+z)\big)_{z=z(y)}.$$

This is a bifurcation equation not containing the parameter ε.

Using topological methods and methods of complex function theory, Böhme was able to solve (13) and to prove the following theorems:

(a) Let H be a real Hilbert space. We assume that (9) has a variational structure, that is

$$\forall x \in H\colon Lx+Nx = F'(x), \qquad F(0) = G(0) = 0,$$
$$x+Mx = G'(x).$$

Then μ_n is a bifurcation point; (9) has for fixed small $r > 0$ at least two solutions (x_i, ε_i) with

$$F(x_i) = r^2, \qquad \varepsilon_i(r) \to 0 \qquad \text{as } r \to 0,\ i = 1, 2.$$

If the operators N, M have the property

$$(*) \qquad N(-x) = -N(x), \qquad M(-x) = -M(x),$$

then there are $2m(\mu_n)$ such solutions $(m(\mu_n) = \dim N(\mu_n L - I))$.

If N, M have a real analytic structure, then there are two (or $2m(\mu_n)$ in the case $(*)$) analytic branches containing $(0, \mu_n)$.

This theorem generalizes results due to Krasnoselskiĭ [24], Berger [4, 5], Naumann [34].

(b) Let H be a real or complex Hilbert space. If $m(\mu_n)$ is odd, then μ_n is a bifurcation point.

(c) Let H be a complex Hilbert space. If N, M are analytic operators, then μ_n is a bifurcation point. The set of all small solutions is diffeomorphic to an analytic set in C^{m+1}, $m = m(\mu_n)$.

If the equation $\mu_n(Lx+Nx) = x+M(x)$ has no nontrivial solutions with small norm, then there exists a number $\bar{s} \geqslant 2^m - 1$ such that, for fixed small (complex) ε, (9) allows $\bar{s}$ nontrivial solutions (x, ε) with small x.

These theorems are also true in Banach spaces equipped with a positive bilinear functional; they are also applicable in the case of unbounded operators. In this connection we refer the reader to [7].

4. Conditions Preventing Second Bifurcation of Branches

Let $x(\mu)$ be a continuous branch in Figure 1. It is possible for that another branch to bifurcate from the branch considered (second bifurcation). Our aim is to find conditions preventing second bifurcation. We sketch our idea as follows. Let us consider

$$(14) \qquad f(x(\mu), \mu) \equiv \mu Lx(\mu) - x(\mu) + Nx(\mu) = 0.$$

If

$$(15) \quad f'_x(x(\mu_0), \mu_0) \neq 0,$$

then from the implicit function theorem it follows that there exists a unique solution of (14) in a neighbourhood of $(x(\mu_0), \mu_0)$, i.e. second bifurcation is impossible. In order to guarantee condition (15) we must have a great deal of information about the spectral properties of (14).

Conditions preventing second bifurcation in the case of the theory of Hammerstein integral equations with oscillation kernels were formulated by Pimbley [37].

As another example we consider the Hartree equation of the helium atom

$$R^3: \ -\frac{1}{2}\Delta u(x) - \frac{2u(x)}{|x|} + u(x)\int \frac{u^2(y)\,dy}{|x-y|} = \lambda u(x)$$

(λ energy) with the auxiliary condition

$$||u||^2 = \int u^2(y)\,dy = 1.$$

A solution of this equation is called a *self-consistent field.* By restricting the Hartree equation to the subspace of spherically symmetric functions we obtain

$$(16) \quad L_2(0, \infty): \left(-\frac{1}{2}\frac{d^2}{dr^2} - \frac{2}{r} + q_x(r)\right) x(r) = \lambda x(r),$$

$$x(0) = 0,$$

$$q_x(r) = 4\pi\left(\frac{1}{r}\int_{s<r} x^2 ds + \int_{s<r} \frac{x^2}{s}\, ds\right).$$

The following general theorem is due to Reeken [40]. It is a generalization of results due to Bazley, Zwahlen [3].

Let G be a bounded or unbounded region of R^n. Consider an equation of the form

$$Bu + Cu = \lambda u$$

in the real Hilbert space $L^2(G)$, where B and C have the following properties:

(1) B is self-adjoint and bounded below, the lower bound μ_0 being an isolated eigenvalue. The resolvent $(B-\mu I)^{-1}$ is strictly positive

for $\mu < \mu_0$; that is, nonnegative functions are transformed into strictly positive functions by $(B-\mu I)^{-1}$.

(2) The nonlinear operator C is generated by an m-linear form $C(u_1, u_2, \ldots, u_m)$ defined on $D(B)$; that is, $C(u) = C(u, u, \ldots, u)$ and this m-linear form is such that $C(u, \ldots, u, h) = q_u h$, where q_u is a bounded measurable function.

(3) There is a self-adjoint operator S which has the same domain as B such that

(a) $||C(u_1, \ldots, u_m)|| \leqslant K_0 ||u_1||_S ||u_2||_S \ldots ||u_m||_S$ for $u_i \in D(B)$, $||u||_S := \alpha||Su|| + \beta||u||$, $\forall u \in D(S)$, $\alpha > 0$, $\beta \geqslant 1$ and suppose that $||C(u)|| \leqslant K_1 ||u||^{m-1} ||u||_S$, where α can be chosen as small as one wishes, independently of K_1.

(b) C is relatively compact with respect to S; that is, from $||u_i||_S < \text{const}$ follows the existence of a subsequence u_i, such that $C(u_{i'})$ converges strongly.

(c) $C'(u)$ is symmetric and relatively compact with respect to S and

$$\langle C'(u)h, h\rangle > \langle q_u h, h\rangle > 0, \qquad \forall u, h \in D(B)$$

and $h \neq 0$, $u > 0$.

(d) $B + q_u$, with $u \in D(B)$, where q_u is regarded as a multiplicative operator, has no eigenvalues in the interior of the essential spectrum of B, whose infimum shall be denoted by μ_e, which is not supposed to be an isolated point of the essential spectrum.

Under these assumptions, one gets the following result:

(a) There exists an interval $\langle \mu_0, \mu_c \rangle$, $\mu_c \leqslant \mu_e$ of solutions u_λ, everywhere positive, where $||u_\lambda||$ and $||u_\lambda||_S$ are continuous functions of λ.

(b) $||u_\lambda||$ is monotonically increasing with λ and $||u_\lambda||_S$ and $||u_\lambda||$ tend to infinity if $\mu_c < \mu_e$. If $\mu_c = \mu_e$, then $||u_\lambda||$ and $||u_\lambda||_S$ may either both stay bounded or both diverge.

(c) There are no other positive solutions for $\mu_0 < \lambda \lessdot \mu_c$.

The Hartree equation (16) is a special case of this theorem with

$$B = -\tfrac{1}{2} d^2/dr^2 - 2/r, \qquad S = -d^2/dr^2,$$

$$\mu_0 = -2, \qquad \mu_c = \mu_e = 0.$$

In recent papers Stuart [48, 50, 51] has proved the global properties of the branches emanating from the other eigenvalues $\lambda_n = -\dfrac{2}{n^2}$,

$n = 1, 2, \ldots$ of (16). In [48] he has shown that (16) with the normalization condition

$$4\pi \int_0^\infty u(r)^2 dr = 1$$

has an infinite sequence of solutions $(u_n, \bar{\lambda}_n)$ with the properties

(i) $-\dfrac{2}{n^2} < \lambda_n < -\dfrac{1}{2n^2}$ and

(ii) u_n has exactly $n-1$ zeros (all of which are simple) in $(0, \infty)$ (see also [59, 14, 16, 17]).

5. Bifurcation in the Theory of Variational Inequalities

Many variational problems in elasticity (for example, buckled beams, buckled plates) have the following form:

$$(17) \qquad \lambda g(x_0) - f(x_0) = \min_{x \in K \subseteq X} \big(\lambda g(x) - f(x)\big).$$

If the displacements are restricted (for example supports), then K is a cone (a closed convex set in X, $\alpha \geqslant 0$, $x \in K \Rightarrow \alpha x \in K$). From (17) it follows that

$$(18) \qquad \lambda \langle g'(x_0), x - x_0 \rangle \geqslant \langle f'(x_0), x - x_0 \rangle, \qquad \forall x \in K.$$

This is an eigenvalue problem with respect to a variational inequality. Results concerning such problems have been proved by Miersemann [33] in a recent paper. If $K = X$, then (18) becomes

$$(19) \qquad \lambda g'(x_0) = f'(x_0).$$

The following theorem, due to the author, is a generalization of results due to Krasnoselskiĭ [24], Berger [4], Naumann [34] and Miersemann [33]. Let X be a real Hilbert space and let $U(0)$ be an open neighbourhood of 0. Furthermore, we assume

(a) $f, g\colon U(0) \subseteq X \to R^1$, $f(0) = g(0) = 0$,

(b) $f', g'\colon U(0) \subseteq X \to X^*$, $f'(0) = g'(0) = 0$,

(c) $f''(0), g''(0)\colon X \to X$, linear bounded operators,

(d) $\langle f''(0)x, h \rangle = \langle f''(0)h, x \rangle$, $\langle g''(0)x, h \rangle = \langle g''(0)h, x \rangle$, $\forall x, h \in X$,

(e) g weakly sequentially lower continuous, f weakly sequentially continuous,

(f) $x \to \langle f''(0)x, x\rangle$ weakly sequentially continuous,

(g) $\langle g''(0)x, x\rangle \geqslant \gamma\|x\|^2$, $\gamma > 0$, $\forall x \in X$,

(h) $\exists w \in K$: $\langle f''(0)w, w\rangle > 0$

(f', f'', g', g'' are Fréchet derivatives).

Under these assumptions, the following result holds:

(a) The variational inequality

$$\lambda_0\langle g''(0)x_0, x-x_0\rangle \geqslant \langle f''(0)x_0, x-x_0\rangle, \quad \forall x \in K \tag{20}$$

has a solution $x_0 \in K$, $x_0 \neq 0$, $\lambda_0 > 0$, where λ_0 is the greatest eigenvalue of (20).

(b) The variational inequality

$$\lambda_\varrho\langle g'(x_\varrho), x-x_\varrho\rangle \geqslant \langle f'(x_\varrho), x-x_\varrho\rangle, \quad \forall x \in K \tag{21}$$

has solutions $x_\varrho \in K$, $x_\varrho \neq 0$; $x_\varrho \to 0$, $\lambda_\varrho \to 0$ as $\varrho \to 0$; in addition λ_0 is the greatest bifurcation point of (21).

Under further assumptions it is possible to prove global results, like the compactness, continuity and existence of an unbounded component. These results [68] generalize the global results due to Bazley, Reeken, Zwahlen [2].

In this paper we have not mentioned important topological results on singularities of smooth mappings. In this connection we refer the reader to [1, 13, 29, 31, 47, 52, 53].

Bifurcation in the presence of a symmetry group was investigated in [41], [30].

Bibliography

[1] V. I. Arnold', 'Lektsii o bifurkatsiyakh i versal'nykh semeistvakh', *Uspekhi Mat. Nauk* **27** (1972) 119–184.

[2] N. Bazley, M. Reeken and B. Zwahlen, 'Global Properties of the Minimal Branch of a Class of Nonlinear Variational Problems', *Math. Z.* **123** (1971) 301–309.

[3] N. W. Bazley and B. P. Zwahlen, 'A Branch of Positive Solutions of Nonlinear Eigenvalue Problems', *Manuscripta Math.* **2** (1970) 365–374.

[4] M. S. Berger, 'A Bifurcation Theory for Nonlinear Elliptic Partial Differential Equations and Related Systems', in: J. B. Keller and S. Antman, *Bifurcation Theory and Nonlinear Eigenvalue Problems*, Benjamin, New York 1969, 113–216.

[5] M. S. Berger, 'Multiple Solutions of Nonlinear Operator Equations Arising from the Calculus of Variations', *Proc. Sympos. Pure Math.* **18**, Part I, Chicago 1968 (1970) 10–27.

[6] K. Beyer and E. Zeidler, 'Existenzbeweis für Gezeitenwellen mit beliebigen Wirbelverteilungen' (to appear).

[7] R. Böhme, 'Die Lösung der Verzweigungsgleichungen für nichtlineare Eigenwertprobleme', *Math. Z.* **127** (1972) 105–126.

[8] C. V. Coffman, 'On the Bifurcation Theory of Semilinear Elliptic Eigenvalue Problems, *Proc. Amer. Math. Soc.* **31** (1972) 170–176.

[9] M. G. Crandall and P. H. Rabinowitz, 'Bifurcation from Simple Eigenvalues', *J. Funct. Anal.* **8** (1970) 321–340.

[10] M. G. Crandall and P. H. Rabinowitz, 'Bifuraction, Perturbation of Simple Eigenvalues and Linearized Stability', *Arch. Rat. Mech. Anal.* **52** (1973) 161–180.

[11] J. Cronin, 'Fixed Points and Topological Degree in Nonlinear Analysis', Providence R. I.: *Amer. Math. Soc. XII* (1964).

[12] E. N. Dancer, 'Global Solution Branches for Positive Mappings', *Arch. Rat. Mech. Anal.* **52** (1973) 181–192.

[13] J. Eells, 'A Setting for Global Analysis', *Bull. AMS* **72** (1966) 751–807.

[14] G. Fonte, R. Mignani and G. Schiffrer, 'Solutions of the Hartree–Fock Equations, *Comm. Math. Phys.* **33** (1973) 293–304.

[15] W. M. Greenlee, 'Remarks on Branching from Multiple Eigenvalues', Nonlin. Probl. Phys. Sci. Biology, Proc. Batelle Summer Instit., Seattle 1972, Lecture Notes Math. **322** (1973) 101–121.

[16] K. Gustafson and D. Sather, 'A Branching Analysis of the Hartree Equation', *Rend. Mat.* **4** (1971) 723–734.

[17] K. Gustafson and D. Sather, 'Large Nonlinearities and Monotonicity', *Arch. Rat. Mech. Anal.* **48** (1972) 109–122.

[18] J. P. Keener and H. B. Keller, 'Perturbed Bifurcation Theory', *Arch. Rat. Mech. Anal.* **50** (1973) 159–175.

[19] J. B. Keller and S. Antman (eds.), *Bifurcation Theory and Nonlinear Eigenvalue Problems*, Benjamin, New York 1969.

[20] H. B. Keller and W. F. Langford, 'Iterations, Perturbations and Multiplicities for Nonlinear Bifurcation Problems', *Arch. Rat. Mech. Anal.* **48** (1973) 83–108.

[21] K. Kirchgässner, 'Multiple Eigenvalue Bifurcation for Holomorphic Mappings', *Contibutions to Nonlinear Functional Analysis*, Academic Press, New York 1971, 69–99.

[22] R. Kluge, 'Zur Existenz und Realisierungsweise von Bifurkationselementen', *Math. Nachr.* **42** (1969) 173–192.

[23] U. Kosel, 'Über die Anzahl von Verzweigungslösungen der Nulllösung Hammersteinscher Integralgleichungen in der Nähe eines zweifachen Eigenwertes', *Beiträge zur Analysis* **3** (1972) 135–145.

[24] M. A. Krasnosel'skiĭ, *Topological Methods in the Theory of Nonlinear Integral Equations*, Pergamon Press, Oxford–London–New York–Paris 1964.

[25] M. A. Krasnosel'skiĭ, *Näherungslösungen von Operatorgleichungen*, Akademie-Verlag, Berlin, 1973 (transl. from Russian).

[26] V. M. Krasnosel'skiĭ, 'Investigation of the Bifuraction of Small Eigenfunctions in the Case of Multidimensional Degeneration', *Soviet Math. Dokl.* **11** (1970) 1609–1613.

[27] A. Langenbach, 'Parameterabhängige Gleichungen', *Beiträge zur Analysis* **4** (1972) 9–15.

[28] A. Langenbach, 'Lösungsbifurkation und differenzierbare Zweige von Eigenfunctionen', *Monatsber. Deutsche Akademie der Wiss.* **13** (1971) 656–664.

[29] H. I. Levine, 'Singularities of Differentiable Mappings', Proc. of Liverpool Singularities Symposium, Springer Lecture Notes **192** (1971) 1–89.

[30] B. V. Loginow and V. A. Trenogin, 'The Use of Group Properties for the Determination of Multiparameter Families of Solutions of Nonlinear Equations', *Mat. Sb.* (*N.S*) **85** (1971) 440–454.

[31] J. N. Mather, 'Stability of C^∞-Mappings VI, the Nice Dimensions', Proc. of Liverpool Singularities Symposium, Springer Lecture Notes **192** (1971).

[32] B. V. Melamed, K zadache o vetvlenii reshenii nelineinogo analiticheskogo uraveneniya', *Doklady Akademii Nauk SSSR* **145** (1962) 531–533.

[33] E. Miersemann, 'Verzweigungsproblem für Variationsungleichungen', *Math. Nachr.* (to appear).

[34] J. Naumann, 'Variationsmethoden für Existentz und Bifurkation von Lösungen nichtlinearer Eigenwertprobleme I, II', *Math. Nachr.* **54** (1972) 285–296; **55** (1973) 325–344.

[35] R. D. Nußbaum, 'The Fixed Point Index and Fixed-Point Theorems for k-Set Contractions', Thesis, Chicago University 1969.

[36] A. Pazy and P. H. Rabinowitz, 'On a Branching Process in Neutron Transport Theory', *Arch. Rat. Mech. Anal.* **51** (1973) 153–164.

[37] G. H. Pimbley jr., 'Eigenfunction Branches of Nonlinear Operators and Their Bifurcations', Springer Lecture Notes in Math. **104** (1969).

[38] P. H. Rabinowitz, 'A priori Bounds for Some Bifurcation Problems in Fluid Dynamics', *Arch. Rat. Mech. Anal.* **49** (1973) 270–285.

[39] P. H. Rabinowitz, 'Some Global Results for Nonlinear Eigenvalue Problems, *J. Funct. Anal.* **7** (1971) 487–513.

[40] M. Reeken, General Theorem on Bifurcation and Its Application to the Hartree Equation of the Helium Atom', *J. Math. Phys.* **11** (1970).

[41] D. Ruelle, 'Bifurcation in the Presence of a Symmetry Group', *Arch. Rat. Mech. Anal.* **51** (1973) 136–152.

[42] B. N. Sadovskii, 'Predel'no kompaktnye operatory i uplotnyayushchye operatory', *Uspekhi Mat. Nauk* **1** (1972) 82–146.

[43] D. Sather, Branching of Solutions of an Equation in Hilbert Space, *Arch. Rat. Mech. Anal.* **36** (1970) 47–64.

[44] D. Sather, 'Branching of Solutions of a Class of Nonlinear Equations', Univ. of Colorado (preprint).

[45] D. Sather, 'Nonlinear Gradient Operators and the Method of Lyapunov–Schmidt', Univ. of Colorado (preprint).

[46] D. H. Sattinger, 'Stability of Bifurcating Solutions by Leray–Schauder Degree', *Arch. Rat. Mech. Anal.* **43** (1971) 154–166.

[47] S. Smale, 'Topology and Mechanics I, II', *Inventiones Math.* **10** (1970) 305–331; **11** (1970) 45–64.

[48] C. A. Stuart, 'Existence Theory for the Hartree Equation', *Arch. Rat. Mech. Anal.* **51** (1973) 60–69.

[49] C. A. Stuart, 'Some Bifurcation Theory for k-Set Contractions', *Proc. London Math. Soc. III. Ser.* **27** (1973) 531–550.

[50] C. A. Stuart, 'Global Properties of Components of Solutions of Nonlinear Second Order Ordinary Differential Equations on the Half-Line', Battelle Advanced Studies Center, Geneva, June 1973.

[51] C. A. Stuart, 'An Example in Nonlinear Functional Analysis: the Hartree Equation', Battelle Advanced Studies Center, Gevnea, August 1973.

[52] R. Thom, 'The Bifurcation Subset of a Space of Maps, Manifolds', Amsterdam 1970, Springer Lecture Notes **197** (1971) 202–208.

[53] R. Thom, *Stabilité structurelle et morphogenèse*, Benjamin 1972.

[54] J. W. Thomas, 'A Bifurcation Theorem for k-Set Contractions', *Pac. J. Math.* **44** (1973) 749–756.

[55] M. M. Vainberg and V. A. Trenogin, *Theorie der Verzweigung von Lösungen nichtlinearer Gleichungen*, Akademie-Verlag, Berlin 1973 (transl. from Russian).

[56] M. M. Vainberg and V. A. Trenogin, 'The Methods of Lyapunov and Schmidt in the Theory of Nonlinear Equations and their Further Development, *Russian Math. Surveys* **17** (1962) 1–60.

[57] J. Vogel, 'Untersuchungen über die Anzahl der Verzweigungsgleichungen bei Hammersteinschen Integralgleichungen im zweidimensionalen Verzweigungsfall der trivialen Lösung', *Nova Acta Leopoldina* **203**, Joh. Ambrosius Barth, Leipzig, 1971.

[58] C. T. C. Wall (ed.), *Proceeding of Liverpool Singularities Symposium*, Springer Lecture Notes **192** (1971) 209.

[59] J. Wolkowisky, 'Existence of Solutions of the Hartree Equations for N Electrons', *Indiana Univ. Math. J.* **22** (1972) 551–568.

[60] W. Wegner and K. Wiedemann, 'Existenz differenzierbarer Zweige von Eigenfunktionen', Diss., Humboldt-Universität 1973.

[61] E. Zeidler, 'Über eine Klasse freier Randwertprobleme der ebenen Hydrodynamik', *Ellipt. Dgl.* **I**, 191–203, Schriftenreihe der Institute für Mathematik der Deutschen Akademie der Wiss. Berlin, Akademie-Verlag 1970.

[62] E. Zeidler, 'Beiträge zur Theorie und Praxis einer Klasse freier Randwertaufgaben', Akademie-Verlag, Berlin 1971.

[63] E. Zeidler, 'Existenzbeweis für conoidal waves unter Berücksichtigung der Oberflächenspannung', *Arch. Rat. Mech. Anal.* **41** (1971) 81–107.

[64] E. Zeidler, 'Existenzbeweis für permanente Kapillar-Schwerewellen mit allgemeinen Wirbelverteilungen', *Arch. Rat. Mech. Anal.* **50** (1973) 34–72.

[65] E. Zeidler, 'Zur Verzweigungstheorie und zur Stabilitätstheorie der Navier–Stokesschen Gleichungen', *Math. Nachr.* **52** (1972) 167–205.

[66] E. Zeidler, 'Existenz einer Gasblase in einer Parallel- und Zirkulationsströmung unter Berücksichtigung der Schwerkraft', *Beiträge zur Analysis* **3** (1972) 67–95.
[67] E. Zeidler, 'Existenzbeweis für asymptotische Wirbelwellen', *Beiträge zur Analysis* **3** (1972) 109–134.
[68] E. Zeidler, 'Lokale und globale Verzweigungsresultate für Variationsungleichungen', *Math. Nachr.* (to appear).

AN ALTERNATIVE APPROACH TO CONSTRUCTIVE $\lambda\Phi_4^4$ QUANTUM FIELD THEORY*

T. BAŁABAN

Institute of Mathematics, Warsaw University, Warsaw, Poland

R. RĄCZKA

Institute for Nuclear Research, Warsaw, Poland

I. Introduction

We shall consider in this review article the explicit construction of interacting quantum scalar fields $\hat{\Phi}(x)$ which satisfy the nonlinear relativistic wave equation in four-dimensional space-time.[1]

In the present work we elaborate the canonical quantization of the classical nonlinear relativistic field theory. We first show that if the initial Cauchy data for the classical field $\Phi(x)$ satisfying the equation

$$(1.1) \quad (\Box + m^2)\Phi(x) = \lambda\Phi^3(x), \qquad \lambda < 0,$$

are properly chosen, then under the Poisson brackets the interacting field $\Phi(x)$ and the asymptotic fields $\Phi_{\text{in}}(x)$ and $\Phi_{\text{out}}(x)$ are canonical, i.e. $(\Pi(x) \equiv (\partial_t \Phi)(x))$

$$(1.2) \quad \begin{aligned} \{\Phi(t, \mathbf{x}), \Pi(t, \mathbf{y})\} &= \delta^{(3)}(\mathbf{x}-\mathbf{y}), \\ \{\Phi(t, \mathbf{x}), \Phi(t, \mathbf{y})\} &= \{\Pi(t, \mathbf{x}), \Pi(t, \mathbf{y})\} = 0, \end{aligned}$$

and

$$(1.3) \quad \{\Phi_{\substack{\text{in}\\ \text{out}}}(x), \Phi_{\substack{\text{in}\\ \text{out}}}(y)\} = \Delta(x-y\colon m).$$

This implies that the classical evolution operator $U(\tau, \tau_0)$, the Möller scattering operators $U(-\infty, \tau)$, $U(\tau, \infty)$ and the S-operator $S = U(-\infty, \infty)$ are canonical transformations.

In addition we show that for the considered class of Cauchy data the generators of the Poincaré group, P_μ and $M_{\mu\nu}$, $\mu, \nu = 0, 1, 2, 3$, are equal to those associated with the free asymptotic fields Φ_{in} or Φ_{out}, i.e.

$$(1.4) \quad P_\mu = P_\mu^{\text{in}} = P_\mu^{\text{out}}, \qquad M_{\mu\nu} = M_{\mu\nu}^{\text{in}} = M_{\mu\nu}^{\text{out}}.$$

The fields $\Phi(x)$, $\Phi_{\text{in}}(x)$ and $\Phi_{\text{out}}(x)$ transform covariantly under the same representation $U_{(a,\Lambda)}$ of the Poincaré group. In particular we have

$$(1.5) \quad \{\Phi, P_\mu\} = \partial_\mu \Phi, \qquad \{\Phi, M_{\mu\nu}\} = (x_\mu \partial_\nu - x_\nu \partial_\mu)\Phi.$$

The classical S-operator is invariant under the Poincaré group and differs from the identity.

We see therefore that this classical field theory with the canonical formulation satisfies most of the conditions which we usually impose in quantum field theory like e.g. locality, relativistic covariance, and asymptotic conditions.

The next step is the passage from the canonical formulation of field theory to an operator quantization. This consists in the construction of an operator representation of the Heisenberg Lie algebra given by Eq. (1.2). We carry out this program in Sec. VI. Finally in Sec. VII we shall consider the quantum S-operator formalism in the present framework.[1]

Our works represent a continuation of Segal's program of the construction of an interacting quantum field as operators acting in the space of solutions of the corresponding classical nonlinear equations [8]–[11] (see also Streater [12]). The construction of an interacting quantum field is carried out by the quantization of solutions of Eq. (1.1). The alternative program of a direct quantization of dynamical equations (1.1) was considered by Rączka [6].

II. Canonical Formalism

Consider the nonlinear relativistic wave equation

$$(2.1) \quad (\Box + m^2)\Phi(x) = \lambda\Phi^3(x), \qquad \lambda < 0, \qquad x = (t, \boldsymbol{x}) \in R^4,$$

with the initial conditions

$$(2.2) \quad \Phi(0, \boldsymbol{x}) = \varphi(\boldsymbol{x}), \qquad \Pi(0, \boldsymbol{x}) = \pi(\boldsymbol{x}).$$

It was shown by Morawetz and Strauss [8] that for every given Cauchy data (2.2) defined and sufficiently regular on R^3 there exists the unique solution $\Phi(x)$ of Eq. (2.1) and the pair $\Phi_{\text{in}}(x)$ and $\Phi_{\text{out}}(x)$ of the solutions of the free Klein–Gordon equation such that

$$(2.3)\quad \Phi_{\text{in}}(t, \boldsymbol{x}) \underset{t\to-\infty}{\leftarrow} \Phi(t, \boldsymbol{x}) \underset{t\to\infty}{\rightarrow} \Phi_{\text{out}}(t, \boldsymbol{x}),$$

in the energy norm given by the formula

$$(2.4)\quad \|\Phi(t, \cdot)\|_E^2 = \int d^3\boldsymbol{x}[\Pi^2(t, \boldsymbol{x}) + |\nabla\Phi(t, \boldsymbol{x})|^2 + m^2\Phi^2(t, \boldsymbol{x})].$$

It was shown in [1] that the functions $\Phi(\tau, \boldsymbol{x})$ and $\Pi(\tau, \boldsymbol{x})$ on a hyperplane $t = \tau$ belong to the Banach space of initial data and therefore may be used for the construction of a new free field $\Phi_\tau(t, \boldsymbol{x})$. This field is given by the formula

$$(2.5)\quad \Phi_\tau[t, \boldsymbol{x}|\varphi, \pi]$$
$$= \int d^3y \Delta_R(t-t', \boldsymbol{x}-\boldsymbol{y}) \overleftrightarrow{\partial}_{t'} \Phi[t', y|\varphi, \pi]\Big|_{t'=\tau}, \qquad t > \tau,$$

and we have

$$(2.6)\quad \Phi_\tau(t, \boldsymbol{x}) = -\int_\tau^\infty dt' \partial_{t'} \int d^3\boldsymbol{y} \Delta_R(t-t', \boldsymbol{x}-\boldsymbol{y}) \overleftrightarrow{\partial}_{t'} \Phi(t', y)$$
$$= \Phi(t, x) - \lambda \int_\tau^\infty dt' \int d^3y \Delta_R(t-t', x-y) \Phi^3(t', y).$$

Replacing in Eq. (2.5) Δ_R by Δ_A we obtain the integral representation for $\Phi_\tau(t, x)$ field for $t < \tau$. We have

$$(2.7)\quad \Phi_{\substack{\text{in}\\\text{out}}}[t, \boldsymbol{x}|\varphi, \pi] = \lim_{\tau\to\mp\infty} \Phi_\tau[t, \boldsymbol{x}|\varphi, \pi],$$

in the energy norm [4].

The free solutions $\Phi_\tau(t, \boldsymbol{x})$ play an important role in the canonical formalism and in the scattering theory of classical fields.

The Cauchy data φ and π may be used as canonical variables in classical field theory. If $F(\varphi, \pi)$ is a smooth functional in the sense of Gateaux over the Banach space of Cauchy data then the functional derivative $\delta F/\delta\varphi(\boldsymbol{x})$ is defined by means of the Gateaux derivative by the formula

$$(2.8)\quad (\partial_S F)(\varphi, \pi) = \lim_{s\to 0} s^{-1}[F(\varphi+sX, \pi)-F(\varphi, \pi)]$$

$$= \left\langle \frac{\delta F}{\delta\varphi}, X \right\rangle = \int d^3x \frac{\delta F}{\delta\varphi(x)} X(x)$$

where $X(x) \in C_0^\infty(R^3)$. If φ and π are functions which satisfy the smooth initial conditions of Morawetz and Strauss [4] then the functions

$$\varphi+sX,\ \pi \quad \text{or} \quad \varphi,\ \pi+sX, \quad 0 \leqslant s < \infty,\ X \in C_0^\infty(R^3)$$

also represent smooth initial conditions and uniquely define solutions $\Phi[t, x|\varphi+sX, \pi]$ or $\Phi[t, x|\varphi, \pi+sX]$ of Eq. (2.1) (cf. App. I of [2]).

In the canonical formalism presented below an important role is played by the Gateaux derivatives and related variational derivatives $\delta\Phi/\delta\varphi(x)$ and $\delta\Phi/\delta\pi(x)$ of solutions of Eq. (2.1) We have:

PROPOSITION 1. The solutions $\Phi[\cdot|\varphi, \pi]$ of Eq. (2.1) have Gateaux derivatives with respect to φ and π in the topology defined by the energy norm (2.4).

PROOF. Denote by Φ_s and Φ the solutions of Eq. (2.1) determined by the initial conditions $(\varphi+sX, \pi)$ and (φ, π) respectively and let $\vartheta_s = s^{-1}(\Phi_s-\Phi)$. The function ϑ_s satisfies the following equation

$$(\Box+m^2)\vartheta_s = \lambda(\Phi_s^2+\Phi_s\Phi+\Phi^2)\vartheta_s,$$

and the initial conditions $\vartheta_s(0, x) = X(x)$, $(\partial_t\vartheta_s)(0, x) = 0$. Because Φ has finite F-norm given by Eq. (1) of App. I, the condition (2) of App. II of [2] is satisfied. Hence by virtue of Lemma I of App. II of [2] one obtains

$$\|\vartheta_s(t, \cdot)\|_E^2 \leqslant c^2 \int (|\nabla X(x)|^2+m^2X^2)d^3x.$$

Denote by ϑ_0 the solution of the equation

$$(2.9)\quad (\Box+m^2)\vartheta_0 = V\vartheta_0,$$

with $V(t, x) = 3\lambda\Phi^2(t, x)$ and initial conditions $\vartheta_0(0, x) = X(x)$ and $(\partial_t\vartheta_0)(0, x) = 0$. We show that $\vartheta_s \to \vartheta_0$ at $s \to 0$, in the energy norm (2.4). For $\vartheta_s-\vartheta_0$ we have the following equation and initial conditions

$$(2.10)\quad (\Box+m^2)(\vartheta_s-\vartheta_0) = 3\lambda\Phi^2(\vartheta_s-\vartheta_0)+3\lambda(\Phi_s+2\Phi)(\Phi_s-\Phi)\vartheta_s,$$

$$(\vartheta_s-\vartheta_0)(0, x) = 0, \quad [\partial_t(\vartheta_s-\vartheta_0)](0, x) = 0.$$

Hence by virtue of Eq. (3) of App. II we have

$$(2.11)\quad ||(\vartheta_s-\vartheta_0)(t,\cdot)||_E \leqslant c_1|\lambda|\left|\int_0^t ||[(\Phi_s+2\Phi)(\Phi_s-\Phi)\vartheta_s](\tau,\cdot)||\,d\tau\right.$$
$$\leqslant c_2|\lambda|(\sup|\Phi_s|+\sup|\Phi|)\sup|\Phi_s-\Phi|\left|\int_0^t ||\vartheta_s(\tau,\cdot)||d\tau\right|$$
$$\leqslant c_3|\lambda|\,|t|\,(\sup|\Phi_s|+\sup|\Phi|)\left(\int[\nabla X(x)|^2+m^2X^2(x)]d^3x\right)\times$$
$$\times\sup|\Phi_s-\Phi|.$$

It follows from the continuity of the solution $\Phi[t, x|\varphi, \pi]$ with respect to the initial conditions that $||\vartheta_s-\vartheta_0||_E \to 0$ for $s \to 0$. The function ϑ_0 as the solution of a linear homogeneous equation with sufficiently regular coefficients depends linearly on X: hence $\Phi[t, x|\varphi, \pi]$ is differentiable in the sense of Gateaux. Similarly one proves the existence of the Gateaux derivative for functionals $\Phi[t, x|\varphi, \pi]$ with respect to π. Q.E.D.

REMARK. One may similarly prove the existence of arbitrary order Gateaux derivatives of the functional $\Phi[t, x|\varphi, \pi]$ with respect to the variables φ and π.

The Poisson bracket $\{F, G\}$ of two smooth functionals of the canonical variables φ and π is defined by the formula

$$(2.12)\quad \{F, G\} = \int d^3x\left(\frac{\delta F}{\delta\varphi(x)}\frac{\delta G}{\delta\pi(x)}-\frac{\delta F}{\delta\pi(x)}\frac{\delta G}{\delta\varphi(x)}\right).$$

In particular for δ-functionals $F(\varphi, \pi) = \varphi(x)$ and $G(\varphi, \pi) = \pi(x)$ we obtain from formula (2.12)

$$(2.13)\quad \{\varphi(x),\pi(y)\} = \delta^{(3)}(x-y), \qquad \{\varphi(x),\varphi(y)\} = \{\pi(x),\pi(y)\} = 0.$$

The following theorem shows that the evolution operator $U(\tau, \tau_0)$ for Eq. (2.1) is a canonical transformation. In fact we have

THEOREM 2. The field $\Phi_\tau[t, x|\varphi, \pi]$ satisfies the following commutation relations:

$$(2.14)\quad \{\Phi_\tau(x), \Phi_\tau(y)\} = \Delta(x-y\colon m).$$

PROOF. Let $\alpha(\boldsymbol{x}) \in C_0^\infty(R^3)$ and let $\alpha(t, \boldsymbol{x})$ be the solution of the Klein–Gordon equation satisfying at some $t = t_0$ the initial conditions $\alpha(t_0, \boldsymbol{x}) = 0$ and $(\partial_t \alpha)(t_0, \boldsymbol{x}) = \alpha(\boldsymbol{x})$. We have then

$$\begin{aligned}(2.15)\quad &\int d^3\boldsymbol{x}\,\partial_s \Phi_\tau[t_0, \boldsymbol{x}|\varphi+sX, \pi]\,\alpha(\boldsymbol{x})\\ &= \int d^3\boldsymbol{x}\left\{\int d^3\boldsymbol{y}\,\Delta(t-t', \boldsymbol{x}-\boldsymbol{y})\overleftrightarrow{\partial}_{t'}\,\partial_s\Phi[t', \boldsymbol{y}|\varphi+sX, \pi]|_{t'=\tau}\right\}\times\\ &\times \overleftrightarrow{\partial}_t\,\alpha(t, \boldsymbol{x})|_{t=t_0} = \int d^3\boldsymbol{x}\,\partial_s\Phi(t, \boldsymbol{x})\overleftrightarrow{\partial}_t\,\alpha(t, \boldsymbol{x})|_{t=\tau}.\end{aligned}$$

We now use the fact that the scalar product

$$(2.16)\quad (\alpha, \beta)_t = \int d^3\boldsymbol{x}\,\alpha(t, \boldsymbol{x})\overleftrightarrow{\partial}_t\beta(t, \boldsymbol{x})$$

is t-independent not only for solutions of the Klein–Gordon equation but also for the solution of a more general equation (2.9). Such an equation for $V = 3\lambda\Phi^2$ satisfies the function $\partial_s\Phi$. Take the solution u_τ of this equation satisfying the following initial conditions:

$$(2.17)\quad u_\tau(\tau, \boldsymbol{x}) = \alpha(\tau, \boldsymbol{x}), \qquad (\partial_t u_\tau)(\tau, \boldsymbol{x}) = (\partial_t \alpha)(\tau, \boldsymbol{x}).$$

We have then

$$\begin{aligned}(2.18)\quad &\int d^3\boldsymbol{x}\,\partial_s\Phi_\tau[t_0, \boldsymbol{x}|\varphi+sX, \pi]\,\alpha(\boldsymbol{x}) = (\partial_s\Phi(t, \cdot), u_\tau(t, \cdot))_{t=\tau}\\ &= (\partial_s\Phi(t, \cdot), u_\tau(t, \cdot))_{t=0} = \int d^3\boldsymbol{x}(\partial_t u_\tau)(0, \boldsymbol{x})X(\boldsymbol{x}).\end{aligned}$$

Consequently

$$(2.19)\quad \frac{\delta}{\delta\varphi(\boldsymbol{x})}\int d^3\boldsymbol{y}\,\Phi_\tau[t_0, \boldsymbol{y}|\varphi, \pi]\,\alpha(\boldsymbol{y}) = (\partial_t u_\tau)(0, \boldsymbol{x}).$$

One derives similarly the expression for the functional derivative $\delta\Phi/\delta\pi(\boldsymbol{x})$. Indeed let $\beta(t, \boldsymbol{x})$ be a solution of the free Klein–Gordon equation satisfying the initial condition

$$\beta(t_0, \boldsymbol{x}) = 0, \qquad (\partial_t\beta)(t_0, \boldsymbol{x}) = \beta(\boldsymbol{x}), \qquad \beta(\boldsymbol{x}) \in C_0^\infty(R^3).$$

Then utilizing the same considerations as previously one obtains:

$$\begin{aligned}&\int d^3\boldsymbol{x}\,\partial_s\Phi_\tau[t_0, \boldsymbol{x}|\varphi, \pi+sX]\,\beta(\boldsymbol{x})\\ &= \int d^3\boldsymbol{x}\,\partial_s\Phi[t, \boldsymbol{x}|\varphi, \pi+sX]\,\overleftrightarrow{\partial}_t\beta(t, \boldsymbol{x})|_{t=\tau}.\end{aligned}$$

Defining now v_τ as a solution of Eq. (2.9) with $V = 3\lambda\Phi^2$ satisfying at $t = \tau$ the same initial conditions as $\beta(t, x)$ we obtain

$$(2.20)\quad \int d^3x\partial_s\Phi_\tau[t_0, x|\varphi, \pi+sX]\beta(x) = -\int d^3x v_\tau(0, x)X(x)\,.$$

Hence

$$(2.21)\quad \frac{\delta}{\delta\pi(x)}\int d^3y\Phi_\tau[t_0, y|\varphi, \pi]\beta(y) = -v_\tau(0, x)\,.$$

It follows from Eqs. (2.19) and (2.21) that the Poisson bracket for the functionals

$$(2.22)\quad \int d^3x\Phi_\tau[t, x|\varphi, \pi]\alpha(x) \quad\text{and}\quad \int d^3y\Phi_\tau[r, y|\varphi, \pi]\beta(y)$$

is well defined by formula (2.12), i.e. the considered integral is convergent. We have moreover

$$(2.23)\quad \left\{\int d^3x\Phi_\tau[t, x|\varphi, \pi]\alpha(x), \int d^3y\Phi_\tau[r, y|\varphi, \pi]\beta(y)\right\}$$
$$= \int d^3z[-(\partial_t u_\tau)(0, z)v_\tau(0, z)+u_\tau(0, z)(\partial_t v_\tau)(0, z)]$$
$$= (u_\tau(t', \cdot), v_\tau(t', \cdot))_{t'=0}.$$

Using now the t-independence of the scalar product for solutions of Eq. (2.9) and the definition of u_τ and v_τ one obtains

$$(2.24)\quad (u_\tau(t', \cdot), v_\tau(t', \cdot))_{t'=0} = (u_\tau(t', \cdot), v_\tau(t', \cdot))_{t'=\tau}$$
$$= \int d^3z\alpha(t', z)\overleftrightarrow{\partial}_{t'}\beta(t', z)|_{t'=\tau} = (\alpha(t', \cdot), \beta(t', \cdot))_{t'=\tau}$$
$$= (\alpha(t', \cdot), \beta(t', \cdot))_{t'=t} = -\int d^3x\alpha(x)\beta(t, x)$$
$$= \int d^3x\alpha(x)\left(\int d^3y\Delta(t-r, x-y)\overleftrightarrow{\partial}_r\beta(r, y)\right)$$
$$= \int d^3xd^3y\alpha(x)\Delta(t-r, x-y)\beta(y).$$

This proves Eq. (2.14). Q.E.D.

COROLLARY 1. The interacting field $\Phi(t, x)$ satisfies the following equal time commutation relations

$$(2.25)\quad \begin{aligned} \{\Phi(t, x), \Pi(t, y)\} &= \delta^{(3)}(x-y),\\ \{\Phi(t, x), \Phi(t, y)\} &= \{\Pi(t, x), \Pi(t, y)\} = 0. \end{aligned}$$

PROOF. This follows from Eq. (2.14) and the fact that for $t = \tau$ the values of the field $\Phi_\tau(\tau, \boldsymbol{x})$ and $\Pi_\tau(\tau, \boldsymbol{x})$ coincide with those of $\Phi(\tau, \boldsymbol{x})$ and $\Pi(\tau, \boldsymbol{x})$ respectively for the interacting field. Q.E.D.

We now show that the asymptotic classical fields $\Phi_{\text{in}}(x)$ and $\Phi_{\text{out}}(x)$ are canonical. Indeed we have:

THEOREM 3. The fields $\Phi_{\text{in}}(x)$ and $\Phi_{\text{out}}(x)$ satisfy the following commutation relations

$$(2.26)\quad \{\Phi_{\substack{\text{in}\\ \text{out}}}(x), \Phi_{\substack{\text{in}\\ \text{out}}}(y)\} = \Delta(x-y:m).$$

PROOF. We first prove the uniform convergence of $\partial_s\Phi_\tau[t, \cdot|\varphi+sX, \pi]$ for $\tau \to -\infty$. By virtue of Eq. (2.6) we have:

$$(2.27)\quad \partial_s\Phi_{\tau_1}(t, \boldsymbol{x})-\partial_s\Phi_{\tau_2}(t, \boldsymbol{x})$$
$$= -3\lambda\int_{\tau_1}^{\tau_2} dt'\cdot\int d^3y\Delta_R(t-t', \boldsymbol{x}-\boldsymbol{y})\,\partial_s\Phi(t', \boldsymbol{y})\Phi^2(t', \boldsymbol{y}).$$

Hence

$$\|\partial_s\Phi_{\tau_1}(t, \cdot)-\partial_s\Phi_{\tau_2}(t, \cdot)\|_E$$
$$\leqslant 3|\lambda|\int_{\tau_1}^{\tau_2} dt'\left\|\int d^3y\Delta_R(t-t', \cdot-\boldsymbol{y})\Phi^2(t', \boldsymbol{y})\,\partial_s\Phi(t', \boldsymbol{y})\right\|_E.$$

Using the definition of E-norm and the properties of the Δ function, for $\tau_1 < \tau_2 < t$ we obtain

$$(2.28)\quad \left\|\int d^3y\Delta(t-t', \cdot-\boldsymbol{y})\Phi^2(t', \boldsymbol{y})\,\partial_s\Phi(t', \boldsymbol{y})\right\|_E$$
$$= \|\Phi^2(t', \cdot)\,\partial_s\Phi(t', \cdot)\|_2.$$

Hence using Lemma 1 of App. II we obtain

$$(2.29)\quad \|\partial_s\Phi_{\tau_1}(t, \cdot)-\partial_s\Phi_{\tau_2}(t, \cdot)\|_E \leqslant 3|\lambda|\int_{\tau_1}^{\tau_2} dt'\|\Phi^2(t', \cdot)\,\partial_s\Phi(t', \cdot)\|_2$$
$$\leqslant \frac{3|\lambda|}{m}\int_{\tau_1}^{\tau_2}\sup_{\boldsymbol{x}}\Phi^2(t', \boldsymbol{x})dt' \sup_{t'\in[\tau_1,\tau_2]}\|\partial_s\Phi(t', \cdot)\|_E$$
$$\leqslant \int_{\tau_1}^{\tau_2}\sup_{\boldsymbol{x}}\Phi^2(t', \boldsymbol{x})dt'\|\partial_s\Phi(0, \cdot)\|_E.$$

If follows from the convergence of the integral $\int_{-\infty}^{\infty} \sup_{x} \Phi^2(t, x) dt$ which is a part of the F-norm given by Eq. (1) of App. I that

$$\|\partial_s \Phi_{\tau_1}(t, \cdot) - \partial_s \Phi_{\tau_2}(t, \cdot)\|_E \underset{\tau_1, \tau_2 \to -\infty}{\to} 0$$

uniformly with respect to s in a bounded interval. Hence

$$(2.30)\quad \partial_s \Phi_{\text{in}}[t, \boldsymbol{x}|\varphi + sX, \pi] = \lim_{\tau \to -\infty} \partial_s \Phi_\tau[t, \boldsymbol{x}|\varphi + sX, \pi],$$

in the energy norm. Using definition (2.5) we obtain for $s = 0$

$$(2.31)\quad \int d^3y \frac{\delta \Phi_{\text{in}}[t, \boldsymbol{x}|\varphi, \pi]}{\delta \varphi(\boldsymbol{y})} X(\boldsymbol{y}) = \lim_{\tau \to -\infty} \int d^3y \frac{\delta \Phi_\tau[t, \boldsymbol{x}|\varphi, \pi]}{\delta \varphi(\boldsymbol{y})} X(\boldsymbol{y}).$$

This expression converges in the energy norm: consequently it converges also after integration with a smooth function of variable $\boldsymbol{x}$. Similarly one derives the formula

$$(2.32)\quad \int d^3y \frac{\delta \Phi_{\text{in}}[t, \boldsymbol{x}|\varphi, \pi]}{\delta \pi(\boldsymbol{y})} X(\boldsymbol{y}) = \lim_{\tau \to -\infty} \int d^3y \frac{\delta \Phi_\tau[t, \boldsymbol{x}|\varphi, \pi]}{\delta \pi(\boldsymbol{y})} X(\boldsymbol{y}).$$

The above considerations show that we have convergence for

$$(2.33)\quad \frac{\delta}{\delta \varphi(\boldsymbol{y})} \int d^3\boldsymbol{x} \Phi_\tau[t, \boldsymbol{x}|\varphi, \pi] \alpha(\boldsymbol{x}) \to \frac{\delta}{\delta \varphi(\boldsymbol{y})} \int d^3\boldsymbol{x} \Phi_{\text{in}}[t, \boldsymbol{x}|\varphi, \pi] \alpha(\boldsymbol{x})$$

and

$$(2.34)\quad \frac{\delta}{\delta \pi(\boldsymbol{y})} \int d^3\boldsymbol{x} \Phi_\tau[t, \boldsymbol{x}|\varphi, \pi] \beta(\boldsymbol{x})$$
$$\to \frac{\delta}{\delta \pi(\boldsymbol{y})} \int d^3\boldsymbol{x} \Phi_{\text{in}}[t, \boldsymbol{x}|\varphi, \pi] \beta(\boldsymbol{x}),$$

$\alpha, \beta \in C_0^\infty(R^3)$ in the sense of distributions from $D'(R^3)$.

We shall now analyze these convergences in detail; consider a point $t = t_0$ for which we shall consider the convergence of (2.33): denote by $\alpha(t, \boldsymbol{x})$ and $\beta(t, \boldsymbol{x})$ the solutions of the free Klein–Gordon equation satisfying the initial conditions $\alpha(t_0, \boldsymbol{x}) = 0$, $(\partial_t \alpha)(t_0, \boldsymbol{x}) = \alpha(\boldsymbol{x})$ and analogously for $\beta(t, \boldsymbol{x})$. The functional derivative $\delta/\delta\varphi(\boldsymbol{x}) \times$ $\times \int \Phi_\tau[t_0, \boldsymbol{y}|\varphi, \pi] \alpha(\boldsymbol{y})$ satisfies Eq. (2.19). We now show that the functions u_τ are convergent for $\tau \to -\infty$ in the energy norm, uniformly with respect to t. The limit $u_{-\infty}$ is then the solution of Eq. (2.9) asymptotically convergent at $t \to -\infty$ to $\alpha(t, \boldsymbol{x})$.

By virtue of Lemma 1 of App. II we have

$$(2.35)\quad \|u_{\tau_1}(t,\ \cdot)-u_{\tau_2}(t,\ \cdot)\|_E \leqslant C_1\|u_{\tau_1}(\tau_2,\ \cdot)-u_{\tau_2}(\tau_2,\ \cdot)\|_E = C_1\|u_{\tau_1}(\tau_2,\ \cdot)-\alpha(\tau_2,\ \cdot)\|_E .$$

If we take the solution α_2 of the free Klein–Gordon equation having the same initial conditions at $t = \tau_2$ as the function u_{τ_1} then

$$\|u_{\tau_1}(\tau_2,\ \cdot)-\alpha(\tau_2,\ \cdot)\|_E = \|\alpha_2(\tau_2,\ \cdot)-\alpha(\tau_2,\ \cdot)\|_E = \|\alpha_2(t,\ \cdot)-\alpha(t,\ \cdot)\|_E .$$

We shall now apply the same arguments as in the proof of Eq. (2.29). We have for the function α the following integral representation

$$(2.36)\quad \alpha(t,\boldsymbol{x}) = u_{\tau_1}(t,\boldsymbol{x})-3\lambda\int_{\tau_1}^{\infty} dt'\int d^3y\Delta_R(t-t',\boldsymbol{x}-\boldsymbol{y})\,\Phi^2(t',\boldsymbol{y})u_{\tau_1}(t',\boldsymbol{y}).$$

For the function α_2 we have a similar representation with τ_2 instead of τ_1 in the lower limit of integration. Hence for $\tau_1, \tau_2 < t$ we have

$$(2.37)\quad \alpha_2(t,\boldsymbol{x})-\alpha(t,\boldsymbol{x}) = -3\lambda\int_{\tau_1}^{\tau_2} dt'\int d^3y\Delta(t-t',\boldsymbol{x}-\boldsymbol{y})\,\Phi^2(t',\boldsymbol{y})u_{\tau_1}(t',\boldsymbol{y}).$$

Applying now the same evaluations as in Eq. (2.29) one obtains

$$(2.38)\quad \|\alpha_2(t,\ \cdot)-\alpha(t,\ \cdot)\|_E \leqslant C_0\left|\int_{\tau_1}^{\tau_2} \sup_{\boldsymbol{x}}\Phi^2(t,\ \boldsymbol{x})dt\right| \|u_{\tau_1}(\tau_1,\cdot)\|_E .$$

Consequently, from Eqs. (2.35), (2.36) and the last inequality one obtains

$$(2.39)\quad \|u_{\tau_1}(t,\ \cdot)-u_{\tau_2}(t,\ \cdot)\|_E \leqslant C_2\|\alpha\|_2\left|\int_{\tau_1}^{\tau_2} \sup_{\boldsymbol{x}}\ \Phi^2(t,\boldsymbol{x})dt\right|,$$

from which the required convergence of u_τ follows.

In particular the derivative $(\partial_t u_\tau)(0,\boldsymbol{x})$ is convergent in $L^2(R^3)$ and by virtue of Eqs. (2.19) and (2.33) we have

$$(2.40)\quad \frac{\delta}{\delta\varphi(\boldsymbol{x})}\int d^3y\Phi_{\mathrm{in}}[t_0,\boldsymbol{y}|\varphi,\pi]\alpha(\boldsymbol{y}) = \lim_{\tau\to-\infty}(\partial_t u_\tau)(0,\boldsymbol{x}).$$

The derivation of the functional derivative with respect to π is similar and we point out the main steps only. We have

$$(2.41)\quad \int d^3x\partial_s \Phi_\tau[t_0, x|\varphi, \pi+sX]\beta(x)$$
$$= \int d^3x\partial_s \Phi[t, x|\varphi+sX]\overleftrightarrow{\partial_t}\beta(t, x)\Big|_{t=\tau}.$$

Defining v_τ as above and using Eqs. (2.34) and (2.21) one obtains

$$(2.42)\quad \frac{\delta}{\delta\pi(x)}\int d^3y\Phi_{\rm in}[t_0, y|\varphi, \pi]\beta(y) = -\lim_{\tau\to-\infty} v_\tau(0, x).$$

Here we have convergence in $\|\cdot\|_2$ norm in $L^2(R^3)$.

Using the same steps as in the derivation of formula (2.24) one obtains

$$\left\{\int d^3x\Phi_{\substack{\rm in\\ \rm out}}[t, x|\varphi, \pi]\alpha(x), \int d^3y\Phi_{\substack{\rm in\\ \rm out}}[r, y|\varphi, \pi]\beta(y)\right\}$$
$$= \int d^3x d^3y\, \alpha(x)\Delta(t-r, x-y)\beta(y)$$

which gives the assertion of Theorem 3. Q.E.D.

III. Commutation Relations and Locality

It is evident from Eq. (2.26) that the asymptotic fields $\Phi_{\rm in}$ and $\Phi_{\rm out}$ are local. The free fields $\Phi_\tau(x)$, $-\infty < \tau < \infty$ which are given by Eq. (2.5) are also local, by virtue of Eq. (2.14).

We now derive non-equal time commutation relations for the interacting field $\Phi(x)$ from which locality will follow. Let $\Delta^\lambda[x, y|\Phi]$ denote the Green function of the linear equation

$$(3.1)\quad (\Box+m^2)u(x) = V(x)u(x), \qquad V = 3\lambda\Phi^2,$$

satisfying for $t_x = t_x = r$ the initial conditions:

$$(3.2)\quad \Delta^\lambda[r, x, r, y|\Phi] = 0, \qquad (\partial_t\Delta^\lambda)[r, x, r, y|\Phi] = -\delta^{(3)}(x-y).$$

This function can be written as the following series which after smearing with a test function $\beta(y) \in S(R^3)$ is convergent in the energy norm:

$$(3.3)\quad \Delta^\lambda[x, y|\Phi] = \Delta(x-y) + \sum_{n=1}^{\infty}(3\lambda)^n \iint_{y^0}^{x^0}\cdots\iint_{y^0_{n-1}}^{x^0_{n-1}} \Delta(x-x_1)\times$$
$$\times\ \Phi^2(x_1)\Delta(x_1-x_2)\ldots\Phi^2(x_n)\Delta(x_n-y)\,d^4x_1\ldots d^4x_n.$$

PROPOSITION 4. The interacting classical field $\Phi(x)$ satisfies the following commutation relations

$$\{\Phi(x), \Phi(y)\} = \Delta^\lambda[x, y|\Phi].$$

PROOF. Let $f, g \in C_0^\infty(R^4)$ and $f(r, \boldsymbol{x}) = f_1(r)\alpha(\boldsymbol{x}), g(s, \boldsymbol{y}) = g_1(s)\beta(\boldsymbol{y})$. Then

$$\{\Phi(f), \Phi(g)\}$$
$$= \int drf_1(r)dsg_1(s)\left\{\int d^3\boldsymbol{x}\Phi(r, \boldsymbol{x})\alpha(\boldsymbol{x}), \int d^3\boldsymbol{y}\Phi(s, \boldsymbol{y})\beta(\boldsymbol{y})\right\}.$$

Denoting by u_r like in (2.17) a solution of Eq. (2.9) with initial conditions $t = r$ equal to $(0, \alpha(\boldsymbol{x}))$ and by v_s, a solution of (2.9) with initial conditions on $t = s$ equal to $(0, \beta(\boldsymbol{x}))$ one obtains by virtue of Eq. (2.23) and definition of $\Delta^\lambda[x, y|\Phi]$

$$\left\{\int d^3\boldsymbol{x}\Phi(r, \boldsymbol{x})\alpha(\boldsymbol{x}), \int d^3\boldsymbol{y}\Phi(s, \boldsymbol{y})\beta(\boldsymbol{y})\right\}$$
$$= \big(u_r(t, \cdot), v_s(t, \cdot)\big)_{t=0}$$
$$= \big(u_r(t, \cdot), v_s(t, \cdot)\big)_{t=r}$$
$$= \int d^3x\alpha(\boldsymbol{x})v_s(r, \boldsymbol{x})$$
$$= \int d^3\boldsymbol{x}\alpha(\boldsymbol{x})\int d^3\boldsymbol{y}\Delta(r, \boldsymbol{x}, s, \boldsymbol{y})\beta(\boldsymbol{y}).$$

This implies the formula (3.4).

COROLLARY. The interacting field $\Phi(\boldsymbol{x})$ is local, i.e.

$$\{\Phi(x), \Phi(y)\} = 0$$

for x and y space-like separated.

PROOF. This follows from formula (3.3)

Since v_s satisfies hyperbolic equation of motion (2.9), the initial conditions for v_s and the condition (3.4) imply that $\operatorname{supp} \partial_r v_s(r, \cdot) \cap \cap \operatorname{supp} \alpha(\cdot) = 0$. This by (3.5) implies that (3.3) and consequently (3.1) is satisfied. Q.E.D.

Thus the classical nonlinear field theory equipped with the Lie algebra structure provided by Poisson brackets is a local field theory, in which the interacting field possesses the local asymptotic fields Φ_{in} and Φ_{out}.

IV. Relativistic Covariance

The nonlinear equation (2.1) may be derived from the following Lagrangian density:

$$L(x) = \frac{1}{2}(\Phi_{,\mu}\Phi^{,\mu} + m^2\Phi) - \frac{\lambda}{4}\Phi^4. \tag{4.1}$$

Using the standard technique one derives the following form for the energy-momentum tensor associated with the density (4.1):

$$T_{\mu\nu}(x) = \Phi_{,\mu}(x)\Phi_{,\nu}(x) - g_{\mu\nu}L(x). \tag{4.2}$$

Let σ be a space-like surface in the Minkowski space. Then the integrals

$$P_\mu(\sigma) = \int_\sigma d\sigma^\nu T_{\mu\nu}, \qquad M_{\mu\nu}(\sigma) = \int_\sigma d\sigma^\lambda(x_\mu T_{\nu\lambda} - x_\nu T_{\mu\lambda}), \tag{4.3}$$

are constants of motion. One verifies, using Eq. (2.25), that the quantities (4.3) satisfy the following commutation relations:

$$\begin{aligned} \{P_\mu, P_\nu\} &= 0, \qquad \{M_{\mu\nu}, P_\lambda\} = g_{\nu\lambda}P_\mu - g_{\mu\lambda}P_\nu, \\ \{M_{\mu\nu}, M_{\lambda\varrho}\} &= g_{\mu\varrho}M_{\nu\lambda} + g_{\nu\lambda}M_{\mu\varrho} - g_{\mu\lambda}M_{\nu\varrho} - g_{\nu\varrho}M_{\mu\lambda}, \end{aligned} \tag{4.4}$$

which are the standard commutation relations for generators of the Poincaré Lie algebra.

Let $\Phi_{\text{in}}(x)$ be the free asymptotic field associated with the interacting field $\Phi(x)$. The Lagrangian density for the Φ_{in} field is given by the formula

$$L_{\text{in}}(x) = \tfrac{1}{2}(\Phi_{\text{in},\mu}\Phi_{\text{in}}^{;\mu} + m^2\Phi_{\text{in}}). \tag{4.5}$$

Using the formula (4.2) and (4.3) one can calculate the corresponding generators P_μ^{in} and $M_{\mu\nu}^{\text{in}}$ of the Poincaré group for the free field $\Phi_{\text{in}}(x)$. Similarly one can calculate the generators P_μ^{out} and $M_{\mu\nu}^{\text{out}}$ associated with the free field $\Phi_{\text{out}}(x)$.

The commutation relations of P_μ and $M_{\mu\nu}$ with the field $\Phi(x)$ are directly obtained by using formula (2.25). One obtains

$$\{\Phi(x), P_\mu\} = \partial_\mu\Phi(x), \qquad \{\Phi(x), M_{\mu\nu}\} = (x_\mu\partial_\nu - x_\nu\partial_\mu)\Phi(x). \tag{4.6}$$

The global transformations $(a, \Lambda) \to U_{(a,\Lambda)}$ of the Poincaré group P in the subset $\tilde{F}$ consisting of all solutions of Eq. (2.1) are given by the formula

$$(U_{(a,\Lambda)}\Phi)(x) = \Phi[\Lambda^{-1}(x-a)]. \tag{4.7}$$

The generators P_μ and $M_{\mu\nu}$ of $U_{(a,\Lambda)}$ by virtue of Eq. (4.6) are given by formulae (4.3).

The field Φ_{in} may be expressed as a following scalar functional of

$$(4.8)\qquad \Phi_{\text{in}}(x) = \Phi(x) - \lambda \int \Delta_R(x-y)\Phi^3(y)d^4y.$$

The transformation (4.7) induces the transformation

$$\Phi_{\text{in}}(x) \to (U_{(a,\Lambda)}\Phi_{\text{in}})(x) = \Phi_{\text{in}}[\Lambda^{-1}(x-a)]$$

of the field Φ_{in}. Hence we have

$$(4.9)\qquad U_{(a,\Lambda)} = U^{\text{in}}_{(a,\Lambda)},$$

$$(4.10)\qquad P_\mu = P^{\text{in}}_\mu, \qquad M_{\mu\nu} = M^{\text{in}}_{\mu\nu}.$$

It is instructive to derive the equality (4.10) directly. We show this in detail for the generator P_0. By virtue of Eq. (4.3) for a space-like surface $\sigma(t)$ perpendicular to the time axis, we have

$$(4.11)\qquad P_0(\sigma(t)) = \frac{1}{2}\int_{\sigma(t)} d^3\boldsymbol{x}\Big(\Pi^2(t,\boldsymbol{x}) + |\nabla\Phi(t,\boldsymbol{x})|^2 + m^2\Phi^2(t,\boldsymbol{x}) - \frac{\lambda}{2}\Phi^4(t,\boldsymbol{x})\Big).$$

We shall evaluate the expression (4.11) for $t \to -\infty$. For the interaction term $\lambda\Phi^4$ utilizing the fact that $|\Phi(t,\boldsymbol{x})| < C|t|^{-3/2}$ for large t and that $\int d^3\boldsymbol{x}\Phi^2(t,\boldsymbol{x})$ is smaller than the total energy E we have

$$(4.12)\qquad \lim_{t\to-\infty}\int d^3\boldsymbol{x}\Phi^4(t,\boldsymbol{x}) \leqslant \lim \max_{\boldsymbol{x}} \Phi^2(t,\boldsymbol{x})\int d^3\boldsymbol{x}\Phi^2(t,\boldsymbol{x}) \leqslant \lim_{t\to-\infty} CE|t|^{-3} = 0.$$

Hence by virtue of Eq. (2.3) we obtain:

$$\lim P_0(\sigma(t)) = \lim \tfrac{1}{2}|\Phi(t,\cdot)||_E = \tfrac{1}{2}||\Phi_{\text{in}}(t,\cdot)||_E = P_0^{\text{in}}.$$

Because P_0 is time independent we have $P_0 = P_0^{\text{in}}$. The derivation of Eq. (4.10) for remaining generators may be performed in a similar manner. Using the Yang–Feldman equation

$$(4.13)\qquad \Phi_{\text{out}}(x) = \Phi(x) - \lambda\int \Delta_A(x-y)\Phi^3(y)d^4y$$

and Eq. (3.7) one shows by analogous considerations that

$$(4.14)\qquad U_{(a,\Lambda)} = U^{\text{out}}_{(a,\Lambda)}$$

and

(4.15) $P_\mu = P_\mu^{\text{out}}, \quad M_{\mu\nu} = M_{\mu\nu}^{\text{out}}.$

Clearly using asymptotic properties of Φ one may derive formula (4.15) directly as in case of the Φ_{in} field.

V. Classical Scattering Operator

Theorem 3 implies that the scattering operator defined in the space $\mathscr{F}$ by the formula $S:\Phi_{\text{in}} \to \Phi_{\text{out}}$ is canonical. It was proven in [5] that the scattering operator is Poincaré invariant on $\mathscr{F}$, i.e. for every $(a, \Lambda) \in \mathscr{P}$ we have:

(5.1) $U_{(\alpha\lambda)} S = S U_{(\alpha, \lambda)}.$

It is very interesting that the inverse scattering problem can be solved in the non-linear relativistic field theory for a large class of interactions. In particular in $\lambda\Phi_4^4$ theory we have

THEOREM 5. The coupling constant λ is determined by the scattering operator

$$\lambda = \lim_{\varepsilon \to 0+} \frac{1}{6\varepsilon^4} W\big(S(2\varepsilon\Phi_{\text{in}}),\ S(\varepsilon\Phi_{\text{in}})\big),$$

where W is the Wronskian

(5.2) $W(\Phi, \psi) = \int d^3x (\Phi \partial_t \psi - \partial_t \Phi \psi)(t, x).$

(For the proof cf. Moravetz and Strauss [5].)

The formula (5.2) implies that, for $\lambda \neq 0$, S is the non-linear operator. In fact if S were linear then

$$W\big(S(2\varepsilon\Phi_{\text{in}}),\ S(\varepsilon\Phi_{\text{in}})\big) = 2\varepsilon^2 W(\Phi_{\text{in}}, \Phi_{\text{in}}) = 0.$$

The non-linearity of S implies that scattering is non-trivial in classical theory of self-interacting scalar fields. It should be stressed that in the quantum case this problem is open for the time being. (See Wilson [13] for a discussion of this problem.)

For arbitrary smooth interaction $F(\Phi)$ we have

THEOREM 6. Let $F(\Phi)$ be the interacting term in Eq. (2.1) given by an analytic function defined in the neighbourhood of the origin which is odd and $F'(0) = 0$. Then F is determined by S.

(For the proof cf. Moravetz and Strauss [5].)

Thus the inverse scattering problem is solved completely in classical non-linear field theory.

Since a classical non-linear field theory is a limit $\hbar \to 0$ of a quantum field theory it is useful to know the analyticity properties of classical scattering operator with respect to the coupling constant λ. For a positive coupling constant λ the energy operator

$$P_0 = \int d^3x \left[\tfrac{1}{2}(\Pi^2 + |\nabla\Phi|^2 + m^2\Phi^2)(t, x) - \frac{\lambda}{4}\Phi^4(t, x) \right]$$

consists of two parts with opposite signs. Hence the solution can increase arbitrarily and the asymptotic fields do not exist for all initial data from $\mathscr{F}$. Therefore one cannot expect the analyticity of S with respect to the coupling constant λ for all initial data from $\mathscr{F}$. However, if one restricts oneself to the spave Y defined as the closure of the smooth free solutions in the norm

$$||\Phi||_Y = \sup_{-\infty<t<\infty} (||\Phi(t, \cdot)||_E + (1+|t|^{3/2}) \sup_x |\Phi(t, x)|)$$

then $Y \subset \mathscr{F}$ and for small initial data we have

THEOREM 7. Consider S as an operator $S\colon (\Phi_{\text{in}}, \lambda) \to \Phi_{\text{out}}$ with the domain D,

$$D = \{(\Phi_{\text{in}}, \lambda) | \Phi_{\text{in}} \in Y,\ \lambda \in C^1,\ |\lambda|\ ||\Phi_{\text{in}}||_Y < \eta\}$$

and with range Y. Then S is complex analytic on this domain.

(For the proof cf. Rączka and Strauss [7].)

It can also be shown that S cannot be analytic in λ for all initial data from $\mathscr{F}$. For details see [7].

The above analysis shows that the classical S-operator is a good candidate for the construction of a quantum $\hat{S}$-operator associated with solutions of a quantized version of the dynamical equation (1.1). We consider this problem in Sec. VII of our work.

VI. Construction of Local Interacting and Asymptotic Quantum Fields

We have shown in Sec. II that the algebraic structure of classical non-linear relativistic field theory expressed in terms of Poisson brackets is precisely as that postulated in quantum field theory, e.g. in the formalism of Lehman, Symanzik and Zimmerman. In particular the asymptotic fields Φ_{in} and Φ_{out} are relativistic, local and canonical and the interpolating field Φ is relativistic and local. Moreover, the representation 'interpolating', 'in' and 'out' of the Poincaré group coincide. Hence if one finds an operator representation of the symplectic structure one will lift all desired properties of asymptotic and interacting fields onto the operator level and obtain a model of interacting quantum field theory. We shall now construct this operator representation of the Lie algebra of Poisson brackets.

It will be evident from the next considerations that in case of non-linear field theory the most important role is played by a vector space Ω of functionals over the space $\mathscr{F}$, defined in the following manner.

DEFINITION 1. A functional F over $\mathscr{F}$, belongs to Ω if

(i) $F \in C^\infty(\mathscr{F})$,

(ii) $D^k F[\zeta](\zeta_1, \zeta_2, \ldots, \zeta_k)$ is bounded on bounded subsets of $\mathscr{F}^{k+1}$,

(iii) $D^k\left(\sigma_2 \dfrac{\delta F}{\delta \zeta}\right)(\zeta_1, \zeta_2, \ldots, \zeta_k) \in \mathscr{F}$, $k = 0, 1, 2, \ldots$

If $F, G \in \Omega$ then $\{F, G\}$ is well defined and also belongs to Ω; indeed by virtue of (2.9) one obtains

$$\sigma_2 \frac{\delta}{\delta \zeta}\{F, G\} = \sigma_2 \frac{\delta}{\delta \zeta} DG[\zeta](\zeta_F)$$

$$= D\sigma_2 \frac{\delta}{\delta \zeta} G[\zeta](\zeta_F) + DG[\zeta]\left(\sigma_2 \frac{\delta}{\delta \zeta} \zeta_F\right)$$

$$= D\sigma_2 \frac{\delta}{\delta \zeta} G[\zeta](\zeta_F) - D\sigma_2 \frac{\delta}{\delta \zeta} F[\zeta](\zeta_G),$$

which is an element of $\mathscr{F}$. Similarly for $k = 1, 2, \ldots$ we have:

$$D^k \sigma_2 \frac{\delta}{\delta \zeta}\{F, G\}[\zeta](\zeta_1, \zeta_2, \ldots, \zeta_k)$$

$$= D^{k+1} \sigma_2 \frac{\delta}{\delta \zeta} G[\zeta](\zeta_F, \zeta_1, \ldots, \zeta_k) -$$

$$- D^{k+1} \sigma_2 \frac{\delta}{\delta \zeta} F[\zeta]\,(\zeta_G, \zeta_1, \ldots, \zeta_k),$$

which is also an element of $\mathscr{F}$. Therefore $\{F, G\} \in \Omega$. Similarly $\{\{F, G\}, H\}$ is in Ω if $F, G, H \in \Omega$. Consequently the vector space Ω is a Lie algebra under the Poisson brackets.

We now construct two convenient carrier spaces. We take as the first carrier space a linear space of C^∞ functionals $\psi(\cdot)$ on $\mathscr{F}$ with the topology defined by the system of seminorms

$$(6.1)\quad ||\psi||_{B,m} = \sup_{\zeta \in B} \sup_{\substack{||\zeta_i||_F \leq 1 \\ i=1,\dots,m}} |D^m\psi[\zeta](\zeta_1, \dots, \zeta_m)|, \qquad m = 0, 1, \dots,$$

where B is an arbitrary bounded subset of $\mathscr{F}$. Since this space resembles the Schwartz space $\mathscr{E}(R^n)$ we shall denote it by the symbol $\mathscr{E}(\mathscr{F})$.

The second space $\mathscr{K}(\mathscr{F})$ is the linear space $\Omega \subset \mathscr{E}(\mathscr{F})$ with a topology defined by the seminorms

$$(6.2)\quad ||\psi||_{\mathscr{K}B,m} = \sup_{\zeta \in B} \sup_{\substack{||\zeta_i||_F \leq 1 \\ i=1,\dots,m}} \left\| \sigma_2 \frac{\delta}{\delta\zeta} D^m\psi[\zeta](\zeta_1, \dots, \zeta_m) \right\|_F.$$

We now give the representation of Lie algebra Ω in these spaces. We denote, for the sake of simplicity, by D_F the first-order differential operator given by the formula

$$(6.3)\quad D_F = \int_{R^3} d^3z \left(\frac{\delta F}{\delta\varphi(z)} \frac{\delta}{\delta\pi(z)} - \frac{\delta F}{\delta\pi(z)} \frac{\delta}{\delta\varphi(z)} \right).$$

THEOREM 8. Let F be in Ω. Then the operator $\hat{F}$ associated with a given functional F by the formula

$$(6.4)\quad \hat{F}[\zeta] = F[\zeta] - \tfrac{1}{2} DF[\zeta](\zeta) + iD_F$$

defines the continuous map the spaces $\mathscr{E}(\mathscr{F})$ and $\mathscr{K}(\mathscr{F})$ into itself. If $F, G \in \Omega$, then for ψ in $\mathscr{E}$ or $\mathscr{K}$ we have

$$(6.5)\quad [\hat{F}, \hat{G}]\psi = i\{\hat{F}, \hat{G}\}\psi.$$

(For the proof. cf. [1], Theorem 1.)

One readily verifies using (3.4) that if $\mathscr{F} \in \Omega$ and $\varrho(\cdot)$ is C^∞ then

$$(6.6)\quad \hat{\varrho}(F) = \varrho(F) + \varrho'(F)[\hat{F} - F].$$

The formula (6.6) implies that $\hat{F^n} \neq (\hat{F})^n$ in general. Hence the quantization formula (6.4) applied for a product Φ^n of fields gives some 'renormalization' counter-terms.

Let $\Phi[x|\varphi, \pi]$ be a solution of the dynamical equation (2.1). We begin the construction of a quantum field $\Phi(t, x)$ by quantizing first the free field $\Phi_\tau(t, x)$. Let $\hat{\Phi}_\tau(t, \alpha)$ denote the operator field obtained from

$$\Phi_\tau[t, \alpha|\varphi, \pi] = \int d^3x\alpha(x)\Phi_\tau[t, x|\varphi, \pi], \qquad \alpha \in S(R^3),$$

by formula (3.4). Then we have

THEOREM 9. The operator field $\hat{\Phi}_\tau(t, \alpha)$ for any $\tau \in (-\infty, \infty)$ and $\alpha \in S(R^3)$ is the continuous mapping of the spaces $\mathscr{E}(\mathscr{F})$ and $\mathscr{K}(\mathscr{F})$ into themselves and satisfies on each of these spaces the commutation relations:

$$(6.7) \quad [\hat{\Phi}_\tau(t, \alpha), \hat{\Phi}_\tau(r, \beta)] = i\int d^3xd^3y\alpha(x)\Delta(t-r, x-y)\beta(y).$$

The field $\hat{\Phi}_\tau(t, \alpha)$ is the strongly continuous function of τ and t.

(For the proof cf. [1], Theorem 2.)

REMARK 1. For simplicity of notation in the following we shall write formula (6.7) and similar formulae in the unsmeared form:

$$(6.8) \quad [\hat{\Phi}_\tau(x), \hat{\Phi}_\tau(y)] = i\Delta(x-y).$$

The operator field $\hat{\Phi}_\tau$ plays the basic role in the determination of the quantum evolution operation $\hat{U}(\tau, \tau_0)$ and the quantum scattering operator $\hat{S}$. These problems will be considered in Section VII.

We now describe the quantum interacting field $\hat{\Phi}$ associated with the classical field $\Phi[x, \varphi, \pi]$ by the formula (6.4).

THEOREM 10. The operator field $\hat{\Phi}(t, \alpha)$ is the continuous mapping of the spaces $\mathscr{E}(\mathscr{F})$ and $\mathscr{K}(\mathscr{F})$ into themselves and satisfies in the distribution sense on each of these spaces the commutation relations

$$(6.9) \quad [\hat{\Phi}(x), \hat{\Phi}(y)] = i\hat{\Delta}^\lambda[x, y|\Phi],$$

where $\Delta^\lambda[x, y|\Phi]$ is given by formula (2.12). The map $t \to \hat{\Phi}(t, \alpha)$ is strongly continuous.

(For the proof cf. [1], Theorem 3.)

COROLLARY 1. The field $\hat{\Phi}(x)$ is local, i.e.

$$(6.10) \quad [\hat{\Phi}(x), \hat{\Phi}(y)] = 0 \quad \text{if} \quad (x-y)^2 < 0,$$

and satisfies on $\mathscr{E}(\mathscr{F})$ or $\mathscr{K}(\mathscr{F})$ the canonical commutation relations

(6.11) $[\hat{\Phi}(t, x), \hat{\Pi}(t, y)] = i\delta^{(3)}(x-y),$

$[\hat{\Phi}(t, x), \hat{\Phi}(t, y)] = [\Pi(t, x), \Pi(t, y)] = 0$

PROOF. If $(x-y)^2 < 0$ then by formula (2.12) $\Delta^\lambda[x, y|\Phi] = 0$. Similarly if $t_x = t_y$ then $\partial_{t_y}\Delta^\lambda[x, y|\Phi] = \delta^3(x-y)$.

The formulae (6.11), (6.9) and (3.3) show that the interacting field has the same distributional character as the free field, i.e. they represent the operator-valued dostributions of $S'(R^4)$ type. Let us note, however, that by Theorem 9 $\hat{\Phi}(t, \alpha)\psi$, $\alpha \in S(R^3)$ is the continuous function of t.

One can easily verify that the regularity properties of $\Phi, \Phi_\tau, \Phi_{\text{in}}$ and Φ_{out} fields will not change if we take the initial conditions $\zeta_{\text{in}} = (\varphi_{\text{in}}, \pi_{\text{in}})$ at $t_0 = -\infty$.

(Cf. [1], Remark 1 to Lemma 5 of Appx. A.)

This implies that all assertions of Theorems 8 and 9 remain true also for this case.

We now find an equation of motion for the quantum field $\Phi(x)$. Acting on the field $\Phi(x)$ by the operator $\Box + m^2$ and using Eqs. (2.1) and (6.4) one finds that $\hat{\Phi}(x)$ satisfies the following dynamical equation:

(6.11) $(\Box + m^2)\hat{\Phi}(x) = \lambda\hat{\Phi}^3(x).$

By virtue of Eq. (6.6), the interaction term in Eq. (6.11) is automatically renormalized: consequently, Eq. (6.11) represents a meaningful equality on the space $\mathscr{K}$. It should be stressed, however, that the dynamical equation (6.11) looses its primary meaning as a tool for description of a dynamics of interacting quantum fields: in fact the quantum interacting field is not obtained by a solution of Eq. (6.11) but is constructed independently from the classical solution $\Phi(x)$ by formula (6.4).

It is instructive to apply the present quantization method in the case of the free field equation $(\Box + m^2)\Phi_0(x) = 0$. In this case the solution $\Phi_0[t, x|\varphi, \pi]$ is given by the formula

$$\Phi_0[t, x|\varphi, \pi] = -\int \Delta(t, x-y)\pi(y)d^3y + \int (\partial_t\Delta)(t, x-y)\varphi(y)d^3y.$$

Applying the formula (6.4) one obtains the quantum field $\Phi_0(x)$ which satisfies the following commutation relations:

$$[\hat{\Phi}_0(x), \hat{\Phi}_0(y)] = i\Delta(x-y).$$

Calculating in the standard manner the creation and annihilation operators, one easily verifies that the equation $\hat{\alpha}\psi_0 = 0$ is satisfied by the Poincaré invariant functional $\psi_0(\varphi, \pi) = 1$ and that the n-particle states are represented by polynomials in canonical variables. Restricting the field $\hat{\Phi}_0$ to the irreducible subspace generated from the vacuum by means of creation operators, one obtains a realization which is identical with the conventional Bargmann–Segal representation.

Similarly, the quantization (6.4) of the external field problem,

$$(\Box + m^2)\Phi(x) = V(x)\Phi(x),$$

provides by restriction to the irreducible vacuum sector of $\mathscr{E}(\mathscr{F})$ the conventional theory.

Let $\hat{\Phi}_{\text{in}}(t, \alpha)$ and $\hat{\Phi}_{\text{out}}(t, \alpha)$ be the operator fields obtained from classical solutions $\Phi_{\text{in}}[t, \alpha|\varphi, \pi]$ and $\Phi_{\text{out}}[t, \alpha|\varphi, \pi]$, respectively, by formula (6.4). Then we have

THEOREM 11. For every $\psi \in \mathscr{E}(\mathscr{F})$ or $\mathscr{K}(\mathscr{F})$ in the strong topology of these spaces we have

$$(6.12)\quad \lim_{\tau \to \mp\infty} \hat{\Phi}_\tau(t, \alpha)\psi = \hat{\Phi}_{\substack{\text{in}\\ \text{out}}}(t, \alpha)\psi.$$

The operator fields $\hat{\Phi}_{\text{in}}(t, \alpha)$ and $\hat{\Phi}_{\text{out}}(t, \alpha)$ represent the continuous mappings of the spaces $\mathscr{E}(\mathscr{F})$ and $\mathscr{K}(\mathscr{F})$ into themselves and satisfy on each of these spaces the commutation relations

$$(6.13)\quad [\hat{\Phi}_{\substack{\text{in}\\ \text{out}}}(x), \hat{\Phi}_{\substack{\text{in}\\ \text{out}}}(y)] = i\Delta(x-y).$$

(For the proof cf. [1], Theorems 4 and 5.)

We shall now consider transformation properties of Φ, Φ_{in} and Φ_{out} with respect to the Poincaré group.

Let $\varphi_{\text{in}}(x) = \Phi_{\text{in}}(0, x)$ and $\pi_{\text{in}}(x) = \Pi_{\text{in}}(0, x)$ be the initial conditions for the classical free field $\Phi_{\text{in}}(x)$. Let $\Phi_{\text{in}}(x)$ represent initial conditions at $t_0 = -\infty$ for the interacting field $\Phi(x)$ which satisfies Eq. (2.1.) The map $(a, \Lambda) \to U_{(a,\Lambda)}$ in the Banach space $\mathscr{F}$ given by the formula $(U_{(a,\Lambda)}\Phi_{\text{in}}(x) = \Phi_{\text{in}}(\Lambda^{-1}(x-a))$ defines the continuous representation of the Poincaré group in the space $\mathscr{F}$. The elements

$$(6.14\text{a})\quad (U_{(a,\Lambda)}\Phi_{\text{in}})(0, x) \quad \text{and} \quad (U_{(a,\Lambda)}\Pi_{\text{in}})(0, x)$$

define the element $\zeta_{\text{in}} = (\varphi_{\text{in}}, \pi_{\text{in}})$ after the transformation. We shall denote transformed element (6.14a) by the symbol $U_{(a,\Lambda)}\zeta_{\text{in}}$.

The map $(a, \Lambda) \to \hat{U}_{(a,\Lambda)}$ in the space $\mathscr{E}(\mathscr{F})$ given by the formula

(6.14b) $(\hat{U}_{(a,\Lambda)}\psi)(\zeta_{\text{in}}) = \psi(U^{-1}_{(a,\Lambda)}\zeta_{\text{in}})$

defines the continuous representation of the Poincaré group in $\mathscr{E}(\mathscr{F})$ or $\mathscr{K}(\mathscr{F})$.

We now show the covariance property of the quantum field Φ.

PROPOSITION 12. The field $\hat{\Phi}(x)$ has the following transformation properties relative to the representation $(a, \Lambda) \to \hat{U}_{(a,\Lambda)}$ of the Poincaré group:

(6.15) $(\hat{U}_{(a,\Lambda)}\hat{\Phi}(x)\hat{U}^{-1}_{(a,\Lambda)}\psi)(\zeta_{\text{in}}) = \hat{\Phi}(\Lambda x+a)\psi(\zeta_{\text{in}})$.

(For the proof cf. [1], Proposition 6.)

By virtue of Proposition 2.3, in the classical field theory we have $P^{\text{in}}_\mu = P_\mu = P^{\text{out}}_\mu$ and $M^{\text{in}}_{\mu\nu} = M_{\mu\nu} = M^{\text{out}}_{\mu\nu}$; hence by virtue of (6.4) we obtain

(6.16) $\hat{P}^{\text{in}}_\mu = \hat{P}_\mu = \hat{P}^{\text{out}}_\mu, \quad \hat{M}^{\text{in}}_{\mu\nu} = \hat{M}_{\mu\nu} = \hat{M}^{\text{out}}_{\mu\nu}$.

By Eqs. (6.16) and (6.4) the quantum generators $\hat{P}_\mu$ and $\hat{M}_{\mu\nu}$ are represented by the first-order differential operator only. Consequently, the vacuum state ψ_0 defined by the formula

(6.17) $\hat{P}_\mu\psi_0 = 0, \quad \hat{M}_{\mu\nu}\psi_0 = 0,$

is given in $\mathscr{K}(\mathscr{F})$ by the functional $\psi_0(\varphi, \pi) = 1$. Hence by (5.5) the interacting and the asymptotic quantum fields have the same vacuum ψ_0 in $\mathscr{K}(\mathscr{F})$. The elements of the Wightman domain given by the formula

(6.18) $\psi(f_1, \ldots, f_n) = \prod_{i=1}^{n} \hat{\Phi}(f_i)\psi_0,$

by virtue of Eq. (6.4) are represented by the sums of products of Frechet's derivatives of the classical field Φ.

VII. Quantum Scattering Operator

Let $\Phi_\tau(t, x)$ be a free classical field, whose initial data for $t = \tau$ are determined by the interacting field, i.e.

(7.1) $\Phi_\tau(\tau, x) = \Phi(\tau, x), \quad \Pi_\tau(\tau, x) = \Pi(\tau, x)$.

The time evolution of $\Phi_\tau(t, x)$ is given by the one-parameter group U_t^τ which is generated by the free Hamiltonian H^τ

(7.2) $H^\tau = \frac{1}{2}\int d^3x[\Pi_\tau^2 + |\nabla\Phi_\tau|^2 + m^2\Phi_\tau^2](t, x).$

Indeed by virtue od Eqs. (2.14) we have

(7.3) $\{\Phi_\tau, H^\tau\} = \partial_\tau\Phi_\tau.$

Let $\hat{H}^\tau$ be the quantum operator corresponding to H^τ by virtue of formula (6.4). Then we have:

PROPOSITION 7.1. The global transformation $t \to \hat{U}_t^\tau$ generated by the operator $\hat{H}^\tau$ in the carrier space $\mathscr{E}(\mathscr{F})$ is given by the formula ($\zeta = \zeta_{\text{in}}$)

(7.4) $\hat{U}_t^\tau\psi(\zeta) = \exp\left\{i\int\limits_0^t \overset{\circ}{H}(\Phi[\tau, \cdot\,|(U_{t'}^\tau)^{-1}\zeta])\, dt'\right\}\psi[(U_t^\tau)^{-1}\zeta],$

where $\overset{\circ}{H}{}^\tau = H^\tau - \frac{1}{2}DH^\tau$ and U_t^τ is the classical transformation in the space $\mathscr{F}$ generated by the hamiltonian vector field associated with H^τ.

(For the proof cf. Bałaban and Rączka [3].)

Let $\hat{\Phi}_\tau(x)$ be the quantum field associated with the classical field $\Phi_\tau(x)$. Since the map $\Phi_\tau \to \hat{\Phi}_\tau$ conserves the Lie bracket structure of the Poisson bracket Lie algebra, the field $\hat{\Phi}_\tau$ by virtue of Eq. (6.8) is a free local relativistic quantum field. In particular, by virtue of Eq. (7.3) the time evolution of $\hat{\Phi}_\tau$ is given by the operators $\hat{U}_t^\tau$ defined by Eq. (7.4)

(7.5) $(\hat{U}_{t'}^\tau\hat{\Phi}_\tau(\hat{U}_{t'}^\tau)^{-1})(t, x) = \hat{\Phi}_\tau(t+t', x).$

PROPOSITION 13. The fields $\hat{\Phi}_\tau(t, x)$ and $\hat{\Phi}_{\tau_0}(t, x)$ are connected by the transformation $\hat{V}(\tau, \tau_0)$, i.e.

(7.6) $\hat{\Phi}_\tau(t, x) = \hat{V}(\tau, \tau_0)\hat{\Phi}_{\tau_0}(t, x)\hat{V}^{-1}(\tau, \tau_0),$

given by the formula

(7.7) $\hat{V}(\tau, \tau_0) = \hat{U}^{\text{in}}_{(\tau-\tau_0)}\hat{U}^{\tau_0}_{(\tau_0-\tau)} = \hat{U}^{\tau}_{(\tau-\tau_0)}\hat{U}^{\text{in}}_{(\tau_0-\tau)}.$

The operator $\hat{U}(\tau, \tau_0) \equiv \hat{V}(\tau_0, \tau)$ satisfies the following equation:

(7.8) $\partial_\tau\hat{U}(\tau, \tau_0) = -i\hat{H}_{\text{int}}[\Phi_{\tau_0}(\tau, \cdot)]\hat{U}(\tau, \tau_0),$

where

(7.9) $\hat{H}_{\text{int}}[\Phi_{\tau_0}(\tau, \cdot)] = \frac{\lambda}{4}\int d^3x\hat{\Phi}_{\tau_0}^4(\tau, x).$

Eqs. (7.6)–(7.8) hold in the sense of strong operator topology in $\mathscr{E}(\mathscr{F})$.

(For the proof cf. Bałaban and Rączka [3].)

Equation (7.8) is the evolution equation for the evolution operator $\hat{U}(\tau, \tau_0)$, which in conventional quantum field theory is formally derived by the passage to the 'interaction picture'. It is usually solved by a formal construction of Liouville–Neumann perturbation series, which in four-dimensional space-time is divergent. In the present approach the action of the evolution operator $\hat{U}(\tau, \tau_0)$ in the carrier space $\mathscr{E}(\mathscr{F})$ can be explicitly calculated. Indeed, using the fact that the evolution operator is given as the product of two one-parameter groups of time translation $\hat{U}^{\text{in}}_{(\tau-\tau_0)}$ and $\hat{U}^{\tau_0}_{(\tau_0-\tau)}$, by wirtue of Eqs. (7.7) and (6.14b) one obtains

$$(7.10)\quad \hat{U}(\tau, \tau_0)\psi(\zeta) = \exp\left\{i \int_0^{\tau-\tau^0} \hat{\mathring{H}}(\Phi[\tau_0, |(U^{\tau}_{t'})^{-1}\zeta])\,dt'\right\} \psi[U^{-1}(\tau, \tau_0)\zeta].$$

We derive now the action in the carrier space of the quantum scattering operator. This operator is defined in the space $\mathscr{E}(\mathscr{F})$ by the formula

$$(7.11)\quad \hat{S} = \lim_{\substack{\tau\to\infty \\ \tau_0\to-\infty}} \hat{U}(\tau, \tau_0).$$

THEOREM 14. The quantum scattering operator $\hat{S}$ is given by the formula:

$$(7.12)\quad (\hat{S}\psi)(\zeta) = \psi(S^{-1}\zeta),$$

where S is the classical scattering operator. The operator $\hat{S}$ is invariant under the action of the Poincaré group and satisfies the condition

$$(7.13)\quad \hat{S}^{-1}\Phi_{\text{in}}\hat{S} = \hat{\Phi}_{\text{out}}.$$

(For the proof cf. Bałaban and Rączka [3].)

It follows from formula (5.2) that S is non-trivial in $\lambda\Phi^4$ theory. This implies that the quantum scattering operator (7.12) is also non-trivial.

It was shown is Section III that the classical scattering operator is non-analytic in λ for all initial data from $\mathscr{F}$. Hence by virtue of formula (7.12) the quantum scattering operator is also non-analytic.

VIII. Discussion

A. We formulated the classical theory in the language of Lie algebra, whose commutators are defined in terms of Poisson brackets. In Sec. VI we construct an operator representation of this Lie algebra. We obtain in this manner an interacting, local, relativistic quantum field $\hat{\Phi}(x)$ which satisfies the asymptotic conditions.

B. The canonical formalism discussed in the present paper may be extended to a class of nonpolynomial interactions where $F(\cdot)$ satisfies the conditions

(i) $F(\cdot)$ is an odd analytic function,

(ii) $F'(0) = 0$,

(iii) $|F(u)u^{-5}| \to 0$ as $|u| \to \infty$.

The extension of the present results may be proven by using in the proofs of Theorems 2 and 3 the corresponding results for a classical nonlinear relativistic wave equation with an analytic nonlinear term [4].

Acknowledgments

The authors would like to thank Professors I. Białynicki-Birula, M. Flato R. Streater and S. Woronowicz for useful discussions and valuable suggestions. The authors are particularly grateful to Professor I. E. Segal for the illuminating discussions of problems of quantization of nonlinear relativistic classical field theory.

References

* Revised version January 1976.

[1] Sections I, II, and IV represent part of our work [2]. Section III contains a new derivation of commutation relations for interacting fields. Section V is based on Moravetz and Strauss work [5] and the unpublished results of Rączka and Strauss [7]. Section VI contains review of results of work [1] and Section VII is based on our work [3].

[2] See Wilson [13]; cf. also Wilson and Kogut [14], where the triviality of renormalized $\lambda\Phi_4^4$ theory for small bare coupling constant is proven.

Bibliography

[1] T. Bałaban, K. Jezuita and R. Rączka, 'Second Quantization of Classical Nonlinear Relativistic Field Theory', Part II: 'Construction of Relativistic Interacting Local Quantum Field', *Comm. Math. Phys.* (1976) (in print).

[2] T. Bałaban and R. Rączka, 'Second Quantization of Classical Nonlinear Relativistic Field Theory', Part I: 'Canonical Formalism', *J. Math. Phys.* **16** (1975) 1475–81.

[3] T. Bałaban and R. Rączka, 'Second Quantization of Classical Nonlinear Relativistic Field Theory', Part III: 'Construction of Quantum Scattering Operator, preprint, Institute for Nuclear Research, 1976.

[4] C. S. Moravetz and W. A. Strauss, *Comm. Pure and Applied Math.* **25** (1972) 1.

[5] C. S. Moravetz and W. A. Strauss, 'On Nonlinear Scattering Operator', *Comm. Pure and Applied Math.* **XXVI** (1973) 47–54.

[6] R. Rączka, 'The Construction of Solution of Nonlinear Relativistic Wave Equation in $\lambda{:}\Phi_4^4{:}$ Theory', *J. Math. Phys.* **15** (1974).

[7] R. Rączka and W. A. Strauss, 'On Analyticity of Solutions of Nonlinear Relativistic Wave Equations in Coupling Constant and Initial Data (in preparation).

[8] I. E. Segal, *J. Math. Phys.* **1** (1960) 468.

[9] I. E. Segal, *J. Math. Phys.* **5** (1964) 269.

[10] I. E. Segal, *Journ. de Math.* **44** (1965), Part I: 71–105, Part II: 107–132.

[11] I. E. Segal, 'La varieté des solutions d'une équation hyperbolique, nonlineaire a ordre 2', III, Lecture Notes College de France, 1964–5.

[12] R. F. Streater, *Comm. Math. Phys.* **2** (1966) 354.

[13] K. G. Wilson, *Phys. Rev.* **D7** (1973) 2911.

[14] K. G. Wilson and J. Kogut, *Phys. Reports* **12C** (1974), Sec. XIII.

PART THREE

GROUP REPRESENTATION IN QUANTUM THEORY

THEORY OF ANALYTIC VECTORS AND APPLICATIONS

M. FLATO

Physique Mathématique, Collège de France, Paris, France and Physique Mathématiqua, Université de Dijon, France

I. Introduction – Physical and Mathematical

When a mathematical-physicist treats a new mathematical formalism, he always has in mind some a priori possible physical applications. As a matter of fact it is in most of the cases the physical intuition which helps him in conjecturing a mathematical theorem – and the mathematical development of the latter which brings new light to the physical situation considered.

The most obvious physical notion connected with the so-called analytic vectors is the notion of an observable in quantum theories. Whether one sticks to the elementary formulations of quantum theories, or whether one prefers more sophisticated formulations, one always makes an a priori assumption concerning the algebra of observables of a given physical situation.

This hypothesis however, contains an important part of the dynamics of the system.

As a simple example of the last statement one can take a Schrödinger equation and define a reasonable C^*-algebra of observables in such a way that the one-parameter group of the time-evolution of the system will be an $*$-automorphism of the algebra only in the trivial case (when the potential is zero).

Also in Wightman quantum field theory, say of a scalar neutral field, the problem of whether the field operators are self-adjoint and therefore candidates for observables (without any additional hypothesis concerning the vacuum!) does not seem to be solved.

It is quite evident that if we do not specify our dynamical hypothesis, we can at best give only necessary conditions for an operator to represent an observable. From the beginning of quantum mechanics it was postulated that an observable should be represented by a self-adjoint operator on a Hilbert space.

The reasons were the following: 1) One wanted to have real eigenvalues and spectral resolution for an obvious physical interpretation, and 2) one wanted the one-parameter group generated by the observable to be represented unitarily on the Hilbert space of states so as to be defined on the whole Hilbert space and to conserve the transition amplitudes.

It has to be mentioned that e.g. for decaying particles it is not at all clear whether all observables (like the Hamiltonian) should really be represented by self-adjoint operators.

If one sticks to observables which are represented by self-adjoint operators (which one usually does!) it was only in the fifties that one remarked that not every self-adjoint operator on the Hilbert space of states corresponds to an observable – this fact being due to the so-called superselection rules.

What is a superselection rule? Similar to a selection rule from atomic spectroscopy, which commutes with a given Hamiltonian, a superselection rule is represented by an operator commuting with *all* observables.

The electric charge operator (as well as the baryon charge, lepton charge, etc...) seems to be a superselection operator. In such a case the Hilbert space of states decomposes into a direct sum (direct integral if the group generated by the superselection operator is non-compact, like in the case of time in non-relativistic quantum mechanics) of coherent subspaces (or sectors) each carrying eigenstates of charge with the *same* eigenvalue.

A state which is a non trivial linear superposition of states in *different* sectors is not physically realizable – though it can be in some cases of non-exact symmetries considered as a mixed state.

Whether charge, lepton number, etc... is or is not a superselection is a question which has to be decided only by experiment!

In any case, since non trivial superselection rules seem to exist, it follows that not every self-adjoint operator on the Hilbert space of states can represent an observable, since the family of all self-adjoint operators is Schur-irreducible.

In what follows we shall not consider superselection rules, only for the sake of simplicity.

One of the fundamental problems of quantum mechanics which was already raised in von Neuman's demonstration of non-existence of hidden variables in quantum mechanics is the following: If A is an observable, namely A is represented by a self-adjoint operator in a Hilbert space and if B is also an observable, not necessarily commuting with A, give at least sufficient conditions such that $\alpha A+\beta B$ will also be an observable for every real α and β.

The same question arises in perturbation theory when we would like to know under which conditions one can assure that given H_0 and H_{int} both essentially self-adjoint on a given domain, $H_0+\lambda H_{\text{int}}$ will also be essentially self-adjoint, at least for small λ.

A closely related question is the following: We know that if we are given a strongly continuous unitary representation of a symmetry group in a second quantized theory (say a Lagrangian theory) we know how to calculate the quantum observables which are conserved in time. Suppose now that we have the inverse problem: Given a family of observables which form a basis of a representation of a finite (or infinite) dimensional real Lie algebra, under what conditions do they come from a unitary representation of a symmetry group?

All these questions will at least be partially answered in this lecture.

Talking about applications of group theory in physics, one should distinguish between symmetry groups coming from geometrical origin and other types of groups like dynamical groups (spectral unifications) internal 'symmetry' groups, classification groups in spectroscopy, etc... A symmetry group which has a geometrical meaning (think of the Poincaré group, for instance) should manifest its global group structure in the physical theory. This means that for not too small distances one has to utilize in relativistic theories finite group transformations, group representations, etc... One should however notice that in particle physics it might very well be that the group aspect is lost and only the Lie algebra aspect remains – especially in high energy strong interactions. After all, we know already from weak interactions that parity as well as even the product of parity by charge conjugation are not conserved in some weak processes.

As to internal 'symmetry' groups (think of SU(3)), dynamical groups, etc... there is no solid reason to impose that the group structure would manifest itself in the theory, and one can very well be satisfied with the Lie algebra aspect – gaining as a supplementary advantage a much richer family of irreducible representations.

As a simple example think of the canonical $(2n+1)$-dimensional nilpotent Lie algebra: $(p_i, q_j) = \delta_{ij}I$ $(i, j = 1, \ldots, n)$, other commutators of which vanish. This Lie algebra, which lies in the foundations of non-relativistic quantum mechanics, has a rich family of irreducible representations. Among them there exist also representations such that every element of the Lie algebra is represented by an essentially skew-adjoint operator, with interesting spectrum, and yet the representation is not integrable to a unitary representation of a group.

It is *only* when passing to a unitary group representation (what is sometimes called the Weyl-trick) that von Neumann was able to prove his famous unicity theorem stating that all irreducible strongly continuous unitary representations of the canonical finite dimensional nilpotent Lie group are physically equivalent (namely projectively equivalent, modulo the choice of the scalar operator representing its center) to the usual Schrödinger representation.

But this restriction to unitary group representations is rather artificial (in this case) and physical examples are known which utilize non-integrable representations of this algebra.

Let us mention briefly some of the mathematical and physical applications of the integrability criteria and of the analyticity (separate and joint) properties to be reviewed later. First of all J. Kisyński from Warsaw applied our criteria to the case of the symplectic group and its applications in quantum mechanics. He also gave a criterion of integrability in the case of Banach space, similar to that given earlier by J. Simon in the case of reflexive Banach space.

Niederle and Kotecky from Prague utilized the integrability criteria to prove that representations, calculated earlier by the Russian school, of the Lie algebras su(p, q) and so(p, q) were integrable. This result from the so(p, q) case was also obtained independently by J. M. Maillard from France.

It will be too long to quote integrability results in particular cases obtained already or in process of preparation, so we limit ourselves to the examples mentioned above.

H. Snellman from Stockholm (and also later in a shorter demonstration B. Nagel and H. Snellman) proved that in Wightman's field theory, in the case that the test functions space is the Schwartz space S, the domain D_0 obtained by the action of the polynomial ring in the field operators on the vacuum contains a dense set of analytic vectors for the Poincaré group. Snellman's demonstration utilizes the conditions which ensure in the unitary case that separate analyticity of a vector for a certain Lie algebra basis implies joint analyticity.

A last we end our introduction with some necessary definitions: A representation of a Lie group G on a topological vector space H will be said continuous if the mapping from $G \times H$ onto H associating to the couple (g, h) the vector $\tau(g) \cdot h$ ($g \in G$, $h \in H$, $\tau(g)$ being the representative of g in the representation τ) is continuous on $G \times H$.

We'll suppose all our representations to be continuous.

A vector $h \in H$ will be called differentiable or analytic if the mapping from G to H which associates to every $g \in G$ the vector $\tau(g) \cdot h$ is so.

Evidently the definitions of an entire vector, of a vector from the class C^k, etc.... are similar.

We can now pass to the second part of the lecture.

II. The Case of One-Parameter Groups

To begin with, we shall give some equivalent definitions for the notion of analytic vectors in the case of one-parameter groups. Our first definition is the specification of the definition we saw above: If the mapping from the real line $\boldsymbol{R}$ to the topological vector space H defined by: $t \to V(t)\varphi$ is analytic in t (here $t \in \boldsymbol{R}$ and $\varphi \in H$) we say that φ is an analytic vector for the one-parameter group $V(t)$.

We suppose from now on that H is a Banach space. If this is the case one has a very simple chacterization of analytic vectors. If A is an operator in H, $\varphi \in H$ is an analytic vector for A (here think of A as the generator of a one-parameter group) iff (if and only if) the series $\sum_{n=0}^{\infty} \frac{t^n}{n!} \|A^n\varphi\| < \infty$ for some $t > 0$. An analogue of the Cauchy estimate gives an equivalent condition for the case of Banach space: φ is analytic for A iff we have $\|A^n\varphi\| \leqslant \lambda^n n!$ for some positive λ.

Another condition can be obtained as follows: let A be a self-adjoint operator, and look at the Schrödinger equation: $i \frac{\partial\psi}{\partial t} = A\psi$. Evidently every ψ_0 in the Hilbert space of initial data can develop into a solution of the equation under the action of the unitary one-parameter group generated by $A: \psi_t = e^{-iAt} \cdot \psi_0$.

Now suppose we transform $t \to it$ or $t \to -it$. The Schrödinger equation will then go to the corresponding heat equations: $\pm \frac{\partial \psi_t}{\partial t} = A\psi_t$. It is evident that if ψ_0 is simultaneously in the domains of self-adjoint operators $e^{\pm tA}$ for some $t > 0$ then $e^{\pm tA}\psi_0$ supply solutions of the equations for small t. What can be trivially proved is also that under the above condition (namely $\psi_0 \in \operatorname{Dom} e^{tA} \cap \operatorname{Dom} e^{-tA}$) ψ_0 will be an *analytic vector* for A. (This condition is a necessary and sufficient condition!)

A more abstract characterization can be given by looking at states on Lie algebras of observables: Let A be a self-adjoint operator. Let also $\{\mathscr{P}(A)\}$ denote the polynomial algebra on A. On $\{\mathscr{P}(A)\}$ we introduce a topology defined by the family of semi-norms $\mathscr{P}_C(A^n) = C^n n!$ with all positive C's. Evidently an *analytic form* on this algebra will be by definition a continuous linear form on the algebra with the above-defined topology. One can study such structures of analytic states (and forms) on any (finite dimensional) Lie algebra and get probably interesting results. However we shall not develop this aspect of the problem in our lecture.

In order to get more intuitive feeling for analytic vectors, we shall introduce the notion of f-regularity: Let H be a Hilbert space, A a self-adjoint operator in H. $\varphi \in H$ will be said f-regular for A, if φ belongs to the domain of $f(A)$ for a given real-valued function f.

Evidently the more rapidly f increases at infinity, the more regular vector has φ to be for the operator A.

EXAMPLES. If $f(X) = \sum_{K=1}^{N} a_K X^K (a_N \neq 0)$ then φ is f-regular for A means that it is of class C^N for the self-adjoint operator A.

If $f(X)$ is any polynomial then φ f-regular is a differentiable vector (class C^∞).

If $f_t(X)$ is the series $\sum_{n=0}^{\infty} \frac{t^n}{n!} X^n$, φ is f_t-regular for a given $t > 0$ means that φ is analytic for A. φ will be entire if φ is f_t-regular for all t.

For a finite linear combination of eigenvectors of A, or for a vector in the closed subspace corresponding to a projector on a compact interval of the continuous spectrum f can be chosen, e. g. as any continuous function. (In this case f can be chosen even as any Borel function.) One can prove that if e.g. A has a pure discrete spectrum, then if φ is f-regular for all continuous functions f, φ is a linear combination of eigenvectors belonging to a compact interval in the spectrum of eigenvalues of A.

One can of course develop this point of view in great detail and obtain many results. We shall, however, not do it here.

Before passing to some classical elementary theorems for the case of one-parameter groups, we shall make a short physical remark: From the different criteria we met for a vector $\varphi \in H$ to be an analytic vector for a self-adjoint operator A follows the following observation: Evidently every eigenfunction of A, or finite combinations of such eigenfunctions, are analytic vectors for A.

In general an analytic vector will be a kind of obvious generalization of the notion of eigenstate. It will be in a way a kind of rapidly decreasing at infinity (regular) combination of eingestates. It will, of course, be very interesting to characterize in more detail the analytic vectors as 'analytic states' from the various points of view mentioned above. Also the connection with the moment problem for observables in analytic states will have to be clarified.

We are now in a position to quote three classical theorems useful for our purposes, concerning one-parameter groups. The most well-known is of course the *Stone theorem*:

THEOREM. Let H be a Hilbert space and let A be a linear operator in it. Then A is self-adjoint if and only if there exists a one-parameter group e^{itA} $(t \in \boldsymbol{R})$ unitary and strongly continuous.

Another well-known theorem, usually referred to as Nelson's theorem is the following one:

THEOREM. A closed symmetric operator is self-adjoint iff it has a dense set of analytic vectors.

This last theorem is very useful in connection with the integrability criterion of Nelson to be mentioned later.

Another interesting result, known for the case of Banach space, is the following:

THEOREM. Let H be a Banach space, D a dense domain in it. Let $V(t)$ be a one-parameter group in H, A its generator, and A_D the restriction of A on D. Suppose now that D is invariant under A_D as well as under the one-parameter group $V(t)$. Then $\overline{A_D} = A$, where by $\overline{A_D}$ we denote the closure of A_D. In the particular case of interest in which H is a Hilbert space and A_D a symmetric operator, this theorem was formulated by Nelson:

THEOREM. H is a Hilbert space, D a dense domain. A is a symmetric operator leaving D invariant. If the unitary one-parameter group exists (A being then the restriction of its generator on D) and leaves D invariant, then A is an essentially self-adjoint operator on D.

We shall now pass to the third part of our lecture.

III. The Case of Finite-Dimensional Real Lie Algebras

There are two interesting aspects in the theory of analytic vectors for finite-dimensional real Lie algebras: the global aspect and the local aspect. In the global aspect we are given a group representation on a Banach space and are asked whether there exists a dense set of analytic vectors for the given representation. On the other hand in the local aspect we begin with a certain type of a Lie algebra representation and then try to find whether this representation is exponentiable to a Lie group representation, and study some related questions.

Though the local aspect utilizes mostly analytic vectors, the global aspect (which will not be developed here in great detail) treats both differentiable and analytic vectors.

So for the sake of completeness we'll say some words about the global aspect of differentiable vectors.

The fundamental theorem here for the case of Banach space representations is due to L. Gårding (later it was extended to more general spaces by F. Bruhat).

GÅRDING THEOREM. Let H be a Banach space, G a Lie group, $\tau(G)$ a continuous representation of G on H. Let $D(G)$ be the space of C^∞ functions with compact support on the group. We now construct the domain $X \subset H$ of vectors $\int_G d\mu(g)\varphi(g)\tau(g)\psi$, for all $\psi \in H$ and $\varphi \in D(G)$, $d\mu(g)$ being the Haar measure on the group. Then the following holds:

1) X is dense in H.

2) X is a domain of differentiable vectors for $\tau(G)$, invariant also under $\tau(G)$.

3) In the particular case when H is a Hilbert space and $\tau(G)$ is unitary, let $X_1, \ldots, X_n$ be the operators on X representing any linear basis of the Lie algebra obtained by differentiating $\tau(G)$ on X. Define on X the Laplace operator $\Delta = X_1^2 + X_2^2 + \ldots + X_n^2$. Then Δ is essentially self-adjoint on X.[1]

Application. Let $\frac{\partial}{\partial X}$, iX, $i\lambda X^2$, iI be a family of skew-symmetric bases of a Lie algebra on $L^2(-\infty, \infty)$ for all $\lambda \in \boldsymbol{R}$. (Evidently the Lie algebra is nilpotent and four-dimensional with the obvious commutation relations). Let U be a strongly continuous unitary irreducible representation of the corresponding Lie group, having the above mentioned bases as bases of the four-dimensional Lie algebra.[2] On the Gårding domain for the representation U in L^2 $(-\infty, \infty)$ the Laplace operator $\Delta = \frac{\partial^2}{\partial X^2} - X^2 - \lambda^2 X^4 - I$ is essentially self-adjoint. But this means that on this domain $-\Delta - I$, *which is the Hamiltonian of the anharmonic oscillator*, is essentially self-adjoint.

We can now pass to the global aspect of analytic vectors. Here we have to limit ourselves to Banach spaces not only for the sake of simplicity but because the density result simply does not work in more general spaces, as one can immediately see from simple examples.

Thanks to the works of Harish–Chandra, Cartier and Dixmier, Nelson, and finally Gårding we know that the following is true:

THEOREM. In a continuous Banach space representation of a real Lie group the analytic vectors form a dense subspace.

Let us pass now to the local aspect of the theory of analytic vectors. To begin with, we should notice that for the case of finite-dimensional real Lie algebras, several inequivalent notions of analyticity exist, and to avoid confusion we shall introduce these notions from the beginning (notion (a) will be the strongest, (c) the weakest).

(a) A vector $\varphi \in H$ is said to be analytic for the whole Lie algebra if for some $t > 0$ and some linear basis $\{X_1, \ldots, X_n\}$ of the Lie algebra, the series $\sum_{n=0}^{\infty} \frac{1}{n!} \sum_{1 \leqslant i_1, \ldots, i_n \leqslant n} t^n X_{i_1} \ldots X_{i_n} \varphi$ is absolutely convergent. In the most important case of interest, namely when H is a Banach space, this condition is equivalent to $||X_{i_1} \ldots X_{i_n}\varphi|| \leqslant C^n \cdot n!$ for some constant $C > 0$.

(b) A vector $\varphi \in H$ is analytic for every element in the Lie algebra (in the sense of analyticity for a single operator as defined before).

(c) A vector $\varphi \in H$ is analytic for every element X_i in a given linear basis of the Lie algebra.

Our main problem of interest is the integrability problem, so we introduce now the following definition:

We say that a real Lie algebra representation is *integrable*, if there exists a continuous representation of the connected and simply connected real Lie group corresponding to our Lie algebra such that the Lie algebra representation is contained in the differential of the Lie group representation.

One should remark that in the case of a strongly continuous unitary representation on a Hilbert space the global density theorem of analytic vectors mentioned above means the existence of a dense set of analytic vectors in the *strongest sense* (a).

This last theorem has a converse discussed by Nelson and completed by Goodman using the Campbell–Hausdorff–Dynkin formula:

Suppose we have a representation on a dense invariant domain D of a real finite-dimensional Lie algebra by skew-symmetric operators. Suppose moreover that in D there exists a dense set of analytic vectors in the strongest sense (a). Then the Lie algebra representation is integrable to a unitary (strongly continuous) group representation.

It was also shown by Nelson that analytic vectors for the Laplace operator $\Delta = X_1^2 + \ldots + X_n^2$ relative to some basis are analytic for the representation. More precisely it was shown that if we have a Lie algebra representation on a dense invariant domain D (by skew-symmetric operators) and if $\varphi \in D$ is an analytic vector for the Laplace operator Δ, then φ is an analytic vector for the whole Lie algebra (sense (a)).

Combining what was said up to now with what we already saw in the one-parameter case, and showing that in the case when Δ is essentially self-adjoint one can extend the Lie algebra representation to the domain $\bigcap_n D(\bar{\Delta}^n)$, where $\bar{\Delta}$ is the closure of Δ and $D(\bar{\Delta}^n)$ is the domain of $\bar{\Delta}^n$, one gets the famous Nelson integrability criterion.

THEOREM. Let T be a representation of a real finite-dimensional Lie algebra on a Hilbert space H by skew-symmetric operators on a dense invariant domain D. Let the Laplace operator Δ relative to some basis be essentially self-adjoint on D. Then T is integrable to a unitary group representation.

This criterion is not always very practical (especially for higher dimensional groups) and not general enough. In particular it is related, as explained before, to the *strongest* notion of analyticity (a).

Intuitively speaking one would have liked to have a more general integrability criterion based on the *weakest* form of analyticity (c).

This result is indeed achieved in [3].

In this work it was proved that under quite general conditions it is enough to have a common invariant dense set of analytic vectors only for a basis of the algebra (sense (c)) to ensure the integrability of the a priori given Lie algebra representation. We shall formulate the result of [3] only for the case of a skew-adjoint Lie algebra representation in a Hilbert space.

The reasons are the following: a) Simplicity and usefulness. b) Though a theorem in [3] was formulated for much more general topological vector spaces, one has of course to put more hypotheses in the general case, and above all to suppose the integrability of the one-dimensional Lie algebras corresponding to a given basis of the Lie algebra.

In the Hilbert space case, this is already ensured in the skew-adjoint case by the theorems we saw for one-parameter groups.

We now formulate the main result of [3]:

THEOREM. Let T be a representation of a real finite-dimensional Lie algebra on a dense invariant domain of vectors, analytic for the skew-symmetric representatives of a linear basis of the algebra, in a complex Hilbert space (weak analyticity in the sense (c)). Then T is integrable to a unique unitary group representation. (The group being connected and simply connected).

To prove such a theorem one needs to rely more heavily on the abstract group structure, rather than utilizing (like in Nelson's theorem) the Campbell–Hausdorff–Dynkin formula. After some structural preliminaries, one takes an algebraic automorphism formula obtained in the preliminaries and topologizes it using duality. One then gets the result after some developments, by utilizing the vector theory of differential equations.

One should stress already at this point that while in the case of Nelson's and Goodman's result for the unitary Hilbert case, an analytic vector for Δ was necessarily analytic for the whole algebra (sense (a)) the situation in our theorem is not similar: we only know from what was said up to now that the existence of a dense invariant set of analytic vectors for the basis implies integrability and therefore by the global theorem implies the existence of (another) dense invariant set of analytic vectors for the whole algebra (sense (a)), which means analytic for the group.

This is about the best we can say on the problem of weak analyticity (sense (c)) implying strong analyticity (sense (a)) in the a priori Lie algebra representation case.

We shall come to this problem later, when we'll assume a *group representation* and ask whether a *given vector* analytic in the sense (c) is necessarily analytic in the sense (a).

In any case, in the weaker statement that a *dense domain* of weak analyticity implies the existence of a *dense domain* of strong analyticity, we find the role played by the real Lie group structure in the integrable case. We can now come back to physical interpretation and especially to the problem of observables. We suppose that our observables are self-adjoint operators and that no non-trivial superselection rules exist.

For an abelian Lie algebra of observables (maximal set of commuting observables) the problem is simple and the answer is known almost from the beginning of quantum mechanics: If X and Y are two observables which commute strongly (their corresponding spectral resolutions commute) then $\alpha X+\beta Y$ are also observables on the intersection domain for all $\alpha, \beta \in \boldsymbol{R}$.

This is the well-known case of *simultaneously-measurable* observables, or what one also calls compatible measurements.

Our theorem can be interpreted as a kind of non-abelian generalization of the latter. We know that in the abelian case (and only in this case) one can have common eigenstates for the whole Lie algebra of observables. In the non-abelian case (like in the case of the Heisenberg algebra) there are 3 interesting possibilities:

(1) We have a Lie algebra representation such that a given basis is represented by essentially skew-adjoint operators on the representation domain but not *all* elements of the Lie algebra are essentially skew-adjoint on that domain.

(2) All elements of the Lie algebra are essentially skew-adjoint, but the representation is not integrable.

(3) The representation is integrable to a unitary one (we'll see later that even in the simplest cases there exist representations of class (2) which are not of class (3)).

We have already seen in the case of one-parameter groups what was the meaning of analytic vectors: they were in a way *close to eigenstates* from several points of view. For every observable we already know that there exists a dense set of analytic vectors. We also know that unless the Lie algebra of observables is abelian, one cannot have common eigenstates for all observables.

We can now analyze the physical meaning of classes (1), (2), and (3). If we have a physical situation in which only the given basis of the Lie algebra has a physical meaning as observables and is measurable by physical experiments, we do not need more than class (1) representations. Even if the closure of all elements in the Lie algebra have to represent observables, one can still utilize representations of class (2) that need not be of class (3).

But in *traditional* quantum mechanics (this goes also to commutation relations representations in quantum field theory) one utilizes unitary group representations, which means class (3) representations.

Though we do not think that this is entirely needed in all situations, we can now interpret the meaning of the utilization of such representations: we begin with a finite-dimensional Lie algebra of skew-symmetric operators such that the basis has a common invariant dense set of analytic vectors, which means that a priori only the basis corresponds to observables. These observables cannot have common eigenstates in general but they can have common analytic vectors which are just common 'close to eigenstates' vectors. Under such condition we propose to call the observables *simultaneously observable*. Our theorem now guarantees that the representation is integrable, which means that we can pass from the basic observables to a *unitary* representation of a *group*, and therefore we can find a domain in which all elements are essentially skew-adjoint – which means that all the Lie algebra elements (after closure and multiplication by i) are observables.

The result of [3] was generalized in [4].

As before, we shall limit ourselves to the unitary Hilbert case, which is of course the most interesting case for applications.

Relying on techniques developed in [3], the following theorem was shown in [4].

THEOREM. Let T be a skew-symmetric representation of a real finite-dimensional Lie algebra g defined on a dense domain D in a Hilbert space H, invariant under $T(g)$. Suppose that there exists a set of *Lie generators* $\{X_1, \ldots, X_n\}$ (this means that g is generated by $\{X_1, \ldots, X_n\}$ by linear combinations of repeated commutators) of g such that D is a domain of analytic vectors for the operators $X_i = T(X_i)$ $(1 \leqslant i \leqslant n)$. Then T is the differential (on D) of a unique unitary representation of the corresponding connected and simply-connected Lie group G on H. In other words, what was proved in [3] for a linear basis is proved in [4] for even a smaller set: a set of Lie generators.

As a trivial consequence of this theorem, one finds the following spectral lemma, which has some physical applications.

LEMMA. Let X_1 and X_2 be two skew-adjoint operators in a Hilbert space H. Suppose moreover that there exists a common dense domain D of analytic vectors for X_1 and X_2 invariant under both of them. If the restrictions of X_1 and X_2 on D generate (by commutators and linear combinations) a finite-dimensional real Lie algebra of operators on D, then X_1+X_2 (defined on $D(X_1) \cap D(X_2)$) is an essentially skew-adjoint operator.

Another question, which we mentioned already before, is the following: Given now a priori a unitary representation of a real finite-dimensional Lie group on a Hilbert space, under what conditions can one say that a *given vector* which is analytic for a *given linear basis* of the corresponding Lie algebra representation (sense (c) above) will be analytic for the whole Lie algebra? (sense (a) above which means here analytic for the group representation).

In [1] we show that there exists at least one linear basis (actually infinitely many bases) for every finite-dimensional real Lie algebra (in the

Hilbert space group representation case) for which every vector analytic separately for this basis will be jointly (for the whole Lie algebra) analytic. Let us now analyze in more detail the main results of [1]. In the first part of this work a result concerning the stability of separate analyticity under the Lie algebra action is obtained.

LEMMA. If T is a representation of a real finite-dimensional Lie algebra g by skew-symmetric operators defined on an invariant domain D in a Hilbert space H, then the set of all analytic vectors for a given arbitrary element of the algebra g is invariant under $T(g)$.

As an immediate conclusion of this lemma one gets a theorem which generalizes the main results of [3] and [4].

THEOREM. Under the hypothesis of the last lemma, suppose moreover that $\{X_1, \ldots, X_n\}$ is a set of *Lie generators* of the Lie algebra g, and that A denotes a set of analytic vectors for $T(X_1), \ldots, T(X_n)$ separately. Then there exists a unique unitary representation of the corresponding connected and simply-connected Lie group (having g as its Lie algebra) on the closure of the smallest set A' containing A and invariant under $T(g)$, the differential of which on A' is equal to T.

This means that we even do not have to suppose the invariance of the set of analytic vectors under $T(g)$.

In the second part of [1] the following results are obtained: At first one proves that given any representation of a compact Lie group then any vector analytic for the closure of the representatives of a given arbitrary linear basis of the corresponding Lie algebra is analytic for the *whole Lie algebra*.

One then proves (utilizing a theorem of F. Browder) that if the Lie algebra g is a *unification* of the Lie algebras g_1 and g_2 (not necessarily without intersection) then given any representation of the corresponding Lie group G on a Banach space, any vector which is analytic for the whole subalgebra g_1 and also for the whole subalgebra g_2 is analytic for the whole Lie algebra g. From this one deduces that in the case of solvable groups every *Jordan basis* is a good basis for our purpose. In other words for every solvable Lie group representation in a Banach space, separate analyticity of a vector relative to a Jordan basis of the Lie algebra implies joint analyticity. (This particular result was first obtained by Goodman.)

Connecting what was said until now, one gets the main result of the second part of [1].

THEOREM. Let G be a real finite-dimensional Lie group. Then there exists a basis $\{X_1, \ldots, X_n\}$ of the corresponding Lie algebra g such that given any representation of G on a Hilbert space, any vector analytic separately for the closures of the representatives of the basis $\{X_1, \ldots, X_n\}$ will be analytic for the *group representation* (which means analytic for the whole Lie algebra, namely jointly analytic).

For a split semi-simple Lie algebra one can construct a very useful family of bases of this type: these are the Cartan–Weyl bases. We now come to another question: Are analytic vectors really *necessary* in order to ensure integrability?

From an example of Nelson one already knows that one can find a two-dimensional abelian Lie algebra and a representation of this real Lie algebra in a Hilbert space such that *every element* of the algebra is essentially skew-adjoint on a common invariant dense domain in the space but the representation is *not integrable.*

The reason is the lack of common analytic vectors for any given two-dimensional linear basis of the algebra. One can ask if in the semi-simple compact case (which has more structure), we really need analytic vectors in order to ensure integrability. This question was treated in [2]. In this work one studies the question of integrability of skew-symmetric representations of compact Lie algebras. In the first part one shows that given a skew-symmetric representation of a compact semi-simple Lie algebra, then if the representatives of the Cartan subalgebra (corresponding to simple roots of the complexified Lie algebra) have on the domain of the Lie algebra representation a total set of common eigenvectors of finite multiplicity, the representation is integrable to a unitary group representation and therefore is completely reducible.

Two remarks should be made at this point: The first is that this result shows that under the conditions stated above the Lie algebra representation *cannot* contain an indecomposable part. The second remark is that one could have thought to deduce from this result that it is possible to find integrability criteria *without* use of analytic vectors. This is of course *not true*, since the hypothesis about common eigenvectors contains our analytic vectors.

In the second part of [2] one indeed confirms the necessity of the existence of analytic vectors for integrability purposes, even in the compact (semi-simple or not) case. The exact result is the following:

THEOREM. Every compact Lie algebra of dimension $n > 1$, has at least one representation in Hilbert space on an invariant domain such that every element of the algebra is represented by an essentially skew-adjoint operator on this domain, every element of some linear basis of the algebra is integrable to a one-parameter *compact* group, but the representation is *not integrable.*

We end this part of our lecture by mentioning two open problems arising from the results mentioned in this part of our talk:

1) Generalize (or give a counterexample to) the main result of the second part of [1] in such a way that *every* linear basis of any real finite-dimensional Lie algebra will imply the same consequence (and not only the special bases utilized in [1]).

2) Prove or give a counterexample to the following conjecture (the answer to which is trivially positive in the compact case!): Given an *irreducible* (in a sense to be precised) representation of a real finite-dimensional Lie algebra on an invariant dense domain in a Hilbert space such that every element of the algebra is represented by an essentially skew-adjoint operator on this domain, then this representation is integrable. Simple examples show that this result is *not true* in the skew-symmetric case. We can now pass to the last part of our lecture.

IV. The Infinite-Dimensional Case: Results and Speculations

This part of our lecture is going to be very short. The reason is very simple: First we were already rather long and secondly very few things are really known in the infinite case, and there is still practically everything to be done in this exciting (but very difficult) domain. Reed and Hegerfeldt constructed weakly analytic (and therefore weakly differentiable) vectors for representations of canonical commutation relations in quantum field theory.

A construction of weakly differentiable vectors (which generalizes the Gårding construction) for a more general class of infinite-dimensional groups, was achieved in [5]. Let us describe very briefly the results of this work.

DEFINITION 1. A *Hilbert Lie group* is a Lie group with a Hilbertian Lie algebra (this means that on its Lie algebra one can put a structure of a Hilbert space).

DEFINITION 2. An *inductively-finite Lie group* is a Lie group the Lie algebra of which is the closure of the union of an increasing sequence of finite-dimensional subalgebras ($g_0 = \bigcup_{n \geq 1} g_n$ and $\bar{g}_0 = g$). We can now formulate the two results of [5].

THEOREM 1. Let $(U, \mathscr{H})$ be a strongly continuous representation on a complex Banach space $\mathscr{H}$ of a Hilbert inductively-finite real Lie group. Then the set $\mathscr{H}_\infty(U, g_0)$ of vectors $\varphi \in \mathscr{H}$ such that the functions $t \to U(\exp(tX))\varphi$ are C^∞ on $\boldsymbol{R}$ for very $X \in g_0$ is dense in $\mathscr{H}$.

THEOREM 2. Under the same hypothesis of Theorem 1, one has:

a) For every $X \in g_0$, $\mathscr{H}_\infty(U, g_0)$ is invariant under the representatives $U_0(X)$ of the Lie algebra g_0 in the representation. $\mathscr{H}_\infty(U, g_0)$ is also invariant under the representatives of the *group representation* having the form $U(\exp(X))$ for every $X \in g_0 \cdot X \to U_0(X)$ is indeed a representation of g_0 on $\mathscr{H}_\infty(U, g_0)$ and $\bar{U}_0(X)$ is the generator of the group $t \to U(\exp(tX))$, $(t \in \boldsymbol{R})$.

b) Two Banach representations of a connected Hilbert finitely-inductive real Lie group are strongly equivalent if and only if the corresponding densely defined Lie algebra g_0 representations on $\mathscr{H}_\infty(U, g_0)$ are so. In other words $\mathscr{H}_\infty(U, g_0)$ is big enough to *characterize* the *Lie group* representation.

We can now ask ourselves the question of what about analytic vectors and integrability in the infinite-dimensional case. We can make only a few remarks concerning this subject:

1) It seems that it is not very difficult to prove integrability in some families of infinite-dimensional Lie algebras if their representations satisfy similar conditions (existence of a dense invariant set of analytic vectors in various different senses, etc...) to the finite-dimensional case. This however, is not very interesting because of the difficulties which one meets in the infinite-dimensional case (and do not exist in the finite-dimensional case), difficulties which will be explained in remarks 2) and 3)

2) Given an abstract infinite-dimensional topological group (even if it is a Lie group) it is not at all an easy matter to know even vaguely how big (is it empty?) can the set of equivalence classes of its unitary

irreducible representations be, under what conditions is a unitary representation completely reducible in such a case, etc...

3) Given a unitary representation of an infinite-dimensional topological group, under what conditions will there exist a dense set of common analytic vectors (in some generalized sense) in the representation space? It is quite evident that if one looks for instance at the group of all unitary operators on a Hilbert space endowed with the strong topology, no common analytic vector will be found for all unbounded skew-adjoint generators of the corresponding one-parameter unitary groups. In the same example if one puts on this group the uniform convergence topology, the group becomes a Lie group and its connected component of the identity is generated by all bounded self-adjoint operators. In such a case every vector is analytic for the generators of the connected component. It is evident, therefore, that in order to find common analytic vectors in some sense, if they exist at all, one should better look for a unitary representation of an infinite 'not too big' real Lie group.

Otherwise even if we get integrability criteria, they might be empty! What is the connection between the infinite case and spectral theory (if any)? Though we do not have a definite answer to this question, we may still make some speculations.

We remember that under the usual conditions mentioned before, if $H = H_0 + H_{\text{int}}$ and if H_0 and H_{int} generate by commutators and linear combinations a *finite-dimensional* Lie algebra, then one can find a suitable domain on which H is self-adjoint.

However in physics this case is very rare!

It is our hope that if H_0 and H_{int} generate infinite-dimensional Lie algebras of *certain types* (to be discovered!), then under some conditions we shall be able to integrate the representation to a unitary representation of a Lie group, and then by redifferentiation to obtain a *new definition* of the 'sum' $H_0 + H_{\text{int}}$ such that its closure will be self-adjoint on some natural domain.

It is also known that in Wightman quantum field theory (think of neutral scalar field) there does not exist a general demonstration (without supposing any additional analyticity properties of the vacuum) of the field operators being self-adjoint as they should.

It is also our hope that once one gives a Lie algebraic formulation of axiomatic quantum field theory, and then develops the theory of analytic vectors for this structure and its representation theory in Hilbert space, this problem of field operators being observables – as well as more complicated problems of dynamics – will be solved.

However these speculations bring us far away from the actual state of affairs, so it seems that this is the right time to end the lecture.

References

[1] As a matter of fact part 3) of the theorem is not included in Gårding's original paper but is a particular case of a more general theorem obtained much later by Nelson and Stinespring.

[2] It is well known that the above Lie algebra representation comes from a unitary group representation. This can also be checked using the criterion we mention later, since a common dense domain of analytic vectors for the basis can be easily constructed.

Bibliography

[1] M. Flato and J. Simon, *J. Funct. Anal.* **13** (1973) 268–76.

[2] M. Flato, J. Simon, and D. Sternheimer, *C. R. A. Sc. Paris* **277** (1973) 939–42.

[3] M. Flato, J. Simon, H. Shellman and D. Sternheimer, *Annales Scient. Ecole Normale Sup.*, 4e série, t. **5** (1972) 423–34.

[4] J. Simon, *Commun. Math. Phys.* **28** (1972) 39–46.

[5] J. Simon, *A Gårding Domain for Representations of Some Hilbert Lie Groups*, Physique-Mathématique, Université de Dijon, preprint, April 1973.

SIMPLE MATHEMATICAL MODELS OF SYMMETRY BREAKING. APPLICATION TO PARTICLE PHYSICS

LOUIS MICHEL

Institut des Hautes Etudes Scientifiques, 91440 Bures-sur-Yvette, France

0. Introduction

There are many approximate symmetries in particle physics. It is tempting to consider them as broken higher symmetries. There has been many simplified models along this lines; in a large subset of them, tentative explanations of the Cabibbo angle are suggested. They are not very convincing.

In this lecture I will first present some mathematical facts relevant to symmetry breaking pertaining to two different approaches. Their application to particle physics seems rather suggestive.

1. What is a Spontaneously Broken Symmetry?

The general expression 'symmetry breaking' covers different physical phenomena. So it might be worthwhile to precise which aspect of symmetry breaking is considered here. We will not consider what could be qualified as 'apparent symmetry breaking'. This is the case for instance of classical systems in the neighbourhood of an unstable equilibrium which possesses a symmetry group G. Such systems, very near from each other,[1] (and not invariant under G) may then evolve to very different states, transformed into each other by G; they may also tend to the same state invariant only under the subgroup H of G. In both cases however the symmetry of each system has increased rather than decreased since the initial state had a smaller symmetry.[2]

The problem which interests us is of a broader nature. For instance, in the preceding example, we would ask the question: 'why there exist stable equilibria with symmetry group H, strictly smaller than G?' To take a concrete example: although interactions between atoms or ions are invariant under translations and rotations, i.e. G is the Euclidean group, at some temperature and pression the lowest energy state might be a crystal; its state is invariant under a crystallographic group H, strict subgroup of G. By Euclidean transformations, this state is transformed into other states of the same crystal; the complete set of transforms by G is called an *orbit* of G. The interesting problem is not to explain which state of the orbit will appear (this might be due to any heterogeneity such as crystal seed, etc.), but why crystals exist? More generally, which subgroups H of the Euclidean group G can be symmetry groups of equilibrium states? As we will see, one can answer this question.

To summarize, we say that a symmetry is spontaneously broken when for a physical problem invariant under a group G there exist solutions (which can be grouped into orbits of G) which are only invariant under a strict subgroup of G. We shall omit from now on the adverb 'spontaneously'!

The mechanism of symmetry breaking is well understood; it appears in statistical mechanics when one goes to the thermodynamics limit for systems for which one has rigourous solutions; it also appears in quantum field when one performs the renormalization. For quantum field theory we refer to early examples with perturbative renormalization[3] and recent examples in the lectures of Glimm and Jaffe at this conference [14], [11]. There are even more examples in statistical mechanics, e.g. models of spontaneous magnetization.[4] It is also a criterion that broken symmetries are well understood when one can predict that they cannot occur, as Dobrushin and Schlossmann [37] have recently proven for a large class of 2-dimensional lattice models invariant[5] under a compact connected Lie group G.

The description of broken symmetry is very simple and natural when one uses the mathematical frame of C^*-algebra. This frame covers both classical and quantum statistical mechanics, and quantum mechanics and quantum field theory. The physical states are positive linear forms on A; in the dual A^* of A, they form a convex set whose extremal points are the pure states. Let G be a locally compact group of automorphisms of A. Let Φ be a G-invariant state and $\mathscr{H}_\Phi$, π_Φ the corresponding Hilbert space and representation of A obtained by the Gelfand–Naimark–Segal construction. When Φ is not a pure state, π_Φ is reducible. From the assumption of asymptotic abelianness[6] one proves[7] [18] that Φ is an integral over a subset θ of the pure states

$$(1) \qquad \Phi = \int_{\theta} \Psi d_{\mu}(\Psi)$$

where $d\mu(\Psi)$ is a G-invariant measure (normalized to $\int_{\theta} d\mu = 1$) and that the representation π_{Φ} is factorial (i.e. $\mathscr{H}_{\Phi} = \int_{\theta}^{\oplus} \mathscr{H}_{\Psi} d\mu(\Psi)$ and all irreducible representations π_{Ψ} on $\mathscr{H}_{\Psi}$ are unitary equivalent). The symmetry is broken. Indeed each state Ψ is only invariant under a strict subgroup G_{Ψ} of G, and the automorphisms $g \in G$, $g \notin G_{\Psi}$ of A are not unitarily implementable on $\mathscr{H}_{\Psi}$. Finally one also proves that the only G-invariant subsets of θ are either of μ-measure one or of μ-measure zero.

When there are G-invariant μ-measure zero subsets, Φ is a ergodic transitive state. The classification of such sets for the Euclidean group is still to be done. When there are no G-invariant μ-measure zero subsets, Φ is called a transitive state. The classification of such states is obtained by finding all isomorphic classes of G-orbits carrying a finite G-invariant measure. This has been done in [18] for the Euclidean group. Outside the crystallographic groups, in three dimensions (230), and those in two dimensions to which are added continuous translations in the third direction (17), and the extension of translations by discrete subgroups of $SO(3)$ (oriented homogeneous material, as ferromagnet, infinite number), there are two infinite classes of helicoidal symmetries with rational or irrational rotation angle (helimagnetic states, cholesteric liquids, etc.).[8, 9]

It is more simple to classify 'possible symmetry breakings' than to prove that they dynamically occur (or do not occur); for a given system the latter requires the study of the action of G on the set of pure states. However there also exists in physics simplified models which predict correctly the possible symmetry breaking; I think particularly of the Landau theory of phase transitions.[10] Since these transitions are reversible it deals not only with symmetry breaking, but also with enlargement of symmetry. However enlargements of symmetry seemed more natural than symmetry breaking (cf. the pioneer work of Curie in this domain [6]).

It is time now to expose the work I did with L.A. Radicati these last few years on two mathematical models of symmetry breaking.

2. Two Mathematical Models of Symmetry Breaking

2.1. SMOOTH ACTION[11] OF A COMPACT LIE GROUP G ON A MANIFOLD M

When a group G acts on a set M, we denote by G_m the isotropy group (or little group) of $m \in M$:

$$(2) \qquad G_m = \{g \in G, g \cdot m = m\}$$

and by $G(m)$ the G-orbit of m:

$$(3) \qquad G(m) = \{m' \in M, \quad \exists g \in G, m' = gm\}.$$

The isotropy groups of elements of the same orbit are conjugated $m' = g \cdot m \Leftrightarrow G_{m'} = gG_m g^{-1}$.

There is a natural definition of isomorphy class of G-orbit; these classes are in bijective correspondence with conjugation classes of subgroups of G. In the action of G in M we call layer the union of all isomorphic orbits: elements $m' \in M$ whose isotropy group is conjugated to G_m form the layer $S(m)$. By inclusion up to a conjugation there is a natural order on the conjugation classes of subgroups of G (we denote by $\{H\}$ the class of $H \subset G$); it is customary to use the reverse order on the isomorphy classes of orbits. Indeed, when G is a Lie group:

$$(4) \qquad \dim G = \dim G_m + \dim G(m)$$

so the smaller is the subgroup G_m the larger is the orbit $G(m)$. To be on the same orbit is an equivalence relation for the elements of M; the quotient is called the *orbit space*; we denote it by M/G and by π the canonical projection $M \xrightarrow{\pi} M/G$.

For smooth action of compact Lie groups G on manifolds everything is beautiful. The isotropy groups are closed Lie subgroups. Orbits and layers are submanifolds, layers are strata (see Thom's lecture), the continuous map π is open, closed, proper. There is a maximal layer S_0 ($\sim$ minimal isotropy group) which is open dense.[12]

Consider some examples: a) M is the five-dimensional phase space of three distinct particles with fixed total energy momentum $\boldsymbol{p}$; G is the little group of $\boldsymbol{p}$ for the Lorentz group, G is isomorphic to $O(3)$; M/G is the Dalitz plot; there are two layers, whose images by π are the interior and the boundary of the Dalitz plot.

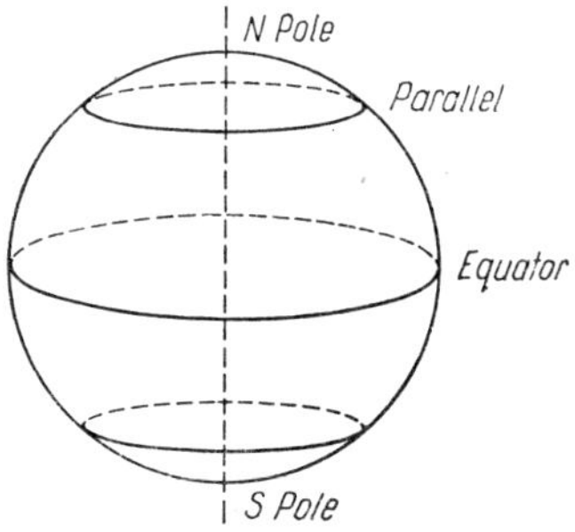

Fig. 1

b) M is S_2 and G is $O(2)$ which includes the inversion through the origin; the rotation axis defines poles and equator on the sphere. The orbits of the open dense layer are the two parallels of same N–S latitude λ: $0 < \lambda < 90°$. There are two other layers of one orbit each: the poles and the equator.

c) M is S_7, the unit sphere of the octet space, i.e. the space of the adjoint representation of $SU(3)$; $G = \text{Aut}\, SU(3)$, isomorphic to the semi-direct product $SU(3) \square Z_2$; in physics, outer automorphisms of $SU(3)$ an generated by charge conjugation. We can use Figure 1 again; the orbit of the open dense layers contain 2 connected components of dimension 6. The equator is a connected orbit of dimension 6; the orbit represented by the two poles contain 2 connected components of dimension 4, the corresponding isotropy group is $U(2)$.

There is a G-invariant Riemann metric on M (take any Riemann metric on M, average it by the group with the Haar measure) and M/G is a metric space. For any globally G-invariant submanifold Ω of M, there is a tubular neighbourhood $U \supset \Omega$, such that $\forall m' \in U$, there exists a unique $m \in \Omega$ such that distance mm' is minimum. The retraction map $r(m') = m$ is equivariant: $\forall g \in G, r(g \cdot m') = g \cdot r(m') = g \cdot m$; hence $G_{m'} \subset G_m$. This proves the (take $\Omega = G(m)$) [25]

THEOREM 1. For every $m \in M$, there is a neighbourhood U such that $\forall m' \in U$, G_m is larger (not necessarily strictly!) than $G_{m'}$.

Taking local geodesic coordinates in m, one sees that the local action of G_m is linear: this is also the linear orthogonal representation D_m of G_m on $T_m(M)$ the tangent vector space at m; D_m is orthogonal and fully reducible; $T_m(S(m))$ and $T_m(G(m))$, the tangent spaces to the layer and the orbit are invariant subspaces. Theorem 1 proves that the subspace of fixed points of G_m on $T_m(m)$ is included in $T_m(S(m))$. For a G-invariant vector field $m \to v_m \in T_m(M)$, $\forall g \in G_m, gv_m = v_m$; hence [25]:

THEOREM 2. A G-invariant vector field on M is at each m tangent to the layer $S(m)$.

Consider the set $\mathscr{F}$ of all G-invariant real valued smooth functions on M, i.e. $f \in \mathscr{F} \Rightarrow \forall g \in G, f(g \cdot m) = f(m)$: f is constant on each orbit; so its gradient is orthogonal to the orbit and if the orbit is isolated in its layer (i.e. $\pi(G(m))$ is an isolated point of $\pi(S(m))$) then $\operatorname{grad} f = 0$. The converse is also true; hence if we call 'critical orbits' those orbits on which the gradient of every $f \in \mathscr{F}$ vanishes, we have[13]

THEOREM 3. The critical orbits are the orbits isolated in their layer.[14]

The interest of this theorem for the study of symmetry breaking is obvious. If the physical G-invariant problem is a variational problem, whatever the function to be varied, the symmetry is broken on critical orbits which are not fixed points, whatever the function to be varied.

While our first mathematical model deals with the smooth action of compact Lie groups on real manifolds, the second model considers linear action of any group on real or complex finite dimensional vector spaces.

2.2. G-INVARIANT ALGEBRAS

Consider two linear representations D and D_1 on two vector spaces $\mathscr{E}$ and $\mathscr{E}_1$. We denote by $\operatorname{Hom}(\mathscr{E}, \mathscr{E}_1)$ the set of linear maps from $\mathscr{E}$ to $\mathscr{E}_1$ (it is a vector space) and $\operatorname{Hom}(\mathscr{E}, \mathscr{E}_1)^G$ those maps invariant by G:

$$\forall g \in G, \quad f_0 D(g) = D_1(g)_0 f; \tag{5}$$

$\dim H(\mathscr{E}, \mathscr{E}_1)^G_{\mathscr{E}} > 0$ if D and D_1 have in common some irreducible representations in their reduction (one says they are not 'disjoint'). As a particular case, consider the action of G on $\mathscr{E}_1 = \mathscr{E} \otimes \mathscr{E}$ defined by the tensor representation $D_1 = D \otimes D$; then each map $f \in$

$\in \text{Hom}(\mathscr{E} \otimes \mathscr{E}, \mathscr{E})^G$ defines an algebra on $\mathscr{E}$ for which G is a group of automorphism. (This algebra is generally not associative). One can decompose the tensor product $\mathscr{E} \otimes \mathscr{E}$ into its symmetric and antisymmetric part $\mathscr{E} \otimes \mathscr{E} = (\mathscr{E} \overset{S}{\otimes} \mathscr{E}) \oplus (\mathscr{E} \overset{A}{\otimes} \mathscr{E})$; the corresponding algebras $\in \text{Hom}(\mathscr{E} \overset{S}{\otimes} \mathscr{E}, \mathscr{E})^G$ and $\in \text{Hom}(\mathscr{E} \overset{A}{\otimes} \mathscr{E}, \mathscr{E})^G$ are respectively symmetric and antisymmetric. Ex.: G is a simple Lie group of the type B_n, C_n, D_n; D is the adjoint representation; then $\dim \text{Hom}(\mathscr{E} \otimes \mathscr{E}, \mathscr{E})^G = l = 1$ and the corresponding algebra is the Lie algebra, while $\dim \text{Hom}(\mathscr{E} \overset{S}{\otimes} \mathscr{E}, \mathscr{E})^G = d = 0$. For the Lie algebra of the type A_n $(n \geqslant 2)$ (e.g. $SU(n+1)$, $SL(n+1)$), $d = 1$ and the corresponding symmetric algebra for $SU(3)$ is well known to physicists. In all physical examples we shall meet below $l \leqslant 1$, $d \leqslant 1$ so the corresponding algebra is unambiguously defined; e.g. the symmetric algebra for the $(3, \bar{3}) \oplus (\bar{3}, 3)$ representation of $SU(3) \times SU(3)$ is studied in details in [26]. When we need to consider one of these algebras we shall denote by $_T$ the corresponding law, i.e.:

(6) $\quad f(a \otimes b) = a_T b.$

In quantum mechanics or quantum field theory, invariance under a group G leads to consider 'Tensor Operators',[15] $T \in \text{Hom}(\mathscr{E}, \mathscr{L}(\mathscr{H}))^G$ where $\mathscr{H}$ is the Hilbert space of states, $\mathscr{L}(\mathscr{H})$ the vector space of linear operators on $\mathscr{H}$, and $\mathscr{E}$ is a finite dimensional vector space on which acts a representation D of G. When D is irreducible, T is an irreducible tensor operator. $\mathscr{E}$ is called the *variance* of the tensor operator. Let $a, b \in \mathscr{E}$, $T_1, T_2 \in \text{Hom}(\mathscr{E}, \mathscr{L}(\mathscr{H}))^G$, then

(7) $\quad a \mapsto T_1(a) + T_2(a)$ is a tensor operator of variance $\mathscr{E}$,

(8) $\quad a \otimes b \mapsto T_1(a) T_2(b)$ is a tensor operator of variance $\mathscr{E} \otimes \mathscr{E}$.

Any polynomial equation

(9) $\quad P(T_1, \ldots, T_2) = 0$

involving tensor operators of the same variance $\mathscr{E}$ defines a polynomial equation in the tensorial algebra $\mathscr{T}$ on $\mathscr{E}$ (i.e. $\mathscr{T} = \bigoplus_{n=0}^{\infty} (\overset{n}{\otimes} \mathscr{E})$) and also an equation $\mathscr{P} = 0$ on the quotient $\mathscr{U}$ of $\mathscr{T}$ modulo the equivalence relations of the type (6) due to the G-invariant algebras. (When this algebra is the Lie algebra, $\mathscr{U}$ is its enveloping algebra.) Generally we want equation $P = 0$ to be satisfied for any $a \in \mathscr{E}$; however for the solutions of $\mathscr{P}(a) = 0$, $a \in \mathscr{E}$, equation $P = 0$ is trivially satisfied and this leads to symmetry breaking 'along the direction' a and the symmetry group is reduced to the isotropy group G_a. For quadratic equations $P = 0$, corresponding for example to simple bootstrap models, and often, for higher degree equations, the idempotent ($\alpha \neq 0$) or nilpotent ($\alpha = 0$) of the algebra

$$(10) \qquad a_T a = \alpha a$$

are solutions of $\mathscr{P}(a) = 0$.

This Part 2.2 can be generalized to include multilinear algebras; for example trilinear algebras with G as group of automorphisms are elements of $\mathrm{Hom}(\mathscr{E} \otimes \mathscr{E} \otimes \mathscr{E}, \mathscr{E})^G$. In some physical applications discussed in [30], idempotents and nilpotents of trilinear algebras appear.

3. Application to Particle Physics

Since the $SU(3)$ symmetry for hadrons was recognized by Gell–Mann and Ne'eman, we know that the hypercharge Y, the electric charge Q, the Cabibbo direction of weak-current $C_{\pm} = C_1 \pm iC_2$, the weak hypercharge[16] Z define directions (i.e. unit vectors) in the octet space: the space of the adjoint representation of $SU(3)$. It is to be noted[17] (and earlier papers quoted there) that the unit vectors y, q, z belong to the critical orbit represented by the poles of Figure 1 while c_1 and c_2 belong to the critical orbit of the equator. We also remark that y, q, z are idempotents and c_+, c_- are nilpotents of the (complex) $SU(3)$ invariant symmetrical algebra defined on the octet space.[18,19]

These remarks extend naturally to

$$(11) \qquad G = (SU(3) \times SU(3)) \,\square\, (Z_2(P) \times Z_2(C))$$

where the charge conjugation C is an outer automorphism for each $SU(3)$ factor while the parity operation P exchanges them (this is an interesting interplay between geometrical invariance in space time and internal hadronic symmetry). In the action of G on S is the unit sphere of the adjoint representation space, there are 12 layers; five of them contain only critical orbits; one contains y, q, a second contains z, a third contains c_+, c_-, a fourth contains what is likely to be the direction of the

CP-breaking interaction in K^0 decay; what is the role of the fifth one?[17]

G-invariance is broken by semi-strong interaction in a different linear representation space; this might be the $(3,\bar{3}) \oplus (\bar{3}, 3)$ representation[20] as first suggested in [12]. There are groups containing P and C whose connected component is $SU(3)$ (the diagonal $SU(3)$) or $SU(2) \times SU(2) \times \times U(1)$: chiral invariance for pions). They correspond to each of the two types of critical orbits, and also respectively to an idempotent and to a nilpotent of the symmetric algebra [26], [27], [5]. However, in nature, semi-strong interactions break G up to the subgroup $H = (U(2) \square Z_2 \square (C)) \times Z_2(P)$. (It is not the isotropy group of the open dense layer.)[21] One can also extend the study of critical orbits and idempotents or nilpotents for the reducible representations interesting for the semi-strong symmetry breaking. Pegoraro, Subba Rao [33] and then Darzens [7] have shown that the physically interesting schemes appear when H is the isotropy group of a critical orbit.

More recently, the group G has been extended to $(SU(4) \times SU(4)) \square \square (Z(P) \times Z_2(C))$ either for having a unified theory of weak and electromagnetic interactions [4], to explain the absence of strangeness changing neutral currents [13] or to include the four basic leptons in the scheme [13], [3], [8]. In a recent preprint Mott [30] has shown that all physical directions of symmetry breaking in these papers correspond both to critical orbits and to idempotents or nilpotents of the involved symmetrical algebra.[22]

Conclusion

Isospin was introduced by Heisenberg [17] in 1932, immediatly after the discovery of the neutron. For the last forty years, 'internal symmetry' have appeared richer and richer and the interplay of the different types of interactions with internal symmetry is fascinating. Of course I believe that complete understanding of 'internal symmetry' breaking will require to solve the hard dynamical problems of the particle interactions. The concepts of 1) critical orbits, 2) idempotents and nilpotents of canonical algebra, seem very useful for the study of this subject. At least they show that many simplified models made to understand the subject do not shed much light on it, because their predictions are not at all specific but are mere consequences of general theorems. Finally we remark that in particle physics, the symmetry breaking is never maximal (i.e. symmetry is not broken up to the minimal isotropy group of the open dense layer). When the symmetry breaking occurs along a direction of critical orbit, it is minimal (isotropy groups of critical orbits are largest of any completely ordered chain of conjugated classes of isotropy group which contains them).

References

[1] Even if the experimentalist prepares them as similar as possible, there are fluctuations which make them different.

[2] Apparent symmetry breaking does also raise interesting problems. R. Thom [39] in his lecture will also distinguish several types of symmetry breaking and study interesting cases which I do not consider here.

[3] Earliest references are B. W. Lee [21], B. W. Lee and J. L. Gervais [22], K. Symanzik [38] and for an early review Lectures at the Cargese Summer School 1970 (Gordon and Breach).

[4] An historical reference is R. Peierls [34]. I am no competent for giving the large list of publications for the last ten years.

[5] For more details: Define an action of G on a manifold X. So G acts (diagonally on X^{Z^2}. The potential U is bounded, has finite range and is invariant under G, and there are some conditions of non-degeneracy. The proof use the limit theorem for random variables on Lie groups.

[6] This assumption, introduced in [39], is: $\forall a, b \in A$, $\forall \Phi A^*$ $\Phi([a, \alpha_t(b)]) \to 0$ as $G \ni t \to \infty$, e.g. if G is the Euclidean group, t is a translation going to infinity).

[7] See D. Kastler, G. Loupias, M. Membkhout and L. Michel [18]. Pioneer works on the same subject are: D. Kastler and D. W. Robinson [19], D. W. Robinson and D. Ruelle [35], D. Ruelle [36], S. Doplicher, D. Kastler and E. Størmer [10].

[8] Of course not all these symmetry classes, in infinite number, are known to occur in Nature, this is already the case of some of 230 crystallographic classes. But it seems to me interesting that man has for instance made borane molecules with the dodecahedron, icosahedron symmetry (e.g. anion $B_{12}H_{12}{}^{2-}$) which has never been observed in natural molecules.

[9] We denote by $\mathscr{B}$ the set of all closed subgroups of the Euclidean group into which the Euclidean symmetry may be broken. For the Euclidean group E (and this is a generalization of a Mostow result for solvable Lie groups) $B \in \mathscr{B}$ is equivalent to 'the homogeneous space E/B is compact'. Crystallographic groups are the discrete subgroups of E which belong to $\mathscr{B}$. There is a natural topology on the space $\mathscr{F}$ of closed subgroups of a locally compact group (See e.g. Bourbaki, Vol. VI, Chap. 8, § 5 which generalizes a work of Macbeath and Swierczkowski). $\mathscr{F}$ is compact. In the case of the Euclidean group, $\mathscr{B}$ is open in $\mathscr{F}$.

[10] It is succently exposed in the Landau-Lifschitz [20] and the Lubarskii [24] text books. For improvements and an up to dated review see Birman [2].
[11] For classical review of the subject see [31], [28]. To avoid details we assume smoothness, i.e. C^∞; most results are valid with weaker hypothesis; Mostow, Palais theorem use C^1. Palais [32] proved that for a compact M, C^1 action is equivariant to C^∞ action.
[12] Remarkable results, not used here, are those of Mostow. For C^1 action, if M is compact, the number of layers is finite; if the number of layers is finite, there is an embedding of M is a finite dimensional vector space on which the action of G is linear orthogonal.
[13] See L. Michel [25]; for more details see: Proceedings of 3rd GIFT Seminar in Theoretical Physics, Madrid (1972) 49–131.
[14] We can also give conditions implying that all G-invariant vector fields vanish on an orbit isolated in its layer. The points of $G(m)$ which have G_m as isotropy group form a submanifold diffeomorphic to the group $H_m = N(G_m)/G_m$ where $N(G_m) = \{h \in G, hG_mh^{-1} = G_m\}$ is the normalizer of G_m in G. Let H_m^0 the connected component which contains the identity. So two such conditions are: 1) $H_m^0 = \{e\}$, 2) The Euler characteristic of the orbit $\chi(G(m)) \neq 0$.
[15] Which are not operators on the Hilbert space of states, but linear functions with operator value.
[16] It is generally admitted that non-leptonic weak interaction is invariant under a $\mathscr{U}(2)$ group, corresponding to the Q-spin of Cabibbo's original paper.
[17] See L. Michel and L. Radicati [26] and earlier references quoted there. The canonical symmetric algebra of the $(6, \bar{6}) \oplus (\bar{6}, 6)$ and the $(8, 8)$ are studied by A. Chakrabarty and C. Darzens [5].
[18] Gell-Mann has denoted d_{ijk} the structure constants of this symmetrical algebra.
[19] It is not clear that physicists should use only the compact form $SU(3)$ and not its complexified form $SL(3, C)$. Indeed the weak currents are not Hermitian; we also know that complexification of the Poincaré group has been fruitful for the study of analytic properties of field theory and CPT invariance. In the adjoint action of $SL(3, C)$ the set $\mathscr{S}$ of semisimple elements contains two orbits, one open dense in $\mathscr{S}$ and one exceptional which contains y, q, z; the set $\mathscr{N}$ of nilpotent elements also contains two orbits, one open dense in $\mathscr{N}$ and the exceptional one which contains c_+, c_- [1].
[20] Remark that the representation is irreducible in the real and it is the real canonical algebra which is considered here.
[21] Indeed the minimal subgroup is $U(1) \times U(1)$.
[22] Similarly, the semi-strong breaking, which is in a small strata, is near a critical orbit corresponding to chiral symmetry and to a nilpotent of a symmetric trilinear algebra (there is no bilinear algebra associated to the $(4, \bar{4}) \oplus (\bar{4}, 4)$ representation).

Bibliography

[1] L. Abellans, *J. Math. Phys.* **13** (1972) 1064.
[2] J. L. Birman, 'IInd Colloquium on Group Theory Methods in Physics', Nijmegen, 1973.

[3] B. J. Bjorken and S. L. Glashow, *Phys. Lett.* **11** (1964) 255.
[4] J. D. Bjorken and C. H. Llewellyn Smith, *Phys. Rev.* **D7** (1973) 887.
[5] A. Chakrabarty and C. Darzens, *Ann. Phys.* **69** (1972) 193.
[6] P. Curie, *Journal de Physique* (*3ème série*) **3** (1894) 393.
[7] C. Darzens, *Ann. Phys.* **76** (1973) 236.
[8] P. Dittner, A. Eliezer and T. K. Kuo, *Phys. Rev. Lett.* **30** (1973) 1274.
[9] S. Doplicher, D. Kastler and D. W. Robinson, *Comm. Math. Phys.* **3** (1966) 1.
[10] S. Doplicher, D. Kastler and E. Størmer, *J. Functional Analysis* **3** (1969) 419.
[11] A. Jaffe, this volume, p. 33.
[12] M. Gell-Man, R. S. Oakes and B. Renner, *Phys. Rev.* **175** (1968) 2195.
[13] S. L. Glashow, J. Iliopulos and L. Maiani, *Phys. Rev.* **D2** (1970) 1285.
[14] J. Glimm, this volume, p. 15.
[15] A. Guichardet and D. Kastler, *J. Math. Pures et Appliquées* **49** (1970) 349.
[16] R. Haag, D. Kastler and L. Michel, Marseille preprint 1969.
[17] W. Heisenberg, *S. Phys.* **77** (1932) 1.
[18] D. Kastler, G. Loupias, M. Mebkhout and L. Michel, *Comm. Math. Phys.* **27** (1972) 195–222.
[19] D. Kastler and D. W. Robinson, *Comm. Math. Phys.* **3** (1966) 151.
[20] L. D. Landau and E. M. Lifschitz, *Statistical Physics*, § 136.
[21] L. D. Landau and E. M. Lifschitz, *Quantum Mechanics*, Ch. 14.
[22] B. W. Lee, *Nucl. Phys.* **B9** (1969) 649.
[23] B. W. Lee and J. L. Gervais, *Nucl. Phys.* **B12** (1969) 627.
[24] C. Ya. Lubarskii, *Application of Group Theory in Physics*, Chap. VII.
[25] L. Michel, *C. R. Acad. Sc. Paris* **272** (1971) 433.
[26] L. Michel and L. Radicati, 'Breaking of the $SU(3) \times SU(3)$ Symmetry in Hadronic Physics', in: *Evolution of Particle Physics*, Academic Press, New York 1970, p. 191.
[27] L. Michel and L. Radicati, 'Properties of the Breaking of Hadronic Internal Symmetry', *Ann. of Phys.* **66** (1971) 758–783.
[28] D. Montgomery, *Differential Analysis*, Bombay Colloquium 1964.
[29] G. D. Mostow, *Ann. Math.* **65** (1957) 513 and 432.
[30] R. E. Mott, 'Algebraic Properties of $SU(4) \times SU(4)$ Symmetry Breaking', Preprint, Queen Mary College, London.
[31] R. S. Palais, *Memoirs Ann. Math.* **36** (1960).
[32] R. S. Palais, *Ann. J. Math.* **92** (1970) 748.
[33] Pegoraro and J. Subba Rao, *Nucl. Phys.* **B44** (1972) 221.
[34] R. Peierls, 'On Ising Model of Ferromagnetism', *Proc. Camb. Phil. Soc.* **32** (1936) 477.
[35] D. W. Robinson and D. Ruelle, *Ann. Inst. Henri Poincaré* **A6** (1967) 299.
[36] D. Ruelle, *J. Functional Analysis* **6** (1970) 116.
[37] Iu. Sukhov, private communication, unpublished.
[38] K. Symanzik, *Comm. Math. Phys.* **16** (1970) 48.
[39] R. Thom, this volume, p. 293.

SINGULAR PERTURBATION METHOD FOR EVOLUTION EQUATIONS IN BANACH SPACES

JANUSZ MIKA

Institute for Nuclear Research, Warsaw, Poland

Abstract

In the paper the singular perturbation method is applied to linear evolution equations in Banach spaces containing a small parameter multiplying the time derivative. Outer and inner asymptotic solutions are formulated and the sense in which they converge to the exact solution is rigorously defined. It is then shown that the sum of the two asymptotic solutions converges uniformly to the exact solution. Possible applications to various physical situations are indicated.

Introduction

A large variety of time-dependent physical systems are described by evolution equations with linear operators satisfying certain requirements in properly chosen Banach spaces. For such systems the semigroup and quasi-semigroup theory developed by Hille and Phillips [2], Yosida [9], and Kato [3] can be used to prove the existence theorems and to supply a convenient formulation of the corresponding solutions.

In many physical systems described by evolution equations a time relaxation constant is very large and the system changes rapidly in time. In other words this means that the time derivative in a corresponding evolution equation is multiplied by a small positive parameter. In such cases the application of the perturbation procedure with respect to that parameter leads to the singular perturbation method.

The motivation to study the singular perturbation method comes from two important reasons. Firstly, the detailed information on the asymptotic behavior of the solution may considerably contribute to the better understanding of the physical properties of the system. Secondly, the application of the perturbation procedure for practical calculations may remove the computational difficulties introduced by the small parameter appearing in the equation.

Many linear operators encountered in mathematical physics have been shown to be infinitesimal generators of strongly continuous semigroups or quasi-semigroups in properly chosen Hilbert or Banach spaces. In particular, to such class belong linear operators describing such physical phenomena as particle transport, particle diffusion, heat transfer, quantum field behavior for some practically important potentials, etc.

The first results concerned with the asymptotic behavior of evolution equations in Banach spaces were obtained by Krein [4] who considered the zeroth order perturbation method for a single equation of evolution. The author extended Krein's results to the systems of equations of evolutions for which only one time derivative is multiplied by a small parameter [5], [6] and developed the general perturbation method for single equation [7]. In this article the results of the last paper will be presented without proofs. It will be also indicated that the present approach gives different and, in some sense, considerably simpler results than those obtained by the standard perturbation procedure as applied, for instance, to the linear transport equation [1].

Evolution Equations in Banach Spaces

Let E be a certain Banach space with a norm denoted as $\|\cdot\|$. A function $G(t)$ with the domain $[0, \infty)$ whose values are bounded operators on E into itself, is called a strongly continuous semigroup if the following conditions are satisfied:

(1)
- (i) $G(t+s) = G(t)G(s)$, $0 \leqslant s$, $0 \leqslant t$;
- (ii) $G(t) = I$ (identity operator on E);
- (iii) for each $x \in E$ the function $G(t)x$ with the domain $[0, \infty)$ and the values from E is strongly continuous on $[0, \infty)$.

The operator $Ax = \lim\limits_{t\to\infty} \frac{1}{t}\big(G(t)x - x\big)$ with the domain $D(A)$ consisting of all $x \in E$ such that the limit exists in the sense of the norm in E, is called an infinitesimal generator of a strongly continuous semigroup $G(t)$.

The important Hille–Yosida theorem states that a necessary and sufficient condition that a closed operator A from E into itself with a domain $D(A)$ dense in E be an infinitesimal generator of a strongly continuous semigroup $G(t)$ of bounded operators such that $||G(t)|| \leqslant \exp(\omega t)$ for some real ω and each $t \in [0, \infty)$, is that the resolvent $(\lambda I - A)^{-1}$ exists for each $\lambda > \omega$ as a bounded operator defined for all $x \in E$ and satisfying the inequality $||(\lambda I - A)^{-1}|| \leqslant (\lambda - \omega)^{-1}$.

For any strongly continuous semigroup $G(t)$ and for each x belonging to the domain $D(A)$ of its infinitesimal generator A the function $G(t)x$ with the values from E is strongly continuously differentiable on $[0, \infty)$ such that

$$\frac{d}{dt}\big(G(t)x\big) = G(t)Ax = AG(t)x. \tag{2}$$

From this it follows that the evolution equation

$$\frac{d}{dt}\big(x(t)\big) = Ax(t) + q(t) \tag{3}$$

with the initial condition

$$x(0) = \eta \subset D(A) \tag{4}$$

has a unique, strongly differentiable solution if A is an infinitesimal generator of a strongly continuous semigroup $G(t)$ and $q(t)$ is a function with the values from E, strongly continuously differentiable on $[0, \infty)$. This solution is given by the formula

$$x(t) = G(t)\eta + \int_0^t dsG(t-s)q(s) \tag{5}$$

where the integral is understood as a strong limit of Riemann sums.

It is seen that if A is a generator of a semigroup $G(t)$ then $\frac{1}{\varepsilon}A$ where $\varepsilon > 0$, is a generator of a semigroup $G\left(\frac{t}{\varepsilon}\right)$. Thus the evolution equation

$$\varepsilon \frac{d}{dt}\big(x_\varepsilon(t)\big) = Ax_\varepsilon(t) + q(t) \tag{6}$$

with the initial condition analogous to (4) has the solution

$$x_\varepsilon(t) = G\left(\frac{t}{\varepsilon}\right)\eta + \frac{1}{\varepsilon}\int_0^t ds\, G\left(\frac{t-s}{\varepsilon}\right)q(s). \tag{7}$$

A function $U(t, s)$ whose domain is the triangle $0 \leqslant s \leqslant t \leqslant t_0$, where $0 < t_0 < \infty$, and whose values are bounded operators on E into itself, is called a strongly continuous quasi-semigroup if the following conditions are satisfied:

(8)
- (i) $U(t, s) = U(t, \tau)U(\tau, s)$, $0 \leqslant s \leqslant \tau \leqslant t \leqslant t_0$;
- (ii) $U(t, t) = I$, $0 \leqslant t \leqslant t_0$;
- (iii) for each $x \in E$ the function $U(t, s)x$ with the values from E, is strongly continuous on the triangle $0 \leqslant s \leqslant t \leqslant t_0$.

A function $B(t)$ whose domain is the interval $[0, t_0]$ and the values are closed operators from E into itself, is called an infinitesimal generator of a strongly continuous quasi-semigroup, if the following conditions are satisfied:

(9)
- (i) for all $t \in [0, t_0]$ the domains of the operators $B(t)$ are independent of t and the common domain $D(B)$ is dense in E;
- (ii) for each $\tau \in [0, t_0]$ the operator $B(\tau)$ is an infinitesimal generator of a strongly continuous semigroup $G_\tau(t)$ such that for each $t \in [0, \infty)$ $\|G_\tau(t)\| \leqslant \exp(\alpha_\tau t)$, $\sup_{\tau \in [0, t_0]} \alpha_\tau = \alpha < 0$;
- (iii) for each $x \in D(B)$ the function $B(t)x$ with the values from E is strongly continuously differentiable on $[0, t_0]$.

For any strongly continuous quasi-semigroup $U(t, s)$ and for each x belonging to the common domain $D(B)$ of its inifinitesimal generator $B(t)$ the function $U(t, s)x$ is strongly continuously differentiable on the triangle $0 \leqslant s \leqslant t \leqslant t_0$ such that

$$\begin{aligned} &\frac{\partial}{\partial t}\big(U(t, s)x\big) = B(t)\,U(t, s)x, \\ &\frac{\partial}{\partial s}\big(U(t, s)x\big) = -U(t, s)\,B(s)x. \end{aligned} \tag{10}$$

From this it follows that the evolution equation

$$(11) \qquad \frac{d}{dt}\big(x(t)\big) = B(t)x(t)+q(t)$$

with the initial condition

$$(12) \qquad x(0) = \eta \in D(B)$$

has a unique, strongly differentiable solution if $B(t)$ is an infinitesimal generator of a strongly continuous quasi-semigroup $U(t, s)$ and $q(t)$ is a function with the values from E, strongly continuously differentiable on $[0, t_0]$. This solution is given by the formula

$$x(t) = U(t,0)\eta + \int_0^t ds U(t, s)q(s)$$

where the integral has the same meaning as in (7).

If $B(t)$ generates a strongly continuous quasi-semigroup $U(t, s)$, then $\frac{1}{\varepsilon}B(t)$ where $\varepsilon > 0$, generates a strongly continuous quasi-semigroup $U_\varepsilon(t, s)$. The evolution equation

$$(13) \qquad \varepsilon\frac{d}{dt}\big(x_\varepsilon(t)\big) = B(t)x_\varepsilon(t)+q(t)$$

with the initial condition analogous to (12) has the solution

$$(14) \qquad x_\varepsilon(t) = U_\varepsilon(t, 0)\eta + \frac{1}{\varepsilon}\int_0^t ds U_\varepsilon(t, s)q(s).$$

The quasi-semigroup $U(t, s)$ generated by $B(t)$ satisfying the conditions (9), can be expressed for each $x \in E$ by a multiplicative integral

$$U(t, s)x = \lim G_{\tau_n}(t_{n+1}-t_n)G_{\tau_{n-1}}(\tau_n-\tau_{n-1}) \dots G_{\tau_1}(t_2-t_1)$$

where $s = t_1, t_2, \dots, t_{n+1} = t$ are the points of division of the interval $[s, t]$ and $t_i < \tau_i < t_{i+1}$. The limit is understood in a strong sense for $\max_i(t_{i+1}-t_i) \to 0$. From this it follows that the quasi-semigroup $U(t, s)$ satisfies the inequality

$$\|U(t, s)\| \leqslant \exp\big(\alpha(t-s)\big)$$

where the constant α was defined in (9). Similarly, for the quasi-semigroup $U_\varepsilon(t, s)$ one has

$$\|U_\varepsilon(t, s)\| \leqslant \exp\left(\frac{\alpha(t-s)}{\varepsilon}\right). \tag{15}$$

A comprehensive analysis of evolution equations with time-independent and time-dependent operators is given in [4].

Asymptotic Solutions to Evolutions Equations

The assumptions (9) concerning the functions $B(t)$ and the assumptions concerning $q(t)$ made in the previous section, allow for the analysis of the zeroth order asymptotic solutions to the evolution equation (see [4]–[6]). However, when considering higher order asymptotic solutions, one needs some more restrictive assumptions for both functions $B(t)$ and $q(t)$.

In this section it will be assumed that $B(t)$ besides (9) satisfies the following additional condition:

(9a) For each $t \in [0, t_0]$ and $x \in D(B)$ the element $B(t)x$ can be written as $B(t)x = B_0x + B_1(t)x$, where B_0 is a closed operator independent of t with the domain $D(B)$ dense in E and $B_1(t)$ is a function whose values are bounded operators on E into itself, $(N+1)$ times uniformly continuously differentiable on $[0, t_0]$.

Similarly, it will be assumed that the function $q(t)$ with values from E is $(N+1)$ strongly continuously differentiable on $[0, t_0]$.

The above assumptions are not particularly restrictive and apply for many equations in mathematical physics, as for instance, linear transport equation or linear quantum field theory equations with cutoff potentials.

The above additional assumptions allow to integrate by parts $(N+1)$ times the integral term in (14) by making use of the formulas for differentiation of the quasi-semigroup (10) and of the fact that, according to (9), the inverse $B^{-1}(t)$ of $B(t)$ exists for each $t \in [0, t_0]$. The result is

(16) $$\frac{1}{\varepsilon}\int_0^t ds U_\varepsilon(t,s)q(s) = -U_\varepsilon(t,0)\bar{x}_\varepsilon^{(N)}(0) + \bar{x}_\varepsilon^{(N)}(t) - \varepsilon^N \int_0^t ds U_\varepsilon(t,s)\frac{d}{ds}\left(\bar{x}_N(s)\right).$$

The function $\bar{x}_\varepsilon^{(N)}(t)$ defined over the interval $[0, t_0]$ and with the values from E is called an outer asymptotic solution of order N and is given by the formulas

(17) $$\bar{x}_\varepsilon^{(N)}(t) = \sum_{n=0}^{N} \varepsilon^n \bar{x}_n(t),$$
$$\bar{x}_0(t) = -B^{-1}(t)q(t), \qquad \bar{x}_n(t) = B^{-1}(t)\frac{d}{dt}\left(\bar{x}_{n-1}(t)\right),$$
$$n = 1, 2, \ldots, N.$$

The last term in (16) satisfies the inequality

$$\varepsilon^N \left\| \int_0^t ds U_\varepsilon(t,s)\frac{d}{ds}\left(\bar{x}_N(s)\right) \right\| \leqslant M\varepsilon^{N+1}$$

which follows from (15) and from the fact that the function $\frac{d}{ds}\left(\bar{x}_N(s)\right)$ is uniformly bounded on $[0, t_0]$. From this it is seen that the difference between the exact solution $x_\varepsilon(t)$ to the evolution equation (13) and the function

(18) $$\bar{x}^{(N)}(t) + U_\varepsilon(t,0)\left(\eta - \bar{x}^{(N)}(0)\right)$$

tends to zero as ε^N with $\varepsilon \to 0$, uniformly with respect to t over the interval $[0, t_0]$.

The above property of (18) explains why the function $\bar{x}_\varepsilon^{(N)}(t)$ is referred to as the outer asymptotic solution. In fact, it is seen from (15) that the second term in (18) tends to zero rapidly when $\varepsilon \to 0$ for all t except the vicinity of $t = 0$. To describe its properties more rigorously, the following definition will be introduced: An abstract function $\varphi_\varepsilon(t)$ defined over the intervals $0 \leqslant t \leqslant t_0$ and $0 < \varepsilon \leqslant \varepsilon_0$ with the values from E or from the Banach space $L(E, E)$ of bounded operators on E into itself, is said to be tending to zero β-nearly uniformly on $(0, t_0]$

with $\varepsilon \to 0$ if for each $\delta > 0$ and $\beta \in (0, 1)$ there exist t_1 and ε_1 such that $0 < t_1 < t_0$ and $0 < \varepsilon_1 < \varepsilon_0$ and for each $\varepsilon \in (0, \varepsilon_1]$ and $t \in [\varepsilon^{\beta} t_1, t_0]$ there holds the inequality $\|\varphi_\varepsilon(t)\| < \delta$. ($\|\cdot\|$ denotes the norm in E or $L(E, E)$, respectively). If the above statements are valid for each $\beta \in (0, 1]$ then $\varphi_\varepsilon(t)$ is said to be tending to zero nearly uniformly. The function $\psi_\varepsilon(t)$ is said to be tending β-nearly or nearly uniformly on $(0, t_0]$ to the function $\psi(t)$ if the difference between these two functions tends to zero, respectively, β-nearly or nearly uniformly.

As an example take for $a > 0$ the function $f_\varepsilon(t) = \varepsilon^{\gamma} \exp\left(-\dfrac{at}{\varepsilon}\right)$ with the values from the one-dimensional Euclidean space R^1. It is seen that $f_\varepsilon(t)$ tends to zero with $\varepsilon \to 0$ uniformly on $[0, t_0]$ if $\gamma > 0$, nearly uniformly on $(0, t_0]$ if $\gamma = 0$, and β-nearly uniformly on $(0, t_0]$ if $\gamma < 0$.

With the above introduced definition, it is seen that the quasi-semigroup $\varepsilon^{\gamma} U_\varepsilon(t, s)$ tends to zero nearly uniformly on $(0, t_0]$ for $\gamma = 0$ and β-nearly uniformly for $\gamma < 0$. In other words, the exact solution $x_\varepsilon(t)$ to the evolution equation (13) tends to the outer asymptotic solution $\bar{x}^{(N)}(t)$ of order N $(N > 0)$ as ε^N with $\varepsilon \to 0$ β-nearly uniformly on $(0, t_0]$. For $N = 0$ one has a nearly uniform convergence.

The last step in the analysis of the asymptotic behavior of $x_\varepsilon(t)$ is to obtain the Nth order asymptotic form of the term $U_\varepsilon(t, 0)\left(\eta - \bar{x}^{(N)}(0)\right)$. This form will be called the inner asymptotic solution of order N and denoted by $U_\varepsilon^{(N)}(t, 0)\left(\eta - \bar{x}^{(N)}(0)\right)$. The function $U_\varepsilon^{(N)}(t, s)$ will be said to approximate the quasi-semigroup $U_\varepsilon(t, s)$ up to the order N if $U_\varepsilon(t, s)$ is tending to zero nearly uniformly on $(0, t_0]$ and for each $x \in D(B)$ the function $\varepsilon^{-N}\left(U_\varepsilon(t, s)x - U_\varepsilon^{(N)}(t, s)x\right)$ is tending to zero uniformly with respect to t on the interval $[s, t_0]$ for any $s \in [0, t_0)$.

By making use of the definition of β-nearly convergence one can show that the function $U_\varepsilon^{(N)}(t, s)$ can be defined in the following way. For any $\mu \in D(B)$ and $s \in [0, t_0)$ consider the system of evolution equations with respect to t with operators independent of t

$$(19) \qquad \varepsilon \frac{d}{dt}\left(y_\varepsilon^{(n)}(t, s)\right) = B(s) y_\varepsilon^{(n)}(t, s) + \sum_{k=1}^{n} \frac{1}{k!}\left(\frac{t-s}{\varepsilon}\right)^k \times$$
$$\times \frac{d^k}{ds^k}\left(B(s)\right) y_\varepsilon^{(n-k)}(t, s), \qquad n = 0, 1, \ldots, N$$

with the initial conditions

$$(20) \qquad y_\varepsilon^{(0)}(s, s) = \mu, \qquad y_\varepsilon^{(1)}(s, s) = \ldots = y_\varepsilon^{(N)}(s, s) = 0.$$

From the properties of the operator $B(s)$ and of the inhomogeneous terms it follows that the system (19) has a unique solution $\{y_\varepsilon^{(0)}(t, s; \mu), \ldots y_\varepsilon^{(N)}(t, s; \mu)\}$. Now the function $U_\varepsilon^{(N)}(t, s)$ is defined such that for any $\mu \in D(B)$

$$(21) \qquad U_\varepsilon^{(N)}(t, s)\mu = \sum_{n=0}^{N} \varepsilon^n y_\varepsilon^{(n)}(t, s; \mu).$$

The inner asymptotic solution of order N is obtained from (19) and (20) upon substitution of $s = 0$ and $\mu = \eta - \bar{x}_\varepsilon^{(N)}(0)$.

Since the difference between $U_\varepsilon(t, s)x$ and $U_\varepsilon^{(N)}(t, s)x$ for any $x \in D(B)$ tends, according to the definition of $U_\varepsilon^{(N)}(t, s)$, to zero as ε^N with $\varepsilon \to 0$ uniformly with respect to t, in (18) one can substitute $U_\varepsilon(t, 0)$ by $U_\varepsilon^{(N)}(t, 0)$ and still the whole expression will be tending uniformly to zero. Therefore, the asymptotic solution of order N to the evolution equation (13), tending uniformly to zero with $\varepsilon \to 0$ as ε^N, has the form

$$(22) \qquad x_{\varepsilon,\mathrm{asym}}^{(N)}(t) = \bar{x}_\varepsilon^{(N)}(t) + U_\varepsilon^{(N)}(t, 0)\left(\eta - \bar{x}_\varepsilon^{(N)}(0)\right)$$

where the outer asymptotic solution of order N is defined by (17) and the inner asymptotic solution of order N by (19) and (20) with $s = 0$ and μ substituted by $\left(\eta - \bar{x}_\varepsilon^{(N)}(0)\right)$.

All the results obtained in this section are obviously valid also for the operator A independent of time.

Conclusions

The analysis of the singular perturbation method given in the previous section shows that the accuracy of the asymptotic solution (22) depends essentially on the ratio of the parameters α and ε with the provision that α is negative since otherwise the whole analysis would not be valid. It is an advantage of the presented approach that the role of both parameters is clearly demonstrated.

In many applications one is not interested in the behavior of $x_\varepsilon(t)$ near $t = 0$. For such cases the outer asymptotic solution gives an exceptionally good approximation if only the ratio ε/α is sufficiently small.

However, if the effect of initial conditions cannot be neglected and the detailed description of the behavior of $x_\varepsilon(t)$ in the so-called boundary layer is needed, then the equations (19) have to be solved for $s = 0$ and $\mu = \eta - \bar{x}_\varepsilon^{(N)}(0)$. Such an approach has still some advantages over the direct method of solving the evolution equation (13) since the boundary layer is normally rather thin and in the rest of the interval $[0, t_0]$ one has to solve the stationary equation and this is usually much less time-consuming than solving the time dependent equation. Besides, the equations (19) include time-independent operators and thus are much easier to solve than the original evolution equation.

Another important feature of the present approach is that it may be perhaps used for introducing new approximate physical models by serving as a perturbation procedure complementary to the existing ones. Such a possibility in quantum field theory has been indicated to the author by Rączka [8].

It is to be noted that the singular perturbation method has been already applied to the evolution equations and, in particular, to the neutron transport equation [1]. The formal approach described in [1] is based, however, on purely intuitive argument and does not give any quantative estimations of the range of validity of the method. Moreover, the final asymptotic solution differs from that derived in this paper and it consists of three terms instead of two. The outer asymptotic solution obtained in [1] is the same as that given by (17) but the inner asymptotic solution is defined in [1] differently so that an additional term, the intermediate asymptotic solution, has to be subtracted from the sum of the inner and outer asymptotic solutions to give the total asymptotic solution. The intermediate solution is defined as an outer asymptotic solution for the inner asymptotic solution and, at the same time, as the limit of the outer asymptotic solution for small values of t. Thus, in the intermediate region the matching procedure for the outer and inner asymptotic solutions is needed and the whole method is then called the matched asymptotic expansion method.

The singular perturbation method presented in this paper does not contain any matching procedure. The convergence of the exact solution to the asymptotic one is clearly defined and the effect of the parameters ε and α on the rate of convergence shown. Finally, the equations to be solved to find the inner asymptotic solution are simpler than those given in [1] which may be important in practical calculations. The difference between the here presented approach and that of [1] is discussed in detail in [7].

Reference

* Supported in part by the International Atomic Energy Agency under the Research Contract No. 1236/RB.

Bibliography

[1] W. L. Hendry, 'Solution to the Linear Time-Dependent Neutron Transport Equation with Time-Dependent Source and Cross Sections', *Nucl. Scie. Eng.* **45** (1971) 1–6.

[2] E. Hille and R. S. Phillips, 'Functional Analysis and Semigroups', *Amer. Math. Soc. Colloq. Publ.* **31** (1957).

[3] T. Kato, 'Integration of the Equation of Evolution in a Banach Space', *J. Math. Soc. Japan* **5** (1953) 208–234.

[4] S. G. Krein, *Linear Differential Equations in a Banach Space* (in Russian), Nauka, Moscow 1967.

[5] J. Mika, 'Singular Perturbation Method for Linear Differential Equations in Banach Spaces', in: *International Conference on Integral, Differential and Functional Equations, Bled, Yugoslavia, May 29–June 2, 1973.*

[6] J. Mika, 'Singular Perturbation Method in Neutron Transport Theory', *J. Math. Phys.* **15** (1974) 892–8.

[7] J. Mika, 'Higher Order Singular Perturbation Method for Linear Differential Equations in Banach Spaces', submitted for publication in *Annales Polonici Mathematici.*

[8] R. Rączka, private communication.

[9] K. Yosida, *Functional Analysis*, Springer, Berlin 1965.

GEOMETRIC QUANTIZATION OF SYMPLECTIC MANIFOLDS

D. J. SIMMS

School of Mathematics, Trinity College, Dublin, Ireland

Introduction

The aim of the theory of geometric quantisation is to study the structure of the phase space of a classical mechanical system and to devise procedures for the construction of Hilbert spaces to represent the corresponding quantum mechanical system. In this framework it seeks to establish relationships between canonical transformations on the classical phase space and operators on the Hilbert space.

Attempts in this direction have only been partially successful up to now. In this article we give an account of the method developed by Kostant [8], [9] in the context of a unified theory of representations of Lie groups. Some of the essential ideas were developed independently by Souriau [17]. More recently [10], [4] there has been joint work with Blattner and Sternberg.

In the following, all manifolds are real, finite dimensional, C^∞, Hausdorff and paracompact, and all functions, vector fields etc. are C^∞, and complex valued unless otherwise specified.

Symplectic Structure on the Phase Space

Let X be the configuration space of a finite dimensional classical mechanical system, assumed to be holonomic, conservative, and time independent. Then X is a manifold of dimension n say, and the velocity phase space is represented by the space $T(X)$ of all contravariant vectors on X. The $2n$-dimensional manifold $T(X)$ is called the *tangent bundle* of X.

The momentum phase space, on the other hand, is represented by the space $T^*(X)$ of all covariant vectors on X. This is also a $2n$-dimensional manifold, called the *cotangent bundle* of X. The momentum and velocity phase spaces are related by a diffeomorphism defined by the Lagrangian function of the system, as outlined in the next section.

Let $\pi: T^*(X) \to X$ be the map which assigns to each covector v the point where it is located. Let x^i be a coordinate system on X with domain V, let $p_i(v)$ denote the components of a covector v in this coordinate system, and let $q^i(v)$ denote the coordinates of the point $\pi(v)$ where v is located. Then the functions $(p_1, \ldots, p_n, q^1, \ldots, q^n)$ form a coordinate system on $M = T^*(X)$ with domain $\pi^{-1}V$. The 1-form $\sum p_i dq^i$ on $\pi^{-1}V$ is independent of the choice of coordinates x^i on X and therefore there is a unique 1-form α on M which equals $\sum p_i dq^i$ on $\pi^{-1}V$ for any choice of coordinates x^i with domain V. The exterior derivative $\omega = d\alpha$ is a 2-form on M which in $\pi^{-1}V$ is equal to $\sum dp_i \wedge dq_i$. Thus ω is a non-singular, real, closed ($d\omega = 0$), differential 2-form on M.

A manifold M together with a non-singular, real, closed differential 2-form ω on M is called a *symplectic manifold*. The 2-form ω is called a *symplectic form*.

Poisson Brackets and the Classical Equations of Motion

On a symplectic manifold (M, ω) we can define a Lie algebra structure on the space $C^\infty(M)$ of all smooth complex valued functions on M, in the following way. For each (complex) covector field u on M, let $\tilde{u}$ denote the unique vector field which satisfies

$$\omega(\tilde{u}, v) = \langle u, v \rangle$$

for all vector fields v. For each function $\varphi \in C^\infty(M)$ we write $\xi_\varphi = \widetilde{d\varphi}$ and we define the *Poisson bracket* of two scalar fields φ and ψ to be the function

$$\{\varphi, \psi\} = \xi_\varphi(\psi).$$

This operation makes $C^\infty(M)$ into a complex Lie algebra.

Let $U(M)$ denote the space of complex contravariant vector fields on M; this is a Lie algebra under the usual Lie bracket of vector fields. The map $\varphi \to \xi_\varphi$ is a Lie algebra homomorphism from $C^\infty(M)$ to $U(M)$. We call ξ_φ the *Hamiltonian vector field generated by* φ.

If, with respect to a coordinate system x^i, ω has components ω_{ij} with inverse matrix ω^{ij}, and if u has components u_i, then $\tilde{u}$ has components

$u^i = \sum \omega^{ji} u_j$ and

$$\{\varphi, \psi\} = \sum \omega^{ij} \frac{\partial \varphi}{\partial x^i} \frac{\partial \psi}{\partial x^j}$$

in the domain of x^i.

If X is the configuration space of a classical system then the kinetic energy T, the potential energy V, and the Lagrangian $L = T - V$ are functions on the velocity phase space $T(X)$. We assume that the mapping $u \to \tilde{u}$ from $T(X)$ to $T^*(X)$ given by

$$\langle \tilde{u}, v \rangle = \frac{d}{dt} L(u + tv)|_{t=0}$$

is a diffeomorphism. It is called the *Legendre transformation* and is discussed in Abraham [1], § 18. Under this map the function $T+V$ on $T(X)$ corresponds to a function H, the *Hamiltonian*, on $T^*(X)$. The classical motion of the system corresponds to the integral curves of the vector field ξ_H on the symplectic manifold $T^*(X)$.

The Classical Systems of Souriau. Invariance Groups

In the previous section we have seen that the momentum phase space $M = T^*(X)$ is a symplectic manifold and the motion of the system can be entirely described in terms of the symplectic form ω and the Hamiltonian function H. It has been proposed by Souriau in his book [17] that, rather than requiring the momentum phase space M to be of the form $T^*(X)$ for some configuration space X, we should allow for more general types of symplectic manifolds as models for classical momentum phase space. For example, he considers a pair $(M, \omega_{s,m})$ where M is the space of all pairs (l, J) where l is a future-oriented time-like straight line in Minkowski space and J is a unit vector orthogonal to l, and $\omega_{s,m}$ is a certain symplectic form which is invariant under the action of the Poincaré group. Since the Poincaré group acts transitively on M we may regard $(M, \omega_{s,m})$ as a model for the momentum phase space of a classical elementary relativistic particle. The translation part of the Poincaré group defines Hamiltonian vector fields on M of the form ξ_{P_i} where $P_i \in C^\infty(M)$. The homogeneous part defines Hamiltonian vector fields $\xi_{M_{ij}}$ where $M_{ij} \in C^\infty(M)$. The functions P_i and $W_i = \sum \varepsilon_i^{jkl} P_j M_{kl}$ then represent the four-momentum and the polarisation, and satisfy $P^2 = m^2$ and $W^2 = s^2m^2$. We may regard the system as one with mass m and spin s. The space M in this example is diffeomorphic to $\boldsymbol{R}^6 \times S^2$ and is not of the form $T^*(X)$ for any configuration space X. Moreover the symplectic form $\omega_{s,m}$ is not of the form $d\alpha$ for any 1-form α, (in other words ω is not an exact form).

For a given Lie group G we say that (M, ω) is a *homogeneous symplectic G-manifold* if ω is a symplectic form on M and an action of G on M is given which acts transitively on M and which leaves ω invariant. Then, as suggested by Bacry [3] and as illustrated by the above example of Souriau, a classical system is said to be an *elementary system relative to an invariance group* G if its momentum phase space is a homogeneous symplectic G-manifold.

For a given connected Lie group G, all homogeneous symplectic G-manifolds can be found by using an idea of Kirillov [7]. Such a manifold is always a covering space of an orbit of the action of G on the dual of a central extension of the Lie algebra of G by a 1-dimensional algebra. Moreover any such covering space is a homogeneous symplectic manifold, where the symplectic form corresponds to the bracket in the algebra. These central extensions play an analogous role in classical mechanics to the role of ray (or projective) representations in quantum mechanics. Using these techniques, Souriau [17] has found models for the elementary relativistic and Galilean systems, and Rawnsley [11, 12] has done the same for the de Sitter systems.

For a very thorough treatment of homogeneous symplectic manifolds, see the Springer article by Kostant [9].

Connections on Hermitian Line Bundles

Let (M, ω) be a symplectic manifold representing the momentum phase space of a classical system. We wish to consider a class of wave functions in order to quantise the system. To maintain a suitable level of generality, we will not simply consider complex valued functions $s: M \to \boldsymbol{C}$, but rather functions of the type

$$s: M \to \bigcup_{x \in M} L_x$$

where $s(x) \in L_x$ for all x, and each L_x is a 1-dimensional complex Hilbert space with inner product $(\cdot, \cdot)$.

In order to introduce notions of differentiability, we give $L = \bigcup_{x \in M} L_x$ the structure of a vector bundle with fibre L_x over x. This means that L is a manifold and an open cover $M = \bigcup_{i \in I} V_i$ and differentiable maps $s_i : V_i \to L$ exist with $s_i(x) \in L_{x'}$, and $s_i(x) = c_{ij}(x) s_j(x)$ where $c_{ij} \in C^\infty(V_i \cap V_j)$ and $|c_{ij}| = 1$. We call a map $s: M \to L$ *a section of* L if $s(x) \in L_x$ for all x, and we denote by $\Gamma(L)$ the complex vector space of all C^∞ sections of L. L itself is called a *Hermitian line bundle over* M. We denote by $\pi: L \to M$ the map which projects L_x onto x.

If $\varphi \in C^\infty(M)$ and $s, t \in \Gamma(L)$ we define $\varphi s \in \Gamma(L)$ and $(s, t) \in C^\infty(M)$ by

$$(\varphi s)(x) = \varphi(x) s(x) \quad \text{and} \quad (s, t)(x) = \big(s(x), t(x)\big).$$

In order to be able to consider the derivative along a vector field ξ of a section s of a Hermitian line bundle L, we introduce the notion of a connection on L. A *connection* ∇ *on a Hermitian line bundle* L is a map which assigns to each complex vector field ξ on M a linear operator

$$\nabla_\xi : \Gamma(L) \to \Gamma(L)$$

such that

(i) $\nabla_{\xi+\eta} = \nabla_\xi + \nabla_\eta$,

(ii) $\nabla_{\varphi\xi} s = \varphi \nabla_\xi s$,

(iii) $\nabla_\xi(\varphi s) = (\xi\varphi) s + \varphi \nabla_\xi s$,

(iv) $\xi(s, t) = (\nabla_\xi s, t) + (s, \nabla_\xi t)$

for all $\xi, \eta \in U(M)$, all $s, t \in \Gamma(L)$ and $\varphi \in C^\infty(M)$. We call $\nabla_\xi s$ the *covariant derivative* of the section s along the vector field ξ with respect to the connection ∇.

Each connection ∇ on a Hermitian line bundle L defines a real closed 2-form on M denoted by $\operatorname{curv} \nabla$ and called the *curvature of* ∇, such that the operator

$$[\nabla_\xi, \nabla_\eta] - \nabla_{[\xi, \eta]}$$

on $\Gamma(L)$ is equal to multiplication by the function $2\pi i(\operatorname{curv} \nabla)(\xi, \eta)$ for all $\xi, \eta \in U(M)$.

A real closed 2-form ω on a manifold M is said to be *integral* if its periods over closed surfaces in M are all integers or, more precisely, if ω defines an integral de Rham cohomology class. For a given real closed 2-form ω on M there exists a Hermitian line bundle L with connection ∇, such that $\text{curv}\nabla = \omega$, if and only if ω is an integral 2-form. If, moreover, M is simply connected then L and ∇ are essentially uniquely determined by ω.

A symplectic manifold (M, ω) is said to be *quantisable* if ω is an integral 2-form. A Hermitian line bundle L with connection ∇ over a quantisable symplectic manifold (M, ω) is called a *quantum bundle* for the symplectic manifold if $\text{curv}\nabla = \omega$.

For a thorough treatment of these ideas see Kostant's Springer article [9]. These concepts also play a major role in Souriau's book [17].

Polarisations

Let (M, ω) be a $2n$-dimensional symplectic manifold. Then the complexification M_x^C of the tangent space M_x to M at x is a $2n$-dimensional complex vector space. By a *complex distribution* F on M we mean a function which assigns to each $x \in M$ a vector subspace F_x of M_x^C. We say that F is *involutive* if the set of vector fields

$$U_F(M) = \{\xi \in U(M) | \xi_x \in F_x \text{ for all } x \in M\}$$

is a Lie subalgebra of $U(M)$. We say that F is *maximally isotropic* with respect to ω if $\omega_x(F_x, F_x) = 0$, and $\dim F_x = n$, for all $x \in M$. For any complex distribution F on M we have complex distributions $\bar{F}$, $F \cap \bar{F}$, and $F + \bar{F}$ on M given by $x \to \bar{F}_x$, $x \to F_x \cap \bar{F}_x$, and $x \to F_x + \bar{F}_x$, where $\bar{F}_x$ denotes the complex conjugate of F_x in M_x^C. If we write $D_x = F_x \cap \bar{F}_x \cap M_x$ then $F_x \cap \bar{F}_x = D_x^C$ and $D: x \to D_x$ is a real distribution on M. If F is involutive and if $F_x \cap \bar{F}_x$ has constant dimension k (say) then D is also involutive (integrable) and defines a k-dimensional foliation of M. We then denote by M/D the quotient space of M by the foliation. A complex involutive maximally isotropic distribution F, such that $F \cap \bar{F}$ has constant dimension, is called a *polarisation* of the symplectic manifold (M, ω), and (M, ω, F) is called a *polarised symplectic manifold*. We call F a *real polarisation* if $F = \bar{F}$; in which case D is n-dimensional and gives a Lagrangian foliation of M. We call F a *Kähler polarisation* if $F \cap \bar{F} = 0$.

For example, when $M = T^*(X)$ is the momentum-phase-space corresponding to a configuration space X, we have a polarisation such that F_x is spanned by the momentum directions $\dfrac{\partial}{\partial p_i}$ at each point x.

This is a real polarisation, and the quotient M/D can be naturally identified with the configuration space X. A choice of polarisation of a symplectic manifold (M, ω) corresponds in this way to a *choice of configuration space*.

A typical example of a Kähler polarisation is given by the Riemann sphere M with complex coordinate z, symplectic form $\omega = \dfrac{idz \wedge d\bar{z}}{(l+z\bar{z})^2}$ and polarisation F spanned at each point by $\dfrac{\partial}{\partial \bar{z}}$.

Line Bundles of Half-Forms

Let (M, ω, F) be a polarised symplectic $2n$-dimensional manifold. For each $x \in M$ let B_x^F denote the set of ordered basis (frames) of the complex vector space F_x. The general linear group $GL(n, C)$ acts simply transitively on the right on B_x^F and $B^F = \bigcup_{x \in M} B_x^F$ is a right principal $GL(n, C)$ bundle over M, called the *frame bundle of* F. There is a unique double covering σ: $ML(n, C) \to GL(n, C)$ and the covering group $ML(n, C)$ is called the *complex metalinear group*.

A *metalinear frame bundle* for F is a right principal $ML(n, C)$ bundle $\tilde{B}^F$ over M together with a double covering σ: $\tilde{B}^F \to B^F$ such that the diagram

$$\begin{array}{ccc} \tilde{B}^F \times ML(n, C) & \to & \tilde{B}^F \\ \downarrow & & \downarrow \\ B^F \times GL(n, C) & \to & B^F \end{array}$$

commutes, where the vertical maps are given by the double coverings and the horizontal maps by the group actions. An element of $\tilde{B}_x^F$ is called a *metalinear frame* of F_x. The relation of metalinear frames to ordinary frames in this theory is analogous to the relation of spinor frames to orthonormal frames in Riemannian geometry.

A section β: $V \to \tilde{B}^F$ over an open set V is called a *metalinear F-frame field on* V. Such a section is said to be *Hamiltonian* if there exist $\varphi_1, \ldots, \varphi_n \in C^\infty(V)$ such that

$$\sigma\beta(x) = \left(\xi_{\varphi_1}(x), \ldots, \xi_{\varphi_n}(x)\right)$$

for all $x \in V$.

Let $\chi: ML(n, C) \to C$ be the unique holomorphic function such that

(i) $\chi(g)^2 = \det \sigma(g)$,

(ii) $\chi(1) = 1$

for all $g \in ML(n, C)$. Let L_x^F denote the set of all complex valued functions v on $\tilde{B}_x^F$ such that

$$v(bg) = \overline{\chi(g^{-1})}v(b)$$

for all metalinear frames b and all $g \in ML(n, C)$. Since $ML(n, C)$ acts simply transitively on $\tilde{B}_x^F$ we see that L_x^F is a 1-dimensional complex vector space and $L^F = \bigcup_{x \in M} L_x^F$ is called the *complex line bundle of half-F-forms.*

A C^∞ section $\nu: M \to L^F$ is called a *half-F-form* on M and the space of such sections is denoted by $\Gamma(L^F)$. A half-F-form ν is said to be *Hamiltonian at* $x \in M$ if there exists a Hamiltonian metalinear F-frame field β on a neighbourhood V of x such that $\nu_y(\beta(Y)) = 1$ for all $y \in V$. There is a unique way of defining for each $\xi \in U_F(M)$ a linear operator ∇_ξ on $\Gamma(L^F)$ such that for all $\xi, \eta \in U_F(M)$, all $\varphi \in C^\infty(M)$, and $\nu \in \Gamma(L^F)$ we have

(i) $\nabla_{\xi+\eta} = \nabla_\xi + \nabla_\eta$,

(ii) $\nabla_{\varphi\xi}(\nu) = \varphi(\nabla_\xi \nu)$,

(iii) $\nabla_\xi(\varphi\nu) = (\xi\varphi)\nu + \varphi(\nabla_\xi \nu)$,

(iv) $(\nabla_\xi \nu)(x) = 0$ if ν is Hamiltonian at x.

For each $x \in M$ we can identify $(F+\bar{F})_x$ with the space of all complex valued linear functions on M_x/D_x by means of the map $v \mapsto \omega(v, \cdot)$, where $D_x = F_x \cap \bar{F}_x \cap M_x$. In this way we can identify the exterior power $\Lambda^{2n-k}(F+\bar{F})_x$ with the space $\Lambda^{2n-k}(M_x/D_x)^*$ of complex valued alternating $(2n-k)$-forms on M_x/D_x where k is the dimension of D_x.

We denote by $|\Lambda^{2n-k}(M_x/D_x)^*|$ the set of complex multiples of the absolute values of functions belonging to $\Lambda^{2n-k}(M_x/D_x)^*$ and we denote by $|\Lambda^{2n-k}(M/D)^*|$ the corresponding complex line bundle over M. We define a map $\langle \cdot, \cdot \rangle$:

$$\Gamma(L^F) \times \Gamma(L^F) \to \Gamma(|\Lambda^{2n-k}(M/D)^*|)$$

by

$$\langle \nu_1, \nu_2 \rangle(x) = \nu_1(b)\overline{\nu_2(b)}|\omega^{n-k}(v_1 \ldots v_{n-k}\bar{v}_1 \ldots \bar{v}_{n-k})|^{-1/2} \times$$
$$\times |u_1 \wedge \ldots \wedge u_k \wedge v_1 \wedge \ldots \wedge v_{n-k} \wedge \bar{v}_1 \wedge \ldots \wedge \bar{v}_{n-k}|$$

where b is any element of $\tilde{B}^F_x$ such that $\sigma(b) = (u_1, \ldots, u_k, v_1, \ldots, v_{n-k}$ and where $(u_1, \ldots, u_k)$ is a basis for $(F \cap \bar{F})_x$. This definition is independent of the choice of b since if b' is another such element of $\tilde{B}^F_x$ then $b' = bg$ with $g \in ML(n, C)$ and $\sigma : ML(n, C) \to GL(n, C)$ maps g to $\begin{bmatrix} A & * \\ 0 & B \end{bmatrix}$ (say). Thus if we put b' in place of b in the definition of $\langle v_1, v_2\rangle(x)$ we see that the effect is to multiply it by

$$\begin{aligned} &\overline{\chi(g^{-1})}\,\chi(g^{-1})|\det B \det \bar{B}|^{-1/2}|\det A \det B \det \bar{B}| \\ &= |\chi(g^{-1})^2 \det A \det B| \\ &= 1 . \end{aligned}$$

Quantisation

Let M be a manifold with quantisable symplectic form ω, and let F be a polarisation of ω. Let L^ω be a Hermitian line bundle with connection ∇ having ω as curvature form. Let L^F be a complex line bundle of half-F-forms. For each $\xi \in U_F(M)$ let ∇_ξ be the unique linear operator on $\Gamma(L^\omega \otimes L^F)$ such that

$$\nabla_\xi(s \otimes v) = (\nabla_\xi s) \otimes v + s \otimes (\nabla_\xi v)$$

for all $s \in \Gamma(L^\omega)$ and $v \in \Gamma(L^F)$.

Write

$$\Gamma^F = \{\psi \in \Gamma(L^\omega \otimes L^F) | \nabla_\xi \psi = 0 \text{ for all } \xi \in U_F\}$$

and let $\langle \cdot, \cdot \rangle$ be the unique sesquilinear map

$$\Gamma^F \times \Gamma^F \to \Gamma(|\Lambda^{2n-k}(M/D)^*|)$$

such that

$$\langle s_1 \otimes v_1, s_2 \otimes v_2\rangle_x = (s_1, s_2)_x \langle v_1, v_2\rangle_x$$

for all $s_i \in \Gamma(L^\omega)$, $v_i \in \Gamma(L^F)$, $x \in M$.

Now we assume that the quotient M/D of M by the foliation D is a Hausdorff manifold such that the projection $\pi : M \to M/D$ gives a surjection $M_x \to (M/D)_{\pi(x)}$. (A removal of this restriction would considerably widen the scope of the theory.) Then M_x/D_x may be identified with $(M/D)_{\pi(x)}$ and in this way the space of densities on the $(2n-k)$-dimensional manifold M/D may be identified with a subspace of

$\Gamma(|\Lambda^{2n-k}(M/D)^*|)$. The image of the above sesquilinear map lies in the subspace, and we write

$$\mathscr{H}_0^F = \left\{\psi \in \Gamma^F \mid \int_{M/D} \langle \psi, \psi \rangle < \infty \right\}.$$

We then have a pre-Hilbert structure $(\cdot, \cdot)$ on $\mathscr{H}_0^F$ given by

$$(\psi_1, \psi_2) = \int_{M/D} \langle \psi_1, \psi_2 \rangle.$$

We denote by $\mathscr{H}^F$ the Hilbert space obtained by completion of $\mathscr{H}_0^F$.

In order to quantise a real classical function $\varphi \in C^\infty(M)$ as an operator on $\mathscr{H}^F$, we consider the Hamiltonian vector field ξ_φ generated by φ. If this real vector field is complete then it defines a 1-parameter group σ_t of diffeomorphisms of M and a unitary map of $\mathscr{H}^F$ onto $\mathscr{H}^{\sigma_t F}$. Under certain conditions there is an integral transform introduced, by Blattner, Kostant, and Sternberg in [4], and discussed by Gawędzki in [6] and in this volume, which maps $\mathscr{H}^{\sigma_t F}$ into $\mathscr{H}^F$. The infinitesimal generator of the composition

$$\mathscr{H}^F \to \mathscr{H}^{\sigma_t F} \to \mathscr{H}^F$$

is taken to be the quantised form of φ.

Applications

The following applications of the ideas outlined above have been made. Renouard [13] has applied Kostant's quantisation procedure to the elementary classical relativistic and Galilean systems of Souriau. He showed that the procedure yields the usual representations of the Poincaré and Galilean groups associated with elementary relativistic and Galilean particles in quantum mechanics. Rawnsley [11] carried out the same programme for the elementary de Sitter systems. Elhadad in [5] has studied polarisations of the phase space of the Kepler problem. Simms in [14] considered the set of classical trajectories of fixed energy E for the Kepler problem, and showed that for every choice of polarisation F the dimension of the resulting Hilbert space $\mathscr{H}^F$ is equal to the multiplicity of the energy level E in the Schrödinger equation. In [15] Simms considered a special polarisation of the phase space of the harmonic oscillator which contains the classical trajectories, and showed that the resulting energy levels and multiplicities are the same as for the Schrödinger equation.

Bibliography

[1] R. Abraham, *Foundations of Mechanics*, Benjamin 1967.
[2] R. Arens, 'Classical Lorentz Invariant Particles', *J. Math. Phys.* **12** (1971) 2415.
[3] H. Bacry, I. A. S. preprint, Princeton 1966.
[4] R. J. Blattner, 'Quantisation and Representation Theory', in: *Harmonic Analysis on Homogeneous Spaces* (edited by C. C. Moore), A. M. S. Proc. Sym. Pure Math. XXVI, 1974.
[5] J. Elhadad, 'Sur l'interpretation en geometrie symplectique des etats quantiques de l'atome d'hydrogene', *Convegno di Geom. Simp. e Fis. Mat., INDAM, Rome 1973*, to appear in Symposia Math. Series, Academic Press.
[6] K. Gawędzki, 'Geometric Quantisation Kernels' (to appear).
[7] A. A. Kirillov, 'Unitary Representations of Nilpotent Lie Groups', *Russian Math. Surveys* **17** (1962) 53.
[8] B. Kostant, *Proceedings of the United States–Japan Seminar in Differential Geometry*, Kyoto 1965.
[9] B. Kostant, *Quantisation and Unitary Representations*, Lectures in Modern Analysis and Applications III (edited by C. T. Taam), Springer Lecture Notes in Mathematics **170** (1970).
[10] B. Kostant, 'Symplectic Spinors', *Conv. di Geom. Simp. e Fis. Mat., INDAM, Rome 1973*, to appear in Symposia Math. Series, Academic Press.
[11] J. Rawnsley, 'De Sitter symplectic spaces and their quantisations', *Proc. Cambridge Phil. Soc.* (to appear).
[12] J. Rawnsley, 'Orbits for inhomogeneous orthogonal groups', Preprint, Ist. di Fis. Theor., Napoli 1974.
[13] P. Renouard, 'Variétés symplectiques et quantification', These, Orsay 1969.
[14] D. J. Simms, 'Geometric Quantisation of Energy Levels in the Kepler Problem', *Conv. di Geom. Simp. e Fis. Mat., INDAM, Rome 1973*, to appear in Symposia Math. Series, Academic Press.
[15] D. J. Simms, 'Geometric Quantisation of the Harmonic Oscillator with Diagonalised Hamiltonian', *Proc. of 2nd Int. Coll. on Group Theoretical Methods in Physics*, Nijmegen 1973.
[16] J. Śniatycki, 'Bohr–Sommerfeld Conditions in Geometric Quantisation', Research paper No. 216 (1974), Dept. of Math. Calgary.
[17] J.-M. Souriau, *Structure des systèmes dynamiques*, Dunod 1970.

GEOMETRIC QUANTIZATION – AN EXTENDED PROCEDURE

K. GAWĘDZKI

Department of Mathematical Methods in Physics, Warsaw University, Warsaw, Poland

In the present lecture we shall use all notions introduced and notations adopted in the D. J. Simms lecture in this volume. We shall refer to this lecture as to [0] and all other references will concern the reference list contained there.

Introduction

If one specifies the quantization construction described in [0] in the simplest case when $M = \boldsymbol{R}^2$ (we shall use x and p as coordinates in $\boldsymbol{R}^2$), $\omega = dp \wedge dx$, and we take the polarization F generated by the vector field $\frac{\partial}{\partial p}$, one arrives at the usual position representation of quantum mechanics. Thus H^F is in fact $L^2(\boldsymbol{R}, dx)$. The Kostant quantization procedure described shortly in [0] is well established in the case when the hamiltonian flow of the function to be quantized (i.e. the flow of the hamiltonian vector field defined by it) preserves the polarization. There are not so many functions like that in the considered case. Only functions φ at most linear in p are admitted. As the corresponding quantized operators $\hat{\varphi}$ are first order differential operators one can give them a standart kernel representation used for all pseudo-differential operators:

$$(1)\qquad (\hat{\varphi}f)(y) = \frac{1}{2\pi}\int e^{ip(y-x)}a_\varphi(x,p)\,dp\wedge dx,$$

where $f \in C_0^\infty(\boldsymbol{R}) \subset L^2(\boldsymbol{R}, dx)$.

The function a_φ, called the *symbol of the operator* $\hat{\varphi}$, is closely connected to φ. More exactly

$$(1')\qquad a_\varphi = \varphi + \frac{i}{2}\,\frac{\partial^2\varphi}{\partial x\partial p}.$$

If one takes another polarization, say this generated by $\frac{\partial}{\partial x}$ one gets the momentum representation of quantum mechanics. The class of functions whose hamiltonian flows preserve the polarization changes. These are now functions at most linear in x. Using the Kostant prescription one can quantize such functions now. But the position and the momentum representations of quantum mechanics can be 'translated' one into another by the Fourier transform. Thus the 'momentum' quantization prescription can be also obtained in the $L^2(\boldsymbol{R}, dx)$ space (by superposing the Fourier transform, the quantized operator in $L^2(\boldsymbol{R}, dp)$ and the inverse Fourier transform). *For the quantized operators obtained in this way the kernel representation* (1) *and* (1′) *holds in unchanged form.* Thus one can treat (1) and (1′) as an (at least formal) generalized prescription for quantization of functions which are combinations of a part at most linear in x and another one at most linear in p. The aim of this lecture is to show how one can obtain kernel representations of the type of (1) and (1′) in the case of a general polarized symplectic manifold. We shall not go into technicalities and refer the reader to [6] for the detailed constructions.

Distinguished Kernels

First we shall define a general notion which corresponds to the $e^{ip(y-x)}$ part of the kernel under the integral sign in (1). We need some extra geometric structure to be introduced.

Let M (connected, simply connected), ω, F, L, L^F, H^F be as in [0]. Suppose that we are given another polarization F' which is transversal to F, i.e. at any point $m \in M$ $F_m \cap F'_m = \{0\}$. Let us consider the bundle $L^F \otimes L^{F'}$. We have a natural isomorphism

$$(2) \qquad (L^F \otimes L^{F'}) \otimes (L^F \otimes L^{F'}) \simeq (L^F \otimes L^F) \otimes (L^{F'} \otimes L^{F'}).$$

But the bundle $L^F \otimes L^F$ (as one can easily see from the definition of L^F in [0]) is naturally isomorphic to the bundle $\Lambda^n(M)^*/F$ of n-forms on M vanishing when contracted with a vector from F. Thus

$$(3) \qquad (L^F \otimes L^F) \otimes (L^{F'} \otimes L^{F'}) \simeq \left(\Lambda^n(M)^*/F\right) \otimes \left(\Lambda^n(M)^*/F'\right) \simeq \Lambda^{2n}(M)^*,$$

the last isomorphism being generated by the external multiplication. Composing isomorphisms (2) and (3) we get

(4) $\quad (L^F \otimes L^{F'}) \otimes (L^F \otimes L^{F'}) \simeq \Lambda^{2n}(M)^*.$

Thus $L^F \otimes L^{F'}$ is a 'square root' of the bundle $\Lambda^{2n}(M)$. The last bundle is trivial as $\overset{n}{\omega}$ is its non-vanishing section. As M has been assumed to be connected and simply connected there exist exactly two sections of $L^F \otimes L^{F'}$ such that their tensor square goes into $\overset{n}{\omega}$ under the isompor-phism (4). We shall choose and fix one of them and shall denote it by $(\overset{n}{\omega})^{1/2}$.

We shall need besides the bundle L^ω also its dual $(L^\omega)^*$. One can provide $(L^\omega)^*$ with a connection ∇^* dual to ∇, i.e. such that if $s \in \Gamma(L^\omega)$, $s^* \in \Gamma\big((L^\omega)^*\big)$, and ξ is a vector field on M then

$$\xi(\langle s|s^*\rangle) = \langle \nabla_\xi s|s^*\rangle + \langle s|\nabla_\xi^* s^*\rangle,$$

where $\langle \cdot | \cdot \rangle$ denotes the canonical bilinear pairing between sections of L^ω and $(L^\omega)^*$.

If ξ is a vector field from $U^{F'}(M)$ then we can define in a similar way besides the operator ∇_ξ acting in $\Gamma(L^\omega \otimes L^{F'})$ an operator ∇_ξ^* in $\Gamma\big((L^\omega)^* \otimes L^{F'}\big)$.

Now let

$$S := \bigcup_{(m, m') \in M \times M} (L^\omega \otimes L^F)_m \otimes \big((L^\omega)^* \otimes L^{F'}\big)_{m'}.$$

S can be easily given a structure of a complex one-dimensional vector bundle over $M \times M$. Let K be a section of S. For each $m \in M$ $K(m, \cdot)$ can be viewed as a section of $(L^\omega)^* \otimes L^{F'}$ (if we do not care about the ambiguity in normalization), and similarly for each $m' \in M$ $K(\cdot, m')$ can be viewed as a section of $L^\omega \otimes L^F$. Moreover if we consider K restricted to the diagonal $\Delta := \{(m, m) \in M \times M\}$ we can view it as a section of $L^\omega \otimes L^F \otimes (L^\omega)^* \otimes L^{F'}$ or simply of $L^F \otimes L^{F'}$. Keeping this identifications in mind we give

DEFINITION. A non-zero section K of S is called a *distinguished kernel* if

1. for each $m \in M$ and each $\xi \in U^{F'}(M)$,

$$\nabla_\xi^* K(m, \cdot) = 0,$$

2. for each $m' \in M$ and each $\xi \in U^F(M)$,

$$\nabla_\xi K(\cdot, m') = 0,$$

3. $K|_{\Delta}$ is proportional to $(\overset{n}{\omega})^{1/2}$.

One can show that a distinguished kernel exists and is unique up to a constant in the following general cases:

A. F and F' are real and $\pi \times \pi' : M \to M/D \times M/D'$ is a diffeomorphism, $\pi \times \pi'(m) := (\pi(m), \pi'(m))$, where π and π' are canonical projections of M onto M/D and M/D' respectively;

B. F is of Kähler type, $F' = \bar{F}$, and M is a homogeneous space for a group of holomorphisms.

Remark. Kernels of a similar type for real polarizations where first introduced by Blattner, Kostant and Sternberg (see [4, 10]) when they examined a generalized Fourier transform intertwining the Hilbert spaces obtained for different polarizations. Our kernels should be rather viewed as kernels composed of two B.K.S. kernels.

Suppose we are given a distinguished kernel K. If $\psi \in H_0^F$, $m' \in M$, we can pair ψ and $K(\cdot, m')$ using the sesquilinear pairing $\Gamma^F \times \Gamma^F \to \Gamma(|\Lambda^{2n-k}(M/D)^*|)$ defined in [0] and try to integrate the resulting tensor density over M/D. The only difference, comparing with the definition of the scalar product in H^F, is that $\int_{M/D} \langle \psi, K(\cdot, m') \rangle$, if exists, then in fact is an element of $((L^{\omega})^* \otimes L^F)_{m'}$, not a number. Let $H_c^F := \{\psi \in H_0^F$: projection of suppt$\psi$ onto M/D is compact$\}$. We shall need some regularity properties of the distinguished kernels we shall use (see [6] for slightly weaker conditions).

Definition. A distinguished kernel K will be called *regular* if

1. for each $\psi \in H_c^F$ and each $m' \in M$ $\langle \psi, K(\cdot, m') \rangle$ is integrable and the map

$$M \ni m' \mapsto \int_{M/D} \langle \psi, K(\cdot, m') \rangle \in ((L^{\omega})^* \otimes L^F)_{m'} \subset (L^{\omega})^* \otimes L^F$$

is a smooth section of $(L^{\omega})^* \otimes L^F$;

2. for each $\chi \in H_c^F$

$$M \ni m' \mapsto \frac{\chi(m') \otimes \int_{M/D} \langle \varphi, K(\cdot, m') \rangle}{(\overset{n}{\omega})^{1/2}(m')} =: \{\psi, \chi\}(m') \in \mathbf{C}$$

is a function from $L^1(M, \overset{n}{\omega})$;

3. the sesquilinear form

$$H_c^F \times H_c^F \ni (\psi, \chi) \mapsto \int_M \{\psi, \chi\}\tilde{\omega} \in \mathbf{C}$$

is norm continuous.

These regularity conditions are always satisfied in the general case B mentioned above and in many special cases.

DEFINITION. A regular distinguished kernel is called a *reproducing kernel* if the sesquilinear form

$$H_c^F \times H_c^F \ni (\psi, \chi) \mapsto \int_M \{\psi, \chi\}\tilde{\omega} \in \mathbf{C}$$

coincides with the one defined by the scalar product.

One can show that if M is a homogeneous space for a group G of diffeomorphisms preserving ω and F then there exists a projective unitary representation U of this group in H^F. One can prove

PROPOSITION. Suppose that the general case A mentioned above holds or that M is compact. Let K be a reproducing kernel and φ a real function on M whose hamiltonian flow preserves F. Let $\xi_\varphi = \xi_\varphi^1 + \xi_\varphi^2$, where ξ_φ^1 takes values in F and ξ_φ^2 in F'. Let $\hat{\varphi}$ be the operator assigned to φ by the Kostant quantization procedure defined on the domain $D_{\hat{\varphi}}$. Then for $\psi \in H_c^F$ and $\chi \in H_c^F \cap D_{\hat{\varphi}}$

$$(\psi, \hat{\varphi}\chi) = \int_M \left(\varphi + \frac{i}{2}\frac{\pounds_{\xi_\varphi^2}\overset{n}{\omega}}{\overset{n}{\omega}}\right)\{\psi, \chi\}\overset{n}{\tilde{\omega}}, \tag{5}$$

if the integral on the right hand side exists.

We shall consider the general case A more carefully. One can show that in this case if there exists a reproducing kernel then it defines an isometric operator $V: H^F \to H^F$ which intertwines both Kostant procedures of quantization of functions whose hamiltonian flows preserve both F and F'.

PROPOSITION. Suppose that the general case A holds. Let K be a reproducing kernel and let φ be a real function on M whose hamiltonian flow preserves F. Let ψ, $\chi \in H_c^F$ and suppose that $V\chi \in D_{\hat{\varphi}'}$, where $\hat{\varphi}'$

is the Kostant quantization of φ acting in $H^{F'}$. Then

$$(6)\qquad (V\psi, \hat{\varphi}' V\chi) = \int_M \left(\varphi + \frac{i}{2}\,\frac{\pounds_{\xi_\varphi}^2 \overset{n}{\omega}}{\overset{n}{\omega}}\right) \{\psi, \chi\}\, \overset{n}{\omega},$$

if the integral on the right-hand side exists.

Comparing (5) and (6) we see that one can use (5) as (at least formal) definition of an extended quantization procedure which, when we deal with the general case A, seems reasonable at least for combinations of functions whose hamiltonian flows preserve either F or F'.

SYMMETRIES GAINED AND LOST*

R. THOM

Institut des Hautes Etudes Scientifiques, Bures-sur-Yvette, France

I. Transversality, Genericity, Structural Stability and All That

I.1. TRANSVERSALITY: THE IMPLICIT FUNCTION THEOREM, GENERALIZED

Let N be Euclidean n-space, with coordinates $(x_1, x_2, \ldots x_n)$, P – Euclidean p-space with coordinates $(y_1, y_2, \ldots, y_p)$. A system of p smooth functions $f_j(x_i)$, $j = 1, 2, \ldots, p$, $i = 1, 2, \ldots, n$, defines a smooth map $f: N \to P$, by the equations: $y_j = f_j(x_i)$. N is called the *source* of the map, P the *target*. We suppose $n \geqslant p$. A point x of N is a *regular* point of the map f iff at x the jacobian matrix $D(f_j)/D(x_i)$ is of maximal rank p. If x is regular, then, in a neighborhood U of x, the level variety $f^{-1}[f(x)]$ is germ of a smooth manifold, of codimension p: the implicit function theorem states that, in a neighborhood of x, the p functions $f_j, j = 1, 2, \ldots, p$ may be taken as local coordinates. The regular points of f form an open set in N; any point x' of the complementary closed set C, has the property that the jacobian matrix is at x' of rank $\leqslant p-1$. The point x' will be called a *critical* point of f, C is the *critical* set of f. The image set $f(C) \subset P$ will be called the set of *singular values*; this set may not be closed in P; but if the map f is *proper* (i.e. the counter-image of a compact $K \subset P$ is compact in N), then $f(C)$ is closed. The complementary set $P - f(C)$ is called the set of *regular values* of the map f. This set is not empty. In fact the classical theorem of Sard states that the set of singular values $f(C)$ has zero p-dimensional measure in P. Let y be a regular value of f; then its counter-image $f^{-1}(y)$ consists only of regular points. Hence, it is a smooth $(n-p)$-manifold $W(y)$.

If y is a *regular* value, and if f is *proper*, then for all maps f' sufficiently near f in the C^r topology, y is also a regular value (continuity of the first order derivatives $\partial f_j/\partial x_i$ on a compact in N), and the corresponding counter-image $f'^{-1}(y)$ is isotopic to $f^{-1}(y)$ in N. The topological stability of the counter-image of a regular value for small perturbations of the (proper) map is one of the fundamental properties of these regular values.

Transversal maps. Let V be an embedded manifold of codimension q in P; a map $f: N \to P$ is called *transversal on* V, if for any couple of points $x \in N$, $y \in V$ such that $y = f(x)$, the direct sum $f_* T_x(N) \otimes T_y(V)$ of the image vector space $f_* T_y(N)$ and the tangent space $T_y(V)$ is the whole tangent space $T_y(P)$. A simple adaptation of the preceding result shows:

(i) If $f: N \to P$ is transversal on V, then the counterimage $f^{-1}(V)$ in N is a smooth manifold $W(f)$ of codimension q = codimension of V.

(ii) If f is proper, and f' sufficiently near f in the space $L(N, P)$ of C^r maps from N to P, then f' is also transversal on V, and $W(f')$ is isotopic to $W(f)$.

(iii) Any map $g: N \to P$ can be approximated in $L(N, P)$ by a transversal map.

Because of (ii) and (iii), the proper transversal maps form an open dense set in $L(N, P)$; for this reason, the property of transversality on a submanifold is said to be *generic*. (If we relax the condition of properness, the property of transversality holds only for a Baire set in $L(N, P)$; one may reobtain an open dense set if one uses a stronger topology, in $L(N, P)$, the so-called Whitney topology.)

The property of transversality is obviously equivalent to the notion of 'general position', so frequently found in the old Italian algebraic geometry.

I.2. STRATIFIED SETS AND TRANSVERSAL MAPS ON THEM

The preceding definition of transversal mappings requires V to be smooth. What happens if V has singularities as for instance in the case of a real algebraic variety? One may then *stratify* the variety V, that is, decompose V into a finite union of disjoint embedded manifolds (*the strata* X_k of V). One imposes, for the gluing of the strata together, some conditions of continuity for the tangent planes, the so-called Whitney (A, B) conditions: for instance, if the stratum Y is in the boundary of X, then if y is a point of Y, x_i a sequence of points of X tending to y, then $\lim T_{x_i}(X)$ contains $T_y(Y)$. (These conditions are shown to hold almost anywhere in the singular locus of an algebraic set). This being done, a map $f: N \to P$ is *transversal* on V iff it is *transversal on each stratum* X_k of V. In that case, the inverse image $f^{-1}(V) = A(f)$ is itself a stratified set in N, and the same deformation theorems hold for $A(f)$ as for the previous manifold $W(f)$. In other words: for almost any map $f: N \to P$ the counter-image $f^{-1}(V)$ is a stratified set, the topological type of which is invariant with respect to small deformations of f.

An example of a stratified set. Consider the general equation of degree k in one complex variable

$$z^k + u_1 z^{k-1} + \dots + u_s z^{k-s} + \dots + u_k = 0,$$

the coefficients u being complex indeterminates. The discriminant variety $G(u) = 0$ is the set of points where the corresponding equation has at least one multiple root. To any point in G, we associate the partition π of k defined by the multiplicities $i_1, i_2, \dots, i_s, \sum_j i_j = k$, of the roots of the corresponding equation. It is easily seen that the locus of points of G admitting the same partition (up to order!) form a manifold X_π. The union of such X form a stratification of G. To any stratum X there corresponds a well defined conjugacy class of subgroups of S_k, the symmetry group of permutations of k objects (these are the Galois groups of the corresponding equations). One obtains that way a natural family of subgroups of S_k, which form (in some sense we do not wish to define here) the 'universal unfolding' of the trivial action (identity) of S_k. (Problem: what is the relation of these S_π to the Young diagrams used in physics?)

I.3. GROUP ACTIONS

Let G be a Lie group acting smoothly on a differentiable manifold M. The *orbit* of a point $m \in M$ is the set of points of the form $g \cdot m$, $g \in G$; it is a manifold, which may not be properly (locally closed) embedded in M when G is not compact (example: the irrational geodesic on the flat torus). The *stabilizer* N_m of m is the subgroup of G leaving m fixed; the stabilizers N_m, $N_{m'}$ of two points in the same orbit are conjugate through an inner automorphism s of G, such that $m' = s \cdot m$. From an infinitesimal point of view, the action defines, for any point $m \in M$, a homomorphism of the Lie algebra L_G into the Lie algebra X_m of all germs of vector fields around m. The image of this homomorphism is the tangent plane to the orbit $G \cdot m$, the kernel is the Lie algebra of the stabilizer $L(N_m)$.

(a) *The compact case.* When G is compact, and acts in a compact manifold M, great simplifications arise: by averaging on G, one may form a Riemann metric invariant under G, hence G acts only through isometries. As a result, any orbit θ admits an invariant tubular neighborhood $T(\theta)$, and there exists an equivariant retraction r: $T(\theta) \to \theta$. The stabilizer subgroup N_m, $m \in \theta$, acts in the normal bundle to θ at m by an orthogonal representation (topologically equivalent, via the exponential map, to the local action of the stabilizer).

This implies the fact that there are only a finite number of conjugacy classes which may occur in such an action. Let F be the set of points m where N_m belongs to a given conjugacy class; this set is an invariant set, and, in the neighborhood of any of its points, F consists of the normal geodesic manifold spanned by the eigenvectors of value one (fixed points of the stabilizer representation). Hence F is a manifold, locally closed, which is fibered into orbits. The decomposition of M into strata F (each stratum corresponding to a conjugacy class) is a stratification in the previous sense: for all 'normal stars' on two points belonging to the same connected component of a stratum are isometric, as defined by conjugate orthogonal representations. These strata have locally semialgebraic models; the quotient space itself M/G may be given a stratified structure.

It has been proved by Palais that such a compact action (G compact in M compact) is 'rigid', that is any nearby action h' to an action h is topologically equivalent to h via an homeomorphism k of M on itself $h' = k \circ h$. This shows that the space Hom(G, Diff$M$) of all actions of G in M consists of components which are orbits under Diff$M$.

The singularities of G-invariant real functions (or maps) are of a completely different nature from the generic singularities of functions without any symmetry constraint. The result (presented here by L. Michel [7]) on the necessary vanishing of the differential df of a G-invariant function f on a 'critical' orbit is a striking example of this phenomenon. It shows that the use of 'genericity' arguments is conclusive only if we know the totality of the constraints to which our function (or more generally the mathematical object we wish to make generic) is subjected.

(b) *The non compact case.* Little is known about actions of non compact groups, except for the two fundamental examples $G = Z$, and

$G = R$, which give rise to the classical dynamical systems theory (discrete = diffeomorphisms, and flows). Of fundamental importance, in that case, is the concept of 'structural stability'. For instance, a flow X in a manifold M is said to be *structurally stable*, if for any flow X' sufficiently near X (in the C^1-topology), there exists an homeomorphism $k = M \to M$, which carries any trajectory of X into a trajectory of X'.

At the origin of the theory of structural stability, it was hoped that structurally stable flows (or diffeomorphisms) were open and dense in the C^1 Banach space of flows $X(M)$, (resp. $D(M)$) on M. Unfortunately, this hope was betrayed (except for $D(S^1)$ and $X(M^2)$), and the recent history of this very lively theory (due to S. Smale and his 'school') points to an ever increasing pathology. Nevertheless, a few simple facts remain: the 'generic' gradient flows are structurally stable, and they admit a structurally stable stratification defined by the so called *stable* and *unstable manifolds* (transversal each to other) associated to the singular points. Moreover, if there exists in M an oriented hypersurface H, such that the flow X is entering H, then the flow X certainly admits attractors in H, and the same property holds true for all flows X' sufficiently near X. Moreover, any attractor (closed invariant set which is the limit set of all nearby trajectories) admits a local Lyapunov function (i.e. a function increasing along the trajectories); hence under perturbation an attractor may not increase too much in size. The difficulty lies in controlling the topological changes undergone by an attractor during a deformation of the field. This is the subject known as bifurcation theory which is well known only in case of gradients (theory of 'elementary catastrophes'). The notion of attractor is important, because it expresses in a geometric way the idea of an 'asymptotic regime' of a dynamical system. It is useful to recognize whether a given attractor is 'locally structurally stable', as this property implies topological equivalence with the attractor of any nearby perturbed field, and, besides the simplest ones, the point and the generic closed trajectory, a lot of others, of very complicated, Cantor-like nature, are known to exist (cf. [17]).

The theory of classical flows, already so difficult, is still outstandingly clear in comparison with the theory of Hamiltonian systems, for which practically nothing is known from a qualitative point of view, except the negative result (which physicists are very reluctant to consider) that ergodicity is not, for such systems, a generic property.

I.4. LOCAL MODELS FOR GROUP ACTIONS

Suppose we are given a discrete group π, with finite presentation, i.e. generators $s_1, \ldots, s_k$, relations $W_j(s_i) = e$. The set of representations of π in the linear group $GL(m; C)$ may be obtained as follows: to any generator s_j we associate a matrix $M_j \in GL(n; C)$; in the space R^{kn^2} of coefficients of these k matrices, we consider the algebraic set defined by writing that the products $W_j(M_i), \ldots$ are the identity matrix. This shows that the set of representations of degree n of π is an algebraic set (on which $GL(n)$ acts by inner automorphisms). It is particularly important to consider the neighborhood of the identity action; it has a stratified structure, each stratum corresponding to a conjugacy class of stabilizers (if π is finite).

In the same way, let A be a Lie algebra defined by generators and relations; if M is a manifold, let $X(M)$ be the (infinite dimensional) Lie algebra of vector fields in M; the preceding construction defines the set $\mathrm{Hom}(A, X(M))$ as an 'algebraic' set in an infinite dimensional space. We may localize this construction to the classification of actions of A around a fixed point; here we take as target the Lie algebra $\mathscr{L}(0)$ of all germs of vector fields around $0 \in R^m$.

As usual in such cases, it is useful to consider first the Taylor expansion at 0 of such an homomorphism. In modern terminology one defines mappings of A into the *space of jets* $J^r(X(m))$ which satisfy the relations of A in J^{r-1}. Here again one gets, for each r, algebraic sets which form a projective limit. The family of stabilizer subalgebras which appear in this construction, and the corresponding algebraic (pro-)set may be called the *universal unfolding* of the constant action of A in R^n.

As a special case, we define a generic zero of a vector field in this way: 0 is a generic zero of X, if the matrix of the linear coefficients in the Taylor expansion of X at 0 has its eigenvalues outside the imaginary axis (*hyperbolic* singular point). It is well known, that, in such case, the local action is topologically equivalent to the linear part: around $0 \in R^n$ is direct sum of two manifolds transversal at 0, W^-, W^+; on W^-

the flow is attracting towards 0 (stable manifold), on W^+, it is dilating. (These manifolds W are those used to define the stratification associated to a gradient field). In the same way, for a closed trajectory C, we take a local hyperplane section H: a trajectory leaving a point $p \in H$ meets H for the first time at p': the correspondance $p \to p'$ defines the 'Poincaré–Floquet' diffeomorphism of C; if this diffeomorphism is generic at $0 = C \cap H$, i.e., if all the eigenvalues of the linear part are outside the unit circle, then the closed orbit C is *hyperbolic*. Hyperbolic fixed points and hyperbolic closed orbits are structurally stable.

II. Why Symmetries Break Down

II.1. BROKEN SYMMETRIES AND RESONANCES

A typical example. Let T^2 be the two-dimensional torus defined by the quotient $\boldsymbol{R}^2/Z \times Z$, with coordinates $x, y \,(\text{mod}\, 1)$. Let $dy = c\,dx$ the linear differential system with irrational slope c; the integral curves are geodesics each of which is not structurally stable. Hence one may expect the following 'natural' evolution: first, the slope c varies till it becomes a rational number p/q near to c; then, all the orbits of the parallel flow of slope p/q are closed, and it degenerates toward a structurally stable flow: a so-called *Morse–Smale* flow, with a finite number s of attracting closed trajectories, an equal number of repelling trajectories, and the other trajectories spiralling from a repelling cycle to an attracting one. The original flow was invariant with respect to translations parallel to the x, y directions. The final flow is no longer invariant: it is only approximately invariant. Let us consider the differential systems $dy/dx = f(x, y)$ as imposing a condition on maps from $0x$ as source space to $0y$ as target space; the translation group acts on both $0x$ and $0y$, and the original flow was invariant under this action. Any solution of $dy/dx = c$ is also invariant under the translation $x' = x+h$, $y' = y+ch$. The same remains true after c has been replaced by p/q. But this time the x-translation factorizes through the subgroup $x \to x+k/q$. At the end stage, there are no more Lie group symmetries for the global flow (except possibly if the breaking occurs with rotational symmetry).

In this example, one sees that 'broken symmetry' may have four different meanings:

(a) The first meaning is perhaps the most classical: It applies when a differential problem is invariant under the action of a Lie group G, and there exist solutions which are invariant only under a subgroup G' of G. In such a case, if $s(x)$ is a G'-invariant solution, then all $g \cdot s(x)$, for $g \in G$ are also solutions; there exists a continuum of solutions parametrized by the homogeneous space G/G'.

It is rather exceptional, that for a G-invariant problem, any solution is effectively G-invariant. As in this first example, this may be impossible only because of dimensionality reason (the dimension of the source space being lesser than the dimension of the full symmetry group G).

(b) The first change (primary symmetry breaking) from slope c to slope p/q exemplifies the second meaning of 'broken symmetry': this occurs when the G action on solutions may be factorized through a quotient group K usually with compact fundamental domain (here the circle group quotient of R by $x \to x+q$). In such a case, it would be more accurate to speak of 'degenerating symmetry', as the action of the symmetry group of any solution factorizes by a normal subgroup.

(c) The last change (secondary symmetry breaking = degeneracy to a 'Morse–Smale' field) is the true symmetry breaking. But the final solutions (the attractors) have also the K-symmetry. (That K, in this example, is a compact group, may play a fundamental role in the fact that the attractor solutions do have this extra symmetry, while the general solution has only the G symmetry, G non compact).

(d) As we shall see in the next chapter (on macroscopic physics), it happens frequently that the secondary breaking is itself symmetrical, modulo a subgroup G'' of G (in our case, the translation $x \to x+1/s$, s the number of attracting trajectories). In fact (but this is quite likely a very special phenomenon due to dimension two) any degeneracy admits such a rotational symmetry, after having applied a global diffeomorphism to T^2. This is our last meaning of 'breaking symmetry' and it is probably the most usual.

Recall that the first meaning applies to a *fixed* differential system, while the three others apply to a *perturbation* of a differential system. The 'symmetry breaking' discussed at length in this meeting on elementary particle theory is obviously of the first type; it may be only a way of looking at things, and most quantum physicists would discard with utmost energy the idea that such a symmetry breaking could be induced by a 'real' physical process. This should not be a justification for not knowing what happens in the macroscopic symmetry breaking of the other three types, particularly the last one.

Resonances. The preceding example may also be looked at in the following way: let two linear oscillators E_1, E_2 with periods T_1, T_2 respectively, be given. If T_2/T_1 is irrational, then the flow on the product torus is of the type considered earlier; it is structurally unstable, and any coupling between the two oscillators (not Hamiltonian) will result in the apparition of a resonance, that is, the two periods will change in such a way as to get a rational ratio; the choice of the phase of the periodic process obtained in that way is a further symmetry breaking. It is quite clear that the smallest deviation from linearity for one of the oscillators may lead to such a behavior.

Moreover, the preceding example may be completed as follows: let X be a flow on T^2, $x(t)$, $y(t)$ a trajectory of the flow: such a trajectory may always be lifted to the universal covering R^2 of T^2, with parametrization $x(t)$, $y(t)$. Then the ratio $y(t)/x(t)$, for t tending to $+\infty$, has a limit, which is the same for all trajectories, the *rotation number* of the flow. (It is the slope c for our original flow). It may be then shown that:

(a) The rotation number is a continuous function of the flow in $X(T^2)$.

(b) Any flow with irrational rotation number, of class C^r, $r > 2$, is isomorphic to the linear flow (Denjoy's theorem).

This shows that in a global deformation of the flow X(a path in $X(T^2)$), in which the rotation number varies monotonically from a to b, the flow has to be isomorphic to the linear flow on a set which has the power of the continuum, despite the fact that, for a rational rotation number, the flow is structurally stable, and stays there for some interval of time. Hence *symmetry breaking may be a relatively easily reversible process.* (It is believed that, in an analytic deformation of X, the set where the flow is irrational has positive measure ...)

II.2. NON GENERICITY OF FIRST INTEGRALS

The problem of broken symmetries may be generalized to the problem of stability of 'first integrals'. Recall that in a classical dynamical system $(M; X)$, a function $F: M \to R$ is a first integral if it is invariant under the flow, i.e. $dF(X) = 0$. Although local first integrals always exist near a regular point of the flow, a global first integral (in a compact manifold M) is not structurally stable. As soon as the flow X has recurrent trajectories, then *Pugh's closing lemma* applies: any recurrent trajectory of X can be approximated by closed trajectories of a field X' nearby X; as these closed trajectories are hyperbolic, there cannot be any first integral for X'. (Recall that the Poincaré diffeomorphism of an hyperbolic closed trajectory is 'spiralling' towards this trajectory in the stable, or unstable manifold.) Despite this negative result, the notion of first integral still plays a very important role in mechanics and physics. This is due to the introduction of the Hamiltonian formalism.

II.3. FIRST INTEGRALS IN HAMILTONIAN FORMALISM

Let M^{2n} be a symplectic manifold, α its fundamental two-form. Let $F: M^{2n} \to R$ be real-valued smooth function; the symplectic gradient of F (notation of J. M. Souriau) is the vector field X defined by: $i(X) \cdot \alpha = dF$, i – interior product. If F is taken as Hamiltonian, then X is the associated field (given by Hamilton–Jacobi equations). We recall that F itself is first integral of X, as

$$i(X) \cdot dF = i(X) \cdot i(X) \cdot \alpha = \alpha(X, X) = 0$$

and

$$d[i(X) \cdot \alpha] = ddF = 0 = L_X \alpha,$$

which shows that the field X leaves invariant α, hence its n exterior product (Liouville measure). (Here L_X denotes the Lie derivative with respect to X.)

Suppose we are given a vector field Y which leaves invariant α and the Hamiltonian H; then the flow X, symplectic gradient of H, leaves invariant the closed one-form $i(Y) \cdot \alpha$.

First prove that $i(Y) \cdot \alpha$ is closed. For

$$d[i(Y) \cdot \alpha] = i(Y) \cdot d\alpha + L_Y \cdot \alpha = 0$$

as $L_Y \alpha = 0$ means that the flow Y leaves invariant the symplectic structure. Moreover, as $i(Y) \cdot \alpha$ is closed,

$$L_X[i(Y) \cdot \alpha] = d[i(X) \cdot i(Y) \cdot \alpha] = d[\alpha(Y, X)],$$

and using the fact that Y leaves H invariant:

$$0 = i(Y) \cdot dH = i(Y) \cdot i(X) \cdot \alpha = \alpha(X, Y).$$

Suppose a Lie group G acts in M^{2n} via *symplectomorphisms* and leaves H invariant; if g_k is a system of generators for its Lie algebra $\mathscr{L}_G$, then g_k have images g_k' in M and the one-forms $i(g_k')\alpha$ may at least locally be integrated as differentials of functions g_k^*; these functions locally span the dual $\mathscr{L}_G^*$ of the Lie algebra of G, which form in that way a system of local first integrals. In case where M^{2n} is the covector bundle of a configuration space V^n, and if G acts only in V, with derived action in $T^*(V) = M$ then $\mathscr{L}_G^*$ is a system of global first integrals.

This way of obtaining first integrals for hamiltonian systems with symmetries is very general (Noether's theorem). Here again one might suspect that the presence of a symmetry for a hamiltonian system is highly non generic, and that it may be destroyed by a small perturbation of the Hamiltonian.

Of course, this is true. But a theorem due to Arnol'd shows that, under a weak hamiltonian perturbation, the increase of the first integral, averaged over the orbit of G, *supposed compact*, is zero. For instance let z be a cyclic parameter, Z the conjugate variable generating the dual Lie algebra (i.e. kinetic momentum for the rotation z). Let $W(z, Z; q, p)$ be the perturbation hamiltonian. The Z-component of its symplectic gradient w is given by

$$i(w) \cdot \alpha(z) = \alpha(w, z) = dW(z)$$

and the integral of the form $dW(z)$ over the circle trajectory of z is zero. This result is very interesting, because it shows that despite the fact that a symmetry by a compact Lie group acting on an Hamiltonian system might be only approximate, nevertheless the corresponding system of first integrals (kinetic momenta) still keeps a kind of statistical existence.

II.4. SPLIT SYSTEMS

DEFINITION. A dynamical system $(P; Z)$ is *split*, if there exist two dynamical systems (M_1, X_1), (M_2, X_2) such that $(P; Z)$ is isomorphic to the direct product $(M_1 \times M_2; X_1 + X_2)$. Given a dynamical system $(P; Z)$ one may try to decompose it as direct product of dynamical systems, which are themselves irreducible with respect to splitting. It is not known whether any two such decompositions are necessarily isomorphic; even for germs of vector fields around a point, and even at the purely formal level of their Taylor expansion, this algebraic problem seems to be entirely open.

We shall be interested here in the structural stability of the 'split' situation: is any system near a split system itself a split system? One may first consider the case where all factors are structurally stable. When all factors are generic gradients, then the product is itself a generic gradient; hence, in that case, the split situation is structurally stable. This seems to be the only positive result one might hope for: as soon as two factors have closed trajectories (or recurrence) then *resonances* by perturbation are unavoidable, and this destroys the split situation (like for the torus field considered in II. 1). Strangely enough, the 'discrete' (diffeomorphisms) case seems to have more stable splittings. For instance the product of two Anosov diffeomorphisms (known to be structurally stable) is itself an Anosov diffeomorphism (hence structurally stable).

II.5. THE MEAN FIELD IN A SPACE OF APPROXIMATE FIRST INTEGRALS

Let $p: M \to Q$ be a proper fibration of the phase space M of a dynamical system on a base space Q; we admit that this fibration defines a system of approximate first integrals: the flow X is of the form $X^0 + Z$, where X^0 is tangent to the fibers, and the 'horizontal component' Z small with respect to X^0 (at least in general). The average on a fiber $p^{-1}(q)$ of the vector Z: $\int_{p^{-1}(q)} Z\, d\mu$ will be called the *mean field* induced by X on the space Q of first integrals.

Of course, the motion of the projection by p of any trajectory m_t in M is not – in general – a trajectory of the mean field. If the vertical dynamic X_0 in a fiber has an attractor then it is better to average the horizontal component only on this attractor, and not on the full fiber. If there are several attractors in competition, the mean field may have several determinations at a point $q \in Q$, also in competition. In the work of Arnold's one find estimates of the error made by using the mean field instead of the true motion.

The case of a 'symplectic' fibration. In the fibration $M \to Q$, M and Q are symplectic manifolds, such that the fundamental 2-form α of M

may be, at any point of M, split into the sum $p(\beta)+\gamma$, where β is the fundamental form of Q, and γ a 2-form of rank $\dim M - \dim Q$, which induces on each fiber a symplectic structure.

Suppose the dynamic on M is given by a hamiltonian H of the form $H(m) = H^0(m) + W(m, q)$, where H^0 depends only on vertical coordinates f_i (such that $\langle p^*(\beta), f_i \rangle = 0$) and W is a small perturbative 'horizontal' hamiltonian. Then the 'horizontal' component Z is the symplectic gradient of W with respect to the horizontal structure $p^*(\beta)$; because of the linearity of the symplectic gradient, one may average Z on the (compact) fiber by averaging W, and taking the symplectic gradient of the averaged W in m. Hence, in that case, the mean field is itself a hamiltonian field defined by an averaged hamiltonian $\tilde{W}$.

Replacing W by its fiber-average $\tilde{W}$ defines a new dynamics on M, which obviously maps fiber into fiber. As each of these displacements are a global symplectic transformation they leave invariant the Liouville measure (defined by γ) of the fibers. In fact this Liouville measure is in general constant.

II.6. COUPLINGS

When two initially disjoint dynamical systems are allowed to interact, then, in general as we saw, new resonances ensue. It is of course very difficult to specify without further hypothesis, which resonance will actually appear. Moreover in case of hamiltonian systems, we can say a bit more.

(a) The sum of the two hamiltonians has to be constant

$$H_1 + H_2 = C.$$

(b) Suppose the two systems exchange energy by swift, catastrophic processes, which spread the Liouville measure ergodically in the hypersurface $H = C$. (This will be called 'thermodynamical coupling'.)

Call $j_1(h_1)$ the Liouville measure of the energy hypersurface (more precisely the derivative dm/dh, where m is the standart Liouville measure inside the hypersurface $H = h$ (set where $H = h$)). Each of the hamiltonians H_1, H_2 may vary from 0 to C. If $H_1 = h$, then $H_2 = C - h$, and the corresponding Liouville measure is the product:

$$j_1(h) \cdot j_2(C-h).$$

Putting $S_i = \text{Log} j_i$, $i = 0, 1$, we have to maximize the sum $S_1(h) + S_2(C-h)$; if we write $1/T_i = dS_i/dh$, the maximum will be obtained for

$$T_1 = T_2,$$

hence for the equality of the 'formal' temperatures of the two systems.

As mean field on the h-axis, we may suppose that the evolution is defined by the gradient of the entropy $S_1 + S_2$; for dynamics of the linear oscillator type (or more generally $j(h) = h^\alpha$, $\alpha > 0$), this leads to the heat equation $dT/dt = K\Delta T$ (for a field of nearby oscillators).

Suppose now the two hamiltonian systems (M_i, X_i) admit the same symmetry group G; if interaction takes place through hamiltonians which are also invariant by G, then, as first integrals, we have, not only the energy, but also the 'kinetic momenta' defined on the dual of the Lie algebra $\mathscr{L}_G^*$: we have invariance of the sum of the two vector momenta $J_1 + J_2$; hence we have to maximize an entropy on an energy interval $0 \leqslant h \leqslant C$ multiplied by the vector space $\mathscr{L}_G^*$; here again, in absence of further hypothesis, we could suppose that the 'mean field' is defined by the gradient of the entropy. For equilibrium, we have to write equality of two Pfaffian forms (replacing the 'formal temperature'). Suppose the two systems geometrically identical, this expresses that if we want a unique equilibrium, the graph of the $S(q)$ function cannot have two points with parallel tangent planes, hence this graph has to be convex.

In Chapter IV (on quantum mechanics), we shall deal with the case where the Lie algebra $\mathscr{L}$ has itself a symplectic structure: the fibration over $\mathscr{L}$ is a symplectic fibration for the interaction, hence we have a mean field which is the *symplectic gradient* of the entropy.

III. How Symmetries Break Down: Examples in Macroscopic Physics

III.1. NATURAL SYMMETRIES

It has been observed, very long ago, that many natural phenomena exhibit symmetries of mysterious origin. In two (old) classic books: 'On Growth and Form' by d'Arcy Thompson, and 'Symmetry', by Hermann Weyl (two books which are still worth reading now, as this topic does not seem to have attracted much attention), one may find a lot of material and examples. Adding to that material some elements found in more recent technical treatises (geomorphology, in particular), I give below a list (probably very incomplete) of such symmetry phenomena:

1. *Fluid Mechanics.* 1° Symmetry breaking of symmetrical initial data. Experience of the cylindrical bathtub. (The Coriolis effect in the determination of the sense of rotation of water does not seem to be predominant.)

Fall in free air of an horizontal resting disk (the center of gravity follows a spiralling curve).

2° Couette–Taylor flow (see, for instance, Feynmans Lectures on Physics, vol. 2).

3° Bénard's phenomenon. (Hexagonal pattern of convection cells in a fluid heated from below.)

4° 'Swell'. Rolling waves in a channel with constant slope.

5° Rayleigh's jet. (Diadem pattern of jets found after the fall of a heavy ball in water.)

6° Hehlmholtz experiment on the diffusion of an ink drop in water. (Jellyfish-like formations: See d'Arcy Thompson.)

7° Periodic flames, 'singing' flames.

8° In meteorology: periodic formations of clouds (Lee waves). Huge cyclonic patterns observed from satellites.

2. *Astronomy.* 1° Spiral nebulae.

2° Periodic emissions of some stellar bodies,

3. *Physico-Chemistry.* 1° Dendritic growth, and global symmetries of crystal formations. (Example: hexagonal symmetry of snow flake.)

2° Polarization of liquid crystals.

3° Periodic deposits in some chemical reactions: Liesegang rings.

4. *Geomorphology.* 1° Periodicity of some erosive patterns (in sandy slopes or in cliffs).

2° 'Cauliflower' structure of stalagmites.

3° Pattern of dunes, of sandy river beds.

4° Cusps of beaches.

5. *Biology.* 1° Bilateral symmetry of most animals (echinoderms to vertebrates).

2° Metamerie: Periodic longitudinal structure of animals like worms, centipeds (annelids); periodic structure of the vertebrate axis (in Somites, or Vertebrae).

3° Patterns of hairs, or feathers (phaneres) on the skin.

4° Symmetries of leaves (phyllotaxis) and of flowers in botanics.

5° At the 'ultrastructure' level: Fibril network structure observed in many tissues (example: collagen).

In all such phenomena, one meets – in principle – the following situation: whereas the initial data have apparently for symmetry group a Lie group G, the final state of affairs has only a subgroup G' of G as symmetry group. The symmetry group G is reduced to one of its subgroups, which contradicts the 'Curie principle': Any symmetry of the causes is to be found in the effects.

This evolution seems as more paradoxical, that, generally the increase of entropy in a system results into an increase of the homogeneous character of the system (think, for instance, of the mixing by diffusion of the gases contained in symmetric containers, when they are allowed to mix by opening a tap on the pipe connecting them).

In such problems, most physicists will have a somewhat contradictory attitude. In quantum mechanics, the physicist accepts without any reluctance indeterminism; but in macroscopic physics, he is a strong believer of determinism: He would say, in our case, that the initial data were only approximately symmetric (mod G), and that this initial dissymmetry, amplifying with time, will lead to an end state which has only the symmetry defined by the subgroup G' of G. Such an answer, obviously, cannot be refuted. But it cannot either be fully confirmed by experiment. Moreover, it does not explain the stability of the G'-symmetry obtained at the end, stability which may be quite remarkable in some experimental circumstances. This is why the physicist should not worry too much about the metaphysical problem of determinism, but try, instead, by a specific modeling, to determine what kind of constraints may apply to such symmetry breaking phenomena.

There is no point in trying to construct a formal deterministic model of the process, as any formal process satisfies the Curie principle: Any automorphism of hypotheses is to be found in the consequences. Perhaps mathematics itself escapes the sterility of formal deduction by allowing itself, at the beginning, the 'axiom of choice': For a non empty set E, the full symmetry group is the group $S(E)$ of all permutations of the elements of E. After having chosen – marked – an element a of E, the couple (E, a) has only as symmetry group the subgroup of $S(E)$ which leaves a fixed.

I believe that it is only by getting some qualitative understanding of the process taking place, that we may hope to specify the statistical constraints applying to the end state, and so to reduce the part of practical indeterminism (structural unstability) which is unavoidable in such symmetry-breaking processes.

III.2. THE BASIC MODEL: COMPETITION OF SINGULARITIES

A geomorphological example. We consider, at the initial time $t = 0$, a sandy slope in form of an inclined plane; we suppose that a gentle rain is steadily falling on this slope. At the top of the hill there will be a very great number of minute brooklets; then, going down the slope, these brooklets will capture each other, thus forming approximately periodic treelike configurations. At the bottom of the slope, we shall have a small number of brooks, separated by approximately equal distance L, the length L is the 'wave-length' of the process. Here, the initial translational symmetry (along coordinate x) is broken by an obvious physical factor: the erosive power of each brook. Where water started to run, there is a sink in the ground, and this sink tends to deepen because of the eroding effect of the running water. We shall study this process in a rough qualitative way.

Let x_i denote the abscissa of the watershed (the divide) separating the basin of the $(i-1)$th stream from the basin of the ith, at the bottom of the slope. We shall suppose that the erosive power of a stream is at first approximation proportional to the width of its basin, of the form $c \cdot (x_{i+1} - x_i)$ for the ith stream. Then the divide x_i may vary, according to the relative strength of the eroding powers of the ith and $(i+1)$th stream acting on both sides of the divide. This leads to the differential equation:

$$dx_i/dt = c \cdot (x_i - x_{i-1}) - c \cdot (x_{i+1} - x_i),$$

$$dx_i/dt = c \cdot (2x_i - x_{i-1} - x_{i+1}).$$

Hence, the equipartition of basins, defined by

$$x_i = \frac{x_{i-1}+x_{i+1}}{2},$$

is a stationnary state. But we have to check the stability of this solution. To do that, we shall use the simplified model of two streams sharing the $[-x, +x]$ interval. Let us suppose that the coefficient c depends only on the width x of the basin. Let u be the abscissa of the divide, near 0. We then have the differential equation:

$$\begin{aligned} u' &= c(x+u)\cdot(x+u)-c(x-u)(x-u) \\ &= [c(x)+uc'(x)](x+u)-[c(x)-uc'(x)](x-u) \\ &= [2c(x)+2xc'(x)]u+\ \ldots \end{aligned}$$

Equipartition will be stable if

$$[c(x)+xc'(x)] < 0.$$

That such a condition is not unnatural may be seen from the following considerations: the eroding power of a stream depends not only on the size of its basin, but also on its slope, which determines the speed of the running water. For an older stream, whose profile is very near the

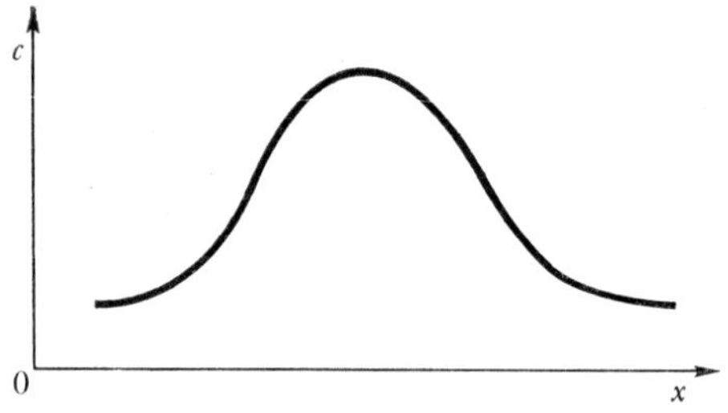

Fig. 1

exponential equilibrium profile, the slope at the end is very small. Hence it may well happen that the eroding power diminishes, while the width of the basin gets larger. (See Fig. 1 for such a function $c(x)$).

If such is the case, the wave-length L of the process is (for a slope of infinite width) the smallest x such that

$$c(x)+xc'(x) < 0 .$$

If we had boundary conditions (for instance a circular mound of circumference A) then L would be the greatest divisor of A for which $c+xc'$ is negative.

As rough as it is, this model describes, I believe, the profound mechanism of many symmetry breakings. In the initial homogeneous situation, there is continuous forming of many little singularities which break the symmetry. But these singularities are unstable and disappear. Now, due to variation of some external parameter, these singularities may become stable; they enter into competition, capture each other, till the extension of a basin causes a diminishing of its eroding power. Then the situation stabilizes through equipartition of the basins.

In the same way, the 'cauliflower' shape of stalagmites (Fig. 2) may be explained as follows: on the upper half of the stalagmite, convex

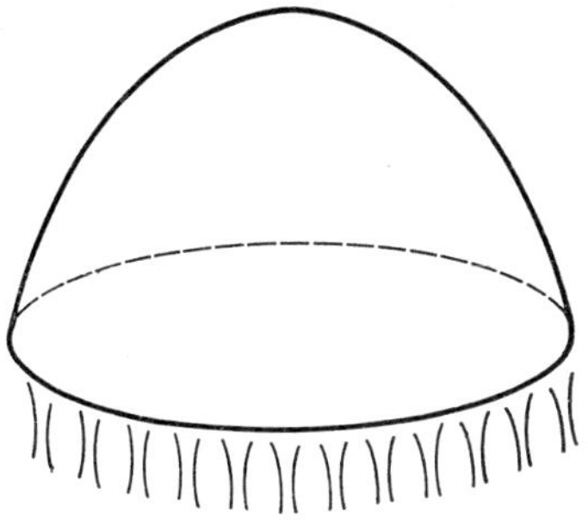

Fig. 2

up, any accidental brooklet is unstable, because it tends to deposit dissolved material, and then to raise up its bed; on the lower half, in contrary, such a stream becomes stable, because the material laid down in vertical '*drapés*' facilitates the running down of water.

I would not wish to convey the idea, that this process of competition between isomorphic singularities is the universal explanation for all symmetrical natural patterns. It is, certainly, of very general nature.

But there is little doubt that some symmetries may require a more sophisticated origin. In particular, biological morphologies may be determined by specific factors of genetic origin. For instance it is known that the sense of rotation of some shells of the Nautilus type is determined by the genotype of the mother. If we cut a mammal egg at the two blastomeres stage, the left half will give rise to a normal embryo, the right half will give rise in 50% of the cases to an embryo with *situs inversus* (heart to the right, liver to the left); hence there exists in the egg a preexisting dissymetry, at least in potentialities. The five-order symmetry, so frequently found in biology, (the fingers of the hand, the petals of the rose) quite likely requires a different explanation from the six-order symmetry, found in crystals, or in equipartition lattices, like in the Bénard phenomenon.

III.3. COMPETITION BETWEEN SYMMETRY-TYPES

Mathematically, the problem of broken symmetries may be described in the following, slightly paradoxical manner: we are given on the $0x$ axis (or on a space E) a constant function. We wish to approximate this function by a non constant one. Is there any canonical way of doing that, especially if we want to preserve some symmetry, or qualitative homogeneity for the approximation? Perhaps Norbert Wiener's 'homogeneous chaos' is the only known attempt to define such a process. If we consider the possibility of breaking the translation symmetry by a discrete subgroup, then these subgroups form a continuous family parametrized by the length k of the fundamental domain. The Fourier transform of the function $f(x)$: $\hat{f}(k)$, such that

$$f(x) = \int \hat{f}(k) \exp ikx \, dk$$

defines, in some sense, the probability that the f function has to degenerate with respect to the k-subgroup, and knowing all these probabilities enables one to reconstruct the function. Can this fact be generalized to other situations?

Suppose an ensemble of processes may be parametrized by the points of a manifold M, and that a Lie group G acts on this manifold; to each point $a \in M$ we associate its stabilizer K_a. Let us suppose that, as in the compact case, all points a with stabilizers in the same conjugacy class (or more generally, in the same isotopy class in G) form a submanifold, a stratum S_K. In this geometric model, if the system has effective G-symmetry, it is initially in the stratum S_G; breaking the symmetry to a subgroup K of G means that the representative point leaves S_G to enter a stratum S_K of the star of S_G. One sees that, in principle, the system has the choice among all strata of the star in the ways of breaking its symmetry. In the case above, of the Fourier transform, the function – being constant – is initially in the stratum S_T of all translations: this a stratum of infinite codimension, which may be looked as the transversal intersection of all strata $S_{(k)}$, corresponding to the subgroup $x \to x+k$. Here degeneracy may be guided by the relative force of the Fourier coefficients. There are cases where the initial stratum is of finite codimension; then generally, the symmetry jumps abruptly while decreasing the codimension of the stratum. This may correspond (see the paper by Ruelle–Takens on the onset of turbulence) to bifurcations of the dynamical system describing the evolution in M.

After the choice of the subgroup K has been made, then one has to describe the precise way the secondary symmetry breaking occurs.

In case of local actions of G (defined by local homomorphisms of the Lie algebra $\mathscr{L}_G$ (mod $\mathscr{L}_K$) in the normal vector fields to the stratum S_K) one is given, for compact K, a field of orthogonal representations of $\mathscr{L}_G$ (mod $\mathscr{L}_K$) in the normal bundle to S_K (the metric being the invariant metric); this field of representations has to satisfy the invariant differential operators defined on S_K by the adjoint G-action. This representation completely defines the incidence of S_K on S_G, which is locally a cone having as base space the homogeneous space H.

To define the 'secondary symmetry breaking' we need an auxiliary dynamical system in the homogeneous space $G/K = H$. Generally, one has to consider the differential operators on H which are invariant under the G-action. Gradients of such solutions (for instance, harmonic functions for the Laplacian), are favorite agents for secondary symmetry breaking. Like in the first examples of resonance on T^2, the remaining K-orbits after the resonance will be defined by maxima of such functions (or, more generally, by critical points). The auxiliary field defined on H which is itself function of the point in S_K describes the 'bifurcation' by which the dynamical system leaves S_G to enter S_K, hence the secondary symmetry breaking.

In case where $\dim G = \dim K$ one gets an homomorphism of the discrete group G/K into the transversal vector fields to S_K. For a closed orbit of a flow, this homomorphism is the Poincaré–Floquet diffeomorphism already described.

IV. Quantum Mechanics as Translational Symmetry-Breaking

IV.1. THE NOTION OF LOCAL PHASE

Let U be a domain of space-time and suppose that some physical process takes place in U; for any subdomain $A \subset U$, there is associated a dynamical system (M_A, X_A), such that if B is a subdomain of A, there exists an homomorphism g_{BA} of (M_A, X_A) onto (M_B, X_B) (i.e. a surjective smooth map $M_A \xrightarrow{g} M_B$, such that if $b = g(a)$, the flow $X_B(b)$ is the image of the vector $X_A(a)$). We have the transitivity relation $g_{CA} = g_{CB} \circ g_{BA}$ for $C \subset B \subset A$. Let u be a point of U; we suppose that the inductive limit (M_A, X_A) for A tending to u may be defined, and is still a reasonable classical dynamical system of hamiltonian type (M_u, X_u). If we associate to any point u of U its local dynamic, we define that way a field of local dynamics, or more briefly, a *metabolic* field.

Suppose then that the medium filling the domain U admits (from the phenomenological point of view) a pseudo-group Γ of symmetries. This means that any transformation γ of Γ acts smoothly in the field of local dynamics, equivariantly with respect to the projection π onto the base space U. In particular, the infinitesimal transformations of the pseudo-group Γ act in any neighborhood, however small, hence also in the punctual dynamics (M_u, X_u). Such a medium filling U will be called a 'local phase' with pseudo-group of invariance Γ.

In this sense, a 'phase transition' developping inside U is to be considered as a local breaking of the Γ-symmetry; in general, the new phase will admit a new pseudo-group of symmetry Γ'. Quite frequently (not always, as phase transitions may be reversible) the new pseudo-group Γ' is a sub-pseudo-group of the old one. Transitions of type 'condensation' like gas $\to$ liquid, liquid $\to$ solid, are of this type.

If the local dynamics is defined by a Hamiltonian $H(x_i, y_i)$, x, y conjugate variables, it is very natural to suppose that a phase transition may occur only for local physical conditions for which this local Hamiltonian is critical; crossing a critical point may affect the ergodicity properties of the flow. As an illustration, passing a critical point of index one may join two disjoint connected componets of the energy hypersurfaces. Suppose 0 is such a critical point then the local Hamiltonian H may be written:

$$H = -x^2 + y^2 + Q(x_i, y_i),$$

where Q is a quadratic positive definite form. Suppose there exists a canonical transformation which reduces H to

$$H = -x^2 + y^2 + \sum (x_i^2 + y_i^2).$$

This may not be the case, although there exists always a diffeomorphism which reduces Q to a quadratic form. If it would not be the case, we would reason on the two jet of H instead of H itself. The 'spherical' orthogonal character of the quadratic form $\sum (x_i^2 + y_i^2)$ expresses the fact that this (local) dynamic does not admit any natural splitting. This is – very likely – the classical translation of the notion of infinitely divisible representations of currents groups (cf. [14]).

Any canonical transformation leaving H invariant has to preserve such decomposition.

Now the group of infinitesimal translations acts in the (x_i, y_i) space in conserving together the metric $\sum (x_i^2 + y_i^2)$ (because they conserve H) and the fundamental symplectic form $\sum dx_i \wedge dy_i$. Hence they leave invariant the unitary structure defined by the complex coordinates $x_i + iy_i$. Now such an Abelian Lie algebra of unitary transformations has to be of even rank and symplectic (H. Weyl's theorem). Let call p_i, q_i the corresponding generators, $\sum dp_i \Lambda dq_i$ the associated symplectic form. The group of infinitesimal translations being a (local) symmetry group for the local dynamics, there exists a surjective map $\pi\colon M_u \to \mathscr{L}(p_i, q_i)$ defining the associated first integral (kinetic momenta). But this map is not, in general, a symplectomorphism.

For instance, in a local chart around the q-orbit, let $\bar{q}$, $\bar{p}$ be conjugate coordinates to q, p respectively in the original symplectic structure;

for this structure, the induced fundamental form is $\alpha = dp \wedge d\bar{p} + dq \wedge d\bar{q}$; to make the mapping of the space p, q; $\bar{p}$, $\bar{q}$, onto the (p, q)-plane a symplectomorphism for the 2-form $dp \wedge dq$ we have to introduce a new symplectic structure, defined by $\beta = dp \wedge dq + d\bar{p} \wedge d\bar{q}$.

Here, we have to assert something about the coupling between local oscillators of the metabolic field. The local dynamics (M_u, X_u) has to be isomorphic to its product with itself, quotiented by the relations obtained when we write that two local oscillators are in equilibrium by their interaction contact. Here we suppose that these interactions are mediated via interaction Hamiltonians, which have the following property: the associated Jacobi field is a Hamiltonian field with respect to structure (β) (this obviously expresses the condition for lifting the corresponding infinitesimal transformations in the projective group to the unitary group) and its rotates structure (α). This being supposed, the contact interaction of any two nearby oscillators of the field is described by vector fields which are Hamiltonian with respect to β. Hence, with respect to the β-structure, there exists an interaction hamiltonian, and for the β-structure, the projection to the (p, q)-plane, as first integrals, is a symplectomorphism. As the fiber is (theoretically) compact, and of constant Liouville measure, we may average this Hamiltonian on the fibers: we get on the (p, q)-plane a mean field, which is a Hamiltonian field for the $dp \wedge dq$ structure.

As this mean field has to be invariant under Euclidean dilatations: $q, p \to \lambda q, \lambda p$, it is necessarily linear, hence given by a quadratic β-Hamiltonian. By the self-coupling condition, which applies also to the averaged Hamiltonians, the curves describing the constancy of the Liouville measure for this oscillator have to be convex for the q-direction (and also the p-direction); hence they have to be given by a positive definite Hamiltonian, hence (up to an affine transformation) by $H = p^2 + q^2$.

Here again, the mean field dynamics, being orthogonal and symplectic, is unitary. Hence the field of local dynamics may be averaged on its first integrals of translation, thus giving a field of unitary representations (rotation $\exp it$) in $\mathscr{L}(p_i, q_i)$.

Now a state of the metabolic field is given by some density of Liouville measure for the α-structure. Moreover, the possibility of getting a new phase at a point $u \in U$, is directly proportional to the measure m on the energy hypersurfaces which are just below the threshold 0: under stochastic perturbations, this measure has in fact a great chance to be pushed across the threshold, in the other phase.

But the global Liouville measure is the same for the α-structure and for the β-structure, as

$$(dp \wedge d\bar{p} + dq \wedge d\bar{q})^2 = -(dp \wedge dq + d\bar{p} \wedge d\bar{q})^2.$$

Hence by averaging the Liouville measure m along the (compact) fibers of the symplectomorphism π: $M_u \to \mathscr{L}(p, q)$ we get the standard Liouville measure (area) in the (p, q)-plane.

As a result, the probability density on the 'vanishing' sphere of the critical point will be described, after averaging, by a probability density on the $\mathscr{L}(p_i, q_i)$ Lie algebra. By averaging this density in $\mathscr{L}(p, q)$, we get the usual wave function. As by self-coupling, densities at opposite points in the (p, q) plane may cancel, one understands the fundamental linear character of quantum mechanical formalism.

Observe that in the projection π, the Liouville measure lying far away from the kernel of π on the unit sphere stands a better chance of being projected far away from 0 in $(0pq)$. Hence there is a direct connection between $\sum (x_i^2 + y_i^2)$ (the energy) and the modulus of the radius vector of the projection in $\mathscr{L}(p, q)$.

The possibility of creating a nucleus of the new phase at a point $u \in U$ may be evaluated by taking a small sphere around u, and integrating the density measure on all points of the sphere on the (p, q)-plane. Letting the radius of the sphere tend to zero amounts to taking the Laplacian of the wave function vector, averaged locally. As the density of velocity vectors is also given by $dm/dt = im$, this linear relation extends by integration, and gives rise to Schrödinger equation. Taken globally, on all points, this equation expresses the fact that the mean field is in local equilibrium through contact interactions. Two local oscillators exchange Liouville measure, exactly like in the heat transfer two local oscillators exchange heat proportionally to the difference of temperature: $dT/dt = k(T_2 - T_1)$.

The field of Liouville measure in $\mathscr{L}(p, q)$ is only a statistical description of a family of symmetry-breakings from full Euclidean invariance (D), to only rotational SO(3)-invariance. The quotient of the Lie algebra $D/\mathrm{SO}(3)$ identifies with the Lie algebra of the translation group $\mathscr{L}_{\mathscr{C}}$. Hence the 'secondary' symmetry breaking is expressed by a function on the homogeneous space R^3 having its value in $\mathscr{L}_{\mathscr{C}}$, more precisely in the tangent space of $\mathscr{L}_{\mathscr{C}}$.

Hence quantum mechanical formalism slightly generalizes the general construction for symmetry-breaking; it expresses how *surface effects* affect the probability of appearance of a new phase... (such surface effects would be expressed by boundary conditions imposed on the Schrödinger equation).

I shall leave here the interesting questions of adapting this approach to specific quantum mechanical features, as the number of particles, the spin, the statistics. For the spin, just observe that we have only a field of homomorphisms of Lie algebras, hence one may get the two-fold covering of SO(3) as well as SO(3) itself. Odd spin expresses fundamentally a change of sign in the self-coupling interaction when one follows a closed one parameter subgroup of SO(3), in its representation in $\mathscr{L}_{\mathscr{C}}$.

Finally, we may turn to our original model and ask: why not explain quantum mechanics by a 'competition of singularities'? Here is a first attempt in that direction.

IV.2. COMPETITION OF SINGULARITIES IN ENERGY QUANTIFICATION

Let us go back to the sandy slope of III.2. Suppose now the x-axis represents energy E. Let us suppose that the same kind of ramifying process (generalized catastrophes, in my terminology) affects energy, as it affected rainwater on the hill. The equation describing the motion of a watershed was (taken here, on the $-x$, $+x$ interval):

$$du/dt = [c(x)+x\cdot c'(x)]u+\ldots$$

If x, u are energies, then c has dimension t^{-1}, and the coefficient c' dimension $t^{-1}E^{-1}$, hence the inverse of an action. This leads us to a model identifying the c' coefficient to the inverse of Planck's constant.

Suppose, like before, that the two streams whose basins are separated by zero have equal basin width x. Let $c(x)$ be the eroding power of a stream of width x; we admit that the two basins work so to speak independently each from the other, so that the separation point at 0 undergoes two separate 'virtual' motions: a displacement u_1 due to the $x<0$ stream, a displacement u_2 due to the second $x>0$ stream;

then, u_1, u_2 are solutions of the differential equations

$$du_1/dt = c(x+u_2)\cdot(x+u_1) - c(x-u_2)\cdot(x-u_1),$$
$$du_2/dt = c(x-u_1)\cdot(x+u_2) - c(x+u_1)\cdot(x-u_2),$$

the second equation is obtained by replacing u_1 by $-u_1$ in the 'natural' equation for du_2/dt; we get a linear system, which leads, for the onset of stability $c(x) = 0$, to a periodic circular motion such that $w^2 = x^2c'^2$, $w = xc'$, $x = (c')^{-1}v$, $E = hv$ in the quantic interpretation.

Of course, one may object that the 'Ansatz': replacing u_1 by $-u_1$ in the du_2/dt equation is highly arbitrary. It points out to the very strange properties of the energy looked as a 'fluid'. The only possible geometric interpretation would be as follows: a virtual 'brooklet' is formed at the threshold of the two basins, and is displaced alternatively along the u_1, u_2 displacements; but, as it is locally a 0(1)-phase, it has the possibility of reversing its local orientation: this one may interpret in two ways: either geometrically by the periodic formation of cusp like thresholds between basins like in Fig. 3: where then a displacement

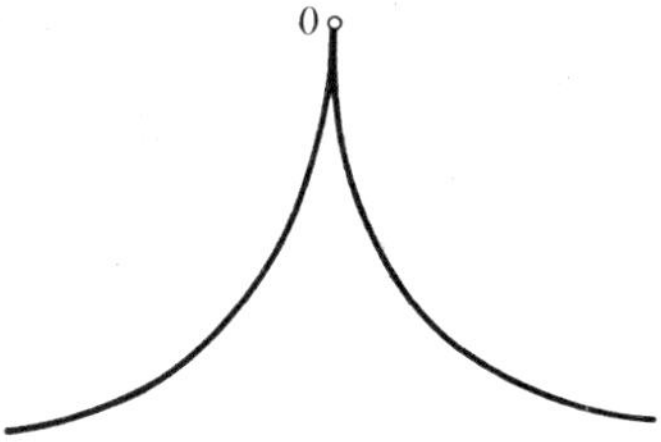

Fig. 3

positive for a basin has to be considered as an increase for the other basin.

Or anthropomorphically, as a sort of optimal strategy devised by a player P who plays alternatively against two players 1, 2. To player 1, P owes u_1, to player 2, he owes u_2. Then if the chances of win against 1 are given by

$$\frac{du_1}{dt} = ku_2$$

those to win against 2 will be given by the opposite formula

$$\frac{du_2}{dt} = -ku_1 .$$

Hence this will be an harmonic oscillator game ...

Reference

* Reprint of the 'Proceedings of the III GIFT Seminar in Theoretical Physics', Universidad de Madrid, 22–29 March, 1972, GIFT. 14/72, p. 1–48.

Bibliograghy

[1] R. Abraham and J. Marsden, *Foundations of Mechanics*, Benjamin, 1967.

[2] R. Abraham and J. Robbin, *Transversal Mappings and Flows*, Benjamin, 1967. (See also the original papers quoted there.)

[3] V. I. Arnol'd, *Sov. Math. Dokl.* **6**, 331–337.

[4] E. Artin, *Modern Higher Algebra (Galois theory)*, Courant Institute of Math. Sciences, 1947.

[5] D'Arcy Thompson, *On Growth and Form*, abridged edition, Cambridge Univ. Press, 1961, first edition 1915.
H. Weyl, *Symmetry*, Princeton Univ. Press, 1952. Related articles of both authors can be found in the book *The World of Mathematics* by J. R. Newman, Simon and Schuster, 1956.

[6] K. Jänich, Lecture Notes in Mathematics, vol. **59**, Springer-Verlag, 1968.

[7] L. Michel, Lectures delivered at the Haifa Summer School, 1971.

[8] J. Milnor, *Topology from the Differentiable Viewpoint*, Univ. Press, Virginia, 1965.

[9] R. Palais, *Bull. Am. Math. Soc.* **67**, 362–364.

[10] C. Pugh, 'The Closing Lemma for Hamiltonian Systems', Intern. Congress of Mathematicians, Moscow, 1966.

[11] D. Ruelle, and F. Takens, *Comm. Math. Physics* **20**, 167–192.
D. Ruelle, and F. Takens, *Comm. Math. Physics* **23**, 343–344.

[12] S. Smale, 'Differentiable Dynamical Systems', *Bull. Am. Math. Soc.* **73**, 803.

[13] J. M. Souriau, *Structure des systèmes dynamiques*, Dunod, 1970.

[14] R. F. Streater, 'Current Commutation Relations, Continuous Tensor Products and Infinitely Divisible Group Representations', *Rendiconti di Sc. Inst. di Fisica E. Fermi* **XI**, 247–63.

[15] R. Thom, 'Ensembles et Morphismes Stratifiés', *Bull. Am. Math. Soc.*, **75**, 240–284.

[16] R. Thom, *Stabilité structurelle et morphogénèse*, Benjamin, 1973.

[17] R. Williams, *Topology* **6**, 473–487.

PART FOUR

QUANTUM MECHANICS AND PARTICLE PHYSICS

RELATIVISTIC COMPOSITE SYSTEMS AND EXTENSIBLE MODELS OF FIELDS AND PARTICLES

A. O. BARUT

University of Colorado, Boulder, Colorado 80302, U.S.A.

I. Introduction

It is appropriate to recall that exactly seventy years ago Henri Poincaré in his 1904 St. Louis address entitled *L'état actuel et le future de physique mathématique*, at the International Congress of Arts and Sciences, spoke about the electromagnetic structure of the electron. The problem is still with us and my talk is essentially on the same subject.

At the turn of this century one of the most fundamental problems of physics was to understand the mass and structure of a charged particle in terms of electromagnetic fields, hence reduce mechanics to electromagnetism. Wien, Abraham, Lorentz, Poincaré and many more have contributed[1] and the goal was an 'electromagnetic world picture'. The climax of these efforts was the famous Rendiconti paper of Poincaré which contains among other results the Lorentz group, the relativistic action principle, the relativistic theory of gravitation, relativistic energy considerations, etc.[1] Einstein's paper on special relativity submitted about the same time as that of Poincaré went around the problem of reducing mechanics to electromagnetism. Instead he elevated mechanics to a relativistic level to be an equal partner of electromagnetism. He was not interested to reduce mass to an electromagnetic energy. After Einstein every physical theory could be made 'relativistic'. Clearly the program of the 'electromagnetic world picture' is a more fundamental approach to Nature which obviously was much ahead of its time. Only sporadic investigations have been devoted after that to the problem of extended models of particles. Two important examples are one by Einstein himself, who in 1919 [9] considered the possibility of gravitational forces holding the particle together, and the other by Dirac (1962) [7], who considered a pulsating spherical shell as a model of the electron.

Unifying attempts in physics are always very attractive and always actual, and so in particular a unification of a charged particle and its electromagnetic field. The so-called 'dualistic point of view' of introducing field and matter separately, and their interactions, (or separate coupled Dirac and Maxwell fields in the language of quantum field theory) is the generally accepted point of view. But the 'unitarian point of view' seems to be logically more superior. Here particles are viewed either as singularities of fields, or, on the other extreme, there are only particles interacting-at-a distance, and no fields. The difficulties of point particles in quantum field theory on the one hand, and the detailed empirical information that is now available about the internal structure of the proton, on the other hand, lead us to consider again extended models of elementary particles.

Concerning the structure of hadrons we seem to be confronted with a remarkable puzzle: Proton shows considerable internal structure, when probed with electrons and neutrinos, expressed by charge and magnetic moment distributions, excited states and structure functions, but no constituents have so far been identified. This puzzle has naturally lead to a lot of ideas and speculations which have been reviewed elsewhere [2]. Here we wish simply to state that a truly extended model of proton would of course immediately solve the puzzle. The extended structure does not need to have further constituents as we shall see. The attitude of physicists concerning the nature of hadrons is gradually shifting, in the last decade or more, from a description by means of coupled or self-coupled fields towards first understanding the internal structure of a single hadron treated as an 'atom' or a 'molecule'.

II. Relativistic Description of Extended Structures

We describe a point structure or an extended structure in Minkowski space M by a world line $Z^{\mu}(\tau)$, or world sheet $Z^{\mu}(\tau, \sigma)$, or a world tube $Z^{\mu}(\tau, \sigma_1, \sigma_2)$, etc. Let $Z^{\mu}(\sigma_{\alpha})$, in general, denote this structure. Here σ_{α} are labels parametrizing the points of the object; $\sigma_0 = \tau$ is a time-like, σ_i three space-like labels, they may be equilibrium positions or initial coordinates. They are the (Poincaré) *invariant* Lagrange coordinates (sometimes also called Gauss coordinates) of a relativistic continuum [11], [12]: $Z^\mu(\sigma_\alpha)$ thus represents the Minkowski coordinates of the point labelled by σ_α.[2]

It is also useful to consider $Z^\mu(\sigma_\alpha)$ as vector fields over the coordinates σ_α. The coordinates σ play the role of the 'space-time' coordinates of the usual field theory and μ will play the role of the 'internal symmetry index'.

The theory must be invariant under arbitrary transformations, diffeomorphisms, of the labels σ_α. The physics is independent of the choice of σ_α. Thus the formalism carries a great deal of, what one calls, 'gauge freedom', which eventually has to be eliminated. This is the prize one must pay in order to have a relativistic description of extensible objects.

We have thus two sets of coordinates covering the Minkowski space M, Z^μ and σ_α. For bounded objects there will be equations of the type $f(\sigma) = 0$ fixing the dimensionality and the shape of the object or the singularity. For example:

Dimension	$Z^\mu(\sigma_\alpha)$	Boundary	Physical System
1	$Z^\mu(\tau)$	—	mass point
2	$Z^\mu(\tau, \sigma)$	$0 \leqslant \sigma \leqslant 1$	string, loop
3	$Z^\mu(\tau, \sigma_1, \sigma_2)$	$a \leqslant \sigma_1, \sigma_2 \leqslant b$	membrane
	$Z^\mu(\tau, \sigma_1 \sigma_2 \sigma_2)$	$\sigma^2 = \varrho^2$	shell
4	$Z^\mu(\tau, \sigma_1 \sigma_2 \sigma_3)$	$\sigma^2 \leqslant \varrho^2$	ball

More precisely we have a domain D^r, $\sigma \in D^r$, diffeomorphisms φ of D^r; $\varphi : \sigma' \to \sigma$, and an imbedding f of D^r into M^n. $f: \sigma \in D^r \to f(\sigma) \equiv Z^\mu(\sigma) \in M^n$. The tangent space $D^r \times R^r$ is mapped into the tangent space TM^n of M^n.

We describe the further motion of the extensible object by a variational principle on the action S defined by

$$S(f) = \int_{D^r} L[f(\sigma), Tf(e_i)]\, d\overset{r}{\sigma}, \tag{2.1}$$

where $Tf(e_i)$ are the tangent vectors of D^r at the point σ. The Lagrangian L, in other words, is a function $L[Z^\mu(\sigma), \partial Z^\mu/\partial \sigma_\alpha]$ and is invariant under

(i) the group of transformations in M^n (in our case under Poincaré transformations),

(ii) the diffeomorphisms φ of D^r

(cf. Fig. 1).

There is a symmetric induced metric on the surface Σ embedded in M:

$$(2.2)\qquad g_{\alpha\beta} = \frac{\partial Z^\mu}{\partial\sigma^\alpha}\frac{\partial Z_\mu}{\partial\sigma^\beta} = g_{\beta\alpha}.$$

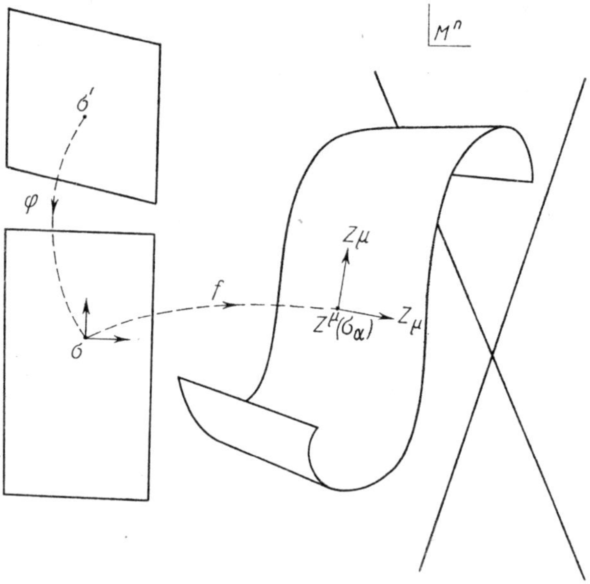

Fig. 1

The surface element is invariant under the choice of the coordinates σ and under Poincaré transformations. Hence the simplest action principle is

$$(2.3)\qquad S(Z) = \lambda \int \sqrt{-g}\, d\overset{r}{\sigma},$$

where g is the determinant of the induced metric $g_{\alpha\beta}$, eq. (2.2) (the minus sign depends on of the signature of the metric). The quantity $\sqrt{-g}$ is a scalar density with respect to σ-transformations. Note that the Einstein's action principle in general relativity is $\int d^4\sigma R\sqrt{-g}$, where R is the scalar curvature. A general factor $\varphi(R)$ could also be introduced in (2.3) under the integral.

The actual motion $Z^\mu(\sigma)$ makes the action $S(Z)$ an extremum, hence for arbitrary variations δZ^μ of $Z^\mu(\sigma)$ we set $\delta S = 0$. Because of the pseudo-euclidean nature of M we get a nontrivial result by minimizing the area; in an euclidean space the minimum of a surface area is of course zero.

$$dS = \lambda \int d\sigma \delta \sqrt{-g} = \lambda \int d\sigma \frac{1}{2} \frac{-\delta g}{\sqrt{-g}} = \lambda \int d\sigma \frac{1}{2} \sqrt{-g}\,\delta g$$

$$= \lambda \int d\sigma \frac{1}{2} \sqrt{-g}\, g^{\alpha\beta} \delta g_{\alpha\beta} = \lambda \int d\sigma \sqrt{-g}\, g^{\alpha\beta} Z_{\mu,\alpha} \delta Z^\mu{}_{,\beta},$$

or, by integrating by parts,

$$\text{(2.4)} \quad \delta S = \lambda \oint d\Sigma_\beta \sqrt{-g}\, g^{\alpha\beta} Z_{\mu,\alpha} \delta Z^\mu - \lambda \int d\sigma [\sqrt{-g}\, g^{\alpha\beta} Z_{\mu,\alpha}]_{,\beta} \delta Z^\mu.$$

If Z^μ is not varied at the boundary of the region of integration in the σ-space, the first term vanishes, and we get from $\delta S = 0$, δZ_μ arbitrary, the 'equations of motion'

$$\text{(2.5)} \quad (\sqrt{-g}\, g^{\alpha\beta} Z_{\mu,\alpha})_{,\beta} = 0,$$

or,

$$\text{(2.5')} \quad P^\beta_{\mu,\beta} = 0,$$

where

$$\text{(2.6)} \quad P^\beta_\mu \equiv \sqrt{-g}\, g^{\alpha\beta} Z_{\mu,\alpha}.$$

From the point of view of the vector fields $Z^\mu(\sigma)$, equations (2.5) are just 'current conservation' laws. We have four conserved currents: $(P_\mu)^\beta$; $\mu = 0, 1, 2, 3$. Of course, they are also the Euler–Lagrange equations:

$$\text{(2.7)} \quad \frac{\partial L}{\partial Z^\mu} - \frac{\partial}{\partial \sigma^\alpha} \left(\frac{\partial L}{\partial Z^\mu_{,\alpha}} \right) = 0,$$

or, the Laplace–Beltrami invariant equations for $Z^\mu(\sigma)$.

Going back to (2.4), and now varying Z^μ at the endpoints as well, we obtain, if the equations (2.5) are valid, the flux equations

$$\text{(2.8)} \quad \oint d\Sigma_\beta P^\beta_\mu = 0, \qquad \mu = 0, 1, 2, 3.$$

Examples

1) $Z_\mu(\tau)$; $\lambda = m$; $S = m\int ds$. There is one value of α. We find

$$(2.9)\quad P_\mu = m\dot{Z}_\mu, \quad \ddot{Z}_\mu = 0, \quad \dot{Z}_\mu(\tau_2) = \dot{Z}_\mu(\tau_1) = \text{const.}$$

2) $Z_\mu(\tau, \sigma)$, $\lambda = M^2$:

$$(2.10)\quad S = M^2 \int d\tau d\sigma [-g_{00}g_{11} + g_{01}^2]^{1/2} = M^2 \int d\tau d\sigma [-\dot{Z}^2 Z'^2 + (\dot{Z}\cdot Z')^2]^{1/2}.$$

Here $\dot{Z}$ and Z' are derivatives with respect to τ and σ, respectively, and α takes two values, 0 and 1, and we have four 2-dimensional 'currents':

$$(2.11)\quad \begin{aligned} P_\mu^0 &= (\dot{Z}_\mu Z'^2 - Z'_\mu(\dot{Z}\cdot Z'))/\sqrt{-g}, \\ P_\mu^1 &= (Z'_\mu \dot{Z}^2 - \dot{Z}_\mu(\dot{Z}\cdot Z'))/\sqrt{-g}, \end{aligned}$$

which are 'conserved'

$$(2.12)\quad \frac{\partial}{\partial\tau} P_\mu^0 + \frac{\partial}{\partial\sigma} P_\mu^1 = 0, \quad \mu = 0, 1, 2, 3.$$

Equation (2.8) becomes

$$(2.13)\quad \int (d\Sigma_0 P_\mu^0 + d\Sigma_1 P_\mu^1) = 0.$$

3) $Z^\mu(\tau, \sigma_1, \sigma_2)$, $\quad \lambda = M^3$, $\quad S = M^3 \int d\tau\, d\sigma_1\, d\sigma_2 \sqrt{-g}$,

$$(2.14)\quad \begin{aligned} S = M^3 \int d\tau\, d\sigma_1\, d\sigma_2 [&\dot{Z}\cdot Z'(\dot{Z}\cdot Z' Z^{*2} - (\dot{Z}\cdot Z^*)(Z'\cdot Z^*)) - \\ &- \dot{Z}^2 (Z'^2 Z^{*2} - (Z'\cdot Z^*)^2) - \dot{Z}\cdot Z^* ((\dot{Z}\cdot Z')(Z'\cdot Z^*) - \\ &- Z'^2(\dot{Z}\cdot Z^*))]^{1/2}, \end{aligned}$$

where Z, Z', Z^* indicate the derivatives of Z with respect to τ, σ_1, and σ_2, respectively. It is straightforward but lengthly to write the 'currents components' P_μ^0, P_μ^1, P_μ^2, and the equations

$$(2.15)\quad \begin{aligned} &\frac{\partial}{\partial\tau} P_\mu^0 + \frac{\partial}{\partial\sigma_1} P_\mu^1 + \frac{\partial}{\partial\sigma_2} P_\mu^2 = 0, \\ &\oint d\Sigma_\beta P_\mu^\beta = 0. \end{aligned}$$

We can also embed a three-dimensional surface (e.g. a spherical shell) into 4-dimensions with the boundary condition $f(\sigma) = 0$, for example.

If we choose the coordinates σ_α such that the boundary condition becomes σ^3 = const, for example, we get

$$(2.16)\quad S = M^3 \int_{\sigma^3=\text{const}} d\tau\, d\sigma_1\, d\sigma_2 \sqrt{-g}\sqrt{-g^{33}}\,.$$

Here g is the 4×4-determinant and g^{33} the contravariant component of g_{33}.

4) In the four-dimensional case, we can introduce two sets of coordinates x^μ, σ_α in the whole Minkowski space. The 'fields' $Z^\mu(\sigma_\alpha)$ can be taught to be defined as elements of a space of functions with compact support, or of a space of rapidly decreasing functions. The difference from the usual field theory is that the Poincaré group acts only on the indices μ, and σ_α are Lorentz-invariant parameters, while the group of diffeomorphisms acts on the indices σ_α – alone. Similarly, for a tensorial property $R_{\mu\nu}(\sigma)$, e.g. curvature of the medium, the group of diffeomorphisms acts on σ, and the Poincaré group on the indices μ, ν. ▼

There is a well-known procedure to derive from the equations of motion, (2.5) or (2.7), local conservation laws with respect to the σ-coordinates, namely those obeyed by the 'energy-momentum tensor'. We multiply (2.7) by $Z^\mu_{,\beta}$:

$$Z^\mu_{,\beta}\left(\frac{\partial L}{\partial Z^\mu} - \frac{\partial}{\partial\sigma^\alpha}\left(\frac{\partial L}{\partial Z^\mu_{,\alpha}}\right)\right) = 0,$$

or,

$$\frac{\partial L}{\partial Z^\mu} Z^\mu_{,\beta} + \frac{\partial}{\partial\sigma^\alpha}(Z^\mu_{,\beta})\frac{\partial L}{\partial Z^\mu_{,\alpha}} - \frac{\partial}{\partial\sigma^\alpha}\left(Z^\mu_{,\beta}\frac{\partial L}{\partial Z^\mu_{,\alpha}}\right) = 0.$$

Hence, if L is a function of Z^μ and $Z^\mu_{,\alpha}$ only (i.e. not explicitly σ-dependent), we get

$$(2.17)\quad \frac{\partial}{\partial\sigma^\alpha}\left(\delta^\alpha_\beta L - Z^\mu_{,\beta}\frac{\partial L}{\partial Z^\mu_{,\alpha}}\right) \equiv \mathring{T}^\alpha_{\beta,\alpha} = 0.$$

In the flat-space vector field case (σ = flat) these would be energy-momentum conservation laws. In general relativity these equations are identities. In the case of our surfaces of 1 to 3-dimensions embedded in M, these lead to *constraint equations*, or coordinate conditions. For $L = \sqrt{-g}$, (2.17) becomes indeed

$$(2.18)\quad (\sqrt{-g}\,\delta^\alpha_\beta)_{,\alpha} - (\sqrt{-g})_{,\beta} = 0.$$

The quantities P^{α}_{μ} are not all independent. In our case, because $P^{\alpha}_{\mu} = \partial L/\partial Z^{\mu}{}_{,\alpha}$ and $L = \sqrt{-g}$ we have the relations

$$(2.19)\quad \begin{aligned} &Z^{\mu,\alpha}P_{\mu,\beta} = 0,\\ &P^{\alpha}_{\mu}P^{\mu\alpha} + Z^{\mu}_{,\beta}Z_{\mu,\beta} = 0 \end{aligned}$$

for each fixed α and β, $\alpha \neq \beta$, (summation over μ, but not over α and β) as can be verified using (2.6). These are thus the so-called *primary constraints* in the generalized Hamiltonian formalism of Dirac [8]. With these relations $\mathring{T}^{\alpha}_{\beta}$ itself vanishes.

In the example of the string, eqs. (2.10)–(2.11), we have the constraints

$$(2.20)\quad \begin{aligned} &P^{0^2} + Z'^2 = 0, \qquad P^0 \cdot Z' = 0,\\ &P^{1^2} + \dot{Z}^2 = 0, \qquad P^1 \cdot \dot{Z} = 0 \end{aligned}$$

to the equations (2.12).

We can impose subsidiary coordinate conditions in order to simplify and reduce the four fundamental equations (2.5). Again in the two-dimensional case, the covariant choice of the *two* coordinate conditions

$$(2.21)\quad \dot{Z}\cdot Z' = 0, \qquad \dot{Z}^2 + Z'^2 = 0$$

simplifies the equation (2.11) to

$$(2.22)\quad P^0_{\mu} = -\dot{Z}_{\mu}, \qquad P^1_{\mu} = Z'_{\mu},$$

hence eq. (2.12) to

$$(2.23)\quad \ddot{Z}_{\mu} = Z''_{\mu}.$$

Note that (2.20) with (2.22) imply (2.21); only two coordinate conditions are involved.

Coming back to the ‘fields’ $Z^{\mu}(\sigma)$, the Poincaré transformations with respect to the μ-index lead actually to 10 conserved ‘currents’ only four of which, P^{α}_{μ}, we have discussed so far. The invariance of $L = \sqrt{-g}$ under $\delta Z^{\mu} = a^{\mu} + \omega^{\mu}_{\nu}Z^{\nu}$ gives 10 quantities

$$(2.24)\quad j^{\alpha}_{(i)} = \frac{\partial L}{\partial Z^{\mu}_{,\alpha}}\,\delta Z^{\mu}_{(i)}, \qquad i = 1, 2, \ldots, 10$$

satisfying

(2.25) $j^{\alpha}_{(i),\alpha} = 0.$

These are $P^{\alpha}_{\mu} = \dfrac{\partial L}{\partial Z^{\mu}_{,\alpha}}$ corresponding to the translations and

(2.26) $J^{\alpha}_{\mu\nu} = P^{\alpha}_{\mu} Z_{\nu} - P^{\alpha}_{\nu} Z_{\mu}, \qquad J^{\alpha}_{\mu\nu,\alpha} = 0,$

corresponding to homogeneous Lorentz transformations. The corresponding global conserved 'charges' can be defined by integrals over the space-like boundaries of the surface

(2.27) $P_{\mu} = \int\limits_{C_{\sigma}} d\Sigma_{\alpha} P^{\alpha}_{\mu}, \qquad J_{\mu\nu} = \int\limits_{C_{\sigma}} d\Sigma_{\alpha} J^{\alpha}_{\mu\nu},$

together with the condition that there are no current fluxes through the time-like boundaries of the surface (Fig. 2):

(2.28) $\int\limits_{C_{\tau}} d\Sigma_{\alpha} P^{\alpha}_{\mu} = 0, \qquad \int\limits_{C_{\tau}} d\Sigma_{\alpha} J^{\alpha}_{\mu\nu} = 0.$

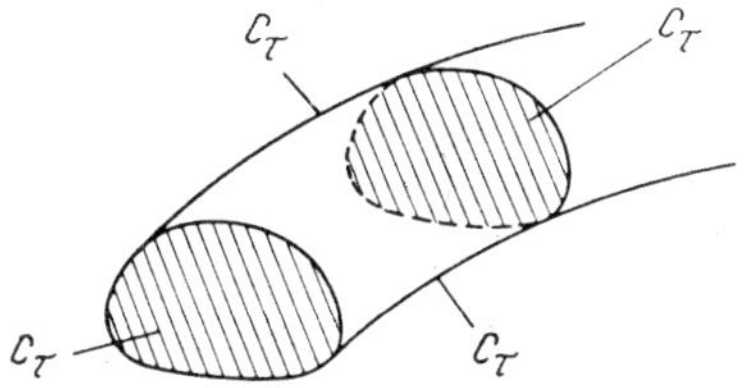

Fig. 2

Generalizations: We may indicate several generalizations and extensions of the above formalism. (1) An extensible object maybe considered in curved spaces, or better yet, in the curvature produced by itself, in a self-consistent manner. (2) As we have emphasized, the Poincaré group acts on the indices μ, but in a global way with respect to σ-coordinates. If we require also a local Poincaré invariance at each point σ, we will be lead to introduce 10 new gauge vector fields, and the σ-derivatives will be replaced by covariant derivatives

$$\partial_{\alpha} Z^{\mu} \to \nabla_{\alpha} Z^{\mu} \equiv \partial_{\alpha} Z^{\mu} + \Gamma^{\mu}_{\nu\alpha} Z^{\nu}.$$

(3) So far we have discussed the classical theory of a single extensible object. We may also consider a field of extensors described by functionals $\Psi(Z^\mu(\sigma_\alpha))$.

III. An Extensible Field ('Aether')

As a first concrete case we consider the extensible object to fill the whole of Minkowski space, i.e. an isotropic medium filling the space-time with mass density ϱ and Young modulus Y such that $c = \sqrt{\varrho/Y}$ is invariant under Lorentz transformations. The points of the medium are labelled by σ_α, and the displacement of the point σ_α we call $Z^\mu(\sigma_\alpha) \equiv A^\mu(\sigma_\alpha)$. Under small diffeomorphisms $\varphi\colon \sigma \to \sigma'$, we can write by expansion

$$A_\mu(\sigma) \to A_\mu(\sigma) + \Lambda_{,\mu}(\sigma). \tag{3.1}$$

The Lagrangian density defined as the difference of kinetic and potential energy densities can be written, in suitable choice of coordinates and units, as

$$L = \tfrac{1}{2} A_\mu^{;\alpha} A^\mu_{,\alpha} \tag{3.2}$$

and leads to the equation

$$\Box A_\mu(\sigma) = 0 \tag{3.3}$$

and to the symmetric energy-momentum tensor

$$T^\alpha_\beta = \tfrac{1}{2}\delta^\alpha_\beta A_\mu^{;\gamma} A^\mu_{,\gamma} - A^{\mu,\alpha} A_{\mu,\beta}, \qquad T^{\alpha,\beta}_\beta = 0. \tag{3.4}$$

The action density (3.2) should be chosen to be invariant under diffeomorphisms so that it will reduce to the intuitive physically interpretable form (3.2) when the σ-coordinates just coincide with the Minkowski coordinates. It will be more complicated than just $\sqrt{-g}$ because in the four-dimensional case this quantity is just equal to one.

The field $A^\mu(\sigma)$ is thus like the Maxwell field, but the Poincaré transformations act only on the indices μ, ν and not on σ-coordinates. Eq. (3.3) describes the wave motions in the medium. The gauge transformations are physically interpreted as the change of parametrization of the points of the medium (aether). The field $A_\mu(\sigma)$ is just the displacement of the aether-point σ which is the observable quantity, subject to freedom of reparametrization. Historically, the aether concept had been abandoned because there is empirically no preferred direction indicating the velocity of the aether. This problem does not arise here. We study and shall quantize *the motions of the medium*, not the structure of the medium itself.

IV. The Electromagnetic String and Extended Singularities of the Electromagnetic Field

The concrete realization of a two-dimensional extensible object that we shall study is the electromagnetic string as an extended singularity of the electromagnetic field. A string as a continuous object made of some kind of matter has been the subject of many other recent studies, also with the problem of hadron structure in mind [15], [10]. The string we introduce is purely a singular electromagnetic field, a distribution [3].

Consider the field of an infinitesimally thin solenoid, $F^M_{\mu\nu}$ satisfying the Maxwell's equations

$$(4.1)\qquad \begin{aligned} F^{M,\nu}_{\mu\nu} &= -4\pi(j^{(e)}_\mu + j^{(s)}_\mu),\\ \tilde{F}^{M,\nu}_{\mu\nu} &= 0, \end{aligned}$$

where j^s_μ is the current producing the field, and j^e_μ the current of other point charges, if they are present, and $\tilde{F}$ is the field dual to F.

$$(4.2)\qquad j^{(e)}_\mu(x) = \sum_i e_i \int ds \dot{Z}_\mu(s)\, \delta\big(x - Z(s)\big).$$

The end points of the solenoid behave like magnetic monopoles. We split the field (4.1) as follows:

$$(4.3)\qquad F^M_{\mu\nu} = F^D_{\mu\nu} + \tilde{\Lambda}_{\mu\nu}$$

where $F^D_{\mu\nu}$ is a non-singular field, part of the field $F^M_{\mu\nu}$ which is exterior to the solenoid, and can be thought to be produced by two magnetic chargers, $+g$ and $-g$, at the positions of the endpoints of the solenoid, and $\tilde{\Lambda}_{\mu\nu}$ is just the remaining singular part. These new fields satisfy the equations

$$(4.4)\qquad \begin{aligned} F^{D,\nu}_{\mu\nu} &= -4\pi j^{(e)}_\mu, & \tilde{\Lambda}^{,\nu}_{\mu\nu} &= -4\pi j^{(s)}_\mu,\\ \tilde{F}^{D,\nu}_{\mu\nu} &= -4\pi k^{(g)}_\mu, & \Lambda^{,\nu}_{\mu\nu} &= 4\pi k^{(g)}_\mu. \end{aligned}$$

The field $F^D_{\mu\nu}$ satisfies the equations of the symmetric electrodynamics of Dirac [5], [6] with $k^{(g)}_\mu$ being the current of the point magnetic charges to be given below. The singular field $\tilde{\Lambda}_{\mu\nu}$ has as current the singular string current, and its dual the current of the magnetic charges.

The Maxwell field $F^M_{\mu\nu}$ is derivable from a potential

$$(4.5) \quad F^M_{\mu\nu} = A_{\nu,\mu} - A_{\mu,\nu} \equiv A_{[\nu,\mu]}.$$

Hence from (4.3)

$$(4.6) \quad F^D_{\mu\nu} = A_{[\nu,\mu]} - \tilde{\Lambda}_{\mu\nu};$$

the singularities in $A_{[\mu,\nu]}$ and $\tilde{\Lambda}_{\mu\nu}$ cancel giving a non-singular $F^D_{\mu\nu}$. Note that $F^D_{\mu\nu}$ is not derivable from a potential, unlike $F^M_{\mu\nu}$, but that (4.6) holds.

The forms of the $\Lambda_{\mu\nu}$ and $k^{(g)}_\mu$ are as follows:

$$(4.7) \quad \Lambda_{\mu\nu}(x) = -4\pi \sum_i g_i \int d\tau d\sigma (\dot{Y}_\mu Y'_\nu - \dot{Y}_\nu Y'_\mu)\, \delta(x-Y),$$

$$(4.8) \quad k^{(g)}_\mu(x) = \frac{1}{4\pi} \Lambda_{\mu\nu}^{;\nu} = \sum g_i \Big[\int ds^{(1)} \dot{Z}^{(1)}_\mu\, \delta(x - Z^{(1)}) -$$

$$- \int ds^{(2)} \dot{Z}^{(2)}_\mu\, \delta(x - Z^{(2)})$$

where $Z^{(1)}_\mu$ and $Z^{(2)}_\mu$ are the two endpoints of the string $Y_\mu(\tau, \sigma)$:

$$Z^{(i)}_\mu(s) \equiv Y_\mu \qquad (s, \sigma = \sigma_i), \qquad i = 1, 2.$$

The potential A_μ in (4.5) can be evaluated and is given by

$$(4.9) \quad A_\mu(x) = \sum e_i \int ds \dot{Z}^i_\mu(s) D(x-Z) +$$

$$+ \sum g_i\, \varepsilon_{\mu\nu\lambda\varrho} \int d\tau d\sigma \dot{Y}^\nu Y'^\lambda \frac{\partial}{\partial x_\varrho} D(x-Y).$$

Indeed, with (4.9) and (4.7) all equations (4.1)–(4.8) are satisfied.

In general, let $F^M_{\mu\nu}$ be a Maxwell-field (i.e. a field satisfying the equations (4.1)) and let $\tilde{\Lambda}_{\mu\nu}$ be its singular part: $\tilde{\Lambda}_{\mu\nu}$ vanishes outside a region D^r. We may even include point singularities (better world-line singularities) in D^r. We subtract the singular parts and define

$$(4.10) \quad F_{\mu\nu} \equiv F^M_{\mu\nu} - \tilde{\Lambda}_{\mu\nu}.$$

We now consider the standard Lagrangian density for the field $F_{\mu\nu}$ which is free of singularities:

$$(4.11)\quad L = \frac{1}{16\pi} F_{\mu\nu}F^{\mu\nu} = \frac{1}{16\pi}(F^{M}_{\mu\nu}-\tilde{\Lambda}_{\mu\nu})(F^{M\mu\nu}-\tilde{\Lambda}_{\mu\nu})$$

$$= \frac{1}{16\pi}(F^{M}_{\mu\nu}F^{M\mu\nu}+\tilde{\Lambda}_{\mu\nu}\tilde{\Lambda}^{\mu\nu}-2F^{M}_{\mu\nu}\tilde{\Lambda}^{\mu\nu}).$$

The second term gives an action of the form

$$(4.12)\quad \frac{1}{16\pi}\int d^4x\tilde{\Lambda}_{\mu\nu}\tilde{\Lambda}^{\mu\nu}$$

$$= \frac{1}{16\pi}\int d^4x\delta^{(4)}(x-Z(\sigma))\delta^{(4)}(x-Z(\sigma'))d^r\sigma d^r\sigma'\lambda_{\mu\nu}\lambda^{\mu\nu},$$

where, as in eq. (4.7), we have expressed $\Lambda_{\mu\nu}$ as

$$(4.13)\quad \Lambda_{\mu\nu}(x) = \int d^r\sigma\delta^{(4)}(x-Z(\sigma))\lambda_{\mu\nu}(Z).$$

We wish to reduce (4.12) to an integral $\int d^r\sigma A$. But the expression for A contains four σ-functions and r-integrations (one of the δ^4-functions integrates with d^4x). We regularize the remaining $(4-r)$ δ-functions. These are just the directions perpendicular to the surface D^r. The regularization means giving a thickness to the surface in perpendicular directions:

$$\delta^{(4-r)} \xrightarrow[\text{regularization}]{} \delta^{(4-r)}_{\text{reg}}.$$

Taking for δ_{reg} Gaussian distributions with a length parameter $a = 1/M$ we obtain the general form

$$(4.14)\quad \frac{1}{16\pi}\int d^4x\tilde{\Lambda}_{\mu\nu}\tilde{\Lambda}^{\mu\nu} \to M^r\int d^r\sigma L^{\text{sing}}.$$

The right-hand side can now be interpreted as the 'mechanical' action of the singular object. Indeed, for the string ($r = 2$), this procedure gives, using (4.7), precisely the action (2.10), namely the invariant area. This is a remarkable result.

The third term in (4.11) has also a remarkable interpretation: Using eq. (4.5) we have

$$-\frac{1}{8\pi}\int d^4x\tilde{\Lambda}_{\mu\nu}F^{M\mu\nu} = \frac{1}{4\pi}\int d^4x\tilde{\Lambda}_{\mu\nu}A^{\mu,\nu}$$

or, by partial integration,

$$(4.15) \quad = -\frac{1}{4\pi}\int d^4x \tilde{\Lambda}_{\mu\nu}^{;\nu} A^\nu = \int d^4x j_\mu^{(s)} A^\mu,$$

where in the last step we have used eq. (4.4). Combining (4.11), (4.14) and (4.15) we have the final result

$$(4.16) \quad I \equiv \frac{1}{16\pi}\int dx F_{\mu\nu} F^{\mu\nu}$$

$$= \frac{1}{16\pi}\int dx F^M_{\mu\nu} F^{M\mu\nu} + M^r \int d^r\sigma L^s + \int dx j_\mu^{(s)} A^\mu.$$

The right-hand side now represents the action of a Maxwell field coupled to a current and the 'mechanical' action of the current. When varied with respect to A^μ, the right-hand side gives the correct Maxwell's equations (4.1) – $j_\mu^{(e)}$ of (4.1) is incorporated here into $j_\mu^{(s)}$ – and furthermore, we will get an interaction term to the right-hand side of the string equation (2.12). For further results see [3].

V. Pulsating Electromagnetic Sphere S^2

The singularity surface is $Z^\mu(\tau, \sigma_1, \sigma_2, \sigma_3)$ with the boundary condition $\sigma_3 = 0$: in spherical coordinates $\sigma_3 = r-\varrho$, $\sigma_1 = \theta$, $\sigma_3 = \psi$. Eq. (4.14) is now the action (2.16)

$$(5.1) \quad M^3 \int_{\sigma_3=0} d\tau d\sigma_1 d\sigma_2 \underset{(3\times 3)}{\sqrt{-g}} \sqrt{-g^{33}}.$$

The contribution of this term to the equation of motion (2.5) is the left-hand side of the following equation of motion

$$(5.2) \quad M^3(\sqrt{-g})^{-1}(\sqrt{-g}\, g^{3\alpha}/g^{33})_{,\alpha} = \tfrac{1}{2} F^M_{\mu\nu} F^{M\mu\nu}$$

which has the geometric interpretation as the total curvature of the surface [7]. The interaction term gives the right-hand side of the equation of motion (5.2) which is the total electromagnetic Lagrangian density just off the surface, $\frac{1}{2}(E^2-B^2)$.

VI. Passage to Hamiltonian Formalism. Quantization

The space of all motions of an extensible object is very large. Hence we first attempt to study and quantize simple motions of the singularity surface such as the rotations of the string, spherically symmetric motions of the shell, etc. Presumably such motions determine states of the system with lowest energies, hence the low-lying discrete excited states of the system when it is quantized. This is not only done for simplicity, but also in order to establish contact with bound-state systems composed of several particles. The idea here is the following. There are classes of motions of the extensible object which are characterized by one or few global coordinates. For example, when one of the endpoint of the string is fixed, a class of motion of the string is specified by the coordinate $\vec{r}$ of the other endpoint relative to the fixed point. Such a motion defines an equivalent one-body problem. Next there is the class of motion of the string which correspond to the relativistic two-body problem for the two endpoints, the interaction being transmitted by the string itself. Similarly for the example of the pulsating sphere, we may consider those motions which involve only the radius ϱ (See Ch. V), i.e. spherically symmetric solutions, then those involving three coordinates ϱ, θ, φ, etc. It would be interesting to classify and characterize all these solutions. In all these cases the partial differential equation of the medium is transformed into an ordinary differential equation for the dynamical variables, i.e. $\varrho(\tau)$.

For example, a spherically symmetric solution of equations (5.2) is obtained by inserting

$$ds^2 = d\tau^2 - (d\sigma^3 + \dot{\varrho}d\tau)^2 - (\sigma^3 + \varrho)^2 d\theta^2 - (\sigma^3 + \varrho)^2 \sin^2\theta \, d\varphi^2$$

into the eq. (5.2) [7]:

$$(6.1) \quad M^3 \left[\frac{d}{d\tau} \frac{\dot{\varrho}}{(1-\dot{\varrho}^2)^{1/2}} + \frac{2}{\varrho(1-\dot{\varrho}^2)^{1/2}} \right] = \int_{\varrho}^{\infty} \frac{1}{2} (E^2 - B^2) \, dV.$$

If the right-hand side of this equation is zero we are studying the free oscillations of the shell. For a purely charged shell, one can put the integrand equal to $\frac{1}{2}\frac{e^2}{\varrho^4}$, the Coulomb energy. Then the equation contains aside from charge e, a single parameter (M^3). It is then possible to pass from (6.1) to a Hamiltonian corresponding to the one-degree of freedom:

$$(6.2) \quad H(\varrho) = (p^2 + M^6 \varrho^4)^{1/2} + \frac{e^2}{2\varrho},$$

where p is the conjugate momentum to ϱ. The parameter M can be determined by the equilibrium radius a of the shell, $M = \frac{1}{a}\left(\frac{e^2}{4}\right)^{1/3}$, or in the electromagnetic picture by the mass of the ground state m: $M = 2m\left(\frac{e^4}{2}\right)^{1/3}$. When the parameter is so fixed there is a unique prediction of the mass of the excited state.

Similarly, in the example of the electromagnetic string which has in addition electric charges e_1 and e_2 at the endpoints, we obtain the non-linear equation of an interacting string (cf. free string (2.12)):

$$(6.3)\qquad \frac{\partial}{\partial\tau} P^0_\mu + \frac{\partial}{\partial\sigma} P^1_\mu = 4\pi\varepsilon_{\mu\nu\lambda\varrho}\dot{Z}^\nu Z'^\lambda (j^{(e)} + j^{(s)})^\varrho,$$

which in the non-relativistic approximation results, for motions relative to one-fixed endpoint, the Hamiltonian

$$(6.4)\qquad H(\vec{r}) = \frac{1}{2m}(p - \mu\vec{D}(\vec{r}, \vec{n}))^2 - \frac{\alpha}{r}$$

where $\vec{r}$ is the relative coordinate of the other endpoint and the other quantities are defined as follows:

$$\alpha = e_1 e_2 + g^2, \qquad \mu = (e_1 + e_2)g \equiv eg,$$

$$(6.5)\qquad \vec{D}(\vec{n}, \vec{r}) = \frac{\hat{r}\times\hat{n}(\hat{r}\cdot\hat{n})}{r[1-(\hat{r}\cdot\hat{n})^2]}.$$

Here g is the magnetic charge associated with the endpoints of the electromagnetic string, and $\hat{n}$ is a fixed direction, a new gauge-coordinate which occurs in the non-relativistic approximation. The Hamiltonian (6.4) is the well-known Hamiltonian of two dyons (particles endowed with both electric and magnetic charges) and so we have established a connection between a certain two-body problem and a class of motion of a string.

It is interesting to note here that the system described by the Hamiltonian (6.4) can have half-odd integer spin values. This is because the angular momentum of the system

$$(6.6)\qquad \vec{J} = \vec{r}\times(\vec{p} - \mu\vec{D}) - \mu\hat{r}$$

when quantized gives

$$(6.7)\qquad \mu = eg = 0,\ \pm\tfrac{1}{2},\ \pm 1, \ldots$$

Thus the fundamental unit of the magnetic charge g_0 is determined by

(6.8) $eg_0 = 1/2.$

Furthermore, the quantization rule (6.7) is equivalent, as can be seen from the electromagnetic string picture, to the quantization of the singular flux inside the string in units of $\frac{h}{2e}$ (as is known from superconductivity). It is important to require that (6.4) is rotationally invariant so that spin is defined. This in turn means that the coordinate $\hat{n}$ in (6.5) must also be rotated as a dynamical variable; it just provides the spin degree of freedom of the system.

We have not only an electromagnetic origin of spin, but also because the value of g is very large (see (6.8)) a natural explanation of the occurrence of strong interactions in nature. These are some of the reasons for using the electromagnetic string as a realistic model of hadrons. These questions have been discussed in detail elsewhere [1].

The essential idea of quantization here, we emphasize, consists in finding classes of motion of the classical system and quantize these motions, and not the medium itself. In other words, we do not make the fields $Z^\mu(\sigma)$ into quantum field operators and field equations into operator equations, but find classical solutions, and quantize the solution manifold.

VII. The Role of Diffeomorphisms of Extensible Objects in Quantization

We have seen in Ch. II that the generalized momenta

(7.1) $P^\alpha_\mu = \partial L/\partial Z^\mu_{,\alpha}$

are not all independent, but satisfy the constraints (2.19). This is already true in the relativistic mechanics of a point particle. From the action $I = m \int ds$, we find

$$p^\mu = \partial L/\partial \dot{Z}_\mu = m\dot{Z}_\mu$$

with the primary constraint $p_\mu p^\mu = m^2$. As a result the Hamiltonian defined by

$$H_0 = p_\mu \dot{Z}^\mu - L$$

vanishes identically. But in such cases the Hamiltonian is not unique, we can add to H_0 any multiple of the constraint equation:

(7.2) $$H_\lambda = H_0 + \lambda(p^2 - m^2)$$

which gives the canonical equations

$$\frac{\partial H}{\partial Z_\mu} = 0, \text{ i.e. } p_\mu = \text{const}, \text{ and } \frac{\partial H}{\partial p^\mu} = 2\lambda p_\mu = \dot{Z}_\mu.$$

The choice of τ is arbitrary; if we choose $Z_0 = \tau$, $\dot{Z}_0 \doteq 1 = 2\lambda p_0$, we get $\lambda = \frac{1}{2}p_0$. That is to say, a definite choice of τ ('gauge') implies a definite value of λ. Similarly in higher dimensions we have for the Hamiltonian density

(7.3) $$H_0 = P^\alpha_\mu Z^\mu_{,\alpha} - L = 0$$

from (2.5) and (2.3). We therefore consider the family of Hamiltonian

(7.4) $$H_\xi = H_0 + \int d\sigma \xi(\sigma) C(\sigma),$$

where $C(\sigma)$ are the constraints (2.19):

$$C^{(\alpha\beta)}(\sigma) = (P^{\alpha 2}_\mu + Z^2_{\mu,\beta}), \qquad \alpha \neq \beta, \text{ fixed},$$

and $\xi(\sigma)$ some weight functions. The Hamiltonian equations are

(7.5) $$\frac{\partial H_\xi}{\partial P^\alpha_\mu} = 2\xi P^\alpha_\mu = Z^{;\alpha}_\mu, \qquad \frac{\partial H_\xi}{\partial Z_\mu} = 2\xi Z^\mu_{,\beta} = \dot{P}^\mu;$$

with $\alpha = 0$, $\beta = 1$ and choosing the 'gauge' $\xi = \frac{1}{2}$, we have $P^0_\mu = Z_\mu$, $\dot{P}^0_\mu = Z''_\mu$, hence

$$\ddot{Z} = Z''_\mu$$

and the corresponding coordinate conditions are again $\dot{Z}_\mu Z'^\mu = 0$, $\dot{Z}^2_\mu + Z'^2_\mu = 0$. The resulting Hamiltonian

(7.6) $$H = \frac{1}{2}\int_{\sigma_1}^{\sigma_2} d\sigma (P^{02}_\mu + Z'^2_\mu)$$

agrees with the non-relativistic form $H = \frac{1}{2}\int_0^L dx \left[p(x)^2 + v^2\left(\frac{\partial q(x)}{\partial x}\right)^2\right]$.

The basic Poisson brackets can be evaluated. For equal times τ one obtains in particular

$$(7.7)\quad [Z_\mu(\sigma), P^0_\nu(\sigma')] = g_{\mu\nu}\delta(\sigma-\sigma')$$

which imply

$$(7.8)\quad [H_\xi, H_\eta] = H_{[\xi,\eta]}$$

where

$$(7.9)\quad [\xi, \eta] = \xi\eta'-\eta\xi'.$$

The family of Hamiltonians are closed under the above algebraic relations. These are the Lie algebraic expressions of the group of diffeomorphism. In a discrete basis

$$\xi(\sigma) = \{e^{-in\sigma}\}$$

with

$$L_n = H(e^{-in\sigma})$$

we obtain the ∞-dimensional Lie algebra

$$(7.10)\quad [L_m, L_n] = (m-n)L_{m+n}.$$

The algebraic structure (7.8)–(7.9) is common to a large number of gauge-type theories [4].

The role of the representations of the groups of diffeomorphisms and of the Poincaré group in a quantized theory is the following: Because these are the symmetry groups of the system, states obtained by applications of the operations of the symmetry groups are physically equivalent. Thus, carrier spaces of irreducible representations represent equivalent states of the same physical system. Distinct systems are given by distinct irreducible representations. The dual space of the Poincaré group is labelled by mass and spin. Thus extensible elementary particles will be characterized by mass, spin and, in addition, those quantum numbers which characterize the irreducible representations of the diffeomorphism group.

VIII. Passage to Dynamical Groups and Infinite-Component Wave Equations

It is known that a Hamiltonian of the type (6.4) which is valid in the rest frame of the system, can be generalized and transformed, for a moving system, into an algebraic infinite-component wave equation of the type

$$(8.1)\quad (J^{\mu}P_{\mu}+K)\Psi(P) = 0 .$$

Here P_{μ} is the total momentum of the extensible object, and J_{μ} and K are infinite-dimensional matrices, elements of the Lie algebra of the dynamical group (e.g. SO (4, 2)) in one of its infinite-dimensional representations [2]. In the rest frame $P_{\mu} \equiv (m, 0, 0, 0)$, eq. (8.1) becomes

$$(8.2)\quad (mJ_0+K)\Psi(0) = 0 .$$

Inserting the realization of J_0 and K as differential operators acting on $L^2(R^3)$ one recovers back eq. (6.4). Depending on the choice of the dynamical group and its representation, eq. (8.1) gives us a wave function in terms of the *global* quantum numbers of the system. For example, eq. (8.1) corresponding to (6.4) gives a solution $\Psi_{njm}(P)$, where n, j, m are the eigenvalues of a basis of the Lie-algebra representation (physically interpreted as principal quantum number, spin and spin-component, respectively).

This type of description is in agreement with our philosophy that we should try to identify global motions of our extensible objects, hence global quantum numbers. Because we have a relativistic description of the extended object, and a definition of the total relativistic momentum

$$P_{\mu} = \int d\Sigma_{\alpha} P_{\mu}^{\alpha}$$

one should be able to obtain directly from the equations of the extensible object, the infinite-component wave equation (8.1), representing the moving object with certain special motions around the center of mass, say. Thus, in the case of the string with spin, when the string moves rigidly, we would expect the Dirac equation

$$(8.3)\quad (\gamma^{\mu}p_{\mu}+K)\Psi(p) = 0 .$$

However, when the string during its motion also rotates we would expect an equation of the type (8.1). Note that (8.3) and (8.1) are algebraically the same equation. The difference is that in (8.3) we use a finite-dimensional representation of the dynamical group (in this case a four-dimensional representation of SO(4, 2)) and in eq. (8.1) we use an infinite-dimensional representation of the same group.

References

[1] See the historical survey by A. I. Miller [13]. Specifically on the theory of the electron see the recent review by F. Rohrlich [14].

[2] Evidently a continuous object can also be embedded in a curved space in which case Z^μ would be the coordinates in this curved space.

Bibliography

[1] A. O. Barut, in: 'Schladming Proceedings', *Acta Phys. Austriaca, Suppl. XI* (1973) 565, and references therein.
[2] A. O. Barut, *Theories of the Proton* (to be published).
[3] A. O. Barut and G. Bronzin, *Nucl. Phys.* **B81** (1974) 477.
[4] A. O. Barut, L. Girardello and W. Wyss, *Lett. N. C.* **4** (1972) 100.
[5] P. A. M. Dirac, *Proc. Roy. Soc.* **A133** (1931) 60.
[6] P. A. M. Dirac, *Phys. Rev.* **74** (1948) 817.
[8] P. A. M. Dirac, *Lectures on Quantum Mechanics*, Yeshiva Univ., New York 1964.
[7] P. A. M. Dirac, *Proc. Roy. Soc.* **A268** (1962) 57.
[9] A. Einstein, *Sitzungsber. d. k. preuss. Akad. Wiss.* (1919) 349.
[10] J. Goldstone, C. Rebbi and C. B. Thorn, *Nucl. Phys.* **B56** (1973) 109.
[11] G. Herglotz, *Ann. d. Phys.* **31** (1911) 393.
[12] G. Herglotz, *Ann. d. Phys.* **36** (1911) 493.
[13] A. I. Miller, *Archives for History of Exact Sciences* **10** (1973) 207–328.
[14] F. Rohrlich, in: 'The Physicist's Conception of Nature', D. Reidel Publ. Co. (1973).
[15] J. M. Souriau, *Compt. Rend. Acad. Sci.* (1972).

QUANTUM MECHANICS IN PARTICLE PHYSICS

A. BÖHM*

Center for Particle Theory, University of Texas, Austin, Tx., 78712, U.S.A.

The general topic of my presentation will be the question of whether and to what extent relativistic quantum mechanics can describe the structure of hadrons in principally the same way as nonrelativistic quantum mechnics describes the structure of atoms. This would mean in particular, whether the mass levels of the hadrons and other properties of one-hadron states like decay rates and form factors can be calculated in analogy to energy levels and transition rates for atoms. I want to discuss this problem using the example of the leptonic and semileptonic decay of the K meson and therewith treat in more detail the decay of hadron states rather than the hadron spectrum (That hadron spectra can be obtained by this method is well known to many people).[1]

The pseudoscalar meson decay, in particular the K-decay is a very suitable example to illustrate the above question. It is kinematically quite simple yet involves large mass differences and therefore exhibits very clearly the fundamental theoretical properties. It is experimentally rather easily accessible and there exists a wealth of experimental data on the decay property; recently it has gained particular interest because two high statistics experiments[2] gave different values for the perhaps most frequently measured parameter of particle physics, the form factor ratio $\xi = \dfrac{f_-}{f_+}$. My personal reason for choosing this topic here is that it will allow me to discuss in detail one particular theoretical assumption which J. Werle told me many years ago and which has never been explicitly made use of, but which will have immediate consequences for the value of ξ.

Before describing the particular assumptions for the relativistic quantum mechanics of a one-hadron system let me recall some of the basic assumptions of quantum mechanics.[3]

Axiom. A physical observable, defined by the prescription for its measurement, is represented by a linear operator in a linear space.

The mathematical image of a physical system is an algebra $\mathscr{A}$ of linear operators in a linear space $\mathscr{H}$. The algebra is generated by some basic physical quantities and the multiplication is defined by algebraic relations.

Examples in nonrelativistic quantum mechanics are: the P_i (representing momentum) and Q_i (representing position) with the algebraic relation $[P_i, Q_j] = \frac{1}{i}\delta_{ij}I$ or the J_i (representing the observable angular momentum) and A_i (representing the Lenz-vector) with the algebraic relations of the enveloping algebra of SO(4).

AXIOM. The state of a quantum mechanical system is described by a Hermitian positive definite operator W (statistical operator, density matrix). The expectation value of the observable A is given by $\varrho(A) = \mathrm{Tr}(AW)$.

If $W(t)$ describes the state of the decaying meson and Λ the projection operator on the subspace of final states for the process $\alpha \to \pi^0 l\nu$, $(\alpha = K^{\pm 0}, \pi^{\pm})$ then the probability for the measurement of the property Λ at the time t is according to this axiom given by $p(t) = \mathrm{Tr}\big(\Lambda W(t)\big)$. The transition rate is $\frac{dp(t)}{dt} = 2\,\mathrm{Im}\,\mathrm{Tr}\big(\Lambda T W(t)\big)$ and the initial decay rate $\Gamma = \left.\frac{dp(t)}{dt}\right|_{t=0}$ is given by

$$(1) \qquad \Gamma = 2\pi \sum_b \sum_{\alpha,\alpha'} \delta(E_{\alpha'} - E_b)\langle b|T|\alpha\rangle\langle\alpha|W(0)|\alpha'\rangle\langle\alpha'|T|b\rangle .$$

Here T is the transition operator and it has been assumed that the process is weak in the sense that T, i.e. the part of the Hamiltonian which is responsible for the decay, causes a negligible level shift.

$|\alpha\rangle$, $|\beta\rangle$ denote the eigenvectors of a complete system of commuting observables (C.S.C.O.).

$|a\rangle$, $|b\rangle$ denote the corresponding free eigenvectors.

$\sum_{\alpha'\alpha}$ means summation (or integration) over all the eigenvalues of the C.S.C.O.

$\sum_b$ means summations over all basis vectors that span the space of final states $\Lambda\mathscr{H}$.

E_b is the energy of the final state, $(E_b = E_{\pi^0} + E_e + E_\nu)$.

In order to proceed we have to make assumptions about the algebra of observables for the one-hadron system. The following three conditions on these assumptions will certainly be generally acceptable: 1) relativistic description, 2) SU(3) classification, 3) $V-A$ transitions. Because of 1) we assume that the generators of the Poincaré group P_μ, $L_{\mu\nu}$ are observables and the enveloping algebra $\mathscr{E}(p)$ extended by parity U_p, time inversion A_T and charge conjugation U_c is a subalgebra of $\mathscr{A}$. Because of 2) we assume that the charges $E_\alpha{}^4$ ($\alpha = \pm 1 \pm 2 \pm 3, 0, 8$) that generate an SU(3) are observables and $\mathscr{E}(\mathrm{SU}(3)_E)$ is a subalgebra of $\mathscr{A}$. Because of 3) in connection with 2) we assume that we have as further observables 8 Lorentz vector operators V_μ^α ($\mu = 0, 1, 2, 3$, $\alpha = \pm 1 \pm 2 \pm 3, 0, 8$) and 8 Lorentz axial vector operators A_μ^α which describe the transitions between one-hadron states.

These three assumptions are rather obvious for a quantum mechanics that is to describe hadrons. The question is how can these 3 assumptions be combined with each other; i.e., what is the relation between the Poincaré group and SU(3), what is the relation between SU(3) and $V-A$, what is the relation between the extended Poincaré group and $V-A$, etc.

If we knew completely the algebra of observables we would of course have the answer to all these questions. As we don't, but rather want to find it, we have to conjecture some properties of the algebra which concern the combinations of the 3 above listed assumptions. Some of these properties are well known and precisely formulated, others are only expressed in vague terms like 'broken SU(3)', or 'octet operators in the symmetry limit'. We will give one possible precise formulation of these vague notions and investigate their consequences.

The assumption that the SU(3) which classifies the particles is a symmetry group, i.e., $\mathrm{SU}(3) = \mathrm{SU}(3)_S$ with

$$(2) \qquad [P_\mu, \mathrm{SU}(3)_S] = 0, \qquad [L_{\mu\nu}, \mathrm{SU}(3)_S] = 0$$

is clearly a highly unrealistic assumption as the masses in an SU(3) multiplet are far from being equal, which would be a consequence of (2). However the relation (which Werle told me many years ago),

$$(3) \qquad [\hat{P}_\mu, \mathrm{SU}(3)_E] = 0, \qquad [L_{\mu\nu}, \mathrm{SU}(3)_E] = 0$$

where

$$\hat{P}_\mu = P_\mu M^{-1}, \qquad M^2 = P_\mu P^\mu$$

does not lead to any immediate contradiction with experimental data. We will therefore assume that the particle-classifying $SU(3)_E$ is a spectrum generating group, i.e., its generators E_α transform from one state to another not only changing the internal quantum numbers isospin I, its component I_3, and hypercharge Y but also the mass in such a way that the experimentally correct mass will go together with the corresponding internal quantum numbers (I, I_3, Y). It is easy [4] to set up algebraic relations between the mass operator M and the SU(3) generators that will give the right mass formula, e.g. it is well known that the assumption that M^2 is the 8th component of an $SU(3)_E$-octet operator will, to a certain extent, do the job. We will, therefore, not discuss this problem but the consequences of the assumption that though M and consequently P_μ does not commute with SU(3), $\hat{P}_\mu = P_\mu M^{-1}$ does.

Turning now to the relation of $V-A$ to the extended Poincaré algebra and $SU(3)_E$ we assume that

$$\text{(4)}\qquad \begin{array}{l} V_\mu^\alpha \text{ are Lorentz vectors,} \\ A_\mu^\alpha \text{ are Lorentz axial vectors.} \end{array}$$

This specifies the relation between $L_{\mu\nu}$, U_p and V_μ^α, A_μ^α. We further assume that V_μ^α and A_μ^α have a definite T-transformation property.[5]

$$\text{(5)}\qquad \begin{aligned} A_T V_\mu^\beta A_T^{-1} &= \varepsilon(\mu) V_\mu^\beta, \\ A_T A_\mu^\beta A_T^{-1} &= \varepsilon(\mu) A_\mu^\beta, \end{aligned}$$

$$\varepsilon(\mu) = -(1-2\delta_{\mu 0}) = \begin{cases} +1 & \text{for } \mu = 0, \\ -1 & \text{for } \mu = 1, 2, 3. \end{cases}$$

In addition to the definite T and P transformation property a definite c-transformation property is assumed.

We postulate

$$\text{(6)}\qquad \begin{aligned} U_c V_\mu^\beta U_c^{-1} &= -V_\mu^{-\beta} \quad (V^{-8} = V^8), \\ U_c A_\mu^\beta U_c^{-1} &= +A_\mu^{-\beta}. \end{aligned}$$

(6) will not be used for the process under consideration and has only been listed for completeness. It is required to relate the decay properties of particles and antiparticles if these are not already given by SU(3).

We do not specify relations between the P_μ and V_μ^β, A_μ^β; these are hoped to be conjectured later from further experimental data (e.g. about the t-dependence of form factors). We will however make in (9) one assumption about these relations.

The fact that the V_μ^α and A_μ^α have been labeled by the same label as the charges E_α – assigning a transition operator to each member of the octet – hints already towards a particular relationship between the V_μ^α, A_μ^α and the particle-classifying SU(3). The simplest assumption about this relation would be that V_μ^α, A_μ^α are $SU(3)_E$ octet operators, i.e.

(7) $$[E_\alpha, V_\mu^\beta] = f^{\alpha\beta\gamma} V_\mu^\gamma \quad \text{or} \quad U V_\mu^\beta U^+ = V_\mu^\gamma U_{\gamma\beta}^8$$

where $U \in SU(3)_E$, $f^{\alpha\beta\gamma}$ are the SU(3) structure constants and $U_{\gamma\beta}^8 = \langle\gamma|U|\beta\rangle$ is the matrix element of U in the 8-dimensional representation.

(7) is one possible precise formulation of the vague assumption that V_μ^β are octet operators of some 'broken' SU(3). This is, however, not the only precise formulation of this vague statement; other possibilities are

(7^p) $$[F_\alpha, \{M^{-p}, V_\mu^\beta\}] = f^{\alpha\beta\gamma}\{M^{-p}, V_\mu^\beta\}, \qquad p = \pm 1, \pm 2, \ldots$$

where M is the mass operator. In the case of SU(3) symmetry, i.e. when (2) holds, (7) and all (7^p) are equivalent.

The same assumption (7^p) we make for the A_μ^β, where p may or may not be different for V_μ^α and A_μ^α. We will first work with the assumption (7), i.e. $p = 0$, and if $p \neq 0$ we have just to replace everywhere V_μ^β by $\{M^p, V_\mu^\beta\}$.

The assumption (7^p) is an ansatz that gives the supression for the K_{l_3} and K_{l_2} decay. It is not essential for the prediction of the form factor ratio ξ, which is a consequence of assumption (3). Though we believe that (7^p) may well have to be altered we will use it here as a model assumption.

Assumption (5) together with (7) (or(7^p)) constitutes the core of the assumption which is usually formulated as V_μ^β and A_μ^β being of first class in the symmetry limit. [5] As a consequence of (7) follows

(8) $$V_\mu^{\beta +} = \omega V_\mu^{-\beta}, \quad |\omega| = 1 \quad (\beta = \pm 1 \pm 2 \pm 3, 0, 8).$$

We will assume $\omega = 1$, i.e. that the [6]

(8′) V_μ^β and A_μ^β are Hermitian.[7]

We need one further assumption that concerns the relations between P_μ and V_μ^α, the analogue of CVC:[8]

The electromagnetic component

$$V_\mu^Q = V_\mu^0 + \frac{1}{\sqrt{3}} V_\mu^8$$

of the octet operator (or any other component) fulfills

(9) $[P^\mu, V_\mu^Q] = 0.$

The analogy between the above assumptions and the usual formulation with currents and axial currents is obvious. Our structure is however entirely within the framework of quantum mechanics and therefore much simpler than the field theoretic structures of current algebra. In a certain sense, these assumptions are – except for few additional specifications – an extraction from the usual formulation of those properties that are really used in the calculation of numbers, as the field theoretic properties are not really made use of in a formulation that extends only to the lowest order of perturbation theory.

To proceed with the calculation of the decay rate in eq. (1) we have to state how the transition operator T is related to the V_μ^α and A_μ^α. We will do this later because it is more convenient to bring first eq. (1) into a more familiar form, by choosing a convenient set of basis vectors or a convenient C.S.C.O.

Usually one chooses as basis vectors the set of generalized eigenvectors of momentum and internal quantum numbers

(10) $|\alpha\rangle = |p, I, I_3, Y, \eta\rangle$

where η denotes some possible further quantum numbers. In the case that (3) and not (2) holds it is however much more convenient to choose for $|\alpha\rangle$ a set of generalized eigenvectors of $\hat{P}_\mu = P_\mu M^{-1}$ and not of P_μ:

(11) $|\alpha\rangle = |\hat{p}, I, I_3, Y, \hat{\eta}\rangle$

which we normalize in a Lorentz invariant way:

(12) $\langle\hat{\alpha}'p'|p\hat{\alpha}\rangle = 2\hat{E}(\hat{p})\,\delta^3(\hat{p}-\hat{p}')\,\delta_{\hat{\alpha}'\hat{\alpha}}$

where

$$\hat{E}(\hat{p}) = \sqrt{1+\frac{\hat{p}^2}{m^2}}, \qquad \hat{\alpha} = (I, I_3, Y, \hat{\eta}).$$

Then $\sum_{\alpha}$ in (1) is given by:

$$\sum_{\alpha} = \sum_{\hat{\alpha}} \int \frac{d^3\hat{p}}{2\hat{E}}.$$

The reason that the choice (11) is more convenient than (10) is the following:

If we calculate the matrix element of V_μ^β in the basis (10) and (11) we can write it as (we ignore for simplicity that there may be additional quantum numbers η like e.g. spin or other charges):

$$(13) \qquad \langle \alpha' p' | V_\mu^\beta | p \alpha \rangle = \langle \alpha' | V^\beta | \alpha \rangle \langle p' | V_\mu | p \rangle,$$

$$(14) \qquad \langle \hat{\alpha}' \hat{p}' | V_\mu^\beta | \hat{p} \hat{\alpha} \rangle = \langle \hat{\alpha}' | \hat{V}^\beta | \hat{\alpha} \rangle \langle \hat{p}' | \hat{V}_\mu | \hat{p} \rangle$$

where we have written the operator V_μ^β in two different ways as a direct product

$$(15) \qquad V_\mu^\beta = V_\mu \otimes V^\beta,$$

$$(16) \qquad V_\mu^\beta = \hat{V}_\mu \otimes \hat{V}^\beta$$

where (15) corresponds to the basis (10), i.e.

$$(17) \qquad |\alpha\rangle = |p, \alpha\rangle = |p\rangle \otimes |\alpha\rangle$$

and (16) corresponds to the basis (11), i.e.

$$(18) \qquad |\alpha\rangle = |\hat{p}, \hat{\alpha}\rangle = |\hat{p}\rangle \otimes |\hat{\alpha}\rangle.$$

(15) is in accordance with the usual description and (16) is the most general form if one assumes that V_μ^β is a purely F-type octet operator.

By assumption (3) $U \in SU(3)_E$ can be written in the basis (18) as

$$(19) \qquad U = 1 \times \hat{U}$$

but in the basis (17)

$$(20) \qquad U = U^{ex} \otimes U^{int} \quad \text{with } U^{ex} \neq 1$$

because U also acts on the momenta p but not on the velocities $\hat{p}$. Consequently all SU(3) dependence of the matrix element in (14) is already contained in the factor $\langle \hat{\alpha}' | V^\beta | \hat{\alpha} \rangle$ and $\langle \hat{p}' | V_\mu | \hat{p} \rangle$ is an SU(3) scalar, whereas for the matrix element (13) both factors have SU(3) dependence, i.e. $\langle p'|V_\mu|p\rangle$ is not an SU(3) scalar. Consequently, if we write (13) and (14) as

$$(13')\quad \langle \alpha' p'|V^\beta_\mu|p\alpha\rangle = \langle\alpha'|V^\beta|\alpha\rangle\{f_+(p'+p)_\mu + f_-(p'-p)_\mu\}$$

and

$$(14')\quad \langle \hat{\alpha}'\hat{p}'|V^\beta_\mu|\hat{p}\hat{\alpha}\rangle = \langle\hat{\alpha}'|\hat{V}^\beta|\hat{\alpha}\rangle\{\mathscr{F}_+(\hat{p}'+\hat{p})_\mu + \mathscr{F}_-(\hat{p}'-\hat{p})_\mu\}$$

(which is a consequence of the Lorentz transformation property of V_μ and $\hat{V}_\mu$) then $\mathscr{F}_\pm$ are $SU(3)_E$ invariants but $f_\pm$ are not.

From the fact that $\mathscr{F}_\pm$ are SU(3) invariants one can derive that as a consequence of (5), (8), and (9) it follows [9]

$$(21)\quad \mathscr{F}_- = 0$$

whereas this cannot be derived for f_-.[10] ((21) is the condition that corresponds to the vanishing of the second class matrix element.)

If the decaying system is prepared to have the internal quantum number

$$\alpha = I, I_3, Y, \quad \text{e.g. is either a } K^\pm \text{ or a } \pi^\pm \text{ meson,}$$

then the statistical operator of the decaying system is given by

$$(22)\quad W(0) = \iint \frac{d^3\hat{p}}{2\hat{E}(\hat{p})}\frac{d^3\hat{p}'}{2\hat{E}(\hat{p}')}|\hat{p},\hat{\alpha}\rangle\langle\hat{\alpha}\hat{p}'|\varphi(\hat{p})\bar{\varphi}(\hat{p}')$$

where $|\varphi(\hat{p})|^2$ is the probability that the momentum value of the decaying particle is $p = m_{(\alpha)}\hat{p}$. ($\varphi(\hat{p})$ corresponds to the wave function in momentum space). For the case that the decaying system has an exact momentum p_A one has[11]

$$(23)\quad |\varphi(\hat{p})|^2 = 2\hat{E}(\hat{p}_A)\,\delta^3(\hat{p}-\hat{p}_A).$$

For the basis vectors $|b\rangle$ of the space of final states in the process $\alpha \to \pi^0 l\nu$ we choose

$$(24)\quad |b\rangle = |\hat{p}_{\pi^0}, \pi^0; e, p_e; \nu, p_\nu\rangle = |\hat{p}_{\pi^0}, \pi^0\rangle \otimes |p_e, e\rangle \otimes |p_\nu, \nu\rangle$$

and correspondingly

$$(24)\quad \sum_b = \sum_{\text{Pol}} \iiint \frac{d^3p_\nu}{2E_\nu}\frac{d^3p_e}{2E_e}\frac{d^3\hat{p}_\pi}{2E_\pi}.$$

If one inserts (24), (22), (11), (12) into (1) one obtains

$$(25)\qquad \Gamma = 2\pi \iiint \frac{d^3p_\nu}{2E_\nu}\frac{d^3p_e}{2E_e}\frac{d^3\hat{p}_\pi}{2\hat{E}_\pi}\times$$
$$\times \int \frac{d^3\hat{p}_\alpha}{2\hat{E}_\alpha}\frac{d^3p'_\alpha}{2\hat{E}'_\alpha}\,\delta(E_\nu+E_e+E_{\pi^0}-E_\alpha)\varphi(\hat{p}_\alpha)\bar{\varphi}(\hat{p}'_\nu)\times$$
$$\times \sum_{\text{Pol}}\langle \nu, e, \pi^0, p_\nu p_e \hat{p}_{\pi^0}|T|\alpha, \hat{p}_\alpha\rangle\langle\hat{p}_\alpha, \alpha|T|\hat{p}_\pi p_e p_\nu, \pi^0 e\nu\rangle .$$

We assume that T conserves total momenta, so its matrix element can be written as a product of the momentum δ-function and a reduced matrix element.

$$(26)\qquad \langle p_e p_\nu \hat{p}_{\pi^0}|T|\hat{p}_\alpha\rangle = \delta^3(p_e+p_\nu+p_{\pi^0}-p_\alpha)\langle\langle e\nu\pi^0 p_e, p_\nu, \hat{p}_\pi|T|\alpha\hat{p}_\alpha\rangle\rangle .$$

With (26) and under the experimental condition that the decaying system has an exact momentum p_A, i.e. that the momentum distribution is given by (23), the decay rate (25) takes the almost familiar form

$$(27)\qquad \Gamma = 2\pi \iiint \frac{d^3p_e}{2E_e}\frac{d^3p_\nu}{2E_\nu}\frac{d^3p_{\pi^0}}{2E_{\pi^0}}\sum_{\text{Pol}}\delta^4(p_\alpha-p_{\pi^0}-p_e-p_\nu)\times$$
$$\times \frac{1}{m_\alpha^2 m_{\pi^0}^2}\frac{1}{2E_\alpha}|\langle\langle e, \nu, \pi^0 p_e p_\nu \hat{p}_{\pi^0}|T|\alpha\hat{p}_A\rangle\rangle|^2$$

where

$$p_\alpha = (E_\alpha, p_A).$$

We now formulate the connection between the transition operator T and the observables of the hadron system. This connection is postulated in analogy to the usual $V-A$ theory for leptonic weak decay.

The reduced matrix element of T is in analogy to the usual $V-A$ assumed to be the product of a leptonic part and a hadronic part, where the leptonic part is given by the usual $V-A$ matrix element for leptons:

$$(28)\qquad \langle\langle e^+\nu\pi^0 p_e p_\nu \hat{p}_\pi|T|\hat{p}_\alpha, \alpha\rangle\rangle$$
$$= \bar{u}(p_\nu)\gamma^\lambda(1-\gamma_5)v(p_e)\langle\langle\pi\hat{p}_\pi|H_\lambda|\hat{p}_\alpha\hat{\alpha}\rangle\rangle .$$

The transition operator in the hadron space H_λ is given by:

$$(29)\qquad H_\lambda = g\sum_{\alpha=\pm1,\pm2}(V_\mu^\alpha + A_\mu^\alpha)$$

where V_μ^α and A_μ^α are the observables discussed above and g is a constant which expresses one strength of the interaction. For decays of the kind $\alpha \to \pi l \nu$ only one component $\alpha = \pm 1$ (strangeness-non-changing) or $\alpha = \pm 2$ (strangeness changing) has a non-zero matrix element. It may be expected that other components of V_μ^α, A_μ^α appear in different decay processes, e.g. electromagnetic decays.[8]

The distinction between the usual Cabibbo model of the $V-A$ theory and the assumption (28), (29) is that the latter assumes a higher universality using only one and the same constant g for the strangeness changing $V_\mu^{\pm 2}$ and strangeness non-changing $V_\mu^{\pm 1}$ instead of a ($g \tan \theta$) for $V_\mu^{\pm 2}$ and g for $V_\mu^{\pm 1}$. However, our V_μ^α are octet operators of a spectrum generating group $SU(3)_E$. The idea behind this is that the relation between the spectrum generating $SU(3)_E$ and the mass operator will result in the Cabibbo suppression.

The reduced matrix element of H_λ is given by (as a consequence of (29), (19), (4) and (5), (8), (9)):

$$\text{(30)} \qquad \langle\langle \pi^0 \hat{p}_{\pi^0} | H_\lambda | \hat{p}_\alpha \alpha \rangle\rangle = g \langle \hat{\pi} | \hat{V}^\beta | \hat{\alpha} \rangle F_+ (\hat{p}^{(\alpha)} + \hat{p}^{(\pi)})_\lambda$$

where F_+ is the SU(3) invariant form factor which is a function of the momentum transfer squared $t = m_{(\alpha)}^2 + m_{(\pi)}^2 - 2m_{(\alpha)} m_{(\pi)} \hat{p}_{(\alpha)} \cdot \hat{p}_{(\pi)}$, and $\langle \hat{\pi} | \hat{V}^\beta | \hat{\alpha} \rangle$ is the SU(3) matrix element of the component of the octet operator $\hat{V}^\beta$ that gives a non-zero contribution (V^{-2} for K^+ and V^{-1} for π^+ decay).

Inserting (28) with (30) into (27) gives

$$\text{(31)} \qquad \Gamma = 2\pi \left| \frac{\langle \hat{\pi} | V^\beta | \hat{\alpha} \rangle g}{m_\pi m_a} \right|^2 \iiint \frac{d^3 p_e}{2E_e} \frac{d^3 p_\nu}{2E_\nu} \frac{d^3 p_\pi}{2E_\pi} \times$$

$$\times \sum_{\text{Pol}} \delta^4 (p_\alpha - p_\pi - p_\gamma - p_e) \frac{1}{2E_\alpha} \left| \bar{u} \gamma^\lambda (1 - \gamma_5) v \left\{ p_\lambda^{(\alpha)} - \right.\right.$$

$$\left.\left. - \frac{1}{2} (1 - \xi^{(\pi, \alpha)}) (p_\lambda^{(\alpha)} - p_\lambda^{(\pi)}) f_+^{(\pi\alpha)} \right\} \right|^2$$

where we have introduced the quantities

$$\text{(32)} \qquad f_+^{(\pi, \alpha)} = \frac{m_{(\alpha)} + m_{(\pi)}}{m_{(\pi)} m_{(\alpha)}} F_+,$$

$$\text{(33)} \qquad \xi^{(\pi, \alpha)} = \frac{m_{(\pi)} - m_{(\alpha)}}{m_{(\pi)} + m_{(\alpha)}}$$

in order that the amplitude in the integral (31) has the familiar form expressed in terms of the usual form factor f_+ and the form factor ratio $\xi = \frac{f_-}{f_+}$ in which the K_{l_3} experiments are analyzed.

The main distinction between our predictions (31), (32), and (33) and the usual expression is the occurrence of mass dependent factors. This, however, one should have expected as it is the precise formulation of the brokenness of SU(3) which is the only essential difference between our quantum mechanical treatment and what is really made use of in the usual treatment. In the symmetry limit in which the masses are equal one obtains the usual symmetry expressions, i.e. that $\xi = 0$ and $f_+^{(\pi\alpha)}$ is an SU(3) scalar, i.e. independent of π, α.

The t dependence of F_+ would follow from the relations between the P_μ and V_μ^β, which we do not know;[12] therefore the t dependence of the K_{l_3} form factor $f_+^{(\pi K)}$ cannot be predicted. However, we predict by (33) the form factor ratio for K_{l_3} decay

$$(34) \qquad \xi^{(\pi K)} = \frac{m_\pi - m_K}{m_\pi + m_K} = -0.57.$$

Further we predict the relative strength of the π_{l_3} and K_{l_3} decay which in the usual treatment is expressed by the vector Cabibbo angle θ_v. In order to facilitate comparison with familiar quantities we introduce the suppression factor S_{l_3} which is defined by:

$$S_{l_3} = \frac{\langle \pi^0 | V^{-1} | \pi^+ \rangle}{\langle \pi^0 | V^{-2} | K^+ \rangle} \sqrt{\frac{\Gamma(K_{l_3})}{\Gamma(\pi_{l_3})} \cdot \frac{\text{Phase space } \pi_{l_3}}{\text{Phase space } K_{l_3}}}.$$

(In the usual notation S_{l_3} corresponds to $f_+(0)\tan\theta_V$.)

From (31) and (32) follows:

$$(35) \qquad S_{l_3} = \frac{m_\pi^2}{m_K^2} \frac{m_{K^+} + m_{\pi^0}}{m_{\pi^+} + m_{\pi^0}} = 0.183.$$

Whereas the prediction of ξ is only a consequence of the relation (3), the prediction of S_{l_3} depends upon the assumption about the relation between V_μ^β and E_α. If this relation is given by (7) then (35) is obtained. If we, however, assume that $\{M^{-1}, V_\mu^\beta\}$ and not V_μ^β is an SU(3)-octet operator, i.e., that (7^p) with $p = -1$ holds, we obtain:

$$(36) \qquad S_{l_3} = \frac{m_\pi}{m_K} = 0.283.$$

The latest experimental value for S_{l_3} is 0.224.[13] The experimental situation with regard to the value of ξ is conflicting, as the result of one high statistics experiment, giving $\xi = 0$, disagrees with another high statistics experiment and most of the other recent experimental data.[2] Since all the assumptions except the precise formulation of SU(3) breaking that we have made use of in the calculation of ξ are well accepted in any theoretical frame for weak leptonic decays, an experimental value of $\xi = 0$ would really be very worrisome (note that ξ is largely independent upon the relation between E_α and V_{μ}^{β}).

The K_{l_2} and π_{l_2} decay is described by A_μ^β. The assumption that A_μ^β is an SU(3) octet operator will lead to results in violent disagreement with experimental data. We therefore assume ($7^{P=-1}$), i.e. that $\{M^{-1}, A_\mu^\beta\}$ is an $SU(3)_E$ octet operator:

$$[E_\alpha, \{M^{-1}, A_\mu^\beta\}] = if^{\alpha\beta\gamma}\{M^{-1}, A_\mu^\gamma\}. \tag{37}$$

By the same arguments as given for the α_{l_3} decay one obtains for the α_{l_2} decay

$$\Gamma = 2\pi \iint \frac{d^3p_e}{2E_l}\frac{d^3p_\nu}{2E_\nu}\,\delta^4(p_{(\alpha)}-p_{(\nu)}-p_{(l)})\frac{1}{2E_\alpha m_\alpha^2}\times \\ \times |\langle\langle l\nu p_l\, p_\nu|T|\alpha, \hat{p}_\alpha\rangle\rangle|^2 \tag{38}$$

where the reduced matrix element is given by:

$$\langle\langle l^+\nu p_l p_\nu|T|\alpha\hat{p}_\alpha\rangle\rangle = \bar{u}(p_\nu)\gamma^\lambda(1-\gamma_5)\,v(p_l)\langle\langle\sigma|H_\lambda|\hat{p}_\alpha\,\alpha\rangle\rangle \tag{39}$$

with

$$\langle\langle\sigma|H_\lambda|\,\alpha\rangle\rangle = gF\langle\sigma|\tilde{A}^\beta|\,\alpha\rangle\,P_\mu^{(\alpha)} \tag{40}$$

σ is the hadronic vacuum state and $\langle\sigma|\tilde{A}^\beta|\alpha\rangle$ is the SU(3) matrix element of the SU(3) octet operator $\tilde{A}^\beta$ which appears in

$$\{M^{-1}, A_\mu^\beta\} = \tilde{A}^\beta \times \tilde{A}_\mu.$$

The suppression factor S_{l_2} for K_{l_2} decay is defined in analogy to the α_{l_3} case by

$$S_{l_2} = \frac{\langle 0|A^{-1}|\pi^+\rangle}{\langle 0|A^{-2}|K^+\rangle}\sqrt{\frac{\Gamma(K_{l_2})\text{ Phase space }(\pi_{l_2})}{\Gamma(\pi_{l_2})\text{ Phase space }(K_{l_2})}}. \tag{41}$$

S_{l_2} corresponds in the usual treatment to $\frac{f_K}{f_\pi} \tan\theta_A$. Inserting (38), (39), (40), into (41) one obtains the following prediction:

$$(42) \qquad S_{l_2} = \frac{m_\pi}{m_K} = 0.283$$

which is in very good agreement with the experimental value[13] of 0.276.

The quantum mechanical description that has been given here may in principle be employed for every weak and electromagnetic decay process which involves transitions between hadrons that can be considered as different states of one and the same hadron system. If different leptons are present the leptonic part of the matrix element will have to be changed appropriately (see e.g. reference [8]) from its form in (28) and (39). An open question is the universality, i.e. to what extent the strength of the decay can be expressed by one constant g and how this strength can be compared if different leptonic parts are involved (as it will occur if one wants to compare the strength of the electromagnetic decay and the weak leptonic decays).

Specific assumptions about the particular structure of the algebra of observables that have been made here may also be only a poor approximation of reality and may have to be improved or altered. In particular any of the assumptions (7^p) about the relations between the V^β_μ, A^β_μ and the $SU(3)_E$ does not seem to completely reproduce the experimental results, though it predicts a suppression of the strangeness changing decay of approximately the right magnitude. One point that has emerged from our discussion is that the assumption that V^β_μ and A^β_μ are octet operators with respect to the particle classifying $SU(3)_E$ may well have to be revised. Though this assumption for V^β_μ gives a reasonable value for the suppression of K_{l_3} decay it does not give the right value for K_{l_2} decay, where one has to use (37). For the application of this procedure to baryon leptonic decay one may have to use a different relation from (7) and ($7^{p=-1}$) or perhaps even an entirely different relation from (7^p).

The relation (3) between the Poincaré group and $SU(3)_E$, though presently not contestable from an experimental point of view, may also be only a rough approximation of reality. But as every theory can give only an approximate description of reality, a model assumption which leads to a better agreement with experimental data may well be an improvement.

References

* Work supported by the U. S. Atomic Energy Commission.

[1] E.g. A. Böhm [5], A. O. Barut [3].

[2] The two high statistics experiments are G. Donaldson *et al.* [11] giving $\xi = 0$ from Dalitz plot analysis; J. Sandweiss *et al.* [16], giving $\xi(0) = -0.655 \pm 0.127$ from polarization analysis. S. Merlan *et al.* [14] find $\xi(0) = -0.57 \pm 0.24$ from Dalitz plot analysis and $\xi(0) = -0.64 \pm 0.27$ from μ polarization analysis. P. Haidt *et al.* [12] find by all three methods values in agreement with (34), however their results favor a linear t dependence of ξ. The $K^0_{\mu_3}/K^0_{e_3}$ branching ratio result of G. W. Brandenburger *et al.* [7] is in disagreement with (34) and also in disagreement with the $\Delta I = 1/2$ rule. I. H. Chiang *et al.* [8] and C. Ankenbrandt *et al.* [2] are in agreement with (34). The $K^0_{l_3}$ Dalitz plot analysis experiments M. G. Albrow *et al.* [1]; E. Dally *et al.* [10] are in agreement with (34). The average value of the older data, L. M. Chounet and J. M. Gaillard [9] is also consistent with (34). Preliminary results of the Aachen-Bary-Brussels-CERN Collaboration also agree with (34).

[3] I will call them axioms, which, however does not mean that I believe in the possibility of a strictly axiomatic formulation.

[4] We use the Cartan–Weyl basis for the SU(3) Lie algebra in which the commutation relations are given by

$$
\begin{aligned}
&[H_i, H_j] = 0, \\
&[E_\alpha, E_\beta] = N_{\alpha\beta} E_\gamma, \\
&[H_j, E_\alpha] = r_j(\alpha) E_\alpha, \\
&[E_\alpha, E_{-\alpha}] = r^i(\alpha) H_i .
\end{aligned}
$$

The root vectors are (in the normalization we use)

$$
\begin{aligned}
r_i(1) &= (1/\sqrt{3})(1, 0), \\
r_i(2) &= (1/2\sqrt{3})(1, \sqrt{3}), \\
r_i(3) &= (1/2\sqrt{3})(-1, \sqrt{3}), \\
r_i(-\alpha) &= -r_i(\alpha),
\end{aligned}
$$

and

$$
\begin{aligned}
N_{\alpha\beta} &= \pm \sqrt{\tfrac{1}{6}} \quad \text{if} \quad r(\alpha)+r(\beta) = r(\gamma) \text{ is also a nonvanishing root vector} \\
&= \text{otherwise; in particular} \\
N_{1,3} &= -N_{3,1} = N_{-3,-1} = -N_{-1,-3} = N_{3,-2}, \\
&= -N_{-2,3} = N_{-2,1} = -N_{1,-2} = N_{2,-3}, \\
&= -N_{-3,2} = N_{-1,2} = -N_{2,-1} = \sqrt{\tfrac{1}{6}} .
\end{aligned}
$$

In the normalization we have used here, the hypercharge is

$$Y = 2H_2$$

and the isospin is

$$I_3 = \sqrt{3}H_1, \quad (I_1 \pm iI_2) = (\sqrt{6})E_{\pm 1}.$$

We also call $H_1 = E_0$, $H_2 = E_8 = E_{-8}$.

Our notation differs from the one conventionally used in physics literature. The connection is

$V_\mu^{\pm 1}$ corresponds to $F_{1\mu} \pm iF_{2\mu}$,
$A_\mu^{\pm 1}$ corresponds to $\overset{5}{F}_{1\mu} \pm i\overset{5}{F}_{2\mu}$,
$V_\mu^{\pm 2}$ corresponds to $F_{4\mu} \pm iF_{5\mu}$,
$A_\mu^{\pm 2}$ corresponds to $\overset{5}{F}_{4\mu} \pm i\overset{5}{F}_{5\mu}$, etc.

[5] The condition of being a first class vector operator in the symmetry limit can be formulated as having a definite T transformation property and being a component of the same octet operator with respect to an SU(3) symmetry group as the Hermitian operator that describes the electromagnetic interactions. V_μ^β being first class with respect to a spectrum generating $SU(3)_E$ may therefore be defined as being an Hermitian[7] $SU(3)_E$-octet operator with definite T transformation property, that has a component V_μ^Q which describes the electromagnetic interaction (or which fulfills $[P^\mu, V_\mu^Q] = 0$).

[6] (5) and (8′) are restrictions which may well have to be generalized to include besides these 'normal' (i.e. fulfilling (5) under A_T) and 'regular' ($\omega = 1$) operators also abnormal (($\varepsilon(\mu)$ replaced by $-\varepsilon(\mu)$ in (5)) and irregular ($\omega = -1$) contributions. C. W. Kim and H. Primakoff [13].

[7] We call an octet operator Hermitian if its components have the same Hermiticity property as the corresponding generators in an unitary representation.

[8] Further relations of this kind have been specified for the calculation of the electromagnetic decay of mesons, e.g. (39) in A. Böhm [6].

[9] From (5) and (8) it follows that either $\mathscr{F}_+$ or $\mathscr{F}_-$ must be zero, from (9) it follows that $\mathscr{F}_- = 0$.

[10] $f_- = 0$ can however be derived in the same way as (21) if (2) holds, and only if (2) holds can $f_- = 0$ be derived if one stays within the framework of the assumptions that are usually made.

[11] This normalizes the statistical operator to $\mathrm{Tr}\, W = 1$.

[12] To conjecture these relations one would in fact proceed in the opposite direction, take the experimental t dependence of a form factor and find the relations such that they result in this t-dependence.

[13] E.g., M. Ross [15].

Bibliography

[1] M. G. Albrow *et al.*, *N. P.* **44B** (1972) 1.
[2] C. Ankenbrandt *et al.*, *P. R. L.* **28** (1972) 1472.

[3] A. O. Barut, in: *Proceedings of the 15th International Conference on High Energy Physics*, Naukova Dumka, Kiev (1972) 454.
[4] A. Böhm, *Phys. Rev.* 158 (1967) 1408.
[5] A. Böhm, in: *Proceedings of the International Conference on Symmetries 1969* (edited by R. Chand), Gordon and Breach, New York.
[6] A. Böhm, *Phys. Rev.* **7** (1973) 2701.
[7] W. Brandenburger *et al.*, *SLAC-Pub. 1212*, March 1973.
[8] I. H. Chiang *et al.*, *P. R.* **D6** (1972) 1254.
[9] L. M. Chounet and J. M. Gaillard, *Phys. Rep.* **40** (1972) 199.
[10] E. Dally *et al.*, *P. L.* **41B** (1972) 647.
[11] G. Donaldson *et al.*, *P. R. L.* **31** (1973) 337.
[12] P. Haidt *et al.*, *P. R.* **D3** (1972) 10.
[13] C. W. Kim and H. Primakov, *Phys. Rev.* **180** (1969) 1502.
[14] S. Merlan *et al.*, Brookhaven Preprint, *BNL-18076*, June 1973.
[15] M. Ross, *Phys. Lettres* **36** (1971) B130.
[16] J. Sandweiss *et al.*, *P. R. L.* **30** (1973) 1002.

ON COMPOSITE GAUGE FIELDS AND MODELS

P. BUDINI

International Centre for Theoretical Physics, Trieste, Italy

P. FURLAN

Istituto di Fisica Teorica dell'Università degli Studi di Trieste, Trieste, Italy

Abstract

After a phenomenological study of the possibility of the existence of massless composite lepton and antilepton states originated by the lepton weak interaction Lagrangian, it is concluded that only one such composite state could exist and be strongly coupled in a unique way to the free left-handed leptons and that its quantum numbers could only be those of the photon ($j = 1,0$; $\Pi C = +1, -1$). In the frame of the unified gauge models with a lower limit for the W meson mass, a condition for its existence is that the W mass be of the order of 100 GeV. It is then shown that the composite massless field has the properties of a gauge field and its gauge transformations are deduced from those of the constituent spinor fields as a consequence of the Wilson expansion of the products of the constituent two-component canonical spinor fields.

This property also applies to the case when the two-component spinor fields build up a degenerate multiplet and the corresponding gauge group is non-abelian. The gauge fields of unified models of weak and electromagnetic interactions could then be considered as composite, with the possibility of computing some of the parameters, such as the coupling constant of the gauge fields, which are otherwise free in these models, and with the advantage of eliminating the infinities of the triangular diagrams.

Parity-conserving quantum electrodynamics is obtained with a method analogous to that used in the unified gauge models; it is finite, and its regularizing length is of the order of magnitude of the square root of the Fermi coupling constant. An example of application to the Salam–Weinberg model is given, together with a rough estimate of $\sin\theta$.

The possibility of a bootstrap mechanism and a dynamical symmetry breaking *à la* Nambu–Jona–Lasinio to substitute the Higgs mechanism is mentioned.

Finally, some possibilities are suggested to explain dynamically the universality of charge quantization.

I. Introduction

If there is some validity, as we believe there is, in the thought often expressed by Dirac [13] that simplicity and mathematical beauty are good guiding principles in searching for the fundamental laws of nature, one would expect that the two-component spinor fields, obeying the most simple first-order linear field equation we know of – the Weyl equation – have a fundamental role in the constitution of matter, as long since advocated by Heisenberg [18].

The particles that most resemble the quanta of these fields are leptons and, among them, massless neutrinos. These are certainly abundant in the universe, but they appear to have a fundamental role only in particular phenomena such as radio-active decays and perhaps some peculiar astrophysical transitions. Actually, they do interact directly or indirectly with every elementary or composite system we know of, but weakly at the energies to which we are normally accustomed.

The question then arises: could it be that, despite their apparent weak interaction at low energy, they nevertheless have a fundamental role in determining the forces and quanta which generate the variety of material phenomena we have learned to know?

To answer this question, we started a phenomenological study on the possible consequences one could draw from the experimentally rather well-established Fermi lepton weak interaction Lagrangian, which is the only Lagrangian containing exclusively two-component left-handed lepton fields.

Starting from this Lagrangian, to which that for the free lepton fields, or better still Weyl fields, could be added, the first question we asked was: which are the forces which this Lagrangian implies between leptons? Obviously they can only arise because of lepton pair exchanges; and of the leptons the lightest neutrinos give the forces of longest range, such that they will be the most effective in giving consequence, if any. These forces have been studied [15], [7], [5], [8]. The most interesting feature is that, while repulsive between two leptons, they are attractive between lepton–antilepton. Furthermore, in the hypothesis of massless neutrinos, their Green function obeys, because of the equations of motion, a continuity equation which limits their divergence to a logarithmic one when computed in perturbation expansion [7], [5], [8].

Their long-range behaviour is well known ($\sim x^{-6}$), it is computable from the local weak Lagrangian and is not dependent on the possible refinement introduced in this Lagrangian to make the theory finite or renormalizable. This instead will determine the behaviour of the Green function at short distances. In view of the continuity equation and with the faith, brought by the development of gauge theories, that the weak Lagrangian can in fact be made renormalizable, we shall suppose that the short-distance behaviour of the Green function is of the Coulomb type ($\sim x^{-2}$) (if the theory is superrenormalizable it will behave as x^{-n} with $n < 2$ or as a power of $\log x^2$). Once the forces are so qualitatively characterized, one must determine at which distance they will pass from the long distance ($\sim x^{-6}$) to the short distance ($\sim x^{-2}$) regime, this distance will be the free parameter entering into the phenomenological study.

In the hypothesis that these forces are sufficient to give rise to composite systems built up of leptons–antileptons, the next step was to study their properties and couplings to the free leptons. We found [7], [5], [8] that if they exist they are strongly coupled to the free fields and have many of the properties of gauge fields; in fact we shall show that they might be identified with the gauge fields. Considering the fundamental role that these fields are considered apt to play nowadays in the constitution of the elementary forms of matter, one is brought to consider the working hypothesis that a Lagrangian of the Fermi type is the basic Lagrangian from which both the elementary systems and their interactions are derived dynamically.

In the frame of this hypothesis, one should then have first multiplets of Weyl leptons (possibly massless). Their interaction should originate both the composite systems and the breaking of the original symmetry (and give masses to both the spinors and the composite fields). It would then be natural to expect that the breaking of the symmetry increases with the complexity of the systems and a hierarchy of 'elementary' systems would arise.

At the first grade of the hierarchy, one would insert the (massive) leptons themselves and the photon as the nearest to the original maximal symmetry: that is, lepton quantum electrodynamics. Hadrons should come after as more complex systems resulting from the composition of more Weyl fields. In general terms one would expect that they still contain some of the features of the fields one started from as, for example, the multiplet structure of the underlying (broken) symmetry.

From this standpoint one should then first tackle the problem of lepton quantum electrodynamics (l.q.e.d.) where the photon plays the role of the 'hydrogen atom' of elementary particles. Certainly the task is not a minor one since, apart from the very tight quantitative constraints in which one must move in l.q.e.d., one must first satisfactorily explain some of its fundamental qualitative features, such as:

a) why spin-one for the photon;

b) why only one and massless photon;

c) why the field representing the photon is a gauge field;

d) why, if the constituent fields are Weyl spinors, parity is conserved in l.q.e.d.;

e) why universal quantization of the electric charge.

In earlier papers we have tried to see if, in the frame of the above hypothesis, and from a phenomenological standpoint (that is, massive leptons interacting through the Fermi Lagrangian), there is an answer to these questions and surprisingly we found that for some of them a non-negative answer seems to be not only possible but sometimes also quite simple. Unfortunately, we had to draw some of the results and conclusions from the use of the Bethe–Salpeter equation in the ladder approximation. We are well aware of the limitations of this mathematical tool, but at the moment for this kind of investigation we know of little better in the mathematical literature. Our attitude has been to extract as much information as possible from the general property of the bound state equations and, whenever the use of approximations such as the ladder was unavoidable, to hope that the qualitative feature of the results will remain also after a more rigorous method is found.

In the present paper we shall briefly summarize the main hypotheses and results of the previous papers and give some new results.

Precisely, we shall show (Secs. II and III) how lepton–antilepton massless bound states could originate from the lepton weak Lagrangian if only the forces generated by neutrino pair exchange continue their x^{-6} long-range behaviour down to distances of the order of 10^{-16} cm, corresponding to the wavelength of the square root of the Fermi coupling constant.[1] These bound states can only have a vector-axial vector coupling to the free spinors, and this result depends on the symmetry properties of the equations and not on the ladder approximation adopted for further calculations.

We shall then show that, if these bound states exist, they can only have the quantum numbers of the photon, that is $j = 1, 0$; $\Pi C = +1, -1$ (space parity Π and charge parity C are not defined separately because of the weak Lagrangian we started from). Besides, we find that there exists only one composite state and no excited ones, or, if these exist and have escaped our computational technique, they will have an exceedingly small coupling constant with the free leptons. The next rather surprising result of this part of the work is that universality of the renormalized coupling constant between the constituent spinors and the composite state (that is, independence from the masses of constituent spinors) seems to originate from the fact that we used chiral-projected (or two-component) fields. In fact, if one excludes the chiral projectors, one immediately obtains a logarithmic dependence of the coupling constant from the mass of the constituent fields (logarithmic singularity for $m \to 0$), as one would have expected *a priori*. If this interesting result will resist a critical analysis of the computational approximations adopted, it could throw a new light on the function of the chiral projectors.

At this stage, before even talking of the l.q.e.d., one has still many points to clarify. The most conspicuous is how to get parity conservation. In fact, since we started from left-handed Weyl fields in the interaction Lagrangian, we obtained a left-handed vertex. To get parity conservation one must then dispose of right-handed leptons and combine the left-handed vertex with a right-handed one to get parity conserving l.q.e.d. One then needs an internal symmetry group for the leptons which will be arranged in a multiplet for this symmetry group. We are aware in fact that this happens in nature for the leptons we know [22]. The tempting thought would be to assign the role of the right-handed lepton to the muon [20], but this way does not seem to be practicable yet, so we postponed the attempt of understanding the problem of parity conservation to that of masslessness of the composite vector state, and, consequently, to that of gauge invariance of the composite field.

In fact, we have supposed that the composite state is massless simply because this makes the equations soluble (and we had in mind the photon), but we know that the masslessness of the photon is necessary if l.q.e.d. must be strictly gauge invariant. So, in order to impose zero mass for the bound state we must show that it is a gauge field. Now, since the bound (vector or axial) field is composed by the spinor fields, its gauge transformation mast depend on that ofthe spinor field and must be deductable from it. The resulting theory must be strictly gauge invariant and then the masslessness of the composite field may be imposed for symmetry reasons.

In order to perform this task, one must first include the previous phenomenological calculations in a Lagrangian formalism and show that from this formalism gauge-invariant field equations can be deducted in such a way that the only gauge transformations are those of the spinor fields, while those of the composite fields follow from them and are such as to make the field equations gauge invariant. This is done in Sec. IV, and it is shown that in fact, independently of the approximations used in solving the bound state equations, the gauge transformation of the vector-axial field may be ascribed to the short-distance behaviour of the products of the spinor fields, according to the general rules established by Wilson [28]. Here again this result is obtained because we have used two-component chiral-projected spinor fields as constituents in the interaction Lagrangian.

In Sec. V we have extended the formalism of the generation of gauge fields to the case when the constituent leptons are members of a multiplet, i.e. to non-abelian gauge transformations of the spinor fields, and, as a consequence, we obtain gauge transformations of the composite vector-axial fields following rules identical to those obtained by Yang and Mills in their famous paper [29].

Once we have shown that the composite fields can be identified with Yang–Mills gauge fields, it is an easy exercise to construct with our mechanism the various gauge models of weak and electromagnetic interactions and to borrow from these models the method for obtaining the conservation of parity in quantum electrodynamics. (One must obviously start from massless leptons to construct the gauge massless fields and then to break the symmetry with the Higgs mechanism.) This is done in Sec. VI, where the Salam–Weinberg [25] model is constructed as an example. With this origin of the gauge models, one avoids the difficulty of the famous triangular [1] anomalies, since triangular diagrams are convergent in this scheme, and also l.q.e.d. is finite since $Z_3 = 0$ and Z_1 and Z_2 are finite. One further advantage is that by the methods given in Secs. II and III one could compute the coupling constant appearing in the vertices and then also the sin θ which gives a measure of the neutral currents in the weak Lagrangian.

This exercise of model building could easily be extended to other models and obviously one could use the method to start a sort of bootstrap procedure, using first the gauge model as a formal mathematical procedure to make the original weak Lagrangian renormalizable; then, giving the intermediate bosons sufficiently high masses, use our method to obtain the composite gauge fields from the weak interaction Lagrangian and eventually re-obtain the model from which one started. In this way one could hope to obtain self-consistency constraints on which to test the various models proposed.

One of the aspects which is for us less satisfactory in this procedure, is the necessity of introducing external new boson fields to break the symmetry and to give mass to the intermediate bosons and to the leptons. From our point of view, we think that the symmetry breaking and the lepton masses should arise from the two-component lepton Lagrangian (or its renormalizable version) from which we started, on a line similar to that originally proposed by Nambu–Jona–Lasinio [23] and recently followed by Coleman and Weinberg, Jackiw and Johnson, and Cornwall and Norton [11], [19], [12].

In the present work, this problem is not tackled but only briefly discussed, together with some speculative considerations on the explanation of the universality of electric charge quantization in Sec. VII.

II. The Vertex

Let us start by taking the current–current weak lepton Lagrangian

(1) $$\mathscr{L}_I(x) = -2\sqrt{2}G_F j_\lambda(x) j^{\lambda +}(x),$$

where

(2) $$j_\lambda(x) = \sum_i :\bar{L}^{(i)}(x)\gamma_\lambda L^{(\nu_i)}(x):$$

and

(3) $$L^{(i)}(x) = \tfrac{1}{2}(1+\gamma_5)\psi^{(i)}(x),$$

$\psi^{(i)}(x)$ represents the Dirac spinor fields of the lepton, and ν_i the corresponding neutrino. The sum in (2) must be extended to all leptons i. To examine if (1) may give rise to lepton composite states, we have used the Bethe–Salpeter equation[2] [7], [5], [8], [10], [17]

$$\frac{i}{(2\pi)^4}\Gamma(X,x) \equiv \left(\frac{1}{2}\vec{\hat{P}}+\vec{\hat{p}}-m\right)\chi_P(x)\left(\frac{1}{2}\vec{\hat{P}}-\vec{\hat{p}}+m\right) \tag{4}$$

$$= -\frac{G_F^2}{2}\gamma^\varrho(1+\gamma_5)\langle 0|T[j_\varrho(x_1)\psi(x_1)\bar{\psi}(x_2)j_\sigma(x_2)]|P\rangle\gamma^\sigma(1+\gamma_5),$$

with

$$X=\frac{x_1+x_2}{2},\qquad x=x_1-x_2,\qquad P=p_1+p_2,\qquad p=\frac{p_1-p_2}{2}$$

and

$$\hat{p}=i\gamma^\mu\frac{\partial}{\partial x^\mu},\qquad \hat{P}=i\gamma^\mu\frac{\partial}{\partial X^\mu},$$

and we defined

$$\chi_P(x_1,x_2)=\langle 0|T[\psi(x_1)\bar{\psi}(x_2)]|P\rangle=\frac{e^{-iP\cdot X}}{(2\pi)^{3/2}}\chi_P(x) \tag{5}$$

as the bound state wave function and $\Gamma_P(x_1, x_2)$ as the corresponding vertex. (For the moment we considered only one lepton and the corresponding neutrino field; further, a Fierz rearrangement was operated on (1) in such a way that from now on $j_\varrho(x)=\bar{\nu}(x)\gamma_\varrho\nu(x)$). It can easily be shown [7], [5], [8], from Eq. (4) that in the Dirac space $\Gamma(X, x)$ has the form

$$\Gamma(X,x)=\gamma^\mu(1+\gamma_5)\mathscr{B}_\mu(X,x), \tag{6}$$

where $\mathscr{B}_\mu(X, x)$ multiplies the unit matrix. This form of the vertex due to the presence of two-component left-handed spinor fields in the Lagrangian (1) greatly simplifies the solution of (4); nevertheless, to attempt an actual solution of it we have to go in the ladder approximation:

$$\langle 0|T[j_\varrho(x_1)\psi(x_1)\bar{\psi}(x_2)j_\sigma(x_2)]|P\rangle\approx\pi_{\varrho\sigma}(x)\chi_P(x_1,x_2), \tag{7}$$

where

$$\pi_{\varrho\sigma}(x)=\langle 0|T[j_\varrho(x)j_\sigma(0)]|0\rangle, \tag{8}$$

and (4) becomes:

$$\frac{i}{(2\pi)^4}\Gamma_P(x) \equiv \left(\frac{1}{2}\vec{\hat{P}}+\vec{\hat{p}}-m\right)\chi_P(x)\left(\frac{1}{2}\vec{\hat{P}}-\vec{\hat{p}}+m\right) \tag{4'}$$

$$= -\frac{G_F^2}{2}\pi_{\varrho\sigma}(x)\gamma^\varrho(1+\gamma_5)\chi_P(x)\gamma^\sigma(1+\gamma_5);$$

$\pi_{\varrho\sigma}$ represents the Green function of the force due to the exchange of lepton loops between the leptons L and $\bar{L}$. Since this force is repulsive between two leptons while attractive between lepton–antilepton, it might give rise to lepton–antilepton bound states. The force will be long range for massless leptons, then in this first phenomenological approach we shall take for $j_\varrho(x)$ the massless neutrino current. In this case, because of the Weyl equation of motion for the neutrino field, $\pi_{\varrho\sigma}(x)$ obeys the equation

$$(9) \qquad \partial^\varrho \pi_{\varrho\sigma}(x) = \partial^\sigma \pi_{\varrho\sigma}(x) = 0$$

and, as a consequence, in momentum space, it has the form:

$$(10) \qquad \pi_{\varrho\sigma}(p) = (p_\varrho p_\sigma - p^2 g_{\varrho\sigma})\pi(p^2),$$

and $\pi(p^2)$ presents only logarithmic divergences even when computed in perturbation expansion starting from the local non-renormalizable current–current Lagrangian (1).

In space-time, $\pi_{\varrho\sigma}(x)$ has the form, valid for $x^2 \neq 0$:

$$(11) \qquad \pi_{\varrho\sigma}(x) = -\frac{2}{\pi^4}\,\frac{2x_\varrho x_\sigma - g_{\varrho\sigma}x^2}{(x^2 - i\varepsilon)^4},$$

and, for the lepton–antilepton neutrino forces, has a behaviour of the type $\sim x^{-6}$ for $x^2 \to \infty$. This behaviour of the Green function will not depend on the possible change brought to the Lagrangian (1) in order to make it renormalizable or simply to regularize it, provided (1) correctly represents the low-energy limit of the lepton weak interaction. This change of the Lagrangian (1) will instead influence the behaviour of $\pi_{\varrho\sigma}(x)$ at $x_\mu \sim 0$ where $\pi_{\varrho\sigma}(x)$ is simply not defined. One could certainly compute $\pi_{\varrho\sigma}(x)$ for $x_\mu \sim 0$ using, for example, an intermediate vector meson model of weak interactions and possibly a renormalizable form of it.[3] For our present tentative approach we shall not undergo such a burdensome and somehow not very significant task and shall suppose instead that for $x_\mu \sim 0$:

$$(12) \qquad \pi_{\varrho\sigma}(x^2) = g_{\varrho\sigma}\frac{M^4}{x^2},$$

where M is a parameter having the dimension of l^{-1} for a renormalizable model, and

$$(13) \qquad \pi_{\varrho\sigma}(x^2) = g_{\varrho\sigma}\frac{M^{6-m}}{x^n} \qquad \text{with } n < 2,$$

or

$$(13') \qquad \pi_{\varrho\sigma}(x^2) = g_{\varrho\sigma}M^6 \ln M/x,$$

for a super-renormalizable theory.

In these forms we have suppressed the terms proportional to $p_\varrho p_\sigma$ in (10). The main justification for this is that these terms could be dropped rigorously in the interaction representation for massless leptons, so that in substance this should be equivalent to terms proportional to the lepton masses, and the position (12) will be more justified in Secs. IV and V where in fact we use massless leptons.

For a renormalizable model we have taken

$$(14) \qquad \pi_{\varrho\sigma}(x; M) = g_{\varrho\sigma}\pi(x^2, M)$$

with the spectral representation

$$(15) \qquad \pi(x^2, M) = i\int_0^\infty d\mu\, g(\mu, M)\Delta_f(x^2, \mu^2),$$

where

$$\Delta_f(x^2, \mu^2) = \frac{i\mu}{4\pi^2}\,\frac{K_1(\mu\sqrt{-x^2})}{\sqrt{-x^2}},$$

which guarantees the behaviour (12) at $x^2 \ll M^{-2}$; and have chosen the spectral function $g(\mu, M)$ such that it correctly represents the behaviour (11) at $x^2 \to \infty$. In order to obtain super-renormalizable behaviour (13) and (13′), one can simply introduce derivatives of Δ_f with respect to μ^2 in (15).

In the computations [10], [17], the results of which will be summarized below, we have chosen:

$$(16) \qquad g(\mu, M) = \frac{1}{(2\pi)^2}\mu^5 e^{-\mu/M}\left(-\frac{\partial}{\partial\mu^2}\right).$$

It can be seen that the parameter M^{-1} represents the length at which the Green function deviates from the long distance behaviour x^{-6} to go to the less singular behaviour. In an intermediate boson theory, M would be connected with the mass of the intermediate boson.

III. Solutions of the B–S Equation

Since we want to study the possible existence of massless bound states, we start from the B–S equation (4′) in ladder approximation. For $P_\mu = 0$, after Wick rotation has been performed in momentum space, we have:

$$(17)\qquad (\gamma\cdot p+im)\chi^{(r)}(p)(\gamma\cdot p+im) = i\frac{G_F^2}{2}\int d^4q\,\tilde{\pi}[(p-q)^2;\, M]\times$$
$$\times\gamma_\varrho(1+i\gamma_5)\chi^{(r)}(q)\gamma_\varrho(1+i\gamma_5)$$

or, for the vertex (4′),

$$(18)\qquad \Gamma^{(r)}(p) = i\frac{G_F^2}{2}\int d^4q\,\frac{\tilde{\pi}[(p-q)^2;\, M]}{(q^2+m^2)^2}\gamma_\varrho(1+i\gamma_5)(\gamma\cdot q-im)\times$$
$$\times\Gamma^{(r)}(q)(\gamma\cdot q-im)\gamma_\varrho(1+i\gamma_5),$$

where $\tilde{\pi}(k^2;\, M)$ is the Fourier transform of (15) with the choice (16) and r denotes a collection of labels (spin, helicity, parity, ...) apt to distinguish the various degenerate bound states having $P_\mu = 0$.

The set $\{\chi^{(r)}(p)\}$ must form the basis for a reducible representation of $O(4)^{\Pi C}$, since $O(4)$ is the little group belonging to $P_\mu = 0$, and Eq. (18) is invariant also under (ΠC) (Π = three-space reflection, C = charge conjugation); therefore we want to expand the B–S amplitudes $\chi^{(r)}(p)$ in a base apt to split their various irreducible components. But we know from Eq. (6) that the vertex has a simpler structure in the Dirac space; therefore, we prefer to analyse it instead of the amplitude.

To this aim we introduce[4] the following four-vector hyperspherical harmonics [4], [21]:

$$(19)\qquad \boldsymbol{Y}^{2j^+2j^-}_{jj_3;n}(\Omega_p) \equiv \sum_{\substack{s,s_3\\ l,m}}\left[\begin{matrix}\frac{1}{2} & \frac{1}{2} \\ s & s_3\end{matrix};\begin{matrix}\frac{n}{2} & \frac{n}{2}\\ l & m\end{matrix}\left|\begin{matrix} j^+ & j^- \\ j & j_3\end{matrix}\right.\right] Y^n_{lm}(\Omega_p)\boldsymbol{e}_{ss_3},$$

where $\{\boldsymbol{e}_{ss_3}\}$, $s = 0, 1, -s \leqslant s_3 \leqslant s$ (boldface symbols denote four-vectors), are four orthonormal four-vectors belonging to the representation $(\frac{1}{2}, \frac{1}{2})$ of $O(4)$, $Y^n_{lm}(\Omega_p)$ are the scalar hyperspherical harmonics, and the square bracket is an $O(4)$ Clebsch–Gordan coefficient explicitly given by

$$\begin{bmatrix} j_1^+ & j_1^- \\ l_1 & m_1 \end{bmatrix} ; \begin{matrix} j_2^+ & j_2^- \\ l_2 & m_2 \end{matrix} \left| \begin{matrix} j^+ & j^- \\ j & j_3 \end{matrix} \right] = (l_1 m_1; \quad l_2 m_2 | j j_3) \times \tag{20}$$

$$\times \sqrt{(2l_1+1)(2l_2+1)(2j^+ +1)(2j^- +1)} \begin{Bmatrix} j_1^+ & j_1^- & l_1 \\ j_2^+ & j_2^- & l_2 \\ j^+ & j^- & j \end{Bmatrix}.$$

In definition (19) (j^+, j^-) define the particular $O(4)$ representation to which $Y^{2j^+2j^-}_{jj_3;n}$ belongs, n is a degenerate quantum number apt to distinguish the various equivalent $O(4)$ representations having the same (j^+, j^-), j gives the total angular momentum content of the particular harmonic and j_3 is its projection. From the 9-j symbol's properties it is easy to see that j^+ and j^- may assume only the two values $|n-1|$ and $n+1$; therefore, once the quantum number n is fixed, it is possible to have only four kinds of four-vector hyperspherical harmonics:

$$\begin{aligned}
&Y^{nn}_{jj_3;n-1}(\Omega_p), && \Pi C = (-)^{n+j}, && n \geqslant 1,\\
&Y^{nn}_{jj_3;n+1}(\Omega_p), && \Pi C = (-)^{n+j}, && n \geqslant 0,\\
&\frac{1}{\sqrt{2}}[Y^{n+1\ n-1}_{jj_3;\ n}(\Omega_p) - Y^{n-1\ n+1}_{jj_3;\ n}(\Omega_p)], && \Pi C = (-)^{n+j+1}, && n \geqslant 1,\\
&\frac{1}{\sqrt{2}}[Y^{n+1\ n-1}_{jj_3;\ n}(\Omega_p) + Y^{n-1\ n+1}_{jj_3;\ n}(\Omega_p)], && \Pi C = (-)^{n+j}, && n \geqslant 1.
\end{aligned} \tag{21}$$

From (21) it is easy to see that the first two belong to equivalent representations of $O(4)^{(\Pi C)}$, so that the harmonics (21) form the basis for three irreducible inequivalent representations of $O(4)^{(\Pi C)}$ (for fixed n).

Once the base vectors are fixed, it is possible to expand the vertex (4′) (written in momentum space for $P_\mu = 0$), in such a way that Eq. (18) decouples into three systems of coupled integral equations, each system corresponding to one irreducible representation of $O(4)^{(\Pi C)}$ (for every fixed n). These equations have been solved using a method based on the spectral representation and solving the integral equations in the first Fredholm approximation of the vertex [3], [2], [16]. The condition of existence of massless bound states establishes for every n a connection (quantum condition) between the Fermi coupling constant

G_F and the free parameter M which, in the kernel $\pi(x^2; M)$ establishes the distance at which the Green function (8) deviates from its long-range behaviour.

Therefore, the kernel $\tilde{\pi}[(p-q)^2; M]$ turns out to be n-dependent through M, since for every irreducible representation of $O(4)^{(IIC)}$ it is possible to establish a relation $M = M(G_F; n)$.

Every solution of Eq. (18) will then belong to a single quantum number n, characterizing, together with j and j_3, the particular bound state. This means that it is possible to have the following three sets of vertices:

$$\Gamma^{(v)}(p)_{njj_3} = \gamma_\mu(1+i\gamma_5)[G_n^{(-)}(|p|)\, Y^{nn}_{jj_3;n-1;\mu}(\Omega_p) + $$
$$+ G_n^{(+)}(|p|)\, Y^{nn}_{jj_3;n+1;\mu}(\Omega_p)],$$

(22)
$$\Gamma^{(a)}(p)_{njj_3} = \gamma_\mu(1+i\gamma_5)\, G_n^{(a)}(|p|) \frac{1}{\sqrt{2}} [Y^{n+1\,n-1}_{jj_3;n;\mu}(\Omega_p) -$$
$$- Y^{n-1\,n+1}_{jj_3\,;n;\mu}(\Omega_p)],$$

$$\Gamma^{(b)}(p)_{njj_3} = \gamma_\mu(1+i\gamma_5)\, G_n^{(b)}(|p|) \frac{1}{\sqrt{2}} [Y^{n+1\,n-1}_{jj_3\,;n;\mu}(\Omega_p) +$$
$$+ Y^{n-1\,n+1}_{jj_3,n,\mu}(\Omega_p)].$$

We wish to stress that this result depends only on the symmetry properties of the integral equation and not on the particular approximations used to solve them. For every sector (v), (a), (b) the relation between M and n is explicitly given by[5]

(23)
$$M^{(v)}(n) = 2\pi\left(\frac{n(n+1)}{6G_F^2}\right)^{1/4},$$
$$M^{(a)(b)}(n) = 2\pi\left(\frac{n+1}{6G_F^2}\right)^{1/4}.$$

Since the B–S equation (17) (or (18)) is a homogeneous integral equation, it gives the hyper-radial functions $G_n^{(k)}(|p|)$ except for a normalization factor. To determine it, the following normalization condition must be imposed:

(24)
$$\frac{1}{4}\,\mathrm{Tr}\int d^4p\, \bar{\chi}^{(r)}(p)\, \chi^{(r)}(p) = -\frac{1}{(2\pi)^4},$$

where

$$\bar{\chi}^{(r)}(p_4, \vec{p}) \equiv \gamma_4\, \chi^{(r)+}(-p_4, \vec{p})\, \gamma_4.$$

Once the explicit (approximate) form of the hyper-radial functions $G^{(k)}(|p|)$ is found, condition (24) throws away all the solutions with $n \geqslant 2$ for sector (v) and $n \geqslant 3$ for sectors (a) and (b), since they give rise to non-normalizable B–S amplitudes. Since the solution $n = 0$ for sector (v) corresponds to a 'repulsive' case (i.e. there is a minus sign in relation (23)), it must be rejected. Therefore, the only acceptable solution for sector (v) is[6]

$$\text{(v)} \qquad n = 1 \qquad (j = 0, 1).$$

As far as the other two sectors are concerned, the only positive norm normalizable solutions correspond to the choices

$$\text{(a)} \qquad n = 2 \qquad (j = 1, 2),$$

$$\text{(b)} \qquad \begin{cases} n = 1 & (j = 1), \\ n = 2 & (j = 1, 2) \end{cases}$$

To decide if these solutions have a physical meaning it remains to evaluate the renormalized effective coupling constants corresponding to the vertices (22). We shall define the value of the vertex (22) on the mass shell of the external leptons, that is for $p^2 = -m^2$, as the renormalized effective coupling constant $g^{(k)}(n; j)$. It turns out that $(k = v, a, b)$

$$\text{(25)} \qquad g^{(k)}(n; j) = k(n) \frac{(mG_F^{1/2})^{n-1}}{I_n^{(k)}(mG_F^{1/2})},$$

where $k(n)$ is some numerical constant depending only on n, and $I^{(k)}(mG_F^{1/2})$ is strictly connected with the normalization integral of (24) and takes on the form

$$\text{(26)} \qquad I_n^{(v)}(mG_p^{1/2}) = \begin{cases} A, & n = 1, \\ \dfrac{1}{mG_F^{1/2}} \sqrt{B - \ln(mG_F^{1/2})}, & n = 2, \\ \dfrac{C}{mG_F^{1/2}}, & n \geqslant 3, \end{cases}$$

$$I_u^{(a)(b)}(mG_F^{1/2}) = \begin{cases} \sqrt{D - \ln(mG_F^{1/2})}, & n = 1, \\ E, & n \geqslant 2, \end{cases}$$

where A, B, C, D, E are finite or infinite (in the non-normalizable cases for which (25) would give $g = 0$) constants independent of m. From relations (25) and (26) it appears that, due to the smallness of the quantity $mG_F^{1/2}$ ($\sim 10^{-6}$), the only physically acceptable solutions correspond to the cases

$$(v),\ (b): \quad n = 1.$$

From relations (25) and (26) it turns out also that $g^{(v)}(1; j)$ is *simply a constant*, while $g^{(b)}(1; 1)$ is logarithmically dependent on the mass of the constituent leptons. Explicitly ($g^{(b)}(1; 1)$ evaluated for the electron)

$$
\begin{aligned}
|g^{(v)}(1; j)| &= 2\pi\sqrt{3}\left[1+\frac{m^2}{M^2}\ln\left(\frac{m}{M}\right)+O\left(\frac{m^2}{M^2}\right)\right] \\
&\simeq 10.9\left[1+O\left(\frac{m^2}{M^2}\ln\left(\frac{m}{M}\right)\right)\right], \\
|g^{(b)}(1; 1)| &= \frac{\pi}{\sqrt{C-\ln\left(\frac{m}{M}\right)}}\left[1+O\left(\frac{m^2}{M^2}\ln\left(\frac{m}{M}\right)\right)\right] \\
&\simeq 0.70\left[1+O\left(\frac{m^2}{M^2}\ln\left(\frac{m}{M}\right)\right)\right]_{m=m_e}.
\end{aligned}
\tag{27}
$$

Taking the limit for $p^2 \to -m^2$ of the physically meaningful vertices of (22) (i.e. cases (v) and (b) with $n = 1$), one obtains

$$
\begin{aligned}
\Gamma^{(v)}_{1jj_3}(p) \underset{p^2\simeq -m^2}{\simeq} g^{(v)}(1; j)\Bigg\{&\Bigg[1-\frac{p^2+m^2}{4m^2}+ \\
&+O\left(G_F^2(p^2+m^2)^2\ln G_F(p^2+m^2)\right)\Bigg]\delta_{\mu\nu}-\Bigg[\frac{p^2+m^2}{4m^2}+ \\
&+O\left(G_F^2(p^2+m^2)^2\ln G_F(p^2+m^2)\right)\Bigg]\times \\
&\times\left(\delta_{\mu\nu}-\frac{4p_\mu p_\nu}{p^2}\right)\Bigg\}e_\nu^{(j,j_3)}\gamma_\mu(1+i\gamma_5),
\end{aligned}
\tag{28}
$$

$$
\begin{aligned}
\Gamma^{(b)}_{11j_3}(p) \underset{p^2\simeq -m^2}{\simeq} g^{(b)}(1; 1)\Bigg[&1-\frac{G_F(p^2+m^2)}{(2\pi)^2 6\sqrt{3}}+ \\
&+O\left(G_F^2(p^2+m^2)^2\ln G_F(p^2+m^2)\right)\Bigg]\frac{1}{2\pi}\vec{e}^{(1j_3)}\cdot\frac{\vec{p}}{|p|}\wedge\vec{\gamma}(1+i\gamma_5).
\end{aligned}
$$

We shall see in the next section that in order to get a local vertex connecting the free states with the composite vector-axial state (which we shall call the local limit) one must integrate in the relative co-ordinates of the composite state. Then in the local limit of the two only $\Gamma^{(v)}(p)$ will survive. In fact it is easy to see [4] that the Fourier transforms of (22) are ($\eta = x_1 - x_2$):

$$\Gamma^{(v)}(\eta)_{njj_3} = \gamma_\mu(1+i\gamma_5)[G_n^{(-)}(|\eta|)\, Y^{nn}_{jj_3;n-1;\mu}(\Omega_\eta) + \tilde{G}_n^{(+)}(|\eta|)\, Y^{nn}_{jj_3;n+1;\mu}(\Omega_\eta)]$$

(22′)

$$\Gamma^{(a)(b)}(\eta)_{njj_3} = \gamma_\mu(1+i\gamma_5)\, \tilde{G}_n^{(a)(b)}(|\eta|)\, \frac{1}{\sqrt{2}}\, [Y^{n+1\, n-1}_{jj_3;n;\mu}(\Omega_\eta) \mp \mp Y^{n-1\, n+1}_{jj_3;n;\mu}(\Omega_\eta),$$

where

$$\tilde{G}_n^{(k)}(|\eta|) \equiv (2\pi)^2(-i)^{n_b} \int_0^\infty \frac{J_{n_b+1}(|\eta||p|)}{|\eta||p|}\, G_n^{(k)}(|p|)|p|^3 d|p|$$

(n_b = orbital q.n.; $n_b = n \pm 1$ for (v), $n_b = n$ for (a), (b)). Remembering that for cases (a) and (b) $n \geqslant 1$, we have

(29)

$$\int d\Omega_\eta Y^{nn}_{jj_3;n+1}(\Omega_\eta) = \int d\Omega_\eta Y^{n\pm 1\, n\mp 1}_{jj_3;n}(\Omega_\eta) = 0,$$

$$\int d\Omega_\eta Y^{nn}_{jj_3;n-1}(\Omega_\eta) = \delta_{n1} \int d\Omega_\eta Y^{11}_{jj_3;0}(\Omega_\eta) = \delta_{n1} \sqrt{2\pi^2}\, e_{jj_3}.$$

Therefore, *the only vertex giving a contribution in the local limit is* $\Gamma^{(v)}(\eta)_{1jj_3}$, and only with its $\tilde{G}_n^{(-)}(|\eta|)$ part,

(30)

$$\int d^4\eta \Gamma^{(k)}(\eta)_{njj_3} = \delta_{kv}\, \delta_{n1}\, \gamma_\mu(1+i\gamma_5)\, (e_{jj_3})_\mu \sqrt{2\pi^2} \int_0^\infty d|\eta||\eta|^3 \tilde{G}_1^{(-)}(|\eta|).$$

We see from (27) that the coupling constant is nearer to the strong than to the electromagnetic coupling which, apart from the limits of validity of our approximations, could be interpreted as an indication that if this is the origin of the electromagnetic vertex one cannot neglect the internal symmetry of the leptons, in analogy withwhat happens in the gauge models of weak and electromagnetic interactions when the couplings with the non-abelian gauge fields are stronger than the electromagnetic coupling.

Before closing this section we wish to point out that for $\Gamma^{(v)}(p)$ the limit $m \to 0$ exists and is finite. This is somewhat surprising since one would rather expect an infra-red divergence.

One could then think that this result depends on the approximation adopted if it were not that, it we perform the same calculation with the same approximation for a vertex when the chiral projector $\frac{1}{2}(1+\gamma_5)$ is eliminated, we obtain in fact the expected infra-red logarithmic divergence; indeed one obtains the result of [10], [17] where the projector was eliminated by hand and the coupling constant was proportional to $\log m$.

This then could mean that the elimination of the infra-red divergence results from the coherent use of two component spinor fields. We intend to analyse this further in the future.

IV. Gauge Invariance

We have imposed zero mass for the bound state and the only reason was that of the simplicity of the integral equations. But if we could insert the field $\mathscr{B}_\mu$ appearing in (6) in a gauge invariant field theory, then masslessness would be imposed by gauge invariance. The field $\mathscr{B}_\mu$ now being composite, we should have that its gauge transformation follows from that of the constituent field $L(X)$, i.e. from

$$(31) \qquad L(X) \to L'(X) = e^{i\varepsilon(X)}L(X),$$

should follow

$$(32) \qquad \mathscr{B}_\mu(X) \to \mathscr{B}'_\mu(X) = \mathscr{B}_\mu(X) - \frac{\partial \varepsilon(X)}{\partial X^\mu}.$$

In order to show if and how this could happen, we must first try to connect the phenomenological calculation of the previous sections with a field theoretical model; that is, to give a Lagrangian formalism from which equations of motion for the fields $L(X)$ ank $\mathscr{B}_\mu(X)$ in interaction could be deduced.

First, one should eliminate the limitations which are implicit in the use of the ladder approximation of the Bethe–Salpeter equation. That is, to sum over all possible irreducible graphs one would obtain from a renormalizable version of the weak Lagrangian. This would be very cumbersome and technically practically impossible; therefore, we shall start from the Lagrangian (for simplicity we leave out lepton indices and summations with respect to them) [6],

$$(33)\qquad \mathscr{L}^{(L)} = \mathscr{L}_0^{(L)} + \mathscr{L}_1^{(L)} = \frac{i}{2}\bar{L}(X)\overset{\leftrightarrow}{\hat{\partial}}_X L(X) - G_F \int \bar{L}(X)\gamma^\mu L(X_1)\times$$

$$\times \bar{L}(X_2)\gamma^\varrho L(X_3) F_{\mu\varrho}(X-X_1, X-X_2, X-X_3)d^4X_1 d^4X_2 d^4X_3,$$

with the condition

$$(34)\qquad \int F_{\mu\varrho}(X-X_1, X-X_2, X-X_3)d^4X_1\, d^4X_2\, d^4X_3 = 2\sqrt{2}g_{\mu\varrho},$$

which guarantees low-energy local behaviour. The Lagrangian must not be interpreted as a non-local Lagrangian but rather as an effective one. The function $F_{\mu\varrho}$ represents the summation over all possible internal virtual lines connecting the fields $L(X)$ in interaction either through the weak Lagrangian (1) or, better still, through its renormalizable intermediate vector boson version.

From (33) one obtains the equation of motion:

$$(35)\qquad i\hat{\partial}L(X) = G_F\int \gamma^\varrho F_{\varrho\sigma}(\xi, \eta, \zeta) L(X+\xi)\bar{L}(X+\eta)\gamma^\sigma \times$$

$$\times L(X+\zeta)d^4\xi d^4\eta d^4\zeta.$$

To connect the present approach with the previous calculation, one has simply to set:

$$(36)\qquad F_{\varrho\sigma} = G_F\,\delta^{(4)}(\xi)\,\delta^{(4)}(\zeta-\eta)\pi_{\varrho\sigma}(\eta),$$

and

$$(37)\qquad \pi_{\varrho\sigma}(\eta) = g_{\varrho\sigma}\pi(\eta^2).$$

Should we go in second quantization, that is, consider the fields in (35) as operators, then $\pi_{\varrho\sigma}$ would represent the causal Green function of neutrino pair exchange and we would further have a chronological operator T in front of the integrand. Eq. (35) then becomes:

$$(38)\qquad i\hat{\partial}L(X) = G_F^2\int \pi(\eta^2)\,T\gamma^\varrho L(X)\bar{L}(X+\eta)\gamma_\varrho L(X+\eta)d^4\eta$$

(here X represents the co-ordinate x_1 of Sec. II, and $T \equiv 1$ in first quantization). We shall, for the sake of simplicity, discuss gauge invariance in Eq. (38) instead of Eq. (35).

Eq. (38) is obviously, at first sight, not invariant with respect to the transformation (31) if the integral on the right-hand side is regular and convergent. To compensate the term $-\big(\hat{\partial}\varepsilon(X)\big)L(X)$ which arises because of the transformation (31) in the left-hand side of (38), one can only hope for the contribution of singularities in the integrand at the point $\eta_\mu = 0$. Let us suppose that near $\eta_\mu^2 = 0$, $\pi(\eta^2)$ behaves as supposed in Sec. III formula (12),

$$\pi(\eta^2) \sim \frac{M^4}{\eta^2}. \tag{39}$$

Substituting this expression in (38) we see that, as far as the contribution of the singularity at $\eta_\mu \sim 0$ is concerned. Eq. (38) becomes scale invariant (this can also be verified *a posteriori* in (93) since $\lambda^2 = M^4 G_F^2$ is adimensional) and we can apply the Wilson [28] expansion to the product of the three L fields near the point $\eta_\mu = 0$:

$$\begin{aligned} &\lim_{\eta_\mu \to 0} T[:\psi^\alpha(X)\overline{\psi}^\beta(X+\eta)\psi^\gamma(X+\eta):] \\ &\equiv \lim_{\eta_\mu \to 0}\left[T\psi^\alpha(X)\overline{\psi}^\beta(X+\eta)\psi^\gamma(X+\eta) - iE(\eta)\frac{\gamma_\mu^{\alpha\beta}\eta^\mu}{\eta^4}\psi^\gamma(X+\eta)\right], \end{aligned} \tag{40}$$

where $E(\eta)$ is either an arbitrary renormalization constant or an arbitrary at the most logarithmically divergent function of η for $\eta_\mu \to 0$ [14], [26].[7]

Let us now perform the gauge transformation (31) on Eq. (38). The left-hand side will become as usual:

$$i\hat{\partial}L = ie^{-i\varepsilon}\hat{\partial}L' + (\hat{\partial}\varepsilon)e^{-i\varepsilon}L'. \tag{41}$$

The right-hand side, as a result of (31), would simply be multiplied by $e^{-i\varepsilon(X)}$ and we would not get compensation of the second term on the right-hand side of (41) unless we get extra contributions from the singular terms of the field products at $\eta = 0$. This can happen; in fact, let us express the field ψ in terms of ψ' according to the correspondent of (31):

$$\begin{aligned} \psi(X) &= e^{-i\varepsilon(X)}\psi'(X), \\ \overline{\psi}(X) &= e^{i\varepsilon(X)}\overline{\psi}'(X), \end{aligned} \tag{31'}$$

and we shall have

$$\begin{aligned} &\lim_{\eta_\mu \to 0}[\psi^\alpha(X)\overline{\psi}^\beta(X+\eta)\psi^\gamma(X+\eta)] \\ &= e^{-i\varepsilon(X)}\lim_{\eta_\mu \to 0}[\overline{\psi}'^\alpha(X)\overline{\psi}'^\beta(X+\eta)\psi'^\gamma(X+\eta)]. \end{aligned}$$

Then let us develop the analysis corresponding to (40) and obtain

$$(42)\qquad \lim_{\eta_\mu\to 0}[:\psi^\alpha(X)\overline{\psi}^\beta(X+\eta)\psi^\gamma(X+\eta):]$$

$$= \lim_{\eta_\mu\to 0}\Big[e^{-i\varepsilon(X)}\psi'^\alpha(X)\overline{\psi}'^\beta(X+\eta)\psi'^\gamma(X+\eta)-$$

$$-iE(\eta)\frac{\gamma_\mu^{\alpha\beta}\eta^\mu}{\eta^4}e^{-i\varepsilon(X+\eta)}\psi'^\gamma(X+\eta)\Big]$$

$$= \lim_{\eta_\mu\to 0}e^{-i\varepsilon(X)}\Big[\psi'^\alpha(X)\overline{\psi}'^\beta(X+\eta)\psi'^\gamma(X+\eta)-$$

$$-e^{-i[\varepsilon(X+\eta)-\varepsilon(X)]}iE(\eta)\frac{\gamma_\mu^{\alpha\beta}\eta^\mu}{\eta^4}\psi'(X+\eta)\Big]$$

$$= \lim_{\eta_\mu\to 0}e^{-i\varepsilon(X)}\Big[\psi'^\alpha(X)\overline{\psi}'^\beta(X+\eta)\psi'^\gamma(X+\eta)-$$

$$-i\left(1-i\frac{\partial\varepsilon}{\partial X_\varrho}\eta_\varrho\right)E(\eta)\frac{\gamma_\mu^{\alpha\beta}\eta^\mu}{\eta^4}\psi'^\gamma(X+\eta)\Big]$$

$$= \lim_{\eta_\mu\to 0}e^{-i\varepsilon(X)}\Big[:\psi'^\alpha(X)\overline{\psi}'^\beta(X+\eta)\psi'^\gamma(X+\eta):-$$

$$-E(\eta)\gamma_\mu^{\alpha\beta}\frac{\eta^\mu\eta^\varrho}{\eta^4}(\partial_\varrho\varepsilon(X))\psi'^\gamma(X+\eta)\Big].$$

In order to test whether the singular term proportional to the arbitrary function $E(\eta)$ may give rise to gauge invariance, we must substitute (42) on the right-hand side of (38). Near $\eta_\mu = 0$ we have supposed that $\pi(\eta^2)$ has the form (39), it is then convenient to divide the domain of integration (whole space-time) on the right-hand side of (38) into two parts: an infinitesimal domain o around the origin where (39) and (42) are both valid and the rest of space-time $V-o$, and we have:

$$(43)\qquad e^{-i\varepsilon(X)}[i\hat{\partial}L'+\hat{\partial}\varepsilon(X)L'] = \lim_{o\to 0}e^{-i\varepsilon(X)}\times$$

$$\times\Big\{G_F^2M^4T\int_o\frac{1}{\eta^2}\gamma_\mu:L'(X)\bar{L}'(X+\eta)\gamma^\mu L'(X+\eta):d^4\eta+$$

$$+G_F^2T\int_{V-o}\pi(\eta^2)\gamma_\mu L'(X)\bar{L}'(X+\eta)\gamma^\mu L'(X+\eta)d^4\eta-$$

$$-\frac{\partial \varepsilon(X)}{\partial X_\varrho} \int_o G_F^2 M^4 E(\eta) \gamma_\sigma \frac{(1+\gamma_5)}{2} \gamma_\mu \gamma^\sigma \frac{(1+\gamma_5)}{2} \times$$

$$\times \psi'(X+\eta) \frac{\eta^\mu \eta_\varrho}{\eta^6} d^4\eta \Big\}.$$

The first two integrals can be summed and in the third we have that, since

$$\gamma_\sigma \frac{(1+\gamma_5)}{2} \gamma_\mu \gamma^\sigma \frac{(1+\gamma_5)}{2} = -2\gamma_\mu \frac{(1+\gamma_5)}{2}, \tag{44}$$

Eq. (43) becomes

$$[i\hat{\partial} L'(X) + \hat{\partial}\varepsilon(X) L'(X)] = G_F^2 T \int \pi(\eta^2) \gamma^\varrho : L'(X) \bar{L}'(X+\eta) \times \tag{45}$$

$$\times \gamma_\varrho L'(X+\eta) : d^4\eta + \gamma_\mu \frac{\partial \varepsilon}{\partial X_\varrho} \lim_{o\to 0} \int_o 2G_F^2 M^2 E(\eta) L'(X+\eta) \frac{\eta^\mu \eta_\varrho}{\eta^6} d^4\eta.$$

It is then clear that gauge invariance is satisfied if

$$\lim_{o\to 0} \int_o 2G_F^2 M^2 E(\eta) \frac{\eta_\mu \eta_\varrho}{\eta^6} d^4\eta = g_{\mu\varrho}, \tag{46}$$

which is well possible considering that the integral is logarithmically divergent at $\eta_\mu = 0$ and the domain of integration is infinitesimal. Then the result of the integration is $g_{\mu\varrho}$ times an arbitrary constant. An appropriate choice of the arbitrary constant E may then satisfy (46), and the last term of the right-hand side of (43) becomes

$$\gamma_\mu \frac{\partial \varepsilon}{\partial X_\varrho} \lim_{o\to 0} \int_o 2G_F^2 M^4 E(\eta) L'(X+\eta) \frac{\eta^\mu \eta_\varrho}{\eta^6} d^4\eta = \hat{\partial}\varepsilon(X) L'(X), \tag{46'}$$

which exactly compensates the gauge term on the left-hand side of (43), which then becomes identical to (38) in the transformed fields $L'(X)$.

It must be stressed that in order to obtain gauge invariance it is necessary to have the form (12) for $\pi(\eta^2)$, that is, to have a Green function of the renormalizable type (otherwise one could not satisfy (46), at least if the limit $o \to 0$ is used)).

In these considerations no essential use was made of particular approximations such as the ladder approximation used in the previous section, in fact it can easily be seen that one could remove condition (37) for the Green function and also (36) and still obtain (46), i.e. gauge invariance.

The problem now is: can we ascribe the compensation of the second term on the left-hand side of (43) to a gauge transformation of a vector (pseudo-vector) boson field? The immediate but formal way would be to write (38) in the form

$$(47)\qquad i\hat{\partial}L(X) = \gamma^{\varrho}A_{\varrho}(X)L(X),$$

where

$$(48)\qquad A_{\varrho}(X) = G_F^2\int \pi(\eta^2)\bar{L}(X+\eta)\gamma_{\varrho}L(X+\eta)d^4\eta,$$

and then for the transformation (31), because of (46),

$$(49)\qquad \gamma^{\varrho}A_{\varrho}(X)L(X) \to \gamma^{\varrho}A'_{\varrho}(X)L'(X)+\gamma^{\varrho}\frac{\partial\varepsilon}{\partial X_{\varrho}}L'(X).$$

However, to interpret $A_{\varrho}(X)$ as a composite field one would need to consider the chain diagram which we do not intend to do in the present paper (see footnote 3).

Let us suppose that because of a force (neutrino pair exchange) represented by the Green function $\pi(\eta)$, a massless lepton–antilepton composite state exists, as discussed in previous sections. In this case

$$(50)\qquad T[\psi_{\alpha}(X)\bar{\psi}_{\beta}(X+\eta)] = H_{\alpha\beta}(X,\eta)$$

represents the composite state operator, and $\langle 0|H_{\alpha\beta}(X,\eta)|P\rangle$ its wave function, and both will rapidly go to 0 for $\eta\to\infty$. Then for these states one could write:

$$\begin{aligned}(51)\qquad & G_F^2\int \pi(\eta^2)T\gamma^{\varrho}L(X)\bar{L}(X+\eta)\gamma_{\varrho}L(X+\eta)d^4\eta\\ &= G_F^2\int \pi(\eta^2)T\gamma^{\varrho}L(X)\bar{L}(X+\eta)d^4\eta\gamma_{\varrho}L(X)+\\ &+G_F^2\int \pi(\eta^2)T\gamma^{\varrho}L(X)\bar{L}(X+\eta)\eta_{\sigma}d^4\eta\gamma_{\varrho}\frac{\partial L}{\partial X_{\sigma}}+\ldots\\ &\simeq G_F^2\int \pi(\eta^2)T\gamma^{\varrho}L(X)\bar{L}(X+\eta)d^4\eta\gamma_{\varrho}L(X).\end{aligned}$$

We shall call the equality (51) the 'local limit'.

Now let us analyse the following term contained on the right-hand side of (51):

$$(52)\qquad \mathscr{R}_{\varrho}(X) \equiv G_F^2 T\int d^4\eta\,\pi(\eta^2)\gamma_{\varrho}:L(X)\bar{L}(X+\eta):$$

Because of the chiral projectors $1 \pm \gamma_5/2$, we have

$$(53) \quad L(X)\bar{L}(X+\eta) = -\frac{1}{2}(\bar{L}(X+\eta)\gamma_\mu L(X))\gamma^\mu \frac{1-\gamma_5}{2}$$

$$= -\frac{1}{4}\operatorname{Tr}[\gamma^\mu\gamma^\sigma L(X)\bar{L}(X+\eta)\gamma_\sigma]\gamma_\mu \frac{1-\gamma_5}{2}.$$

Then

$$(54) \quad \mathscr{R}_\varrho(X) = -\frac{1}{4} G_F^2 T \int d^4\eta\pi(\eta^2) \times$$

$$\times \operatorname{Tr}[\gamma_\mu\gamma^\sigma : L(X)\bar{L}(X+\eta) : \gamma_\sigma]\gamma_\varrho\gamma^\mu \frac{1-\gamma_5}{2}.$$

If we put

$$(55) \quad gB_\mu(X) \equiv \frac{G_F^2}{2} T \int d^4\eta\pi(\eta^2)\operatorname{Tr}[\gamma_\mu\gamma^\sigma : L(X)\bar{L}(X+\eta) : \gamma_\sigma]$$

$$\equiv +G_F^2 T \int d^4\eta\pi(\eta^2) : \bar{L}(X+\eta)\gamma_\mu L(X) :$$

we have

$$(56) \quad \mathscr{R}_\varrho(X) = -\frac{1}{2} gB_\mu(X)\gamma_\varrho\gamma^\mu \frac{1-\gamma_5}{2},$$

and using Eqs. (51), (52) and (56), the right-hand side of Eq. (38) becomes

$$\mathscr{R}_\varrho(X)\gamma^\varrho L(X) = -\frac{1}{2} gB_\mu(X)\gamma_\varrho\gamma^\mu \frac{1-\gamma_5}{2}\gamma^\varrho L(X)$$

$$= gB_\mu(X)\gamma^\mu L(X),$$

and Eq. (38) becomes

$$(57) \quad i\gamma^\mu\partial_\mu L(X) = gB_\mu(X)\gamma^\mu L(X)$$

with $gB_\mu(X)$, a 'scalar' in the Dirac space, given by (55). Gauge invariance of (38) will determine gauge invariance of (57) which, in turn, will mean that because of (31) the field B_ϱ will transform according to:

$$(58) \quad B_\varrho(X) \to B'_\varrho(X) = B_\varrho(X) - \frac{1}{g}\frac{\partial\varepsilon}{\partial X^\varrho}.$$

That is, in the local limit, $B_\varrho(X)$ is a gauge field. The gauge transformation of the field B_ϱ can also be obtained, in the local limit, directly from its definition (55) (see Appx.).

From (51) one easily sees that the local limit is a good limit as long as the wave lengths in $L(X)$ are large with respect to the dimension of the bound state wave function which in our model is of the order of $\sqrt{G_F}$. So that for all known experimental situations it is a good limit (far lower than the present lowest limits of validity of quantum electrodynamics).

If one takes the matrix element of (55) between $\langle 0|$ and $|P\rangle$ one sees that $\langle 0|gB_\mu(X)|P\rangle$ is the integral on the internal coordinates of the vertex $\mathscr{B}_\mu(X, \eta)$ discussed in previous sections. In fact, from Eq. (55)

$$(59)\qquad \langle 0|gB_\mu\,(X)|P, r\rangle = \left\langle 0\left|\frac{G_F^2}{2}\,T\int d^4\eta\pi(\eta^2)\frac{1}{4}\mathrm{Tr}[\gamma_\mu\gamma^\sigma\times \right.\right.$$

$$\left.\left.\times\,(1+\gamma_5)\!:\!\psi(X)\overline{\psi}(X+\gamma)\!:\!\gamma_\sigma(1+\gamma_5)]\right|P, r\right\rangle$$

$$= \frac{1}{4}\mathrm{Tr}\gamma_\mu\frac{G_F^2}{2}\int d^4\eta\pi(\eta^2)\gamma^\sigma(1+\gamma_5)\langle 0|T[:\psi(X)\times$$

$$\times\,\overline{\psi}(X+\eta):]|P, r\rangle\gamma_\sigma(1+\gamma_5);$$

and from Eqs. (5), (6) and (4′) we see that

$$(60)\qquad \langle 0|gB_\mu(X)|P, r\rangle = \frac{1}{4}\mathrm{Tr}\gamma_\mu\frac{G_F^2}{2}\int d^4\eta\pi(\eta^2)\gamma^\sigma(1+\gamma_5)\chi_P^{(r)}(X;\eta)\times$$

$$\times\,\gamma_\sigma(1+\gamma_5) = \frac{(-i)}{(2\pi)^4}\,\frac{1}{4}\mathrm{Tr}\gamma_\mu\int d^4\eta\Gamma_P^{(r)}(X;\eta)$$

$$= -\frac{i}{(2\pi)^4}\int d^4\eta\mathscr{B}_\mu^{(r)}(X;\eta).$$

The gauge vector-axial field appearing in (57) is then the integral on the internal coordinates of the vertex $B_\varrho(X, \eta)$ appearing in the Bethe–Salpeter equation of Sec. II. In spite of the fact that this definition of the composite state field is different from the conventional one [30], we think that it is appropriate since it well represents through (38) the physical fact that the interaction of the composite field with the free states is in effect non-local, the non-locality being given by the wave function of the composite state.

It can easily be seen that the asymptotic field $B_\rho^{\substack{\mathrm{in}\\ \mathrm{out}}}(X)$ defined by (57) obeys a Klein–Gordon equation with the mass equal to that of the composite state. If we wish this equation to be gauge invariant, this mass must then be zero,

$$\Box B_\mu^{\substack{\text{in}\\\text{out}}}(X) = 0. \tag{61}$$

In this case, (57) and (61) will be invariant against the simultaneous gauge transformations (31) and (58), and $B_\mu(X)$ will be an abelian gauge field.

Now we wish to see the connection between the constant g, defined together with the field operator $B_\mu(X)$ in Eq. (55), and the renormalized effective coupling constants $g^{(k)}(n; j)$ of Sec. III.

If we denote the polarization vectors of the vector-axial bound state $|P, r\rangle$, considered as a localized state, by $\varepsilon^{(jj_3)}(P)$ ($j = 0, 1, -j \leqslant j_3 \leqslant j$), we may write

$$\langle 0|gB_\mu(X)|P, jj_3\rangle = g\varepsilon_\mu^{(jj_3)}(P)\frac{1}{(2\pi)^{3/2}}\, e^{-iP.X}. \tag{62}$$

From Eqs. (60) and (62) we then have

$$g\varepsilon_\mu^{(jj_3)}(P) = \frac{(-i)}{(2\pi)^4}\frac{1}{4}\,\mathrm{Tr}\gamma_\mu\int d^4\eta\Gamma_P^{(r)}(\eta), \tag{63}$$

and, taking $P \equiv 0$ and going to euclidean variables,

$$g\varepsilon_\mu^{(jj_3)} = \frac{1}{(2\pi)^4}\frac{1}{4}\,\mathrm{Tr}\gamma_\mu\int d^4\eta\Gamma_g^{(r)}(\eta; P = 0). \tag{64}$$

But from Eq. (30) we have seen that only $\int d^4\eta\Gamma_{g1jj_3}^{(v)}(\eta)$ is different from zero. Thus, taking into account our conventions on the Fourier transform and the results of Sec. III (in particular the existence of $\Gamma^{(v)}(p)$ for $m \to 0$), we have

$$\begin{aligned} g\varepsilon_\mu^{(jj_3)} &= \lim_{p\to 0}\frac{1}{4}\,\mathrm{Tr}\gamma_\mu\frac{1}{(2\pi)^4}\int d^4\eta e^{ip\cdot\eta}\Gamma_{g1jj_3}^{(v)}(\eta) \\ &= \lim_{p\to 0}\frac{1}{4}\,\mathrm{Tr}\gamma_\mu\Gamma_{g1jj_3}^{(v)}(p) = \frac{1}{4}\,\mathrm{Tr}\gamma_\mu g^{(v)}(1; j)\varepsilon_\nu^{(jj_3)}\gamma^\nu(1+i\gamma_5) \\ &= g^{(v)}(1; j)\varepsilon_\mu^{(jj_3)}. \end{aligned} \tag{65}$$

Therefore, the parameter g appearing in definition (55), once Eq. (62) is valid, is nothing but the renormalized effective coupling constant $g^{(v)}(1; j)$ defined in Sec. III, in the limit $m \to 0$.

The connection between phenomenological computation of bound states of the previous sections and the field theoretical approach of the present one is thus completed.

V. Non-Abelian Gauge

Leptons can be grouped into multiplets and, in our scheme, these multiplets will have a fundamental role in determining both the parity conservation in l.q.e.d. and the actual value of the electric charge. We shall suppose that they build up the basis for a representation of an internal non-abelian symmetry group for the weak Lagrangian.

Then let $L(X)$ represent a lepton multiplet and let the weak Lagrangian be

$$\mathscr{L}_{\mathrm{I}} = -G_F^2 \int \pi(\eta^2) \bar{L}(X) \gamma_\mu \vec{\tau} L(X) \bar{L}(X+\eta) \vec{\tau} \gamma^\mu L(X+\eta) d^4\eta, \tag{66}$$

where the components of $\vec{\tau}$, operators in the multiplet space, are the elements of the algebra of the symmetry group. This Lagrangian is invariant with respect to the non-abelian gauge transformation:

$$L(X) = S(X) L'(X) = \exp[-i\vec{\tau} \cdot \vec{\varepsilon}(X)] L'(X), \tag{67}$$

but the wave equation

$$i\hat{\partial} L(X) = G_F^2 \int \vec{\tau} T \pi(\eta^2) \gamma_\mu L(X) \bar{L}(X+\eta) \gamma^\mu \vec{\tau} L(X+\eta) d^4\eta, \tag{68}$$

if transformed with (67), becomes

$$\begin{aligned} S(X)\left[i\hat{\partial} L' + \vec{\tau}\gamma_\mu \frac{\partial \vec{\varepsilon}}{\partial X_\mu} L' \right] \\ = G_F^2 \int \pi(\eta^2) T \gamma_\mu S(X) S^{-1}(X) \vec{\tau} \exp[-i\vec{\tau} \cdot \vec{\varepsilon}(x)] L'(X) \bar{L}' \times \\ \times (X+\eta) \exp[i\vec{\tau} \cdot \vec{\varepsilon}(X+\eta)] \vec{\tau} \gamma^\mu S(X+\eta) L'(X+\eta) d^4\eta, \end{aligned} \tag{69}$$

and gauge invariance is obtained as before from the contribution near $\eta_\mu = 0$ of the expansion of the gauge exponential,

$$\exp\left[i\vec{\tau} \cdot \eta_\mu \frac{\partial \vec{\varepsilon}}{\partial X_\mu} \right] \simeq 1 + i\vec{\tau}\eta_\mu \frac{\partial \vec{\varepsilon}}{\partial X_\mu}, \tag{70}$$

provided

$$(71) \quad 2\int G_F^2 M^4 \frac{\eta_\mu \eta_\rho}{\eta^6} E\vec{\tau}\cdot\vec{\tau}S(X+\eta)L(X+\eta)d^4\eta = g_{\mu\rho}S(X)L(X).$$

The gauge invariance here can also be ascribed to a composite gauge field

$$(72) \quad gB_\mu(X) = G_F^2 \int \pi(\eta^2)T\gamma_\mu\vec{\tau}L(X)\bar{L}(X+\eta)\vec{\tau}d^4\eta,$$

which will interact with the lepton multiplet L in the local limit according to the equation

$$(73) \quad i\hat{\partial}L(X) = gB_\mu(X)\gamma^\mu L(X).$$

But in reality the interaction is non-local, as described by (68).

Then, as seen above, because of the transformation (67), the field $B_\mu(X)$ will transform according to

$$(74) \quad B_\mu(X) \to B'_\mu(X) = S^{-1}(X)B_\mu(X)S(X)+S^{-1}(X)\frac{\partial S(X)}{\partial X^\mu}$$

$$= S^{-1}(X)B_\mu(X)S(X)+i\vec{\tau}\frac{\partial\vec{\varepsilon}(X)}{\partial X^\mu}.$$

and ensures gauge invariance to (73). $B_\mu(X)$ is then an ordinary gauge field of Yang–Mills (we shall not consider the B field Lagrangian). It can be decomposed:

$$(75) \quad gB_\mu(X) = g_1\frac{\vec{\tau}}{2}\cdot\vec{V}_\mu(X)+g_2S_\mu(X),$$

where, applying Eq. (53),

$$(76) \quad g_1\vec{V}_\mu(X) = -G_F^2 T\int d^4\eta\pi(\eta^2):\bar{L}(X+\eta)\gamma_\mu\vec{\tau}L(X):,$$

and

$$(77) \quad g_2S_\mu(X) = \tfrac{3}{2}G_F^2 T\int d^4\eta\pi(\eta^2):\bar{L}(X+\eta)\gamma_\mu L(X):,$$

and g_1 and g_2 are conserved separately, and S_μ transforms as an abelian gauge field once an abelian phase is added in (67).

If we consider the Yang–Mills gauge fields as composite fields, we can then construct the gauge models of weak and electromagnetic interaction, starting from a fundamental weak Lagrangian between Weyl massless spinors, reversing the usual procedure of obtaining the weak Lagrangian from the gauge Lagrangians.

VI. Gauge Models

We shall now give as an example a sketch of the derivation of the Salam–Weinberg model. The model fixes the massless multiplets in lepton space: they are a left-handed massless doublet and a right-handed singlet:

$$L(X) = \frac{1+\gamma_5}{2}\begin{bmatrix}\nu(X)\\ e(X)\end{bmatrix}, \qquad R(X) = \frac{1-\gamma_5}{2}\, e(X),$$

where $\nu(X)$ and $e(X)$ represent the neutrino and electron fields, respectively.

We then postulate that the Lagrangian is given by the sum of the free Lagrangian

$$(78) \qquad \mathscr{L}_0 = \frac{i}{2}\bar{L}\overset{\leftrightarrow}{\hat{\partial}}L + \frac{i}{2}\bar{R}\overset{\leftrightarrow}{\hat{\partial}}R,$$

plus the weak interaction Lagrangian

$$(79) \qquad \mathscr{L}_{\mathrm{I}} = \mathscr{L}^{(1)} + c_2\mathscr{L}^{(2)} + c_3\mathscr{L}^{(3)},$$

where

$$(80) \qquad \mathscr{L}^{(1)}(X) = -\frac{G_F}{\sqrt{2}}\,\bar{L}(X)\gamma^\mu\vec{\tau}L(X)\bar{L}(X)\gamma_\mu\vec{\tau}L(X)$$

is responsible for the known standard weak interaction between electron and neutrino, while

$$(81) \qquad \mathscr{L}^{(2)}(X) = -G_F\bar{L}(X)\gamma^\mu L(X)\bar{L}(X)\gamma_\mu L(X)$$

and

$$(82) \qquad \mathscr{L}^{(3)}(X) = -G_F\bar{R}(X)\gamma^\mu R(X)\bar{R}(X)\gamma_\mu R(X)$$

contribute only to the unknown diagonal weak Lagrangian. c_2 and c_3 are then arbitrary parameters. The Lagrangian $\mathscr{L}_{\mathrm{I}}$ is invariant with respect to an SU(2) non-abelian gauge group times two U(1) abelian gauge groups: one for the left-handed $\mathscr{L}^{(2)}$ and one for the right-handed $\mathscr{L}^{(3)}$. In this way the invariance group of $\mathscr{L}_{\mathrm{I}}$ is $\mathrm{SU}(2)\times\mathrm{U}(1)\times\mathrm{U}(1)$.

Starting from these Lagrangians and iterating them, one arrives at the following effective Lagrangians:

$$
\begin{aligned}
\mathscr{L}_{\mathrm{I}}^{(1)} &= -G_F^2 T \int \bar{L}(X)\gamma^{\mu}\vec{\tau}L(X)\bar{L}(X+\eta)\gamma_{\mu}\vec{\tau}L(X+\eta)\pi(\eta^2)\,d^4\eta, \\
(83) \qquad \mathscr{L}_{\mathrm{I}}^{(2)} &= -\beta G_F^2 T \int \bar{L}(X)\gamma^{\mu}L(X)\bar{L}(X+\eta)\gamma_{\mu}L(X+\eta)\pi(\eta^2)\,d^4\eta, \\
\mathscr{L}_{\mathrm{I}}^{(3)} &= -\gamma G_F^2 T \int \bar{R}(X)\gamma^{\mu}R(X)\bar{R}(X+\eta)\gamma_{\mu}R(X+\eta)\pi(\eta^2)\,d^4\eta,
\end{aligned}
$$

with β and γ arbitrary parameters (depending on c_2 and c_3). The vertex obtained from these Lagrangians will be a sum of three vertices: one right-handed singlet and two left-handed, of which one is a triplet and one a singlet. In the local limit we shall then have

$$
\begin{aligned}
(84) \qquad \mathscr{L}_{\mathrm{eff}} &= g(\beta)\bar{L}(X)\frac{\vec{\tau}}{2}\gamma^{\mu}L(X)\vec{A}_{\mu}(X)+ \\
&+g_2(\beta)\frac{1}{2}\bar{L}(X)\gamma^{\mu}L(X)B_{\mu}^{(2)}(X)+g_3(\gamma)\bar{R}(X)\gamma^{\mu}R(X)B_{\mu}^{(3)}(X),
\end{aligned}
$$

where

$$
\begin{aligned}
g(\beta)\vec{A}_{\mu} &= (1-\beta)G_F^2 T\int d^4\eta\,\pi(\eta^2):\bar{L}(X+\eta)\gamma_{\mu}\vec{\tau}L(X): \\
(85) \qquad g_2(\beta)B_{\mu}^{(2)} &= -(3+\beta)G_F^2 T\int d^4\eta\,\pi(\eta^2):\bar{L}(X+\eta)\gamma_{\mu}L(X): \\
g_3(\gamma)B_{\mu}^{(3)} &= -\gamma G_F^2 T\int d^4\eta\,\pi(\eta^2):\bar{R}(X+\eta)\gamma_{\mu}R(X):
\end{aligned}
$$

are Yang–Mills gauge fields which are subject to gauge transformations which make

$$(86) \qquad \mathscr{L} = \mathscr{L}_0+\mathscr{L}_{\mathrm{I}}^{(1)}+\mathscr{L}_{\mathrm{I}}^{(2)}+\mathscr{L}_{\mathrm{I}}^{(3)}$$

$SU(2)\times U(1)\times U(1)$ gauge invariant. The coupling constants g_2 and g_3 will depend on β and γ and are themselves arbitrary. We can use this arbitrariness to obtainn

$$(87) \qquad g_2(\beta) = g_3(\gamma) = g'.$$

In this way we obtain the Salam–Weinberg effective Lagrangian

$$
\begin{aligned}
(88) \qquad \mathscr{L} &= \mathscr{L}_0+\mathscr{L}_{\mathrm{eff}} = \frac{i}{2}\bar{L}\overset{\leftrightarrow}{\hat{\partial}}L+\frac{i}{2}\bar{R}\overset{\leftrightarrow}{\hat{\partial}}R+g\bar{L}\frac{\vec{\tau}}{2}\gamma^{\mu}L\vec{A}_{\mu}+g' \\
&\times\left[\frac{1}{2}\bar{L}(X)\gamma^{\mu}L(X)+\bar{R}(X)\gamma^{\mu}R(X)\right]B_{\mu}(X),
\end{aligned}
$$

where

$$(89) \qquad B_{\mu}(X) \equiv B_{\mu}^{(2)}(X)+B_{\mu}^{(3)}(X).$$

From this point one can proceed as customary in the gauge models; that is, break the symmetry with the Higgs mechanism and give appropriate masses to both the bosons and the leptons. The electromagnetic potential will then be given by the linear combination

$$(90)\qquad A_\mu(X) = \cos\theta A_\mu^{(3)}(X) - \sin\theta B_\mu(X),$$

and, provided

$$(91)\qquad g\cos\theta = g'\sin\theta = e = \frac{gg'}{\sqrt{g^2+g'^2}},$$

it will interact in the standard way (in the local limit) with the electron field $e(X)$ only:

$$(92)\qquad \mathscr{L}_{\text{eff e.m.}} = e\bar{e}(X)\gamma^\mu e(X) A_\mu(X).$$

Obviously, since $A_\mu(X)$ is the local limit of a composite gauge field, the exact electrodynamical equation of motion is of the type (38) and, consequently, the renormalization constants are $Z_3 = 0$, and Z_1 and Z_2 are finite. The limit of validity of the local quantum electrodynamics expressed by (92) is of the order of $\sqrt{G_F}$, that is, far beyond the present experimental limit. Also, the gauge model thus derived presents the advantage of finiteness. In fact, the known triangular diagrams [1] are finite in the present approach (they are actually hexagonal diagrams and finite in perturbation expansion).

In principle, one could now compute the values of g following the procedure used in Secs. II and III and then fix the free parameter γ in (87) in such a way to make g' satisfy (91) and to give the correct value of e. In other words, one could then compute the value of

$$\sin\theta = \frac{g'}{\sqrt{g^2+g'^2}},$$

which gives the measure of the neutral currents and of the W meson masses. One should actually write a new integral equation for the field $\vec{A}_\mu$ (if not of A_μ), but, in order to give orders of magnitude, one can simply use the result of the previous calculations [9] and find

$$\sin\theta \simeq 0.14,$$
$$M_W \simeq 100 \text{ GeV},$$
$$M_Z \simeq 200 \text{ GeV}.$$

One could now use this procedure to start a bootstrap mechanism and precisely use a gauge model to make the Lagrangian (1) renormalizable, then fix the masses of the intermediate bosons in such a way that massless composite gauge fields are originated from (1). One has to choose the multiplets appearing in the original Lagrangian in such a way that the composite gauge fields are precisely those of the selected gauge model. At this stage one will have an effective gauge invariant Lagrangian formally identical (in the local limit) to that used to make the Lagrangian (1) renormalizable, and the circle of the bootstrap will be closed (the gauge fields which will remain massless after the breaking of the symmetry – usually the electromagnetic field – will obviously remain out of the bootstrap.

This could be applied to different models to try to get self-consistency conditions. One unpleasant feature of this procedure is the introduction of new boson fields to break the symmetry, as requested by the Higgs mechanism. In the present frame it would be much more coherent to use the original weak Lagrangian, not only to generate composite states but also to break the symmetry in the spirit of the idea of Nambu–Jona–Lasinio [23], recently reconsidered by Coleman and Weinberg [11], Jackiw and Johnson [19], and Cornwall and Norton [12]. This would bring us to interesting considerations which are, however, outside the scope of the present work.

VII. Conclusion and Outlook

We have started from a phenomenological study of the weak lepton Lagrangian with the aim of exploring the possibility that it may give rise to massless composite lepton–antilepton states. The result of this study was that, within the reach of the disposable computational instruments and our ability to use them, if that Lagrangian kept its validity up to distances of the order of $\sqrt{G_F}$, then it could give rise to only one composite ground state with $j = 1, 0$; $\Pi C = +1, -1$, that is, with the quantum numbers of the photon strongly coupled, in a unique way, to the free massive leptons.

This uniqueness of the composite state and its quantum numbers was mainly due to the fact that the weak Lagrangian maximally violates parity (it is strictly V–A), that is, it is composed only by chiral-projected or two-component left-handed leptons.

We have then interpreted this as an indication that the Weyl fields might be the fundamental fields and that the fundamental interaction Lagrangian might be a generalization of the weak lepton Lagrangian where the lepton fielsd are simply substituted by massless Weyl or, equivalently, Majorana fields.[8]

If we can take the massless Weyl fields as constituent fields, we can then interpret the vector-axial composite massless field as the strongly coupled gauge fields. We have shown, in fact, how this is possible and how one could in such a way obtain a gauge-invariant Lagrangian in which the gauge transformation of the (composite) axial-vector field derives from that of the component spinor field. Here again, in order to obtain the correct gauge transformation, the chiral projectors were essential, together with the hypothesis that at short distances the forces binding the leptons in the composite vector-axial-state is of the Coulomb type (Eq. (12)) corresponding to a renormalizable theory.

This derivation of the gauge fields from the basic Weyl fields could have a deeper meaning on which we would like to develop some further, even if speculative, considerations.

Despite the limitations of the ladder approximation, let us reconsider Eq. (4′), which for small x_μ and $P_\mu = 0$ (after (7)) will become

$$(93)\qquad (\vec{\hat{p}}-m)\chi_0(x)(m-\overleftarrow{\hat{p}}) = -\frac{\lambda^2}{x^2}\gamma^\rho(1+\gamma_5)\chi_0(x)\gamma_\rho(1+\gamma_5),$$

where $\lambda^2 = G_F^2 M^4$ is an adimensional parameter determined by the condition to have a massless bound state. In Sec. IV the value found for λ^2 was

$$(94)\qquad \lambda^2 = \frac{6}{(2\pi)^4}\, n(n+1),$$

that is, it was determined by the Casimir operator of $O(4)$, the group of symmetry for Eq. (17). It must be stressed that at this stage the Fermi coupling constant has practically disappeared from the dynamical equations.

Let us suppose that for small values of x_μ (high value of p_μ) the masses of the constituent leptons can be dropped. Eq. (93) then becomes

$$\vec{\hat{\partial}}\chi_0(x)\overleftarrow{\hat{\partial}} = \frac{\lambda^2}{x^2}\gamma^\rho\chi_0(x)\gamma_\rho, \tag{95}$$

where the bound state wave function $\chi(x)$ is now defined by

$$\chi_P(x) = (2\pi)^{3/2}\langle 0|TL(x)\bar{L}(0)|P\rangle. \tag{96}$$

Eq. (95) is obviously scale-invariant, but also $O(5)$-invariant (in Euclidean space) because of the so-called accidental symmetry generated by the 'potential' x^{-2}.

Now let us suppose that Eq. (95) is soluble, or at least that the necessary modifications for $x_\mu \gg \sqrt{G_F}$ are not essential in the determination of its left-hand side which, at $p_\mu = 0$, gives the renormalized coupling constant. Then, its solution (the wave function of the composite state) and the coupling constant of the composite field to the free spinors would only be determined by the parameter λ^2, which will in turn depend on the Casimir operator of $O(5)$ and/or $O(4)$, and would further be scale-invariant, that is, independent of the masses of the constituent spinors.

This coupling constant could not yet be the electric charge since the leptons are left-handed, so that one must have a multiplet of leptons and consequently a multiplet of gauge field, and only a linear combination of them could be the photon. But if one could consider the multiplet degenerate (massless spinors), the consideration made for Eq. (95) would remain valid: the solution and the renormalized coupling constant will depend on the invariants of the previously mentioned groups and also on the internal symmetry of the leptons (in this way the electric charge could be smaller than the coupling constant of the original gauge fields, as it happens in gauge theories (see Sec. VI)). This picture could then furnish a basis for the explanation of the universality of the charge and the universality of its quantization. In fact, should one further suppose that for any charged spinor field the interaction with the massless photon field, a mixture of composite lepton–antilepton fields, is always determined by the dynamics at very short distance ($x_\mu < \sqrt{G_F}$) described by an equation similar to (95), which is characterized by the high symmetry envisaged above plus the internal lepton symmetry implicit in the photon, then one would deduce that for all systems the charge depends only on the invariants of those groups and is universally quantized (in opposition to what happens with the mass where we have different dynamical groups and different mass quantizations system by system).

Should this picture be at least partially true, one would then be inclined to think that weak and electric properties of leptons are the first manifestation of the excited state of the matter and that hadron physics arises from the higher complexity of the composite systems. The affinity of weak and electromagnetic properties of leptons advocated by the gauge theories would, in this conception, be connected by their being the nearest to the ground state of matter, whereas for the more complex systems describing the hadrons one would, still expect that the composite systems be labelled according to the symmetry (SU(3) or SU(4)) spanned by the original lepton multiplets, while couplings between the systems are strong and determined by the dynamics of the equations for the vertices [24]. As in the previous case, the symmetry of the dynamics determining the vertices and composite state could be larger than that of the original Lagrangian of the Fermi V–A type. Here one would be inclined to think that the quantum numbers of the composite systems and of their interactions (vertices) are precisely determined by a principle of maximal symmetry of the dynamics which lies at their origin, compatibly with the fundamental interaction of the simplest constituent spinor fields. This would give a dynamical explanation to the fact that in going from weak to strong interactions the symmetry steadily increases, since it is originated by the dynamical nature of the interactions which are such as to maximalize the symmetry.

Only further critical analysis and the use of mathematical methods more rigorous than those now available will determine if these fascinating possibilities have something to do with the reality of nature.

Appendix

Let us consider the transformation properties of the field $B_\mu(X)$ under gauge transformations (31) in terms of the $L(X)$ field, independently of the equation of motion (38). Explicitly,

$$\text{(A.1)}\quad gB_\mu(X) = G_F^2 T\int d^4\eta\pi(\eta^2)\left[\gamma_\mu\frac{1+\gamma_5}{2}\right]_{\beta\alpha} :\bar{\psi}^\beta(X+\eta)\psi^\alpha(X):.$$

The dots notation means that we consider the 'finite' part of this operator product and, since the ψ fields are assumed to be *canonical* fields (dimension $l^{-3/2}$), as proposed by Wilson [28] and Zimmermann [30],[9] we define in a similar way

$$\text{(A.2)} \quad \lim_{\eta_\mu \to 0} T{:}\bar{\psi}^\beta(X+\eta)\psi^\alpha(X){:} = \lim_{\eta_\mu \to 0}\left[T\bar{\psi}^\beta(X+\eta)\psi^\alpha(X) + \right.$$

$$\left. + iE(\eta)\frac{\gamma_\sigma^{\alpha\beta}\eta^\sigma}{\eta^4}\right],$$

where $E(\eta)$ is either an arbitrary renormalization constant or an arbitrary at the most logarithmically divergent function of η for $\eta_\mu \to 0$ [14], [26]. Due to definition (A.2) the product ${:}\bar{\psi}^\beta(X)\psi^\alpha(X+\eta){:}$ does not transform homogeneously under (31); in fact (we omit the T symbol):

$$\text{(A.3)} \quad \lim_{\eta_\mu \to 0}[{:}\bar{\psi}^\beta(X+\eta)\psi^\alpha(X){:}] = \lim_{\eta_\mu \to 0}\left[\exp[i\varepsilon(X+\eta)]\exp[-i\varepsilon(X)] \times \right.$$

$$\left. \times \bar{\psi}'^\beta(X+\eta)\psi'^\alpha(X) + iE(\eta)\frac{\gamma_\sigma^{\alpha\beta}\eta^\sigma}{\eta^4}\right] = \lim_{\eta_\mu \to 0}\left\{[1 + i\eta^\rho\partial_\rho\varepsilon(X)] \times \right.$$

$$\left. \times \bar{\psi}'^\beta(X+\eta)\psi'^\alpha(X) + iE(\eta)\frac{\gamma_\sigma^{\alpha\beta}\eta^\sigma}{\eta^4}\right\} = \lim_{\eta_\mu \to 0}\left\{{:}\bar{\psi}'^\beta(X+\eta)\psi'^\alpha(X){:} + \right.$$

$$\left. + i\eta^\rho\partial_\rho\varepsilon(X)\left[{:}\bar{\psi}^\beta(X+\eta)\psi^\alpha(X){:} - iE(\eta)\frac{\gamma_\tau^{\alpha\beta}\eta^\tau}{\eta^4}\right]\right\}.$$

But

$$\lim_{\eta_\mu \to 0} {:}\bar{\psi}^\beta(X+\eta)\psi^\alpha(X){:}$$

is well defined by hypothesis; therefore

$$\text{(A.4)} \quad \lim_{\eta_\mu \to 0} \eta^\rho{:}\bar{\psi}^\beta(X+\eta)\psi^\alpha(X){:} = 0,$$

and finally we obtain the variation of product (A.2) under gauge transformations (31):

$$\text{(A.5)} \quad \lim_{\eta_\mu \to 0} T[{:}\bar{\psi}^\beta(X+\eta)\psi^\alpha(X){:}]$$

$$= \lim_{\eta_\mu \to 0}\left\{T[{:}\bar{\psi}'^\beta(X+\eta)\psi'^\alpha(X){:}] + \big(\partial_\rho\varepsilon(X)\big)E(\eta)\gamma_\tau^{\alpha\beta}\frac{\eta^\rho\eta^\tau}{\eta^4}\right\}.$$

Let us now consider Eq. (A.1). The integration is performed over the whole space-time, but the operator $T{:}\psi^{\alpha}(X)\bar{\psi}^{\beta}(X+\eta){:}$, once inserted between the $\langle 0|$ and $|P, r\rangle$ states as in Sec. II, gives rise to the bound-state Bethe–Salpeter amplitude

$$\chi_P^{(r)\alpha\beta}(X;\eta) = \langle 0|T[{:}\psi^{\alpha}(X)\bar{\psi}^{\beta}(X+\eta){:}]|P, r\rangle,$$

which, as a function of η, is different from zero only in a very small domain (of size $\leqslant \sqrt{G_F}$) around $\eta = 0$, since it must describe a quasi-localized bound state. Let us indicate this state of affairs by saying that the integration can be extended without appreciable error to a small domain o. Let us now write the identity

$$o = \lim_{o'\to 0}[o - o' + o']$$

and correspondingly split the integration, and we shall obtain

$$\text{(A.6)}\quad gB_{\mu}(X) = \left(\gamma_{\mu}\frac{1+\gamma_5}{2}\right)_{\beta\alpha} G_F^2 T\int_o d^4\eta\,\pi(\eta){:}\bar{\psi}^{\beta}(X+\eta)\psi^{\alpha}(X){:}$$

$$= \left(\gamma_{\mu}\frac{1+\gamma_5}{2}\right)_{\beta\alpha} G_F^2 T\lim_{o'\to 0}\left\{\int_o \frac{M^4}{\eta^2}{:}\bar{\psi}'^{\beta}(X+\eta)\psi'^{\alpha}(X){:}\,d^4\eta\, + \right.$$

$$+(\partial_{\rho}\varepsilon(X))\gamma_{\tau}^{\alpha\beta}\int_{o'}\frac{M^4}{\eta^6}\eta^{\rho}\eta^{\tau}E(\eta)\,d^4\eta\, +$$

$$\left. + \int_{o-o'}\pi(\eta){:}\bar{\psi}'^{\beta}(X+\eta)\psi'^{\alpha}(X){:}[1+\eta_{\rho}\partial_{\rho}\varepsilon(X)]\,d^4\eta\right\},$$

where we supposed that o is small enough to allow the Taylor expansion of the gauge phase factor.

Now let us suppose further that in the whole domain of integration o

$$\text{(A.7)}\quad \eta^{\rho}\partial_{\rho}\varepsilon(X) \ll 1,$$

which not only means that the maximum value of η_{ρ} in o is small but also that the variation of the phase $\varepsilon(X)$ is small in o. In other words, the 'wave length' of the variable phase is large compared with the dimension of the bound state. In this case, the term $\eta^{\varrho}\partial_{\varrho}\varepsilon(X)$ can be dropped from the last integral on the right-hand side of (A.6) since the product ${:}\bar{\psi}'^{\beta}(X+\eta)\psi'^{\alpha}(X){:}$ is not singular in the domain $o-o'$. Then, summing the last integral with the first and defining E in such a way that (see Eq. (46))

$$\text{(A.8)}\quad \lim_{o'\to 0} 2G_F^2 M^4 \int_{o'} \frac{d^4\eta}{\eta^6}\eta_\mu\eta_\varrho E(\eta) = g_{\mu\varrho},$$

we have that

$$\text{(A.9)}\quad gB_\mu(X) = \left(\gamma_\mu \frac{1+\gamma_5}{2}\right)_{\beta\alpha} G_F^2 T \int_o d^4\eta\pi(\eta) : \bar{\psi}'^\beta(X+\eta)\psi'^\alpha(X): +$$

$$+\partial_\mu\varepsilon(X) = gB'_\mu(X)+\partial_\mu\varepsilon(X).$$

That is, $B_\mu(X)$ is a gauge field with the right transformation properties, as given by Eq. (58). It is to be noted that to get local gauge covariance for the composite field, according to (A.7) one must suppose that the variation of the gauge phase inside the composite state is negligible, which is quite reasonable and well represents the concept of 'local limit' expressed in the text.

References

[1] In the frame of the intermediate vector meson models of weak interactions this condition is equivalent to fixing the order of magnitude of the vector meson mass to one or a few hundred GeV. Actually, the experimental lower limit of the vector meson mass is at present far from that value, but seems to increase with experimental progress.

[2] One could also study the possibility of composite systems deriving from chain diagrams generated directly by the Lagrangian (1), as done for example by Nambu–Jona-Lasinio [23] and Thirring [2], [3], [16], but in this case the loops of the chains must be regularized, which is somehow equivalent to introducing a massive intermediate boson into the Lagrangian (1) and thereby postulating the existence of a strong coupling mediated by this meson between the leptons (the corresponding force is strong and short range). In the present paper we wish to show all the possible consequences from the Lagrangian (1) and from the long-range forces it might imply without explicitly introducing new strong interactions.

[3] If one only considers the forces originating from the exchange of massless neutrinos (or basic massless Weyl fields) one could also try to construct a theory in which neutrino and lepton currents are mediated by a massive neutral vector boson. It is known that this theory is renormalizable. One could then let the mass of the neutral vector boson go to infinity and hopefully obtain in such a way a renormalizable version of the theory starting from the local Lagrangian (1).

[4] For the definition of symbols and details of the computations referred to in this section, see P. Furlan [17].

[5] It must be stressed that after the parameter M is determined by (23), the Fermi coupling constant G_F disappears from the equations and does not play any role in determining the dynamics of the bound state and its coupling to the free states.

[6] The solutions $n = 1$; $j = 0,1$; $-j \leqslant j_3 \leqslant j$ are the four components of a four-vector amplitude describing a massless vector-pseudovector boson with the quantum numbres of the photon, i.e. $j = 1,0, \Pi = (-)^j$, $C = -1$. The three $j = 1$ amplitudes correspond to positive-norm (bound) states, while the $j = 0$ 'time' amplitude corresponds to a negative-norm (bound) state. This means that our 'generalized photon' is described covariantly with a formalism *à la* Gupta–Bleuler and, since $P_\mu = 0$, no gauge condition is imposed which is able to keep only the transverse components $j = 1$, $j_3 = \pm 1$.

[7] Had we taken the original Eq. (35) we should also have expanded the product of the two last fields $\bar{\psi}^\beta \psi^\gamma$ in (40) which would have had different arguments, say, $x+\eta$ and $x+\eta+\tau$, and this expansion would give terms proportional to $\gamma_\varrho \tau^\varrho/\tau^4$. We shall see later that the singular terms give finite contributions only if the 'potential' represented by F in (35) is, as (39), quadratically divergent at the origin. Since we have taken the approximation (36), we have assumed that F contains a $\delta^4(\tau)$ function which would eliminate all possible contributions generated by the above-mentioned terms; for this reason and in order to eliminate unnecessary formal complications, in the present approach we have only expanded $\psi^\alpha(X)\, \bar{\psi}^\beta(X+\eta)$.

[8] This Lagrangian is then similar to that originally proposed by Heisenberg and his collaborators [18], [14], [26] in its non-linear spinor theory, and thereafter reconsidered by Nambu and Jona–Lasinio [23] as a basis for spontaneous symmetry-breaking. In order to adhere as much as possible to our phenomenological approach in this version, we have kept in the Lagrangian a dimensional coupling constant of the order of magnitude of the Fermi one, and as a consequence our spinor fields are canonical fields of dimension $l^{-3/2}$.

[9] An analogous method for obtaining the gauge transformation of the non-canonical (dimension $l^{-1/2}$) fields of Heisenberg [18] was used before us by Dürr and co-workers [14], [26].

Bibliography

[1] S. L. Adler, in: *Lectures on Elementary Particles and Quantum Field Theory*, (ed. by S. Deser, M. Grisaru, and H. Pendelton), MIT Press, Cambridge 1970.

[2] K. Baumann, P. G. O. Freund and W. Thirring, *Nuovo Cimento* **18** (1960) 906.

[3] K. Baumann, and W. Thirring, *Nuovo Cimento* **18** 357 (1960).

[4] M. Böhm, H. Joos and M. Krammer, *Nucl. Phys.* **B51** (1973) 397; preprint *DESY* 73/20 (1973).

[5] P. Budini, ICTP, Trieste, preprint IC/72/54, in *Proceedings of the "Neutrino '72" Symposium on Neutrinos*, Balatonfüred, Hungary, 12–17 June, Vol. II, 149.

[6] P. Budini, *Lettere al Nuovo Cimento* **9** (1974) 493.

[7] P. Budini and P. Furlan, *Lettere al Nuovo Cimento* **4** (1972) 305.

[8] P. Budini and P. Furlan, *Nuovo Cimento* **13A** (1973) 937.

[9] P. Budini and P. Furlan, ICTP, Trieste, Internal Report IC/73/86.
[10] P. Budini and P. Furlan, *Nuovo Cimento* **20A** (1974) 1.
[11] S. Coleman and E. Weinberg, *Phys. Rev.* **D7** (1973) 1888.
[12] J. M. Cornwall and R. E. Norton, *Phys. Rev.* **D8** (1973) 3338.
[13] P. A. M. Dirac, in: *From a Life of Physics*, IAEA, Vienna 1969.
[14] H. P. Dürr and N. J. Winter, *Nuovo Cimento* **70A** (1970) 467.
[15] G. Feinberg and J. Sucher, *Phys. Rev.* **166** (1968) 1638.
[16] P. G. O. Freund, *Acta Phys. Austr.* **14** (1961) 445.
[17] P. Furlan, '$O(4)^{\Pi C}$ Analysis of Possible Composite Lepton States Generated by Weak Interactions', *Nuovo Cimento* **24A** (1974) 1.
[18] H. P. W. Heisenberg, *Introduction to the Unified Theory of Elementary Particles*, Interscience Publishers, New York 1966; *Naturwiss.* **61** (1974) 5.
[19] R. Jackiw and K. Johnson, *Phys. Rev.* **D8** (1973) 2386.
[20] E. S. Konopinski and H. M. Mahmoud, *Phys. Rev.* **92** (1953) 1045.
[21] M. Krammer, Internal Report DESY-T73/1 (1973).
[22] R. E. Marshak, 'Lectures on Field-Theoretic Models of Particles', in: *Proceedings of the International Spring School*, Yalta, USSR (1966).
[23] Y. Nambu and G. Jona-Lasinio, *Phys. Rev.* **122** (1961) 345; *Phys. Rev.* **124** (1961), 246.
[24] J. C. Pati and Abdus Salam, *Phys. Rev.* **D8** (1973) 1240; *Phys. Rev. Letters* **31** (1973) 661; *Phys. Rev.* **D10** (1974) 275.
[25] Abdus Salam and J. C. Ward, *Phys. Letters* **13** (1964) 168.
[26] H. Saller, *Nuovo Cimento* **4A** (1971) 404; *Nuovo Cimento* **12A**, (1972) 349.
[27] S. Weinberg, *Phys. Letters* **19** (1967) 1264.
[28] K. G. Wilson, *Phys. Rev.* **179** (1969) 1499.
[29] C. N. Yang and R. L. Mills, *Phys. Rev.* **96** (1954) 191.
[30] W. Zimmermann, *Nuovo Cimento* **10** (1958) 597.

A SHORT REVIEW OF METHODS USED TO COMPUTE BOUND STATES

MARCEL GUENIN

Department of Theoretical Physics, University of Geneva, Geneva, Switzerland

1. Introduction

The purpose of this talk is to make a small (and very incomplete) review of various methods suggested for the discussion and, eventually, the computation of the properties of bound states of particles.

We shall not be interested in the numerical computations, but shall focus our attention on the questions of the principles, picking out some of the typical approaches. Since Arthur Jaffe is also speaking on this subject, we shall completely leave aside the approaches based on the methods of constructive quantum field theory, even if our personal opinion is that these methods offer great potential for future developments.

We first have to define what we mean by bound state problem. The ideal theory should have the following properties: Given a certain number N of 'elementary' fields and the corresponding Hamiltonians describing the various couplings of these fields, one wants a theory which could give:

1) The ground and excited states of a bound state of M given particles.

2) Their quantum numbers, given those of the 'elementary' field.

3) The lifetime of the excited states.

4) The scattering amplitude between an 'elementary' particle and a bound state, or between 2 bound states.

The theory should be relativistically invariant and satisfy the standard properties of quantum field theory, like unitarity and crossing symmetry.

Such a theory does unfortunately not exist. Not only are the proposed equations to tough to be practically solved, they are even in principle not able to solve the problem.

In what follows, we shall make absolutely no claim for completeness, even if the list of references given at the end should represent a fairly large part of the litterature published on the subject during the last few years. In this talk I shall stick to an old tradition concerning raporteur talks, which is that the raporteur doesn't speak of his own work, nor quote himself. That will be rather easy for me since I did not publish anything on that subject and whatever I may have in preparation is far from being ripe enough to be discussed. Being therefore a non-expert on the subject, I apologize in advance for all mistakes or wrong attributions I am certainly going to make, my only excuse is that it may be a good thing if, from time to time, a certain field is reviewed by an outsider.

The first feeling of such an outsider in considering this field is that the problem of bound states is in fact a fantastic one. The dilemma of the specialists in this field is that they can either write beautiful and more or less exact equations, or soluble ones. Ignoring all subtilities, one can roughly divide all attemps into two classes: first the attempts based on some kind of equivalent or effective potentials, thus essentially reducing the problem to a non-relativistic one, secondly the attempts based on the Bethe–Salpeter equation, to which one could add equations of the Weinberg type and others. In fact these approaches are not completely disconnected, since one very often uses quasi-potentials to reduce the Bethe–Salpeter equation to a more tractable form. There exist a few other isolated proposals which can be found in the references and would not fit into the two main categories defined above, they are not without interest, but there is no time to discuss them here.

2. The Equivalent or Effective Potential Approaches

This is historically the oldest approach. In fact it goes back to Yukawa who did show that to first order in perturbative expansion, a Yukawa interaction taken in the limit of no-recoil is equivalent to a Yukawa potential.

This type of approach present the very great advantage that we know very well what is a bound state in a potential. The disadvantage is that it is basically a non-relativistic approach, the relativistic effects appearing as velocity dependent correction terms to a standard type of potential. Nevertheless, this type of approach if suitably used can give fairly good results, especially in the case of weak binding. A good typical modern version is that of Groth and Yennie, but one can also see Todorov, Logunov and Tavkhelidze, Faustov and many others.

The idea is to start from an equation of the form

$$[E-H_e(x_e)-H_p(x_p)]\psi(x_e, x_p) = V(x_e-x_p)\psi(x_e, x_p)$$

in the case of electron-proton system, that is for the hydrogen atom, where $H_e = \boldsymbol{\alpha}_e \boldsymbol{p}_e+\beta m$; $H_p = \boldsymbol{\alpha}_p \boldsymbol{p}_p+\beta M$ the potential being so far not yet defined.

Using the Lipmann–Schwinger equation, the matrix element for the electron-proton scattering to first Born approximation is given by

$$-(2\pi)^4 i\delta(p_1+p_2-p_3-p_4)\frac{mM}{\sqrt{E_1 E_2 E_3 E_4}} \times$$
$$\times u_e^+(p_3, s_3)u_p^+(p_4, s_4)V(\boldsymbol{p}_1-\boldsymbol{p}_3)u_e(p_1, s_1)u_p(p_2, s_2).$$

The corresponding first order amplitude for the relativistic scattering in the CMS would be given in the Coulomb gauge by

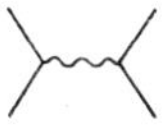

$$ie^2(2\pi)^4\delta(p_1+p_2-p_3-p_4)\frac{mM}{\sqrt{E_1 E_2 E_3 E_4}} \times$$
$$\times u_e^+(p_3, s_3)u_p^+(p_4, s_4)\frac{1}{(\boldsymbol{p}_1, \boldsymbol{p}_3)^2}(1-\boldsymbol{\alpha}_\perp \cdot \boldsymbol{\alpha}_p)u_e(p_1, s_1)u_p(p_2, s_2)$$

where $\boldsymbol{\alpha}_\perp = \boldsymbol{\alpha}-(\boldsymbol{\alpha}\cdot \boldsymbol{k})\dfrac{\boldsymbol{k}}{\boldsymbol{k}^2}$ with $\boldsymbol{k} = \boldsymbol{p}_1-\boldsymbol{p}_3$.

The idea is to identify these two results, since they both come from a first order expansion. It is thus hoped to retain the essential of the relativistic case. This amount therefore to put

$$V(\boldsymbol{p}_1-\boldsymbol{p}_3) = -\frac{e^2}{(\boldsymbol{p}_1-\boldsymbol{p}_3)^2}(1-\boldsymbol{\alpha}_{e\perp}\cdot \boldsymbol{\alpha}_p).$$

This potential is the sum of the Coulomb potential and of what is called the Breit potential. Putting this form of the potential in the equation we started with we get the Breit equation.

The subsequent treatment of the equation does not present difficulties of principle, so that it is the point to make two remarks. The first is to notice that the form given to the potential is gauge dependent. Different potentials, corresponding to different gauges, would give the same first Born approximation to the scattering amplitude, but would yield different results when used in bound states problems. It turns out that this difficulty is not very deep, since one can, in principle, correct the difference by going to the next order of perturbation. The choice made by Grotch and Yennie of the Coulomb gauge, corresponds to making the corrections of the second order minimal. The second remark concerns the fact that the identification of the potential which is made on the mass shell. This feature is found in most of the treatments of the bound states problem: one is nearly always making the extension of-shell of a potential or an amplitude, which is, a priori, 'known' only on-shell. Of course the result is dependent upon the extension chosen, so that the various authors are forced to make some argument to justify their choices. The same problem occurs, by the way, also in the usual treatments of the Bethe–Salpeter equation. We recognize that the arguments given are not always unconvincing, it is however from a fundamental point of vue very unsatisfactory to have to choose between an infinite number of possible extensions with no rule coming from first principles to force the choice.

To sum up, we can say that this type of effective potential approach has the advantage of simplicity and of clarity as far as the hypothesis go, but that one has to keep in mind that besides some problems with the gauge, they require the postulate of the full off-shell potential, which is essentially justified only by plausibility arguments.

Nevertheless, this approach of Grotch and Yennie which we did take as an example is one of the best treatments we know. This is unfortunately not the case of the next example which we are going to discuss rapidly. This is again a potential-type approach, but with no reference to field theory. It is based on Greenberg's harmonic oscillator symmetric quark model, and we follow the treatment as given by Feynman, Kislinger and Ravndal. What they want, is a model which would give equally spaced levels, in the mass squared, hence the idea of starting with an harmonic oscillator, since its levels are equally spaced. The harmonic oscillator, is written as

$$H = \frac{1}{2m}P^2 + \frac{1}{2}m\omega_0^2 X^2.$$

One multiplies by $2m$ and set $m^2\omega_0^2 = \Omega^2$, this gives

$$2mH = P^2 + m^2 X^2$$

the mass (of the quark) disappears on the right-hand side. On the left-hand side, if we add the constant m^2, we get m^2+2mH, or its eigenvalues, $m^2+2mW \cong (m+W)^2$, i.e. the square of the relativistic energies for small W. This leads to consider the operator (for 3 quarks a, b, c)

$$K = 3(p_a^2+p_b^2+p_c^2)+\tfrac{1}{36}\Omega^2[(u_a-u_b)^2+ \\ +(u_b-u_c)^2+(u_c-u_a)^2]+C$$

where C is a constant, 3 and 36 are chosen to simplify the expressions in the computations.

One then consider that $p_a^2 = p_{a0}p_{a0}-p_{a1}p_{a1}-p_{a2}p_{a2}-p_{a3}p_{a3}$ and $u_{a\mu}$ as conjugate position, that is identify $p_a^\mu = i\dfrac{\partial}{\partial u_{a\mu}}$. One then make a change of variables, putting

$$p_a = \tfrac{1}{3}P-\tfrac{1}{3}\xi, \qquad u_a = R-2x,$$

$$p_b = \frac{1}{3}P+\frac{1}{6}\xi-\frac{1}{2\sqrt{3}}\eta, \qquad u_b = R+x-\sqrt{3}\,y,$$

$$p_c = \frac{1}{3}p+\frac{1}{6}\xi+\frac{1}{2\sqrt{3}}\eta, \qquad u_c = R+x+\sqrt{3}\,y,$$

then R, x, y are conjugate to the momenta P, ξ and η and

$$K = P^2-\mathfrak{N}, \qquad \mathfrak{N} = \tfrac{1}{2}\xi^2+\tfrac{1}{2}\eta^2+\tfrac{1}{2}\Omega^2(x^2+y^2)+C.$$

$\mathfrak{N}$ is called the mass-square operator and depends only on the internal motion. It is exactly of the form of the sum of 2 (4-dim.) harmonic oscillators, each giving eigenvalues spaced by Ω.

Everyone in the audience will, however, immediately notice that the spectrum of $\mathfrak{N}$ is not semi-bounded, because of the Minkowski metric. The situation is in fact very similar to the situation for the quantized electromagnetic field. But in the latter case, we have a subsidiary condition which couples the different components and one is making the treatment with an indefinite metric. One could try to make something similar here, but it is not what Feynman and his coworkers are doing, they simply postulate: 'One supposes that only spacelike excited states exist'.

The consequence is that one is essentially back to a non-relativistic treatment with the difficulties of the harmonic potential in the definition of unitarity, and the model cannot predict anything which could not have been said in a non-relativistic treatment. There is a lot more to criticize in this paper, but we think that what we said suffices to show that it is not the miracle answer to the bound state problem.

3. Bethe–Salpeter Type of Equations

The Bethe–Salpeter equation and its generalizations, or equations of the Weinberg type, are from the intellectual point of view much more satisfactory, at least at first sight. It is also a field where the modern techniques of mathematical physics could be profitably used to solve some of the basic problems.

The main drawback of this approach are first that there exists no known exact solution, secondly no general existence proof, thirdly that in order to get numerical results one has to resort to very drastic approximations which reduce considerably the value of the results obtained. Finally, it is allowed to have strong doubts wether even in principle the Bethe–Salpeter equation will provide the answer to the bound state problem, as we shall try to discuss briefly.

Essentially all what was known about the 2-body Bethe–Salpeter equation up to 1969 is very nicely presented in the review article by Nakanishi. An attempt to connect the more rigorous work on composite particles of Nishijima, Haag and Zimmermann to the Bethe–Salpeter equation has been published by Tirapegui. We shall therefore concentrate the very small amount of time available to a few remarks not covered in these two papers.

To fix the ideas, we shall write down the 2-body Bethe–Salpeter equation corresponding to an interaction of the form $\lambda_1 : \psi_1 \psi_1 \phi : + \lambda_2 : \psi_2 \psi_2 \phi :$, where ψ_1, ψ_2, ϕ are taken for simplicity to be hermitian, scalar fields. The introduction of spin can of course be done, that of internal exact symmetries is even simpler, since it amounts to making the coupling parameter in the Bethe–Salpeter equation a function of the Casimir operators of the symmetry group. Graphically, the Bethe–Salpeter equation is then written as

 $(G = G^0 + G^0 IG)$

where  denotes the sum over all 2-particles irreducible graphs, i.e. graphs which cannot be cut in two by breaking two internal lines, in such a way that the two initial external lines lie on one half of the graph and the two final ones in the other (N.B. the difinition of what is to be understood as irreducible graph may become more sophisticated for some other types of interaction). Note also that the definition given for two particle irreducibility presupposes that we know which are the initial and the final lines, this definition is therefore not crossing-symmetric. Also keep in mind that even if we use for this talk the graphical notation, its translation into analytical expression is not always trivial, especially in the 3-body case.

There is not much to be said about the derivation of the Bethe–Salpeter equation which can be done in quite a variety of ways. One can start from the time ordered 4-points function and use the reduction formulas of LSZ; one can also simply compare the perturbative expansion of the right and left hand sides of the equation. Roughly speaking, all derivations are about as rigorous as the latter one, up to some refinements. This is in itself not bad, one can postulate the equation and ask about its solutions.

The Bethe–Salpeter equation contains, however, besides the unknown amplitude, the sum over all irreducible graphs, which is in itself also not known. One has to understand, therefore, that one is in fact dealing with a sequence of equations, indexed by the number of irreducible graphs taken into account. Inversely a knowledge of the full 2-body amplitude would define the sum over all irreducible graphs via the Bethe–Salpeter equation.

A further point of interest is the Wick rotation, which amounts to changing t in $-it$, that is going over to a euclidean theory. This is very advantageous for numerical applications on the one hand, or for existence proofs in some cases. Indeed, for a finite number of irreducible graphs taken into account, the Bethe–Salpeter equation takes the form of a Lippmann–Schwinger equation, and the proofs valid for potentials can be taken over with only slight changes. The major problem is, that there is at least to my knowledge no rigorous general proof of the justification of the Wick rotation. It is not difficult for a particular choice of irreducibles graphs to show that one is not crossing any singularity as one moves the integration path, but it is not sufficient; one should also show that the asymptotic behavior is such as to permit it. To my knowledge, this has never been checked, even not a posteriori.

If we now have a look at how the Bethe–Salpeter equation is applied, we first notice that most of the considerations are made using the ladder approximation, corresponding to the equation

In the particular case where the exchanged boson is of zero mass, an exact solution has been given long ago by Cutkovsky. How good or bad is the ladder approximation is not clear. There have been recently a number of attemps based on the infinite momentum limit which would tend to show that the ladder approximation, at least for bosons, is after all not too bad. We do not want to discuss these arguments here, one has only to remark that by going over to the infinite momentum frame one is loosing the full Poincaré invariance, retaining only invariance under the little group; this might not be too troublesome but clearly much remains to be done rigourously.

One has to be extremely careful that the approximation does not introduce solutions which would not be present for the exact equation. How this can happen can be seen with the hydrogen atom or with the positronium. It is well known that for such systems, the ladder approximation will give a more or less satisfactory spectrum for the excited levels. On the other hand these levels should not appear as solution of the exact Bethe–Salpeter equation. Indeed the same interaction which is responsible for the existence of the bound state is also responsible for the decay of the excited states. As the bound state Bethe–Salpeter equation is an eigenvalue equation for the really stationary states, the excited states should not be stable solutions if one takes the complete Bethe–Salpeter equation, and not only a subset of the irreducible graphs. In this respect I should like to draw the attention to a nearly forgotten paper by Kita and Wakano, who claim to have proven rigourously this statement. I confess that I have not been able to check their proof, but here again a serious, careful and rigorous investigation is badly needed. All this means that the Bethe–Salpeter equation is not capable of predicting the excited levels of bound states, only an approximation which retains from the interaction the binding effects and neglects the part responsible for the decay will do so.

We have been, so far, criticizing some questions of principle. One has to recognize, however, that even if the Bethe–Salpeter equation is far from being a really satisfactory answer to the bound state problem, one has nothing better. It is therefore perfectly natural to have developed computation techniques for solving this equation in concrete cases. The different proposals average about one per paper, but they have nearly always one common aspect, namely to reduce the number of intermediate integrations. This necessitate of course supplementary assumptions, but may just mean going from the impossible to the possible.

An often used trick is the one devised by Blankenbecler and Sugar. One first writes the Bethe–Salpeter equation in the form $T = I + TG^0I$, or, in the ladder approximation

$$T(p, q) = g^2 \frac{1}{(p-q)^2+m^2} - \frac{i}{(2\pi)^4} \int d^4k \, \frac{g^2}{(p-k)^2+m^2} \times$$

$$\times \frac{1}{\left(\frac{1}{2} P+k\right)^2 +1} \, \frac{1}{\left(\frac{1}{2} P-k\right)^2 +1} \, T(k, q),$$

where P is the center of mass momentum, and both particles a and b are supposed to have the same mass, equal to 1. We define $s = -P^2$ and put q on the mass shell, that is $q_0 = 0$, $q^2 = \frac{s}{4} - 1$. The idea is now to replace the product of the two propagators in the kernel by a simpler expression, which should retain the main analytic properties. One defines

$$E(k) = 2\pi \int ds' \, \frac{1}{s'-s} \, \delta\left(1+\left(\frac{1}{2} P'+k\right)^2\right) \delta\left(1+\left(\frac{1}{2} P'-k\right)^2\right),$$

where $P'^2 = -s$ and P' has only a 4-th component. In the center of mass frame, the result of the integration is

$$E(k) = \frac{1}{2} \pi\delta(k_0) \cdot \frac{1}{\sqrt{k^2+1} \cdot (k^2-q^2)} \, .$$

One then write

$$-i \frac{1}{(k^2+1)^2} = E(k) + R(k),$$

where R cannot produce any 2-particles singularities. The Bethe–Salpeter equation can the be written as $T = W + WET$ where $W = V(1-RV)^{-1}$ plays the role of en effective potential. This integral equation only requires the knowledge of T for $p_0 = q_0 = 0$ because of the δ function in E. Notice that E is not unique, one could add any function not singular along the positive cut of s. The simplest approximation, which is the only usually used, is to put $R = 0$. We then have

$$W(p, q) = V(p, q) = g^2 \frac{1}{(p-q)^2 + m^2},$$

in other words, W is exactly the Yukawa potential and our Bethe–Salpeter equation becomes

$$T(p, q) = V(p, q) + \frac{1}{4} \int \frac{d^3k}{(2\pi)^3} \frac{V(p, k)\, T(k, q)}{(k^2+1)^{1/2}(k^2-q^2)}.$$

The only difference between this equation and the Lippmann–Schwinger one in potential scattering is the factor $(k^2+1)^{-\frac{1}{2}}$ in the Green function. This equation is slightly more convergent than the Lippmann–Schwinger equation and satisfies a relativistic (elastic) unitarity condition. However, the square root complicates the analytical structure and the amplitude does not satisfies the Mandelstam representation.

Concerning the generalization of the Bethe–Salpeter equation to 3 particles, there is no problem to formulate the system of equations corresponding to the Faddeev amplitudes, at least as long as one uses the ladder approximation. The general case, with 3 particles irreducibles graphs has been treated by Freedman, Lavelace and Namyslowski. It may be useful to give a graphical representation of these equations, since they don't seem to be very well known and are rederived periodically.

We can write the splitting graphically as follows

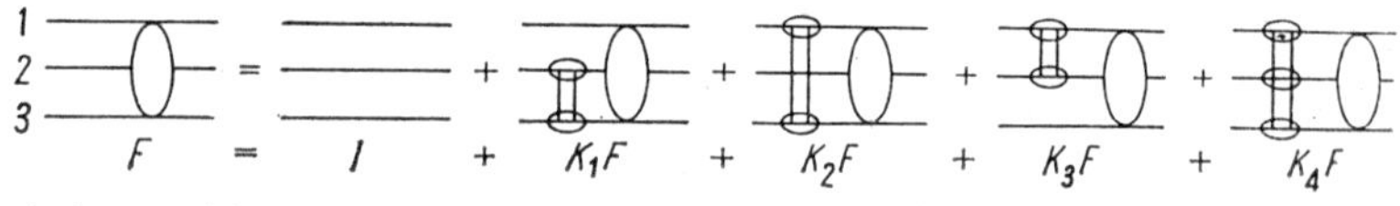

and the Faddeev equations as

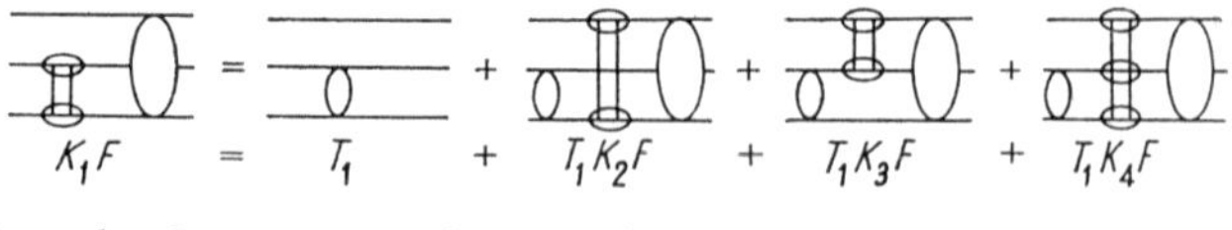

N.B. T_1 is only the connected part of the 2-body amplitude. Of course, and as usual, the equation for bound states is obtained in neglecting the first term.

In matrix form, the equations read

$$\begin{bmatrix} K_1 \\ K_2 \\ K_3 \\ I \end{bmatrix} F = \begin{bmatrix} T_1 \\ T_2 \\ T_3 \\ I \end{bmatrix} + \begin{bmatrix} 0 & T_1 & T_1 & T_1 \\ T_2 & 0 & T_2 & T_2 \\ T_3 & T_3 & 0 & T_3 \\ I & I & I & I \end{bmatrix} \begin{bmatrix} I & 0 & 0 & 0 \\ 0 & I & 0 & 0 \\ 0 & 0 & I & 0 \\ 0 & 0 & 0 & K_4 \end{bmatrix} \begin{bmatrix} K_1 \\ K_2 \\ K_3 \\ I \end{bmatrix} F$$

or

$$\begin{bmatrix} K_1 \\ K_2 \\ K_3 \\ K_4 \end{bmatrix} F = \begin{bmatrix} T_1 \\ T_2 \\ T_3 \\ K_4 \end{bmatrix} + \begin{bmatrix} 0 & T_1 & T_1 & T_1 \\ T_2 & 0 & T_2 & T_2 \\ T_3 & T_3 & 0 & T_3 \\ K_4 & K_4 & K_4 & K_4 \end{bmatrix} \begin{bmatrix} K_1 \\ K_2 \\ K_3 \\ K_4 \end{bmatrix} F.$$

we shall set

$$N = \begin{bmatrix} \begin{bmatrix} 0 & T_1 & T_1 & T_1 \\ T_2 & 0 & T_2 & T_2 \\ T_3 & T_3 & 0 & T_3 \\ I & I & I & I \end{bmatrix} \begin{bmatrix} I & 0 & 0 & 0 \\ 0 & I & 0 & 0 \\ 0 & 0 & I & 0 \\ 0 & 0 & 0 & K_4 \end{bmatrix} \end{bmatrix}.$$

As in the Faddeev approach, the terms in T_i always contain a δ function in the arguments $x_j - x_k$, $j \neq k \neq i$, but a product of the form $T_i T_k$ $(i \neq k)$ or of the form $T_i T_4$ no longer. That makes that the 3rd iteration of the kernel N no longer contains a δ function. That was already true of the 2nd iteration in Faddeev case. Notice, however, that the transcription of thes graphical equations into analytical expressions is not at all trivial.

The N-body case can be treated in a similar way, the splitting being then the one given by Yakobovsky.

4. Conclusion

It is difficult to make a conclusion. We have tried to show through a few examples that both the equivalent potential approach and the Bethe–Salpeter equation approach are far from being satisfactory answers to the bound state problem. It is true that both of these approaches can justly claim a certain number of numerical successes, but from a more fundamental point of vue one is far from understanding what one is doing. A main reason for many of the difficulties is that the entire formalism, be that of potential theory or of quantum field theory, is a formalism constructed to solve the scattering problem and one is lacking a formalism truly adapted to the bound state problem. What this formalism should be is anybody's gess, and it would probably be worthwhile to look for new ideas. Even within the Bethe–Salpeter approach, they remain a very great number of problems to be solved, both mathematical and physical ones.

Bibliography

C. Alabiso, P. Butera and G. Prosperi, 'Variational Padé Solution of the Bethe–Salpeter Equation', *Nucl. Phys.* **B46** (1972) 593–614.

S. Aks, J. Sienicki and B. Varga, 'Quasisecular Renormalization of the Φ^4 Model of Quantum Field Theory', *P. R.* **D6** (1972) 520.

S. Aks, and B. Varga, 'Quasisecular Perturbative Method for Calculating Bound States in Quantum Field Theory', *P. R.* **D6** (1972) 2773.

A. Atanasov, 'On the Bound States of a Three-Particle Nonrelativistic System', *Acta Physica Polonica* **A37** (1970) 337.

J. Ball and F. Zachariasen, 'Asymptotic Behavior of Form Factors of Composite Particles', *P. R.* **170** (1968) 1541.

K. Bardakci and M. Halpern, 'Theories at Infinite Momentum', *P. R.* **176** (1968) 1686.

D. Basu, 'Formal Equivalence of the Hydrogen Atom and Hormonic Oscillator and Factorization of the Bethe–Salpeter Equations', *I. M. P.* **12** (1971) 1474.

S. Biswas, R. Chaudhuri and G. Malik, 'Bound State Problem in a Model Non-polynomial-Lagrangian Theory', *P. R.* **D8** (1973) 1808.

I. Bjorken, 'Asymptotic Sum Rules at Infinite Momentum', *P. R.* **179** (1969) 1547.

I. Bjorken, I. Kogut and D. Soper, 'QED at Infinite Momentum: Scattering from External Field', *P. R.* **D3** (1971) 1382.

I. Bjorken and E. Paschos, 'Inelastic Electron-Proton and γ–Proton Scattering and the Structure of the Nucleon', *P. R.* **185** (1969) 1975.

P. Bogolyubov, 'Equations for Bound States (Quarks)', *Sov. J. of Part. and Nuclei* **3** (1972) 71.

M. Böhm, H. Joos and M. Krammer, 'Relativistic Scalar Quark Model with Strong Binding', *Nuovo Cim.* **7A** (1972) 21.

S. Brodsky, 'Brandeis lectures 1970', Gordon and Breach.

S. Brodsky, F. Close, and J. Gunion, 'Compton Seattering and Fixed Poles in Parton Field-Theoretic Models', *P. R.* **D5** (1972) 1384.

S. Brodsky, F. Close, and J. Gunion, 'Phenomenology of Photon Processes, Vector Dominance and Crucial Tests for Parton Models', *P. R.* **D6** (1972) 177.

S. Brodsky and R. Roskies, 'QED and Renormalization Theory in the Infinite Momentum Frame', *P. L.* **41B** (1972) 517.

E. Blankenbeder and R. Sugar, 'Linear Integral Equations for Multichannel Scattering', *P. R.* **142** (1966) 1051.

S. Chang and S. Ma, 'Feguman Rules and QED at Infinite Momentum', *P. R.* **180** (1969) 1506.

S. Chang and S. Ma, 'Multiphoton Exchange Amplitudes at Infinite Energy', *P. R.* **188**, (1969) 2385.

H. Cheng and T. Wu, 'Theory of High-Energy Diffraction Scattering I', *P. R.* **D1** (1970) 1069.

H. Cheng and T. Wu, 'Theory of High-Energy Diffraction Scattering II', *P. R.* **D1** (1970) 1083.

H. Cheng and T. Wu, 'Impact Picture and Scattering of Deuterons I', *P. R.* **D6** (1972) 2637.

M. Ciafaloni, 'Model Dependence of the Asymptotic Behavior of Form Factors for Relativistic Composite Models', *P. R.* **176** (1968) 1898.

M. Ciafaloni and P. Menotti, 'Asymptotic Behavior of Form Factors for Some Composite Models', *P. R.* **173** (1968) 1575.

R. Cutkosky, 'Solutions of a Bethe–Salpeter Equation', *P. R.* **96** (1954) 1135.

S. Drell and T. Lee, 'Scaling Properties and the Bound-State Nature of the Physical Nucleon', *P. R.* **D5** (1972) 1738.

S. Drell, D. Levy and T. Yan, 'Theory of Deep-Inelastic Lepton–Nucleon Scattering and Lepton-Pair Annihilation Processes I', *P. R.* **187** (1969) 2159.

S. Drell, D. Levy and T. Yan, 'Theory of Deep-Inelastic Lepton–Nucleon Scattering and Lepton-Pair Annihilation Processes II', Deep-Inelastic Electron Scattering', *P. R.* **D1** (1970) 1035.

S. Drell, D. Levy and T. Yan, 'Theory of Deep-Inelastic Lepton–Nucleon Scattering and Lepton-Pair Annihilation Processes III, Deep-Inelastic Electron–Positron Annihilation', *P. R.* **D1** (1970) 1617.

S. Drell and T. Yan, 'Theory of Deep-Inelastic Lepton–Nucleon Scattering and Lepton-Pair Annihilation Processes IV, Deep-Inelastic Neutrino Scattering', *P. R.* **D1** (1970) 2402.

S. Deser, W. Gilbert and E. Sudarshan, 'Structure of the Vertex Function', *P. R.* **115** (1959) 731.

R. Esch, 'On Constructing Massive Particles out of Massless Particless in the Nambu Model', *Can. J. Phys.* **41** (1971) 2833.

H. Ezawa, T. Muta and H. Umezawa, 'An Approach to the Elementarity of Particles', *Progr. Theor. Phys.* **29** (1963) 877.

R. Faustor, 'The Proton Structure and Hyperfine Splitting of Hydrogen Energy Levels', *Nucl. Phys.* **75** (1966) 669.

V. Fainberg, 'On Analytic Properties of Causal Commutators', *Sov. Phys. JETP* **9** (1959) 1066.

P. Federbusch, 'Existence of Spurious Solutions to Many-Body Bethe–Salpeter Equations', *P. R.* **148** (1966) 1551.

E. Fermi and C. Yang, 'Are Mesons Elementary Particles?', *P. R.* **76** (1949) 1739.

R. Feynman, M. Kislinger and F. Ravudal, 'Current Matrix Elements from a Relativistic Quark Model', *P. R.* **D3** (1971) 2706.

D. Freedman, C. Lavelace and Namyslowski, 'Practical Theory of Three-Particle States, II-Relativistic, Spin Zero', *Nuovo Cimento*, **53** (1966) 258.

M. Fortes and A. Jackson, 'Relativistic Effects in Low-Energy Nucleon–Nucleon Scattering', *Nucl. Phys.* **A175** (1971) 449.

S. Fubini and G. Furlan, 'Renormalization Effects for Partially Conserved Currents', *Physics* **1** (1965) 229.

K. Fujimura, T. Kobayashi and M. Namiki, 'Electromagnetic Inelastic Form Factors of Processes $ep \to eN^*$ in a Relativistic Extended Particle Model Based on the Quark Model', *Progr. Theor. Phys.* **44** (1970) 193.

G. Ghicardi and A. Rimini, 'On the Number of Bound States of a Three-Particle System', *Nuovo Cim.* **37** (1965) 450.

C. Gignoux and A. Laverne, 'Three-Nucleon Bound State from Faddeev Equations with a Realistic Potential', *P. R. L.* **29** (1972) 436.

M. Gell-Mann and F. Low, 'Bound States in Quantum Field Theory', *P. R.* **84** (1951) 350.

H. Grotch and D. Yennie, 'Effective Potential Model for Calculating Nuclear Corrections to the Energy Levels of Hydrogen', *Rev. Mod. Phys.* **41** (1969) 350.

J. Gunion, S. Brodsky and R. Blankenbeder, 'Composite Theory of Large Angle Scattering and New Tests of Parton's Concepts', *PL* **39B** (1972) 649.

W. Greenberg, 'Spin and Unitary Spin Dependence in a Paraquark Model of Baryons and Mesons', *P. R. L.* **13** (1964) 598.

C. Garibotti, M. Pellicoro and M. Villari, 'Padé Method in Singular Potentials', *Nuovo Cimento* **66A** (1970) 749.

R. Haag, 'Quantum Field Theories with Composite Particles and Asymptotic Conditions', *P. R.* **112** (1958) 669.

J. Healy, 'New Rigorous Bounds on Coupling Constants in Field Theory', *P. R.* **D8** (1973) 1904.

K. Hepp, 'On the Quantum Mechanical N-Body Problem', *H. P. A.* **42** (1969) 425.

W. Hunziker, 'Proof of a Conjecture of S. Weinberg', *P. R.* **135** (1964) B 800.

R. Haymaker, 'Application of Analyticity Properties to the Numerical Solution of the B–S Equation', *P. R.* **165** (1968) 1790.

M. Ida, 'Integral Representations of Bethe–Salpeter Amplitudes', *Progr. Theor. Phys.* **23** (1960) 1151.

P. Johnson and R. Yaes, 'Crossing-Symmetric Bethe–Salpeter Equation and the Generalized Ladder Graphs', *P. R.* **D4** (1971) 3766.

B. Jouvet and M. Whippman, 'Infinitely Composite Particles and the Condition $Z = 0$', *P. R.* **D4** (1971) 3222.

R. Karplus and A. Klein, 'Electrodynamic Displacement of Atomic Energy Levels III. The Hyperfine Structure of Positronium', *P. R.* **87** (1952) 848.

W. Kaufmann, 'Numerical Solutions of the Bethe–Salpeter Equation', *P. R.* **187** (1969) 2051.

R. Keam, 'Reduction of the Bethe–Salpeter Equation for a Zero-Mass System,' *J. M. P.* **9** (1968) 1462.

R. Keam, 'Perturbation Treatment of the Bethe–Salpeter Equation for Tightly Bound Fermion–Antifermion System', *J. M. P.* **12** (1971) 515.

H. Kita and M. Wakano, 'On the Bethe–Salpeter Equation and the Excited States of the Hydrogen Atom', *Progr. Theor. Phys.* **17** (1957) 63.

J. Kogut and D. Soper, 'QED in the Infinite-Momentum Frame', *P. R.* **D1** (1970) 2901.

C. Lovelace, 'Practical Theory of Three-Particle States I. Norelativistic', *P. R.* **145** (1964) B 1225.

A. Laverne and C. Gignoux, 'A Detailed Analysis of ^{3}H from Faddeev Equations in Configuration Space', *Nucle. Phys.* **A203** (1973) 597.

K. Ladanyi, 'Variational Solutions of the B–S Equation', *Nuovo Cim.* **56A** (1968) 173.

K. Ladanyi, 'Least Squares Calculation of Scattering with the Bethe–Salpeter Equation', *Nuovo Cim.* **61A** (1969) 173.

T. Lee, 'Simple Bound-State Model of the Physical Nucleon', *P. R.* **D6** (1972) 1110.

M. Levine, J. Wright and J. Tjon, 'Solution ofthe Bethe–Salpeter Equation in the Inelastic Region', *P. R.* **154** (1967) 1433.

M. Levine, J. Wright and J. Tjon, 'Effect of Self-Energy Terms in the Bethe–Salpeter Equation'.

A. Logunov and A. Tarkhelidze, 'Quasi-Optical Approach in Quantum Field Theory', *Nuovo Cim.* **29** (1963) 380.

Cz. Mattlioli, 'Ladder Series Convergence and the Bethe–Salpeter Equation', *Nuovo Cim.* **56A** (1968) 144.

C. Majumdar, 'Solution of Faddeev Equations for a One-Dimensional System', *J. M. P.* **13** (1972) 705.

S. Mandelstam, 'Dynamical variables in the Bethe–Salpeter Formalism, *Proc. Roy. Soc.* **A233** (1955) 248.

B. Mc Innis and C. Schwartz, 'Calculation of Scattering with the Bethe–Salpeter Equation', *P. R.* **177** (1969) 2621.

A. Migdal, 'Stability and vacuum and Limiting Field', *Soviet Phys. JETP* **34** (1972) 1184.

J. Mc Kee and P. Ralph (editors), *Three Body Problem in Nuclear and Particle Physics*', North-Holland Publ. Comp., Amsterdam 1970.

N. Nakanishi, 'A General Survey of the Theory of the Bethe–Salpeter Equation', *Progr. Theor. Phys. Suppl.* **43** (1969) 1.

N. Nakanishi, 'Vector-Scalar Sector Solutions to the Spinor Bethe–Salpeter Equation', *J. M. P.* **12** (1971) 1578.

Y. Nambu and G. Jona-Lasinio, 'Dynamical Model of Elementary Particles Based on an Analogy with Superconductivity I', *P. R.* **122** (1961) 345.

Y. Nambu and G. Jona-Lasinio, 'Dynamical Model of Elementary Particles Based on an Analogy with Superconductivity', *P. R.* **124** (1961) 246.

A. Nandy, 'Bound State Problem in the Electrodynamics of Spin-Zero Particles', *P. R.* **D5** (1972) 1531.

K. Nishijimas, 'Formulation of Field Theories of Composite Particles', *P. R.* **III** (1958) 995.

J. Nuttal, 'Asymptotic Form of the Three-Particle Scattering Wave-function for Free Incident Particles', *J. M. P.* **12** (1971) 1896.

H. Pagels, 'Chiral Symmetry Realization in the Quark Model', *P. R. L.* **28** (1972) 1482.

D. Parry, 'Nucleon Resonance Production in High Energy pp Collisions', *Lett. Nuovo Cim.* **4** (1972) 267.

J. Passi, 'Compositeness and Vanishing of Renormalization Constants', *Nucl. Phys.* **B46** (1972) 497.

M. Parkovič, 'Relativistic versus Nonrelativistic Form Factors', *P. R.* **D4** (1971) 1724.

K. Rothe, 'Study of Bound-State Solutions to the Pion-Nucleon B–S Equation', *P. R.* **170** (1968) 1548.

E. Salpeter and H. Bethe, 'A Relativistic Equation for Bound-State Problems', *P. R.* **84** (1951) 1232.

E. Salpeter, 'Wave Functions in Momentum Space', *P. R.* **84** (1951) 1226.

C. Smith, 'A Relativistic Formulation of the Quark Model for Mesons', *Annals of Phys.* **53** (1969) 521.

D. Stojanov and A. Tarkhelidze, 'A Relativistic Three-Body Problem', *Phys. Letters* **13** (1964) 76.

M. Sundaresan and P. Watsan, 'Relativistic Quark Models Based on the Bethe–Salpeter Equation', *Annals of Phys.* **71** (1972) 443.

L. Susskind, 'Model of Self-Induced Strong Interactions', *P. R.* **165** (1968) 1535.

L. Susskind and G. Frye, 'Anomalous and Normal Singularities and the Infinite-Momentum Limit', *P. R.* **165** (1968) 1553.

H. Suura, 'Long-Range-Potential Model of Hadrons', *P. R.* **D6**, (1972) 3538.

C. Schwartz, 'Solution of a Bethe–Salpeter Equation', *P. R.* **137** (1965) B 717.

C. Schwartz and C. Zemach, 'Theory and Calculation of Scattering with the Bethe–Salpeter Equation', *P. R.* **141** (1966) 1454.

S. Schweber, 'The Bethe–Salpeter Equation in Nonrelativistic Quantum Mechanics', *Annals of Phys.* **20** (1962) 61.

A. Staruszkiewicz, 'Canonical Theory of the Two-Body Problem in the Classical Relativistic Electrodynamics', *Ann. Inst. H. Poincaré* **14A** (1971) 69.

E. Stueckelberg and G. Wanders, 'Particule élémentaire et particule composé', *Arch. des Sciences, Genève* **8** (1955) 71.

E. Stueckelberg and G. Wanders, 'Zur Deutung der relativistischen Wellenfunktionen', *H. P. A.* **28** (1955) 353.

L. Susskind, 'Dynamical Theory of Strong Interactions', *P. R.* **154**, (1967) 1411.

I. Schwinger, *Proc. Natl. Acad. Sci.* **37** (1951) 455.

F. Schroeck, 'A Note on Kato's Perturbation Theory', *J. M. P.* **14** (1973) 829.

K. Seto, '*O* (3,1) Harmonics and its Application to B–S Equation', *Progr. Theor. Phys.* **46** (1971).

E. Tirapegui, 'The NHZ Construction and the $Z = 0$ Condition for Compositeness', *Nucl. Phys.* **B34** (1971) 598.

B. Varga and S. Asks, 'Quasisecular Clothing Transformation for the 4-Model of Quantum Field Theory', *P. R.* **D6** (1972) 529.

C. van Winter, *Kgl. Danske Videnskab Selskab, Mat-Fys. Skrifter* **2** (1964).

S. Weidenberg, 'High-Energy Behavior in Quantum Field Theory', *P. R.* **118** (1960) 838.

S. Weidenberg, 'Quasi Particles and the Born Series', *P. R.* **131** (1963) 440.

S. Weidenberg, 'Systematic Solution of Multiparticle Scattering Problems', *P. R.* **133** (1964) B 232.

S. Weidenberg, 'Dynamics at Infinite Momentum', *P. R.* **150** (1966) 1313.

G. Wick, 'Properties of the Bethe–Salpeter Wave Functions', *P. R.* **96** (1954) 1124.

C. Woo, 'Compositeness and Light-Cone Singularities', *P. R.* **D6** (1972) 1127.

O. Yakubovsky, 'On the Integral Equations in the Theory of *N* Particle Scattering', *Sov. J. of Nucl. Phys.* **5** (1967) 937.

A. le Yaohac, L. Oliver and P. Pére, *Nucl. Phys.* **B37** (1972) 552.

W. Zimmerman, 'Some Consequences of Infinite Man Renormalization', *C. M. P.* **8** (1968) 66.

E. Zur Linden and H. Mitter, 'Bound-State Solutions of the B–S Equation in Momentum Space', *Nuovo Cimento* **61B** (1969) 389.

LA CATEGORIE DES CROCS ET LEUR INTERPRETATION

C. PIRON

Section de Physique, Université de Genève, Genève, Suisse

Voici le plan de cet exposé:

1) Pourquoi est-il utile d'introduire la notion de catégorie en physique? Deux exemples en guise de contre-exemples.

2) La catégorie des crocs, exemples, quelques propriétés importantes.

3) L'interprétation physique : Les systèmes de propositions, les observables et les symétries.

1

Le language des catégories a été introduit en mathématique dans le but de décrire les objets de la géométrie en termes indépendant des coordonnées choisies. Pour illustrer l'utilité de cette notion en physique nous allons donner deux exemples de description de systèmes physiques en termes apparament habituelles. L'interprétation quelque peu surprenante que nous en donnerons prouvera à l'évidence l'utilité du language catégorique appliqué aux objets physiques.

Modèle I. On considère l'espace complexe hilbertien $L^2(R^2)$ des fonctions de carré sommable sur R^2 et l'équation de 'Schrödinger':

$$i\partial_t \psi_t = H\psi_t, \qquad \psi_t \in L^2(R^2)$$

avec H l''Hamiltonien' :

$$H = i(x_1 \partial_{x_2} - x_2 \partial_{x_1}).$$

Modèle II. On considère l'espace 'de phase' à deux dimensions:

$$-1 < \eta < +1, \qquad 0 < \varphi < 2\pi$$

(pour $\eta = \pm 1$, on identifiera les différentes valeurs de φ) et les équations 'canoniques' :

$$d_t \eta = -\partial_\varphi H,$$
$$d_t \varphi = +\partial_\eta H$$

avec l'‘Hamiltonien’:

$$H = B_1(t)\sqrt{1-\eta^2}\cos\varphi + B_2(t)\sqrt{1-\eta^2}\sin\varphi + B_3(t)\eta.$$

Le modèle I décrit un oscillateur harmonique classique. Les observables sont données dans ce modèle par des opérateurs multiplicatifs de la forme:

$$\psi(x_1, x_2) \mapsto f(x_1, x_2)\psi(x_1, x_2).$$

En particulier l'énergie est définie par

$$E = f(x_1, x_2) = \tfrac{1}{2}(x_1{}^2 + x_2{}^2).$$

Ainsi l'‘Hamiltonien’ H n'appartient pas à l'algèbre des observables et, bien que le système soit classique, son spectre est discret. De plus $\psi_t(x_1, x_2)$ représente un mélange (et $|\psi_t(x_1, x_2)|$ représente le même mélange!). Les états physiques (états purs) ne sont pas représentables dans $L^2(R^2)$, ce seraient des mesures de Dirac. [1]

Le modèle II décrit un système quantique. Un spin 1/2 dans un champ magnétique $\vec{B}(t)$. A chaque point (η, φ) correspond dans le formalisme habituel un vecteur état de C^2

$$(\eta, \varphi) \mapsto \frac{1}{\sqrt{2}} \begin{bmatrix} e^{-i\varphi/2}(1+\eta)^{1/2} \\ e^{+i\varphi/2}(1-\eta)^{1/2} \end{bmatrix}.$$

Les observables sont des fonctions définies sur les couples de points (η, φ) et $(-\eta, \varphi \pm \pi)$ de l'espace ‘de phase’. Ici aussi l'‘Hamiltonien’ H n'est pas une observable du système.

2. La catégorie des crocs

Pour définir une catégorie, on se donne les objets et les morphismes.

a) *Les objets*

Un croc $\mathscr{L}$ est un ensemble muni de deux structures:

i) Une structure d'ordre partiel définie par la relation:

$$a < b$$

qui fait de $\mathscr{L}$ un treillis complet :

Pour toute famille de $a_i \in \mathscr{L}$, $i \in J$ il existe une borne supérieure $\bigvee_J a_i \in \mathscr{L}$:

$$a_i < x, \quad \forall\, i \in J \Leftrightarrow \bigvee_J a_i < x$$

et une borne inférieure $\bigwedge_J a_i \in \mathscr{L}$:

$$x < a_i, \quad \forall\, i \in J \Leftrightarrow x < \bigwedge_J a_i.$$

ii) Une permutation de $\mathscr{L}$: $a \to a'$ appelée orthocomplémentation qui fait de $\mathscr{L}$ un treillis orthomodulaire:

1°) $a'' = a, \ \forall a \in \mathscr{L}$,

2°) $a < b \Rightarrow b' \vee (b \wedge a') = a'$,

3°) $a \vee a' = I$ et $a \wedge a' = 0, \ \forall a \in \mathscr{L}$.

I et 0 étant respectivement l'élément maximal et minimal de $\mathscr{L}$.

b) *Les morphismes*

Un morphisme $\mu : \mathscr{L}_1 \to \mathscr{L}_2$ est une application d'un croc $\mathscr{L}_1$ dans un croc $\mathscr{L}_2$ telle que:

1°) $\mu(\bigvee_J a_i) = \bigvee_J \mu a_i$

pour toute famille de $a_i \in \mathscr{L}_1, i \in J$

2°) $a < b' \Rightarrow \mu a < (\mu b)'$.

On démontre alors qu'un tel morphisme satisfait en plus les relations suivantes :

3°) $\mu 0_1 = 0_2$.

0_1 et 0_2 étant respectivement l'élément minimal de $\mathscr{L}_1$ et de $\mathscr{L}_2$

4°) $\mu(\bigwedge_J a_i) = \bigwedge_J \mu a_i$,

5°) $\mu(a') = (\mu a)' \wedge \mu I$.

Exemples. L'ensemble $P(\Gamma)$ des parties d'un ensemble Γ est un croc pour la relation d'ordre donnée par l'inclusion et l'orthocomplémentation donnée par le passage au complément. Soient deux ensembles Γ_1 et Γ_2 et une application f d'une partie de Γ_2 dans Γ_1 alors l'image inverse f^{-1} est un morphisme de $P(\Gamma_1)$ dans $P(\Gamma_2)$. Réciproquement tout morphisme de $P(\Gamma_1)$ dans $P(\Gamma_2)$ est de cette forme [4].

L'ensemble $P(H)$ des sous-espaces fermés d'un espace hilbertien H est un croc pour la relation d'ordre donnée par l'inclusion et l'orthocomplémentation donnée par le passage au sous-espace orthogonal. Si d'une manière générale nous désignons par P_a le projecteur orthogonal associe au sous-espace $a \in P(H)$ il est facile de vérifier les propriétés suivantes :

1°) $a < b \Leftrightarrow P_a P_b = P_a$,

2°) $P_{a'} = I - P_a$,

3°) $P_{a \wedge b} = \underset{n \to \infty}{\text{s-lim}} (P_a P_b)^n$.

De même l'ensemble $P(A)$ des sous-espaces associés aux projecteurs d'une algèbre de von Neumann A définie sur H est un croc et de plus l'injection naturelle de $P(A)$ dans $P(H)$ est un morphisme [7], [8].

Une transformation unitaire (ou antiunitaire) de l'espace hilbertien H définit un automorphisme sur le croc $P(H)$ et, réciproquement, tout automorphisme de $P(H)$ peut être défini par une transformation unitaire (éventuellement antiunitaire) de H [12].

Ayant discuté ces quelques exemples reprenons l'exposé général et avant tout donnons quelques définitions.

DÉFINITION. Un croc $\mathscr{L}$ est dit *boolien* (une algèbre boolienne complète) si l'une des deux conditions équivalentes suivantes est réalisée :

1°) $x \wedge (y \vee z) = (x \wedge y) \vee (x \wedge z), \quad \forall x, y, z \in \mathscr{L}$,

2°) $x \vee (y \wedge z) = (x \vee y) \wedge (x \vee z), \quad \forall x, y, z \in \mathscr{L}$.

DÉFINITION. Soit $\mathscr{L}$ un croc, nous dirons que a et $b \in \mathscr{L}$ sont *compatibles* et nous noterons $a \leftrightarrow b$ si l'une des quatre conditions équivalentes suivantes est réalisée [9] :

1°) Le sous-croc engendré par a, b, a' et b' est boolien,

2°) $(a \wedge b) \vee (a \wedge b') \vee (a' \wedge b) \vee (a' \wedge b') = I$,

3°) $(a \vee b) \wedge (a \vee b') = a$,

4°) $(a \vee b') \wedge b < a$.

DÉFINITION. On appelle *centre d'un croc* $\mathscr{L}$, le sous-croc boolien formé des éléments de $\mathscr{L}$ compatibles avec tous les autres.

Soit $\mathscr{L}$ un croc, le segment $[0, a]$, ensemble des $x \in \mathscr{L}$ tels que $0 < x < a \in \mathscr{L}$, est un croc pour l'orthocomplémentation (relative):

$$x \mapsto x' \wedge a.$$

De plus l'injection naturelle de $[0, a]$ dans $\mathscr{L}$ est un morphisme. C'est cette propriété qui justifie le nom de croc abréviation de canoniquement relativement orthocomplémenté.

Soit $\mu:\mathscr{L}_1 \to \mathscr{L}_2$ un morphisme, alors $\ker\mu$, c'est-à-dire l'ensemble des $x_1 \in \mathscr{L}_1$ tels que $\mu x_1 = 0_2$, est un segment $[0, z_1]$. Il est alors facile de montrer que z_1 l'élément maximal de $\ker\mu$ est dans le centre de $\mathscr{L}$. En effet, si $z_1 \in \ker\mu$ alors $(z_1 \vee a_1) \wedge a_1 \in \ker\mu \forall a_1 \in \mathscr{L}_1$ et si z_1 est maximal alors :

$$(z_1 \vee a_1) \wedge a_1' < z_1$$

c'est-à-dire $z_1 \leftrightarrow a_1$, $\forall a_1 \in \mathscr{L}$ [2].

La catégorie des crocs possède un élément nul : le croc $P(\phi)$ des parties de l'ensemble vide. Elle possède également un produit appelé l'union direct: Soit $\mathscr{L}_\alpha$, $\alpha \in \Omega$ une famille de crocs, l'union direct des $\mathscr{L}_\alpha$ est le croc $\bigvee_{\Omega} \mathscr{L}_\alpha$ obtenu de la façon suivante: on munit le produit cartésien des $\mathscr{L}_\alpha$ de la relation d'ordre:

$$\{x_\alpha\} < \{y_\alpha\} \Leftrightarrow x_\alpha < y_\alpha, \quad \forall \alpha \in \Omega$$

et de l'orthocomplémentation:

$$\{x_\alpha\} \to \{x_\alpha\}' = \{x_\alpha'\}.$$

On montre alors que le centre de $\bigvee_{\Omega} \mathscr{L}_\alpha$ est l'union directe des centres des $\mathscr{L}_\alpha$.

Un croc est irréductible s'il n'est d'aucune manière union directe de deux crocs différents de l'élément nul $P(\phi)$. Il en est ainsi si et seulement si le centre de $\mathscr{L}$ ne contient que 0 et I.

3. L'interprétation physique

Avant de construire les crocs qui interviennent en physique, nous voulons définir les questions et les propositions.

DÉFINITION. On appelle question toute expérience conduisant à une alternative dont les termes sont 'oui' et 'non' ('réussie' et 'ratée'). Pratiquement pour définir une question il faut se donner le type de système physique considéré, le ou les appareils, leur mode d'emplois et enfin l'interprétation du résultat.

Si α est une question on peut définir une nouvelle question notée $\alpha^{\sim}$, en échangeant de rôle du 'oui' et du 'non'.

Si $\{\alpha_i\}$ est un ensemble de questions se rapportant toutes au même type de système, on peut définir une nouvelle question notée $\prod_i \alpha_i$ par la prescription suivante:

On choisi une des α_i arbitrairement, on exécute l'expérience correspondante, et on attribue à $\prod_i \alpha_i$ la réponse ainsi obtenue. Il existe une question triviale qui consiste à ne rien mesurer du tout (éventuellement en cassant tout) en énonçant chaque fois la réponse 'oui'.

Quand un système physique donné a été préparé de telle manière que le physicien puisse affirmer avec certitude que l'expérience éventuelle correspondant à la question α conduirait à la réponse 'oui', on dira que α est vraie pour ce système particulier.

DÉFINITION. On dit qu'une question α est plus forte qu'une question β et on note $\alpha < \beta$, si chaque fois que α est vraie on peut affirmer que β est vraie.

Remarque. $\alpha < \beta$ n'est pas une tautologie, c'est l'énoncé d'une loi physique relative au type de système physique considéré.
Une telle relation est transitive, c'est donc une relation de préordre qui induit une relation d'équivalence:

$$\alpha \sim \beta \Leftrightarrow \alpha < \beta \quad \text{et} \quad \beta < \alpha .$$

DÉFINITION. On appelle proposition et on note a la classe d'équivalence contenant la question α.

Une proposition a est dite vraie pour un système donné si l'une des questions $\alpha \in a$ est vraie. Ce fait correspond à une propriété du système physique.

Pour définir le croc $\mathscr{L}$ des propositions associées à un type de système physique donné on procède de la façon suivante: Sur l'ensemble $\mathscr{L}$ des propositions on définit:

1°) Une relation d'ordre partielle:

$$a < b \Leftrightarrow \alpha < \beta \quad \text{pour} \quad \alpha \in a \text{ et } \beta \in b .$$

THÉORÈME. Muni de cette relation $\mathscr{L}$ est un treillis complet.

DÉMONSTRATION. Soit $\{b_i\}$ une famille quelconque de propositions de $\mathscr{L}$ et choisissons un représentant β_i pour chaque b_i. La classe d'équivalence contenant $\prod_i \beta_i$ définit une nouvelle proposition $\bigwedge_i b_i$ qui est la borne inférieure des b_i car $\prod_i \beta_i$ est vraie si et seulement si chacune des

β_i est vraie. L'existence d'une borne supérieure découle immédiatement de l'existence d'une borne inférieure, car on peut poser:

$$\bigvee_i b_i = \bigwedge_m x$$

où m désigne l'ensemble des x qui majorent les b_i:

$$m = \{x | x \in \mathscr{L} \text{ et } b_i < x, \ \forall i\}.$$

2°) Une orthocomplémentation. Pour cela on introduit tout d'abord la notion de complément compatible.

DÉFINITION. Deux propositions a et b sont des *compléments compatibles* l'une pour l'autre:

i) si ce sont des compléments:

$$a \vee b = I, \quad a \wedge b = 0 \quad \text{et,}$$

ii) s'il existe une question α telle que:

$$\alpha \in a \quad \text{et} \quad \alpha^{\sim} \in b.$$

Puis on postule les deux axiomes suivants [11]:

AXIOME C. Pour chaque proposition $a \in \mathscr{L}$ il existe au moins un complément compatible noté a'.

AXIOME P. Si $a < b$ sont des propositions de $\mathscr{L}$ et si a' et b' sont respectivement des compléments compatibles pour a et b alors, le sous-treillis engendré par $\{a, b, a', b'\}$ est distributif.

De ces deux axiomes on déduit l'existence et l'unicité du complément compatible. L'application $a \mapsto a'$ est alors une orthocomplémentation qui fait du treillis $\mathscr{L}$ un croc.

Réciproquement on a le théorème suivant:

THÉORÈME. Si dans un croc $\mathscr{L}$ on interprète l'orthocomplément comme un complément compatible alors les axiomes C et P sont satisfaits.

Il est important de remarquer que les axiomes C et P sont valables aussi bien en physique classique qu'en physique quantique. Ainsi, dans le cas classique le croc des propositions est $P(\Gamma)$ où Γ est l'espace de phase. C'est un croc boolien et ceci caractérise le cas classique. Inversement en mécanique quantique le croc des propositions est $P(H)$ où H est l'espace d'Hilbert $L^2(R^3)$. C'est un croc non boolien et de plus irréductible. Pour comprendre le rôle joué par la distributivité on peut encore faire la remarque suivante : La borne inférieure joue ici le même rôle que le 'et' en logique formelle en effet, $a \wedge b$ est vraie si et seulement si '*a* est vraie' et '*b* est vraie'. Par contre la borne supérieure ne joue pas le rôle du 'ou' car on a seulement:

$$\text{`}a \text{ vraie' ou `}b\text{ vraie'} \Rightarrow \text{`}a \vee b \text{ vraie'}.$$

En fait, on a le théorème suivant:

THÉORÈME. Si pour toute paire de propositions a et $b \in \mathscr{L}$ on a:

$$\text{`}a \vee b \text{ vraie'} \Rightarrow \text{`}a \text{ vraie' ou `}b \text{ vraie'}$$

alors $\mathscr{L}$ est distributif.

L'état d'un système particulier donné est défini par l'ensemble de toutes ces propriétés, c'est-à-dire l'ensemble de toutes les propositions vraies pour ce système. En théorie classique comme en mécanique quantique un tel ensemble est identique avec l'ensemble des propositions moins fortes qu'un atome donné, c'est-à-dire une proposition $p \neq 0$ telle que:

$$x < p \Rightarrow x = 0 \quad \text{ou} \quad x = p.$$

Ainsi les états (purs) du système sont en correspondance biunivoque avec les atomes du croc. Or, par définition, si une proposition a est différente de 0 il doit être possible d'exhiber un système pour lequel '*a* est vraie'. Cette remarque justifie l'axiome suivant:

AXIOME A_1. Le croc des propositions d'un système physique est atomique: pour toute proposition $a \neq 0$ il existe un atome $p < a$.

Enfin l'étude de l'effet d'une 'mesure idéale' sur le système [11] permet de justifier un dernier axiome:

AXIOME A_2. Si $a \neq 0$ et p un atome tel que $p \wedge a' = 0$ alors $(p \vee a') \wedge a$ est aussi un atome.

DÉFINITION. Un croc satisfaitant les axiomes A_1 et A_2 est appelé un *système de propositions.*

Rappelons quelques résultats [9], [11]. Le centre d'un système de propositions est lui-même un système de propositions, il est isomorphe au croc boolien des parties d'un ensemble. Dans le cas classique le centre est le système de propositions tout entier ainsi se trouve justifié l'espace de phase en physique classique. Un système de propositions irréductible est essentiellement le croc $P(H)$ des sous-espaces fermés d'un espace hilbertien [1]. Dans le cas général, où le centre n'est ni $\{0, I\}$ ni le système de propositions tout entier, $\mathscr{L}$ est l'union directe de systèmes de propositions irréductibles c'est-à-dire, essentiellement de la forme $\bigvee_{\Omega} P(H_\alpha)$ où les H_α, $\alpha \in \Omega$ sont des espaces hilbertiens.

En résumé un système de propositions est susceptible de décrire aussi bien un système classique qu'un système quantique.

Une observable est une correspondance entre les propositions définies par l'échelle de l'appareil de mesure et certaines propositions du système. C'est cette correspondance qui nous permet d'affirmer une propriété du système en vu d'une propriété de l'appareil de mesure, d'où la définition mathématique suivante [10]:

DÉFINITION. On appelle *observable* tout morphisme

$$\mu: \mathscr{B} \to \mathscr{L}$$

d'un croc boolien B dans le système de propositions $\mathscr{L}$. En général on impose en plus la condition $\mu I = I$.

Les observables μ_i: $\mathscr{B}_i \to \mathscr{L}$ sont dites compatibles entre elles si l'une des trois conditions équivalentes suivantes est réalisée:

1°) Il existe une observable μ et des morphismes π_i tels que le diagramme suivant est commutatif:

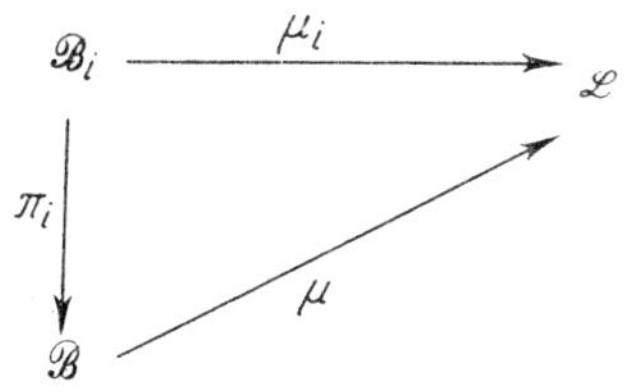

2°) Le sous-croc engendré par les images des μ_i est boolien

3°) Chaque paire de propositions de l'union ensembliste des images des μ_i est une paire compatible.

De telles définitions recouvrent bien les notions habituelles d'observables et d'observables compatibles. En effet, on peut montrer que:

i) Dans le cas classique: $\mathscr{L} = P(\Gamma)$, chaque observable peut être définie comme l'image inverse d'une fonction sur Γ et, inversement chaque fonction définit une observable. Dans ce cas toutes les observables sont compatibles entre elles.

ii) Dans le cas de la mécanique quantique: $\mathscr{L} = P(H)$ avec H un espace hilbertien séparable, à chaque observable correspond un opérateur autoadjoint dont la famile spectrale définit le morphisme et, inversement à chaque opérateur autoadjoint correspond une famille spectrale qui elle-même définit une observable. Deux observables sont compatibles si et seulement si les opérateurs correspondants commutent fortement.

Il est important de remarquer que pour traiter la dynamique du système, on doit, même dans le cas classique, faire jouer un rôle privilégié à un sous-ensemble particulier de l'ensemble des observables (par exemple les fonctions C^K). A plus forte raison, dans le cas quantique, il faut faire le choix d'un sous-ensemble d'observables privilégiées. Nous ne voulons pas discuter ici ce problème qui est l'un des buts de l'exposé de M. Flato [4].

Pour terminer, discutons très brièvement le concept de symétrie [3].

DÉFINITION. On appelle symétrie du système tout automorphisme du système de propositions correspondant.

Si le système de propositions est le croc $\mathscr{L} = \bigvee_{\Omega} P(H_\alpha)$ où pour chaque $\alpha \in \Omega$, H_α est un espace hilbertien sur les complexes, alors les symétries de $\mathscr{L}$ sont données par le théorème suivant qui est une généralisation de celui de E. Wigner:

THÉORÈME. Tout automorphisme de $\bigvee_{\Omega} P(H_\alpha)$ est induit par une permutation des points de Ω:

$$f : \Omega \to \Omega$$

et une famille $\{U_\alpha\}$ d'opérateurs unitaires ou antiunitaires:

$$U_\alpha : H_\alpha \to H_{f\alpha}.$$

Note

[1] Ce formalisme hilbertien est fort utile en théorie de la diffusion classique: voir par exemple: P. D. Làx and R. S. Phillips [6].

Bibliographie

[1] I. Amemiya and H. Araki, *Publ. Research Inst. Math. Science Kyoto University* **A2** (1967) 423.

[2] J. Dixmier, *Les algèbres d'opérateurs dans l'espace hilbertien*, 2ème éd., Gauthier-Villars, Paris 1969, p. 42, corollaire 3.

[3] G. Emch and C. Piron, *J. Math. Phys.* **4** (1963) 469.
[4] M. Flato, dans ce livre, 231.
[5] P. R. Halmos, *Lectures on Boolean Algebra*, D. Van Nostrand Co., Princeton 1963, p. 70, Theorem 5.
[6] P. D. Làx and R. S. Phillips, *Scattering Theory*, Academic Press, New York 1967.
[7] L. H. Loomis, *Mem. Amer. Math. Soc.* **18** (1955).
[8] S. Maeda, *J. of Sc. of Hiroshima Univ.* **A19** (1955) 211.
[9] C. Piron, *Helvetica Phys. Acta* **37** (1964) 439.
[10] C. Piron, *Foundations of Quantum Mechanics IL corso*, Academic Press Inc., New York 1971, p. 274.
[11] C. Piron, *Found of Phys.* **2** (1972) 287.
[12] E. Wigner, *Group Theory*, Academic Press, New York 1959.

BROKEN GAUGE AND SCALE INVARIANCE

L. O'RAIFEARTAIGH

Dublin Institute for Advanced Studies, Dublin, Ireland

Introduction

In this talk we shall be considering some important developments which have taken place recently [7], [9], [18], [27], [30], [31] in two related fields, namely, unified gauge theory (UGT) and renormalization group theory (RGT). I feel somewhat embarrassed at being a rapporteur for these results as I have been only marginally engaged in these fields, but it appears that nobody else will be covering them at this meeting ("Since the wise men have not spoken ..."[1]), and in any case it is no harm to have them reviewed by an outsider. The task of covering these developments has been eased by the appearance of some excellent review articles [7], [9], [18], [27], [30], [31] in the proceedings of last September's Aix-en-Provence Conference and by some recent articles of Weinberg [31] and Collins and Macfarlane [6], and finally by the help of some of my colleagues.[2]

The general idea in both UGT and RGT is to start from a limit in which there are no (or only trivial) masses, exploit the symmetry of that limit, and then introduce the masses as a form of symmetry breaking. This approach is not as obvious and simple as it might appear, because even (or especially) in the symmetric limit there are infrared and ultra-violet divergences, and so the process of taking and then breaking the symmetric limit is quite delicate. Most recent progress has been in this delicate area.

It is sometimes said that one of the characteristic features of a good theory is that it can be summarized on the back of an envelope, i.e. that there are certain key equations, such as Einstein's equations $G_{\mu\nu} = \varkappa T_{\mu\nu}$ or the Schrödinger equation $H = \frac{1}{2}\nabla^2 + V$, which contain the essentials of the theory, the rest being merely the interpretation and development of these equations. In the short time at my disposal, the best I shall be able to do is give the Einstein–Schrödinger equations of UGT and

RGT, together with a little discussion to make them meaningful. Hopefully even the discussion will not go far beyond what can be written on the back of an envelope!

Unified Gauge Theory

The starting point for UGT is that in the symmetric limit the Lagrangian is invariant under a gauge group. For example, for vector mesons, fermions and scalars in interaction, a typical symmetric limit Lagrangian would be

$$L = -\tfrac{1}{4}\operatorname{tr} F^2 + \tfrac{1}{2}(D_\mu\phi)^2 + \bar{\psi}(iD - m_0)\psi + g\bar{\psi}\phi\psi + V(\phi) \tag{1}$$

where

$$\begin{aligned} F = F_{\mu\nu} &= A_{\mu,\nu} - A_{\nu,\mu} + i[A_\mu, A_\nu], \qquad D_\mu\psi = (\partial_\mu - iA_\mu)\psi, \\ A_\mu &= eA^a_\mu I_a, \qquad \phi = \phi^a I_a, \qquad D_\mu\phi = \partial_\mu\phi - i[A_\mu, \phi], \end{aligned} \tag{2}$$

e and g being coupling constants (matrices in the reducible case) and I_a are the generators of a compact Lie group G. The Lagrangian (1) is invariant under the gauge transformation

$$\begin{aligned} A_\mu &\to SA_\mu S^{-1} - iS_{,\mu}S^{-1}, \qquad \phi \to S\phi S^{-1}, \qquad \psi \to S\psi, \\ S &= e^{i\Lambda}, \qquad \Lambda = \Lambda^a(x) I_a, \end{aligned} \tag{3}$$

provided only that m_0 and $V(\phi)$ are chosen to be gauge-invariant. It is well-known that such theories are renormalizable, so that the symmetric limit has no ultraviolet problems (though it has infrared ones).

The question now is how to proceed from the symmetric limit to the physically realistic case in which the A-fields have non-zero masses and the ψ-fields a mass-matrix M which is *not* gauge invariant (and so the fermion current is not conserved). It is well-known that if we simply insert ad hoc values for the meson-masses and for M then (for non-abelian G) the theory becomes non-renormalizable (so at best we trade the infrared divergences for ultraviolet ones!). The key discovery of UGT is that the renormalizability (and unitarity) can be preserved if we introduce the masses in a more delicate way. The way to introduce the masses according to UGT is simply to assume that the ϕ are not conventional fields with $\langle\phi\rangle_0 = 0$, but rather satisfy

$$(4) \qquad \langle\phi\rangle_0 = \frac{M}{g}$$

where M is any required fermion mass matrix.

(Equation (4) is the Einstein or Heisenberg–Schrödinger equation of UGT!) If one then defines the conventional fields $\phi' = \phi - M/g$, rewrites (1) in terms of ϕ' and then drops the primes, one sees by inspection that the mesons acquire the *induced* mass-matrix

$$(5) \qquad \tfrac{1}{2}\operatorname{tr}[M, A]^2$$

In addition, they acquire a new *induced* interaction

$$(6) \qquad \operatorname{tr}[D_\mu\phi][M, A_\mu],$$

which, as we shall see, is crucial for renormalization. Note particularly, that the parameters in (5) and (6) are completely determined by e, g and M. It is this relationship between the parameters in (5) and (6) (especially the meson masses in (5)) and (e, g, M) that constitutes the *unification* of the theory (over and above any unification one might insert by hand in the choice of G, e, g and M).

The real glory of UGT, however, is not the unification which is clearly quite simple, but rather the renormalizability and unitarity. The point is that vector-meson propagators take the form $P_{\mu\nu} = \left(g_{\mu\nu} - (\alpha-1)\dfrac{k_\mu k_\nu}{k^2 - \alpha\mu^2}\right)\Big/(k^2 - \mu^2)$ where α is a gauge parameter, and such propagators for $\alpha \neq \infty$ have poles of negative residue, and for $\alpha = \infty$ lead to divergences which increase with order. At first sight UGT would appear to have not only this problem but also the problem of (unobserved) mass-less Goldstone [11], [12] bosons, because the Ansatz (4) destroys the gauge-invariance of the vacuum. It turns out, however, that in UGT the Goldstone and other difficulties just cancel (like in Tarzan stories, where, attacked to the left by elephants and to the right by buffalo, Tarzan leaps into a tree and lets the herds fight it out!)

To obtain a formed proof that the difficulties cancel, one uses the gauge-invariance of the Lagrangian (which survives the Ansatz (4)) to construct the following two gauges:

1) *Unitary Gauge*. In this gauge S is fixed so that the Goldstone bosons vanish. With S fixed, the propagators for the A-fields come out to be $P_{\mu\nu} = (g_{\mu\nu} - k_\nu k|\mu^2)/(k^2 - \mu^2)$ i.e. the manifestly unitary but not manifestly renormalizable ones ($\alpha = \infty$).
2) *Renormalization gauge*, $\mathscr{L} = \mathscr{L}_R$. In this gauge S is chosen so that the vector meson propagators are $(g_{\mu\nu} - k_\mu k_\nu|k^2)/(k^2 - \mu^2)$ and the theory is manifestly renormalizable. But the Goldstone bosons are not zero and it is not manifestly unitary.

Finally, one uses [25] the equivalence (modulo Faddeev–Popov [8] terms etc.) of the two gauges, i.e.

(7) $$S = S(\mathscr{L}_R) = S(\mathscr{L}_\mu)$$

where S is the S-matrix, to argue that since $S(\mathscr{L}_R)$ is renormalizable and $S(\mathscr{L}_\mu)$ unitary, S is both renormalizable and unitary.

Of course, it is one thing to argue formally in this way, and quite another to check that in practice, e.g. in a Feynman diagram calculation, everything is really as it should be. However, it turns out that for Feynman diagrams everything really *is* as it should be [16], [18], [20], [24]. Furthermore, a close examination shows that what makes $S(\mathscr{L}_\mu)$ renormalizable, and $S(\mathscr{L}_R)$ unitary, in spite of their not being manifestly so, is precisely the induced vector-meson meson interaction (6). In each order of perturbation this term comes to the rescue in restoring the renormalizability of $\mathscr{L}_\mu$ and the unitarity of $\mathscr{L}_R$ respectively.

Finally, we must come down out of the ivory tower and consider applications. First, UGT provides us with a completely new class of renormalizable theories, namely non-abelian ones, and in view of the small number of renormalizable theories available, this in itself is no mean achievement. However, the main practical application of UGT is to weak and electromagnetic interactions, from which, indeed, UGT originated. In those theories the A-fields are identified with the (massless) EM field and (massive) intermediate vector bosons of the weak interactions respectively, and the ψ-field with the leptons, whose masses definitely break the gauge symmetry. The relationship between the (e, g, M) algebra and the known leptons is such that the scheme must also include either (hitherto unobserved) neutral currents or leptons in the theory [31]. And indeed recent experiments[3] indicate the existence of neutral currents.

For EM interactions, perhaps the most interesting possibility opened up by UGT is that some hitherto infinite electromagnetic mass-differences, such as the proton–neutron mass-difference, become finite [31]. Finally, there are speculative attempts to extend UGT to the strong interactions, but the best I can do here is to refer you to Weinberg's preprint [31] for further information here also.

Renormalization Group Theory

For RGT we consider *any* renormalizable theory and begin with the limit in which there are no masses whatsoever. Now if all coupling constants g were scale invariant and there were no infrared or ultraviolet divergences, the scaling properties of the invariant functions (truncated Green's functions, Wilson coefficients [28] etc.) would be trivial, e.g.

$$G(\lambda p, g) = \lambda^d G(p, g), \tag{8}$$

where p represents the momenta, and d (an integer) is the natural dimension of G. However, renormalization modifies this simple picture and what RGT describes is the modification. The starting point is the equation connecting the bare and renormalized Green's functions, namely,

$$\Gamma(p, g, \mu) = Z(g_0, \Lambda, \mu)\Gamma_0(p, g_0, \Lambda), \tag{9a}$$

$$\operatorname*{Lt}_{\Lambda=\alpha} \mu \frac{\partial}{\partial \mu} g(g_0, \Lambda, \mu) = \beta(g), \tag{9b}$$

$$\operatorname*{Lt}_{\Lambda=\alpha} \mu \frac{\partial}{\partial \mu} \operatorname{Log} Z(g_0, \Lambda, \mu) = \gamma(g) \tag{9c}$$

where Λ is the cut-off, Z is the renormalization constant and subscript zero denotes bare quantities. It is important to note that the indicated limits exist only *because of the renormalizability*. [Note also that the derived quantities are re-expressed as functions of g before the limit is taken.] The renormalization point in p-space is taken outside the physical region for reasons of analyticity, and does not appear in $\beta(g)$ and $\gamma(g)$. Eq. (9a) is invariant [1] under the renormalization group [1] of the g_0's and Z's. As an example we consider QED, where, on account of the Ward identity, the only contributions to $g \equiv e$ come from the graphs of Fig. 2, and hence we have the simple explicit expansion [6]

$$(10) \qquad e(e_0, n, \mu) = e_0 + \frac{2e_0^3}{(2\pi)^{n/2}} \sqrt{\frac{4-n}{2}} \int_0^1 xdx(1-x)f(x) + O(e_0^5),$$

$$f(x) = [m^2 - q^2x(1-x)]^{\frac{n-4}{2}} \qquad \text{for } q^2 = -\mu^2, m^2 = 0.$$

Here we have used dimensional regularization, so that $\Lambda \to \infty$ is equivalent to $n \to 4$, and we have taken the renormalization point $q^2 = -\mu^2$ to be spacelike to avoid the cut in the integrand starting at zero. From (10) we see at once that

$$(11) \qquad \beta(e) = \frac{e^3}{12\pi^2} + O(e^5).$$

The crux of RGT is to express equations (9) for $\Gamma(p, g, \mu)$ in a form *which contains no bare quantities*. This is done in two steps. First, from (9) one obtains the equivalent differential equation (Gell-Mann–Low equation [10])

$$(12) \qquad \left[\mu \frac{\partial}{\partial \mu} + \beta(g) \frac{\partial}{\partial g} + d - \gamma(g)\right] \Gamma(p, g, \mu) = 0,$$

for $\Gamma(p, g, \mu)$. Then one solves this equation [3] to obtain the relation

$$(13) \qquad \Gamma(\lambda p, g, \mu) = \lambda^{d-\delta(\lambda)} \Gamma(p, g(\lambda), \mu)$$

where

$$(14) \qquad \begin{aligned} &\lambda \frac{d}{d\lambda} g(\lambda) = \beta(g(\lambda)), \qquad g(1) = g, \\ &\lambda \frac{d}{d\lambda} \delta(\lambda) = \gamma(g(\lambda)), \qquad \delta(1) = 0. \end{aligned}$$

Equations (13), (14) are the 'Einstein or Schrödinger equations' of RGT. Their content is clear. They contain the modifications to the simple scaling law (8) which are induced by the renormalization process. This consists of the replacement of d and g by the λ-dependent functions $\delta(\lambda)$ and $g(\lambda)$, whose form is determined by (14). Indeed, if we regard (13), (14) as the 'Schrödinger' equation of the theory, the quantity $\beta(g)$ plays the role of the 'potential' in that it determines the form of $g(\lambda)$ and hence $\delta(\lambda)$ and the structure of the theory. Thus the analysis of (13), (14) reduces to the analysis of the structure function $\beta(g)$ and of its effect on $g(\lambda)$ and $\delta(\lambda)$ through (14). To date, one has been interested only in making this analysis in the limiting cases $\lambda \to \infty$ (ultraviolet) and $\lambda \to 0$ (infrared) and then only when $g(\lambda)$ does not diverge in the limit. There are then two cases to consider, namely, $g(\lambda) \to g^* \neq 0$ and $g(\lambda) \to 0$.

Case I. $g(\lambda) \to g^* \neq 0$. From (14) we see that $\beta(g(\lambda)) \to 0$, and since there is no reason to assume otherwise, one makes the simplest assumption, namely,

$$(15) \qquad \beta(g(\lambda)) \sim b(g(\lambda)-g), \qquad \delta(\lambda) \sim \delta^*.$$

Inserting these Ansätze in (14) we obtain

$$(16) \qquad (g(\lambda)-g) \sim \lambda^{-|b|}, \qquad \Gamma(\lambda p, g, \mu) \sim \lambda^{d-\delta^*}\Gamma(p, g, \mu).$$

From the second equation we see that in this case we have scaling with *anomalous dimension* δ^*.

Case II. $g(\lambda) \to 0$. In this case the theory is said to be *asymptotically free*. It is the most interesting case, since $g(\lambda) \to 0$ means that *perturbation theory becomes valid* near the limit. In particular, we can check for consistency using perturbation theory, since from perturbation theory and the definitions (9) we have

$$(17) \qquad \beta(g) \sim bg^3, \qquad \gamma(g) \sim cg^2.$$

Inserting the first equation in (17) into (13) we find that for consistency we should have

$$(18) \qquad \underset{\text{(UV-stability)}}{b < 0 \text{ for } \lambda \to \infty}, \qquad \underset{\text{(IR-stability)}}{b > 0 \text{ for } \lambda \to 0}$$

and then

$$(19) \qquad g(\lambda) \sim \frac{1}{\sqrt{|b| \operatorname{Log} \lambda}}, \qquad \delta(\lambda) \sim \frac{1}{(\operatorname{Log} \lambda)^c}.$$

Thus $g(\lambda) \to 0$ rather slowly, and we have scaling only up to log corrections.

The crucial observation is that since $\beta(g)$ reduces to bg^3, the asymptotic structure of the theory is completely determined by b, *which is numerical constant* (*or more generally, a matrix*) *that can be calculated from one-loop diagrams* (we have already calculated it for QED in (11)). The question now is: for which theories is $b \gtrless 0$? It turns out [4], [5] that all conventional abelian-field theories have $b > 0$ (like QED) and hence are IR but not UV stable. The most interesting result, however, is that massless non-abelian gauge fields, in self-interaction, or in interaction with non-exotic fermions, have $b < 0$, and hence are UV stable [13], [22]. The interest of this result is tempered, however, by the fact that it no longer holds when scalar interactions are introduced [14], and as we have seen in UGT, it is precisely the introduction of scalars that provides the masses for such theories. So for the moment the result $b > 0$ for non-abelian fields, and with it the possibility of UV asymptotic freedom, remains in some what of a limbo.

Before leaving the massless limit, I should perhaps mention an important spin-off, namely conformal invariance. One sees either from the Lagrangian [5] or Dyson–Schwinger equations [19] that the scale-invariance of a field theory implies conformal invariance, and this fact has been imaginatively exploited. In particular, it has been shown that conformal invariance determines the 2 and 3 point functions [23] and so reduces a theory to its 'skeleton'. The skeleton theory has been attacked using harmonic analyses over the conformal group [19], [21], [23].

We come now to the legitimacy of the massless limit and the problem of introducing masses. A heuristic argument for the legitimacy of the limit is based on estimates for the behaviour of the massive Green's functions in the Euclidean region (Weinberg's theorem [29]). A more systematic approach is provided by the Callan–Symanzik equation [2], [26], [27]

$$(20)\quad \left[\mu\frac{\partial}{\partial\mu}+\beta(g)\frac{\partial}{\partial g}+d\gamma(g)\right]\Gamma^{(n)}(p, g, \mu)$$
$$= \Gamma^{(n+1)}(p_0 = 0, p, g, \mu),$$

(n = number of external particles) which is derived from a Ward–Takahashi identity and PCAC relation $\partial^{\mu}S_{\mu}^{1} = \theta$ for the dilation current $S_{\mu} = T_{\mu\nu}\chi^{\nu}$. This equation provides the GML equation with a right hand side which gives an estimate for the error in approaching the limit $m \to 0$ and provides the input for $m \neq 0$.

A more recent approach [6], due independently to 't Hooft [15], Weinberg [31] and Jegerlehner and Shroer [17], has been to calculate the renormalization constants Z at $m = 0$, even when $m \neq 0$ in the Green's functions, and to treat the masses in the Green's functions in the same way as the coupling constants. In this approach one obtains the modified scaling equation

$$\Gamma(\lambda p, g, m, \mu) = \lambda^{d-\delta(\lambda)}\Gamma(p, g(\lambda), m(\lambda), \mu), \tag{21}$$

where the *effective* masses $m(\lambda)$ satisfy the differential equation

$$\lambda \frac{d}{d\lambda} m(\lambda) = -[1+\mu(g(\lambda))]m(\lambda), \qquad m(1) = m, \tag{22}$$

analogous to (14). A simple analysis of these equations shows that for both $\lambda \to 0, \infty$, $\theta(\mu|g(\lambda)) \nrightarrow \infty$, we have

$$m(\lambda) \to \frac{\lambda^{-1}}{\lambda^{\mu}}, \qquad m(\lambda) \to \frac{\lambda^{-1}}{(\text{Log}\,\lambda)^{\mu}} \tag{23}$$

in cases I ($g(\lambda) \to g^*$) and II ($g(\lambda) \to 0$) respectively. In particular (for $\mu|(g(\lambda)) > -1$) $m(\lambda) \to \infty$ and $m(\lambda) \to 0$ in the IR and UV limits respectively. In the latter case, if one assumes that the Green's functions can be expanded in powers of $m(\lambda)$, one obtains the corrections to the massless limit result as an expansion in $m(\lambda)$. These corrections are particularly useful if, on account of some symmetry property, the Green's functions vanish in the limit.

References

[1] After coming here, I find that George Sudarshan also is covering UGT. However, his treatment – on the applications of UGT to weak interactions – is orthogonal to mine.

[2] R. Acharya, Z. Horvàth, G. B. Mainland and S. Sen.

[3] CERN Gargamelle Report (1973). See also Barut's remarks following Sudarshan's talk.

Bibliography

[1] N. Bogoliubov and D. Shirkov, *Theory of Quantized Fields*, Interscience, New York 1959.
[2] C. Callan, *Phys. Rev.* **D2** (1970) 1541.
[3] S. Coleman, *Erice Lecture Notes*, 1971.
[4] S. Coleman and D. Gross, *Phys. Rev. Letters* **31** (1973) 851.
[5] S. Coleman and R. Jackiw, *Ann. Phys.* **67** (1971) 552.
[6] J. C. Collins and A. J. Macfarlane, *Phys. Rev.* **D10** (1974) 1201.

[7] R. Crewther, *Proceedings of Second International Conference on Elementary Particles*, Aix-en-Provence, 1973; Suppl. *Journal de Physique* **34**, C1 (1973).
[8] L. Feddeev and V. Popov, *Physics Lettres* **25B** (1967) 29.
[9] R. Gatto, *Proceedings of the Second International Conference on Elementary Particles*, Aix-en-Provence, 1973; Suppl. *Journal de Physique* **34**, C1 (1973).
[10] M. Gell-Man and F. Low, *Phys. Rev.* **95** (1954) 1300.
[11] J. Goldstone, *Nuovo Cim.* **19** (1961) 154.
[12] J. Goldstone, A. Salam and S. Weinberg, *Phys. Rev.* **127** (1962) 965.
[13] D. Gross and F. Wilczek, *Phys. Rev. Letters* **30** (1973) 1343.
[14] D. Gross and F. Wilczek, *Phys. Rev.* **D8** (1973) 3633.
[15] G. 't Hooft, *Nucl. Phys.* **B61** (1973) 445.
[16] G. 't Hooft, *Nucl. Phys.* **B35** (1971) 167.
[17] F. Jegerlehner and B. Schroer, Schladming Lectures, 1973.
[18] B. W. Lee and J. Zinn-Justin, *Phys. Rev.* **D5** (1972) 3155.
[19] G. Mack, *Proceedings of Second International Conference on Elementary Particles*, Aix-en-Provence, 1973; Suppl. *Journal de Physique* **34**, C1 (1973).
[20] G. B. Mainland and L. O'Raifeartaigh, *Phys. Rev.* **D12** (1975) 489.
[21] A. Migdal, *Phys. Letters* **37B** (1971) 98, 386.
[22] H. Politzer, *Phys. Rev. Letters* **30** (1973) 1346.
[23] A. Polyakov, *JEPT Letters* **12** (1970) 381.
[24] D. Ross and J. C. Taylor, *Nucl. Phys.* **B51** (1973) 125; **B5**, 8 (1973) 643; **B5** (1973) 23.
[25] A. Salam and J. Strathdee, *Nuovo Cim.* **11A** (1972) 397.
[26] K. Symanzik, *Comm. Math. Phys.* **18** (1970) 227.
[27] K. Symanzik, *Proceedings of Second International Conference on Elementary Particles*, Aix-en-Provence, 1973; Suppl. *Journal de Physique* **34**, C1 (1973).
[28] K. Wilson, *Phys. Rev.* **179** (1969) 1499.
[29] S. Weinberg, *Phys. Rev.* **118** (1960) 838.
[30] S. Weinberg, *Proceedings of Second International Conference on Elementary Particles*, Aix-en-Provence, 1973; Suppl. *Journal de Physique* **34**, C1 (1973).
[31] S. Weinberg, *Phys. Rev.* **D8** (1973) 3497.

UNITARY UNIFIED FIELD THEORIES*

E. C. G. SUDARSHAN

Center for Particle Theory, University of Texas, Austin, Tx. 78712, U.S.A.

Abstract

This is an informal exposition of some recent developments. Starting with an examination of the universality of electromagnetic and weak interactions we outline the attempts at their unification. The theory of unitary renormalizable self-coupled vector mesons with dynamical sources is formulated for a general group. With masses introduced as variable parameters it is shown that the theory so defined is indeed unitary. Diagrammatic rules are developed in terms of a chosen set of fictitious particles· A number of special examples are outlined including a theory with strongly interacting vector and axialvector mesons and weak vector mesons. Applications to weak interactions of strange particles is briefly outlined.

I. Introduction

Despite the multiplicity of particle species there has been the continuing search for regularity and groupings of fundamental particles and for fundamental conservation laws. We distinguish broadly three families of particles [11]: (1) The leptons, including the electron, the electronic neutrinos and their antiparticles constituting the electronic subfamily; and the muon, the muonic neutrino and their antiparticles constituting the muonic subfamily. (2) The photon constituting a family by itself. (3) The hadrons consisting of the nucleons and the strange baryons and their antiparticles, constituting the baryon subfamily; and the mesons. The question of which resonance states are to be considered as particles is not immediately clear. But it appears definitely clear that all other entities like nuclei and atoms may be considered as being composed of these.

For completeness one should perhaps include the graviton in a separate family for itself.

There are a number conservation laws associated with particle reactions in addition to the traditional conservation laws of energy and momentum. The conservation of electric charge and the conservation of baryon numbers are two of the most familiar and best established.

The two lepton numbers of the two subfamilies also seem to be strictly conserved. In addition to these there are a number of approximate non-Abelian symmetry groups which seem to serve to generate the pattern of particle multiplets. To the degree that we can assume invariance under these non-Abelian groups and to the extent to which we know the manner of breaking of these symmetries we can find interrelationships between physical quantities.

These interrelationships fall short of a satisfactory approach to the dynamics of particle interactions. We need to have a stronger and more precise framework incorporating the mechanism of interactions. The action principle with an associated Lagrangian density in terms of a few primitive fields is a standard framework for this purpose. While at various times people have been exasperated with the many failures of specific Lagrangian structures and specific methods of calculation to the extent to declare field theory to be dead, Lagrangian field theory seems to have regular reincarnations. Much of the problem lies in the difficulty of performing reliable calculations and the richness of the phenomena. It often turned out to be convenient and prudent, if not entirely satisfactory, to deal with limited sets of particles or interactions. Thus we have theories which apply only to strong interactions, or only to the weak interactions of nonstrange particles, or only to the mutual interactions of leptons. Even if we are able to calculate reliably for a small group of processes, we would like to have a formulation of the interaction structure of all species of particles.

A. THE NATURE OF THE UNIVERSALITY OF ELECTROMAGNETIC AND WEAK INTERACTIONS

The electromagnetic interactions involve all the different families of particles and seem to possess several general features. The interaction of the electromagnetic field with all charged particles seems to be governed by the same numerical coupling constant (or multiples of it); and as far as the charged leptons are concerned, the minimal (Dirac) interaction with this coupling seems to be able to yield quantitative explanation of their electromagnetic properties when radiative corrections are taken into account. It is then a reasonable postulate to say that this must be true of the elementary hadrons also; the observed differences coming in from the presence of strong interactions or possibly from the observed hadrons not being elementary entities, or both reasons. In any case the long range Coulomb interaction tells us that the charges of the hadrons and of the leptons are the same. Thus electromagnetism is universal in that it is possibly governed by the same coupling constant; and it couples to both the hadron and the lepton families by an interaction of the form

$$(1.1) \quad e\bar{\psi}\gamma^{\mu}\psi A_{\mu}$$

for all spinor fields.

Unlike spin 1/2 particles we seem to have no stable elementary charged particles with spin 0 or 1. With spin 0 particles however there is only one unique interaction structure; the coupling constant for these particles has the same value. Thus the electromagnetic interaction with charged spin 0 fields is completely determined to have the form

$$(1.2) \quad ei(\varphi^* \partial^{\mu}\varphi - \partial^{\mu}\varphi^*\varphi)A_{\mu}.$$

But for spin 1 particles we have a certain amount of freedom in coupling, corresponding to an anomalous moment. If we had elementary spin 1 charged particles in mesic atoms we would have had direct access to the information. Lacking this we have to obtain this indirectly. We shall see below that in place of a minimal interaction a particular anomalous magnetic moment interaction would have to be included in the formulation of the electromagnetism of charged spin 1 fields so that the general interaction would have the form

$$(1.3) \quad ei(V^{\alpha *}\partial^{\mu}V_{\alpha} - \partial^{\mu}V^{\alpha *}V_{\alpha})A_{\mu} + ei\varkappa V^{\alpha *}V^{\beta}(\partial_{\alpha}A_{\beta} - \partial_{\beta}A_{\alpha}).$$

Similar considerations apply to weak interactions. The lepton as well as the hadrons participate in weak interactions and with the same 'universal' form and coupling constant. Being a richer phenomenon involving a variety of particles the precise formulation of the universality is a difficult task but the numerical agreement of the coupling strengths and the similarity of the beta decay and muon decay interactions is striking leading to the *V–A* chiral coupling structure.

$$(1.4) \quad \frac{G}{\sqrt{2}}\bar{A}\gamma^{\mu}(1+\gamma_5)B \cdot \bar{C}\gamma_{\mu}(1+\gamma_5)D.$$

The leptonic weak interactions of the strange hadrons have a somewhat more complicated form. By reason of the numerical agreement of the coupling strengths we tend to include the nonleptonic strangeness violating decays among weak interactions though we shall not be able to write down any such explicit interaction structures at the present time.

Even within this domain of universal interactions we must note that the *universality* is in the interaction *structure* including the coupling constant and not in the total Lagrangian density. Different charged fields correspond to different masses. They may even possess other couplings which are different. But in the electromagnetic *interaction* term they are all involved in the same manner.

A similar situation was already encountered [15] in the context of chiral four-fermian interactions in which the muon and the electron subfamilies enter on the same footing despite the pronounced mass difference between the muon and the electron. And there is good reason to believe that the same interaction structure is equally valid for the nucleons. Thus our search for universality of interaction structures should not be confused with a search for symmetries of the system: the invariances are broken symmetries, the breaking coming from mass differences in the Lagrangian density.

In searching for universal interaction structures we have to distinguish between primary interactions and effective interactions. The effect of strong interactions alters the effective interaction of a nucleon with the electromagnetic field and its weak interactions. Even for the electron there are nonlinear effects due to its coupling with the electromagnetic field that exhibits itself as radiative corrections. These modifications depend not only on the primary interactions themselves but also on the masses and other dynamical attributes. The radiative corrections to the muon, the electron and to the proton are all different. It is to the *primary interactions* that we must address ourselves and they are the ones that we expect to have *universal* features.

B. UNIFICATION OF ELECTROMAGNETISM AND WEAK INTERACTIONS

Hadrons, by their very nature, do have strong interactions and therefore it would be harder to discern the primary interactions for them. But we have recognition of their participation in electromagnetic and weak interactions with the leptons. We see that both electromagnetic and weak interactions are vectorial in nature, and therefore one is led to look for vector interactions. Since the photon field is a vector field we may look for vector fields associated with weak interactions [10]. If such vector fields are involved, they must be associated with very large mass 'weak bosons' since the four-fermion interactions appear as if they are local interactions. These heavy bosons would also serve to produce the degree of *reduction* in the observed strength of the effective four-fermion interaction in comparison with the effective mutual repulsion of the two electrons. We can attempt to choose the mass of the heavy bosons in such a fashion as to make the photon and the weak heavy bosons the components of a triplet and thus 'unify' weak and electromagnetic interactions.

In creating such a unified theory we must recognize also the differences between the weak and electromagnetic interactions. While the weak interactions are chiral and parity nonconserving, the electromagnetic interactions are vector parity conserving. Secondly, the photon mass appears to be precisely zero and the source of this field is strictly conserved. But neither is true for the weak boson fields. All these suggest that the weak bosons and the electromagnetic field must have other fields also as partners if only the observed leptons are included [16]. The precise number of additional fields is dependent on the specific theory we construct.

This attempt at unification pays other dividends as well: the four-fermion interactions produced infinities characteristic of Lagrangian field theories and these could not be as easily circumvented as in the case of quantum electrodynamics by renormalization. It is desirable to have a renormalizable theory of weak interactions and this is one of the dividends of a unified theory. However, this involves a more systematic study of vector meson theories since the theory of minimally coupled charged vector mesons is not renormalizable. There are also questions of whether vector mesons coupled to nonconserved sources is renormalizable.

A popular method of obtaining renormalizability is to employ a highly symmetric Lagrangian density which involves the coupling of a set of (four) vector bosons which are selfcoupled and which are also coupled to leptonic and hadronic sources. Almost all the masses are assumed to be zero in the Lagrangian density and the 'observed' masses coming about by a spontaneous violation of the symmetry [12].[1] The hope is that such an intrinsic mechanism for generating masses retains the renormalizability that is obtained in the theory of zero mass conserved current theories. One difficulty that obtains for non-Abelian selfcoupled vector mesons is that if the theory is to be renormalizable by conventional methods we must include some fictitious particles in the intermediate states [2], [3]: these particles have fantastic properties like being spin 0 fermions; and they enter only the virtual states. The requirement that spontaneous violations arise requires the existence of scalar mesons in the theory [5], [6], [4]. Their presence leads to further complications.

Having generated unified weak and electromagnetic interactions for leptons, we may ask ourselves two questions: (i) Among hadrons, when we have spin 1 charged particles, either we must assume that they are not participating in primary interactions or that the electromagnetic interactions are renormalizable. But if we want to have a theory involving strongly interacting primary vector bosons some of which are charged, renormalizability of the theory again comes into question. How should we generate the electromagnetic and weak interactions of these primary strong bosons so as to preserve universality on the one hand and renormalizability (and unitarity) on the other? (ii) How can we generate the nonleptonic weak decays of hadrons within such a scheme?

Most of the investigations pertaining to unified theories of weak and electromagnetic interactions start from a Yang–Mills field [19] coupled to scalar fields with spontaneously broken symmetry. In this presentation, we shall follow an alternate path and start with selfcoupled vector mesons with Yang–Mills type highly symmetric interaction, but there are mass terms [7] which are arbitrary except that the mass matrix commutes with electric charge. We will consider successive modifications of this model and then compare it with spontaneously broken symmetry models of similar appearance.

Needless to say a genuinely unified theory should presuppose an understanding of both the existence of the various particle families and the three different kinds of interaction.

II. Theory of Self-Coupled Vector Mesons

Let us choose a group $\mathscr{G}$ with structure constants Γ_{abc} which contains a charge-like generator Q which is such that the electric charge matrix my be written as

$$\hat{Q} = Q + Q_0 \tag{2.1}$$

where Q_0 is invariant under $\mathscr{G}$. Let us consider a set of vector fields W^α_a which transforms as the adjoint representation of $\mathscr{G}$ and a vector field W^α_0 which transforms as a singlet under $\mathscr{G}$. Then consider the Lagrangian density

$$\begin{aligned} \mathscr{L} = \mathscr{L}_l + \mathscr{L}_W = \mathscr{L}_l &- \tfrac{1}{4}\{\partial^\alpha W^\beta_a - \partial^\beta W^\alpha_a - ig\Gamma_{abc} W^\alpha_b W^\beta_c\}^2 - \\ &- \tfrac{1}{4}\{\partial^\alpha W^\beta_0 - \partial^\beta W^\alpha_0\}^2 + \tfrac{1}{2} M_{00} W^\alpha_0 W_{0\alpha} + \\ &+ \tfrac{1}{2} M_{ab} W^\alpha_a W_{b\alpha} + M_{Q0} W^\alpha_Q W_{0\alpha} + \\ &+ g W^\alpha_a L_{a\alpha} + g_0 W^\alpha_0 L_{0\alpha} + \\ &+ \chi_a\{\partial_\alpha W^\alpha_a - ig\Gamma_{aQb} W_{Q\alpha} W^\alpha_b\} + \chi_0 \partial_\alpha W^\alpha_0 + \\ &+ \tfrac{1}{2} n_{ab}\chi_a\chi_b + n_{Q0}\chi_Q\chi_0 + \tfrac{1}{2} n_{00}\chi_0^2. \end{aligned} \tag{2.2}$$

The summation convention over repeated indices is understood. In this expression χ_a and χ_0 are scalar Lagrange multiplier fields transforming respectively as the adjoint representation and the identity representation of $\mathscr{G}$. The coupling constants g and g_0 are chosen real; the 'mass' matrices M are nonnegative and hermitian; the coefficients M_{00}, n_{00}, M_{Q0} and n_{Q0} are chosen real and nonnegative. The lepton Lagrangian density $\mathscr{L}_l$ need not be specified at the present time except to guarantee certain equations of motion for the leptonic currents $L_{a\alpha}$ and $L_{0\alpha}$. Then the variational equations of motion give:

$$\partial_\alpha W^\alpha_a = ig\Gamma_{aQb} W_{Q\alpha} W^\alpha_b - n_{ab}\chi_b - \delta_{aQ} n_{Q0}\chi_0, \tag{2.3}$$

$$\partial_\alpha W^\alpha_0 = -n_{00}\chi_0 + n_{Q0}\chi_Q, \tag{2.4}$$

$$(2.5)\quad \partial^\beta\{\partial_\alpha W_{a\beta}-\partial_\beta W_{a\alpha}-ig\Gamma_{abe}W_{b\alpha}W_{c\beta}\}$$
$$= ig\Gamma_{abc}W_b^\beta\{\partial_\alpha W_{c\beta}-\partial_\beta W_{c\alpha}-ig\Gamma_{cef}W_{e\alpha}W_{f\beta}\}+$$
$$+M_{ab}W_{b\alpha}+\delta_{aQ}M_{Q0}W_{0\alpha}-gL_a-$$
$$-\partial_\alpha\chi_a-ig\Gamma_{Qab}W_{Q\alpha}\chi_b-\delta_{aQ}\Gamma_{Qbc}W_{b\alpha}\chi_c,$$

$$(2.6)\quad \partial^\beta\{\partial_\alpha W_{0\beta}-\partial_\beta W_{0\alpha}\} = -\partial_\alpha\chi_0+M_{00}W_{0\alpha}+M_{Q0}W_{Q\alpha}-g_0L_0\ .$$

From these equations of motion we can deduce the equations of motion for the Lagrange multiplier fields:

$$(2.7)\quad \Box\chi_a+(M_{ab}n_{bc}-M_{Q0}n_{Q0}\delta_{aQ}\delta_{cQ})\chi_c+$$
$$+(n_{00}M_{Q0}\delta_{aQ}-n_{Q0}M_{aQ})\chi_0$$
$$= igM_{ab}\Gamma_{bQc}W_Q^\alpha W_{c\alpha}-ig\Gamma_{abc}M_{cd}W_b^\alpha W_{d\alpha}+$$
$$+igM_{Q0}\Gamma_{abQ}W_b W_{0\alpha}+ig\Gamma_{bQa}n_{Qc}\chi_b\chi_c+$$
$$+ig\delta_{aQ}\Gamma_{cQb}n_{bd}\chi_c\chi_c-ig\Gamma_{bQa}W_{Q\alpha}\partial^\alpha\chi_b+$$
$$+ig\Gamma_{abc}W_{b\alpha}\partial^\alpha\chi_c-ig\delta_{aQ}\Gamma_{cQb}W_{b\alpha}\partial^\alpha\chi_c+$$
$$+g^2\delta_{aQ}\Gamma_{cQb}\Gamma_{bQd}\chi_c W_Q^\alpha W_{d\alpha}-$$
$$-g^2\Gamma_{abc}\Gamma_{dQc}\chi_d W_Q^\alpha W_{b\alpha}-g^2\Gamma_{abQ}\Gamma_{dQe}\chi_d W_{b\alpha}W_{e\alpha}-$$
$$-g\partial^\alpha L_{a\alpha}-ig^2\Gamma_{abc}\Gamma_{b\alpha}W_c^\alpha,$$

$$(2.8)\quad \Box\chi_0+(M_{00}n_{Q0}+M_{Q0}n_{Q0})\chi_0+(M_{00}n_{Q0}\delta_{bQ}+M_{Q0}n_{Qb})\chi_b$$
$$= -g_0\partial^\alpha L_{0\alpha}.$$

In deriving these results we have made repeated use of the antisymmetry of Γ_{abc} in its indices and the Jacobi identity they satisfy. The requirement that the mass matrices do not connect components with different charge labels imply that the quantities $\Gamma_{Qbe}n_{bf}$ and $\Gamma_{Qbe}M_{bf}$ are antisymmetric in e, f

$$(2.9)\quad \Gamma_{Qbe}M_{bf} = -\Gamma_{Qbf}M_{be},\qquad \Gamma_{Qbe}n_{bf} = -\Gamma_{Qbf}n_{be}.$$

We shall also assume that

$$(2.10)\quad M_{Qb} = M_{QQ}\delta_{Qb},\qquad n_{Qb} = n_{QQ}\delta_{Qb}.$$

Under these conditions we can get some simplification in the equation for χ_Q. We get

$$(2.11)\quad \Box\chi_Q+(M_{QQ}n_{QQ}-M_{Q0}n_{Q0})\chi_Q+n_{00}M_{Q0}\chi_0$$
$$= ig(i\partial^\alpha L_{Q\alpha}-g\Gamma_{Qbc}L_{b\alpha}W_c^\alpha).$$

We choose the lepton Lagrangian density $\mathscr{L}_l$

(2.12) $$\mathscr{L}_l = \bar{L}(i\hat{\partial} - g\tau_a \hat{W}_a - f g_0 \hat{W}_0)L + \bar{R}(i\hat{\partial} - g_0 \hat{q}\hat{W}_0)R - (\bar{R}kL + \bar{L}kR)$$

where L and R are the positive and negative chiral components of lepton fields furnishing a representation of $\mathscr{G}$. We have introduced a new coupling parameter f. The matrices τ_a are the matrix realizations of the group $\mathscr{G}$ on the leptons, k the mass matrix and $\hat{q}$ the matrix of electric charges. We choose f such that

(2.13) $$\tau_Q + f = \hat{q}.$$

Thus f is the average charge of the lepton multiplet, possibly fractional. It follows that

(2.14) $$i\partial^\alpha L_{0\alpha} = \bar{R}\tau_Q kL - \bar{L}k\tau_Q R.$$

(2.15) $$i\partial^\alpha L_{a\alpha} - g\Gamma_{abc} L_{b\alpha} W_c^\alpha = \bar{L}\tau_a kR - \bar{R}k\tau_a L.$$

Both these terms would vanish if the lepton masses were to be all zero, but we do not wish to make this drastic simplification.

The Lagrange multiplier fields χ_Q, χ_0 thus satisfy linear inhomogeneous equations. In the absence of lepton terms suitable linear combinations of them obey simple wave equations. The diagonalization of the effective mass matrix for the χ_Q, χ_0 fields becomes much simplified if we choose

(2.16) $$m_{Q0} = 0.$$

We shall adopt this choice in the sequel. The freedom to choose n_{ab} and n_{00} arbitrarily is a parameter freedom ('gauge' freedom) of the theory as we shall see below.

A. QUANTIZATION. UNITARIZED TRANSITION AMPLITUDES

Starting with the Lagrangian density (2.2) we can quantize the Yang–Mills fields W_a^α and the vector field W_0^α. Using equations (2.3) and (2.4) together with (2.9) and (2.16) the Lagrangian density can be rewritten by eliminating χ_a and χ_0 by virtue of the equations

$$\chi_0 = -(n_{00})^{-1}\partial_\alpha W_0^\alpha,$$
$$\chi_a = -(n)_{ab}^{-1}(\partial_\alpha W_b^\alpha - ig\Gamma_{bQc} W_{Q\alpha} W_c^\alpha)$$

in the form

$$(2.17)\quad \mathscr{L} = \mathscr{L}_{l} - \tfrac{1}{4}(\partial^{\alpha} W_{a}^{\beta} - \partial^{\beta} W_{a}^{\alpha} - ig\Gamma_{abc} W_{b}^{\alpha} W_{c}^{\beta})^{2} - \tfrac{4}{1}(\partial^{\alpha} W_{0}^{\beta} - \partial^{\beta} W_{0}^{\alpha})^{2} + \tfrac{1}{2} M_{ab} W_{a}^{\alpha} W_{b\alpha} + M_{Q0} W_{Q}^{\alpha} W_{0\alpha} + M_{Q0} W_{Q}^{\alpha} W_{0\alpha} + \tfrac{1}{2} M_{00} W_{0}^{\alpha} W_{0\alpha} + g W_{a}^{\alpha} L_{a\alpha} + g_{0} W_{0}^{\alpha} L_{0\alpha} - \tfrac{1}{2}(n_{00})^{-1}(\partial_{\alpha} W_{0}^{\alpha})^{2} - \tfrac{1}{2}(\partial_{\alpha} W_{a}^{\alpha} - ig\Gamma_{aQb} W_{Q\alpha} W_{b}^{\alpha})(n^{-1})_{ac}(\partial_{\beta} W_{c}^{\beta} - ig\Gamma_{cQd} W_{Q\beta} W_{\alpha}^{\beta}).$$

For almost all values of the parameters the fields W_{a}^{α} and W_{0}^{α} can be quantized using canonical methods. If we do, we get a positive metric spin 1 degree of freedom and a negative metric spin 0 degree of freedom provided M_{ab} is nonsingular and M_{00} is not zero. (We shall treat the singular cases separately later.) The quantized theory leads to a renormalizable S-matrix but the theory contains transitions to states of negative metric associated with the spin 0 part of the vector field. To restore unitarity we must remove this contribution. If we can arrange to obtain the equations of motion of the Lagrange multiplier fields from a new Lagrangian $\mathscr{L}_{\chi}$ which itself is renormalizable then we could remove these nonunitary contributions from the transition amplitude. We should take this unitarized amplitude as the physical amplitude. Accordingly, if we write down the amplitudes formally as integrals over paths

$$(2.18)\quad \mathscr{A} = \int \exp\left\{i\int d^{4}x(\mathscr{L}_{W}(x) + \mathscr{L}_{l}(x))\right\} d\{W, \chi, L, R\} = \int \exp\left\{i\int d^{4}x \mathscr{L}(x)\right\} d\{W, L, R\}$$

where $\mathscr{L}(x)$ is given by (2.17). If $\mathscr{L}_{\chi}$ is the effective Lagrangian from which (2.7) and (2.8) simplified by (2.14) and (2.15) are obtained, the contribution of the negative metric spin 0 part is given by the factor

$$(2.19)\quad X = \int \exp\left\{i\int d^{4}x \mathscr{L}_{\chi}(x)\right\} d\{\chi\}$$

with the understanding that no χ quanta are to be taken either in the initial or in the final states. Here X^{-2} is the formal determinant of the quadratic form in $\mathscr{L}_{\chi}$ in those cases where the equations of motion for χ are linear in χ. In the general case we still use the notation X for this amplitude. The unitarized scattering amplitude is then given by the expression

$$(2.20)\quad F = \int X^{-1} \exp\left\{i\int (\mathscr{L}(x) + \mathscr{L}_{l}(x)) d^{4}x\right\} d\{W, L, R\}.$$

This amplitude is explicitly renormalizable and it is expected to be unitary.

B. INTRODUCTION OF FICTITIOUS PARTICLES

The unitarizing factor X^{-1} can itself be often displayed as due to a set of fictitious particles [8]. This is a purely mathematical manipulation; and we find it convenient to deal with such fictitious particles for the various special cases studied.

The general theory is now developed once for all. But for practical calculations and direct verification of unitarity, etc., it is better to re-express the content of the theory in terms of rules for calculating amplitudes in perturbation theory.

From (2.7), (2.8), (2.14) and (2.15) we see that a linear combination of χ_Q and χ_0 obeys the free field equation with a suitable mass; by proper choice of parameters this mass may be made equal to zero. We could therefore choose this combination of χ_Q and χ_0 to be zero at will. The corresponding combination of χ_Q and χ_0 then contains no contributions from the negative metric spin 0 part. By proper choice of the mass matrices this combination of W_Q^α and W_0^α could be chosen so as to be mass eigenstates one of which corresponds to zero mass. This component is then identified with the photon field.

The normalization matrices n_{ab}, n_{Q0}, n_{00} are at our control: we have chosen to make them commute with the charge matrix Γ_{aQb}. Much freedom remains in the theory. The changes in these mass parameters correspond to changes in the masses of the χ field which may be identified as 'gauge transformations' in the original sense of the term introduced by Weyl [17], [18]. Since the effect of the χ quanta are removed by division by X these 'gauge' changes should leave physical amplitudes unaltered. Every one of our theories then contains such a gauge group. Special examples of such gauge groups are seen in the parameter independence of the physical transition amplitudes of the models.

III. Models

A. MASSIVE YANG–MILLS FIELD

As the simplest example we take the Yang–Mills field with finite mass [7]. In this case the group $\mathscr{G}$ is $SU(2)$ and the structure constants Γ_{abc} are simply $i\varepsilon_{abc}$. We shall simplify the model by ignoring the singlet and the fermion sources. Accordingly we start with the symmetric Lagrangian density

$$(3.1)\quad \mathscr{L} = -\tfrac{1}{4}(\partial_\mu \vec{W}_\nu - \partial_\nu \vec{W}_\mu + g\vec{W}_\mu \times \vec{W}_\nu)^2 + \tfrac{1}{2}M^2(\vec{W}_\mu)^2 + + \vec{\chi}\cdot\partial^\mu \vec{W}_\mu + \tfrac{1}{2}\xi(\vec{\chi})^2.$$

The equations of motion are

$$(3.2)\quad \partial_\alpha \vec{W}^\alpha = -\xi\vec{\chi},$$

$$(3.3)\quad \partial^\beta(\partial_\alpha \vec{W}_\beta - \partial_\beta \vec{W}_\alpha + g\vec{W}_\alpha \times \vec{W}_\beta) = -g\vec{W}_\beta \times (\partial_\alpha \vec{W}_\beta - \partial_\beta \vec{W}_\alpha + g\vec{W}_\alpha \times \vec{W}_\beta) + M\vec{W}_\alpha - \partial_\alpha \vec{\chi}.$$

From these we deduce

$$(3.4)\quad (\Box + M^2) = g\partial^\alpha(\vec{W}^\beta \times \{\partial_\alpha \vec{W}_\beta - \partial_\beta \vec{W}_\alpha + g\vec{W}_\alpha \times \vec{W}_\beta\} + \partial^\beta(\vec{W}_\alpha \times \times \vec{W}_\beta) = -g\vec{W}^\alpha \times \partial_\alpha \vec{\chi}.$$

Equation (3.4) can be derived from the equivalent Lagrangian with (3.2)

$$(3.5)\quad \mathscr{L}_\chi = -\tfrac{1}{2}\{\partial_\alpha \vec{\chi}\cdot\partial^\alpha \vec{\chi} - M^2\vec{\chi}\cdot\vec{\chi} - g\vec{\chi}\cdot W_\alpha \times \partial^\alpha \vec{\chi}\}.$$

The unitarized amplitude is then

$$(3.6)\quad \int X^{-1}\exp\left\{i\int d^4x \mathscr{L}(x)\right\} d\{\vec{W}^\alpha\}$$

where

$$(3.7)\quad \mathscr{L}(x) = -\tfrac{1}{4}(\partial_\mu \vec{W}_\nu - \partial_\nu \vec{W}_\mu + g\vec{W}_\mu \times \vec{W}_\nu)^2 + \tfrac{1}{2}M^2 \vec{W}_\mu \cdot \vec{W}^\mu - -\tfrac{1}{2}\xi^{-1}(\partial_\mu \vec{W}^\mu)^2$$

and X is given by

$$(3.8)\quad X = \int \exp\left\{i\int d^4x \mathscr{L}_\chi\right\} d\{\vec{\chi}\}.$$

We can think of X^{-1} as due to a formal Lagrangian density $\mathscr{L}_D$ due to a fictitious scalar particle $\vec{D}$ obeying Fermi statistics:

$$(3.9)\quad X^{-1} = \int \exp\left\{i\int d^4x \mathscr{L}_D\right\} d\{\vec{D}\},$$

$$(3.10)\quad \mathscr{L} \sim g\vec{D}\cdot\vec{W}_\alpha x \partial^\alpha \vec{D}.$$

The Feynman rules may be summarized as follows (We take $\xi = 1$ to obtain the simplest form for these rules. We note that if one takes the limit $\xi \to 0$ in (3.1), the theory is *explicitly unitary*):

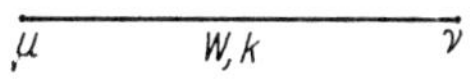

$$-ig_{\mu\nu}(k^2 - M^2)^{-1}$$

$$i(k^2 - M^2)^{-1}$$

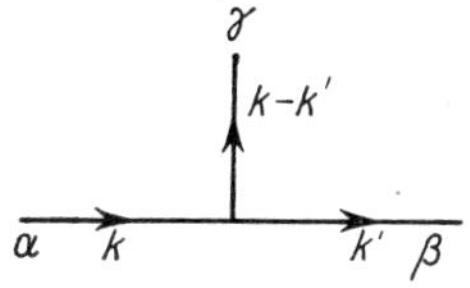

$$-g\varepsilon_{abc}\{(k+k')^{\gamma}g^{\alpha\beta} + (k-2k')^{\alpha}g^{\beta\gamma} + + (k'-2k)^{\beta}g^{\gamma\alpha}\}$$

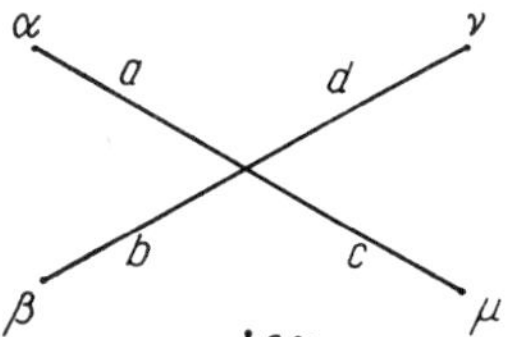

$$ig^2\{\delta_{ac}\,\delta_{bd}(g_{\alpha\nu}g_{\beta\mu} + g_{\alpha\beta}g_{\mu\nu} - 2g_{\alpha\mu}g_{\beta\nu}) + + \delta_{bc}\,\delta_{ad}(g_{\alpha\mu}g_{\beta\nu} + g_{\alpha\beta}g_{\mu\nu} - 2g_{\alpha\nu}g_{\beta\mu}) + + \delta_{ab}\,\delta_{cd}(g_{\alpha\mu}g_{\beta\nu} + g_{\alpha\beta}g_{\mu\nu} - 2g_{\alpha\nu}g_{\beta\mu})\}$$

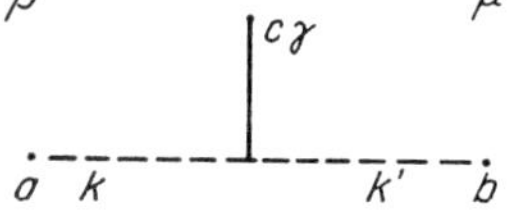

$$\tfrac{1}{2}g\varepsilon_{abc}(k+k')^{\gamma}$$

Direct calculation shows that in the one loop case (i.e., up to fourth order in the coupling constant) the theory is unitary. With proper attention being paid to the statistical weights of the diagrams (and the relevant signs for the fictitious particle loops) we get unitarity in the two loop diagrams also. Of course, this was to be expected from the general theory; and by virtue of the constraint equation in (3.5) we must count $\partial_\alpha \vec{W}^\alpha$ and $\vec{D}$ *as identical particles* in determining the statistical weights.

B. CHARGED BOSON-PHOTON SYSTEM

As a second model we consider the same group but break the mass degeneracy by requiring that the third component have zero mass and that the corresponding Lagrange multiplier field obeys the free wave equation. In this case it is more convenient to use the notation W_μ^+, W_μ^-, A_μ for the three components of the vector isovector field. We write the Lagrangian density [1]:

$$(3.11)\quad \mathscr{L}_W = -\tfrac{1}{2}|\partial_\mu W_\nu^+ - \partial_\nu W_\mu^+ + ie(W_\mu^+ A_\nu - W_\nu^+ A_\mu)|^2 + $$
$$+ M^2 W_\mu^+ W^{-\mu} - \tfrac{1}{4}(\partial_\mu A_\nu - \partial_\nu A_\mu - ie(W_\mu^+ W_\nu^- - W_\mu^- W_\nu^+)\}^2 +$$
$$+ \chi^+(\partial_\mu + ieA_\mu) W^{-\mu} + \chi^-(\partial_\mu - ieA_\mu) W^{+\mu} +$$
$$+ \zeta \partial_\mu A^\mu + \frac{\alpha}{2}\zeta^2 + \chi^+\chi^-.$$

This is invariant under gauge transformations of the second kind:

$$(3.12)\quad \begin{array}{lll} W_\mu^\pm \to W_\mu^\pm e^{\mp ie\Lambda}, & A_\mu \to A_\mu - \partial_\mu \Lambda, & \\ \chi^\pm \to \chi^\pm e^{\mp ie\Lambda}, & \zeta \to \zeta, & \Box \Lambda = 0. \end{array}$$

The Lagrange multiplier fields satisfy

$$(3.13)\quad (\partial_\mu \pm ieA_\mu)^2 \chi^\pm + M^2 \chi^\pm + e^2 W^{\pm\mu}(\chi^\pm W_\mu^\mp - \chi^\mp W_\mu^\pm) = 0,$$

$$(3.14)\quad \Box \zeta = 0.$$

These equations for $\chi^\pm$ are valid if and only if

$$(\partial_\mu \pm ieA_\mu) W^{\mp\mu} + \chi^\mp = 0 \text{ hold,}$$

and these could be derived from

$$(3.15)\quad \mathscr{L}_\chi = -(\partial_\mu - ieA_\mu)\chi^+ \cdot (\partial^\mu + ieA^\mu)\chi^- + M^2 \chi^+ \chi^- -$$
$$-\tfrac{1}{2}e^2(W_\mu^- \chi^+ - W_\mu^+ \chi^-)^2.$$

As before to get the unitary amplitude we compute the amplitude for the equivalent Lagrangian

$$(3.16)\quad \mathscr{L} = -\tfrac{1}{2}|\partial_\mu W_\nu^+ - \partial_\nu W_\mu^+ + ie(W_\mu^+ A_\nu - W_\nu^+ A_\mu)|^2 +$$
$$+ M^2 W_\mu^+ W^{-\mu} - \tfrac{1}{4}\{\partial_\mu A_\nu - \partial_\nu A_\mu - ie(W_\mu^+ W_\nu^- - W_\nu^+ W_\mu^-)\}^2 -$$
$$- |(\partial_\mu - ieA_\mu) W^{+\mu}|^2 - \frac{1}{2\alpha}(\partial_\mu A^\mu)^2$$

with the unitarizing factor

$$(3.17)\quad X = \left(\int \exp\left(i\int d^4x \mathscr{L}_\chi\right) d\{\chi\}\right)^{-1}.$$

Again X^{-1} can be rewritten in terms of fictitious fields. The Feynman rules are given below:

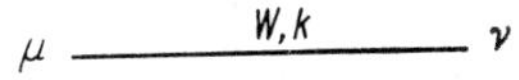

$$-ig_{\mu\nu}/(k^2-M^2)$$

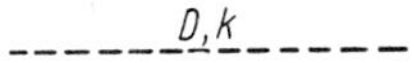

$$i/(k^2-M^2)$$

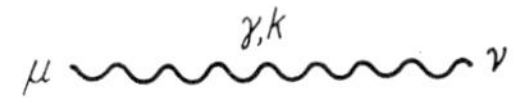

$$-i[{}_{\mu\nu}g-(1-\alpha)k_\mu k_\nu/k^2]/k^2$$

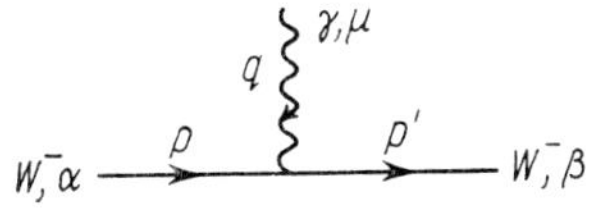

$$ie[(p+p')_\mu q_{\alpha\beta}+2q_\beta g_{\mu\alpha}-2q_\alpha g_{\mu\beta}]$$

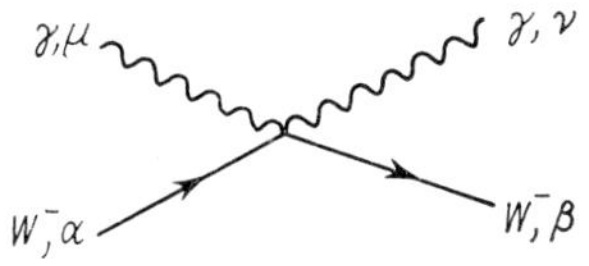

$$-2ie^2 g_{\mu\nu}g_{\alpha\beta}$$

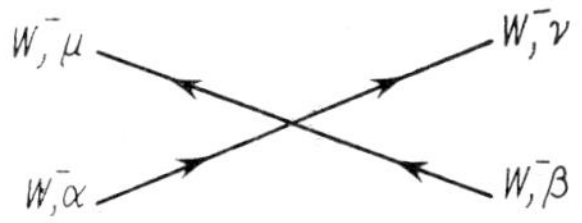

$$-ie^2(g_{\mu\alpha}g_{\nu\beta}+g_{\mu\beta}g_{\nu\alpha}-2g_{\mu\nu}g_{\alpha\beta})$$

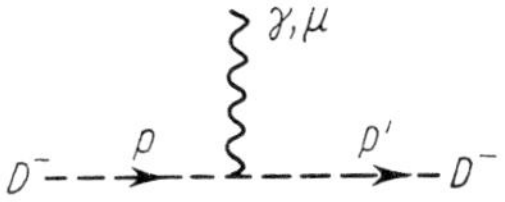

$$-ie(p+p')_\mu$$

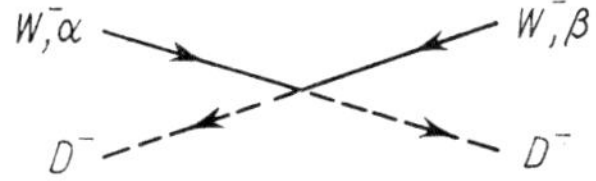

$$2ie^2 g_{\alpha\beta}$$

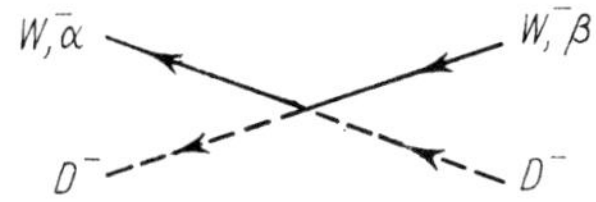

$$-ie^2 g_{\alpha\beta}$$

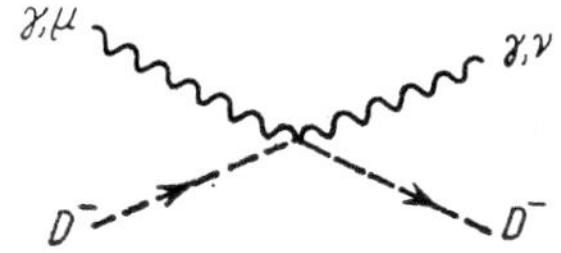

$$2ie^2 g_{\mu\nu}$$

This model has been checked for unitarity to fourth order [1]; and some specific processes to the sixth order (two loop level). Again we see that provided proper statistical weights are attached for the diagrams we recover unitarity.

C. UNIFIED MODEL OF LEPTONS IN INTERACTION

We now consider a simplified but realistic model of a physical system involving a triplet of vector mesons $\vec{W}_{\alpha'}$, a singlet $W_{0\alpha}$ and the leptons μ and ν_μ in interactions. For the Lagrangian density we write

$$\begin{aligned}
\mathscr{L} &= \mathscr{L}_{\text{I}}+\mathscr{L}_{\text{II}}+\mathscr{L}_{\text{III}},\\
\mathscr{L}_{\text{I}} &= -\tfrac{1}{4}(\partial_\alpha \vec{W}_\beta-\partial_\beta \vec{W}_\alpha+g\vec{W}_\alpha \times \vec{W}_\beta)^2-\tfrac{1}{4}(\partial_\alpha W_{0\beta}-\partial_\beta W_{0\alpha})^2+\\
&+\bar{R}(i\hat{\partial}+g_0\hat{B}_0)R+\bar{L}(i\hat{\partial}+\tfrac{1}{2}g\vec{\tau}\cdot\hat{\vec{W}}+\tfrac{1}{2}g_0\hat{B})L,\\
\mathscr{L}_{\text{II}} &= -m_\mu\bar{\mu}\mu-\tfrac{1}{2}M^2\{(W_1^\alpha)^2+(W_2^\alpha)^2\}-\tfrac{1}{2}M_{33}(W_3^\alpha)^2-\\
&-M_{30}W_3^\alpha W_{0\alpha}-\tfrac{1}{2}M_{00}(W_0)^2,\\
\mathscr{L}_{\text{III}} &= \chi_0\partial_\alpha W_0^\alpha+\chi_3\partial_\alpha W_3^\alpha+\tfrac{1}{2}n_{33}\chi_3^2+n_{30}\chi_3\chi_0+\tfrac{1}{2}n_{00}\chi_0^2+\\
&+\chi^-(\partial_\alpha-igW_{3\alpha})W^{+\alpha}+\chi^+(\partial_\alpha+igW_{3\alpha})W^{-\alpha}+\xi^{-1}\chi^+\chi^-.
\end{aligned} \tag{3.18}$$

We choose the matrices M, n such that we could diagonalize them both by choosing

$$\begin{aligned}
gW_3^\alpha &= e(A^\alpha-\cot\theta Z^\alpha),\\
g_0 W_0^\alpha &= e(A^\alpha+\tan\theta Z^\alpha),\\
e &= -(g^2+g_0^2)^{1/2}\cos\theta\sin\theta = -gg_0(g^2+g_0^2)^{-1/2},\\
\cot\theta &= g/g_0.
\end{aligned} \tag{3.19}$$

This choice is manifested in (3.18) by the replacement

$$\begin{aligned}
\mathscr{L}'_{\text{I}} &= -\tfrac{1}{2}W^+_{\alpha\beta}W^{-\alpha\beta}+M^2W^+_\alpha W^{-\alpha}-\tfrac{1}{4}A_{\alpha\beta}A^{\alpha\beta}-\\
&-\tfrac{1}{4}Z_{\alpha\beta}Z^{\alpha\alpha}+\tfrac{1}{2}M_z^2Z_\alpha Z^\alpha,
\end{aligned} \tag{3.20}$$

$$\begin{aligned}
\mathscr{L}'_{\text{II}} &= \bar{\nu}_L i\hat{\partial}\nu_L+\bar{\mu}(i\hat{\partial}-m_\mu)\mu-e\bar{\mu}\hat{A}\mu+\\
&+\frac{e}{\sqrt{2}\sin\theta}(\bar{\nu}_L\hat{W}^+\mu+\hat{W}\bar{\mu}^-\nu_L)+\\
&+\frac{e}{\sin 2\theta}(\bar{\nu}_L\hat{Z}\nu_L)-e\bar{\mu}\left\{\cot 2\theta\frac{1+\gamma_5}{2}-\tan\theta\frac{1-\gamma_5}{2}\right\}\hat{Z}\mu,
\end{aligned} \tag{3.21}$$

$$(3.22)\quad \mathcal{L}'_{\text{III}} = -\tfrac{1}{2}\zeta(\partial_\alpha A^\alpha)^2 - \tfrac{1}{2}\eta(\partial_\alpha Z^\alpha)^2 - -\xi^{-1}|(\partial_\alpha - ie(A_\alpha - \cot\theta Z_\alpha))W^+_\alpha|^2,$$

$$W^+_{\alpha\beta} = (\partial_\alpha W^+_\beta - \partial_\beta W^+_\alpha) + ie(W^+_\alpha A_\beta - W^+_\beta A_\alpha) - -ie\cot\theta(W^+_\alpha Z_\beta - Z_\alpha W^+_\beta),$$

$$(3.23)\quad A_{\alpha\beta} = \partial_\alpha A_\beta - \partial_\beta A_\alpha, \qquad Z_{\alpha\beta} = \partial_\alpha Z_\beta - \partial_\beta Z_\alpha.$$

The equations of motion can be manipulated to yield

$$(3.24)\quad \partial_\alpha A^\alpha_{} + \zeta^{-1}\chi_A = 0,$$

$$(3.25)\quad \partial_\alpha Z^\alpha + \eta^{-1}\chi_Z = 0,$$

$$(3.26)\quad \Box\chi_A = 0,$$

$$(3.27)\quad (\Box + \eta^{-1}M_Z^2)\chi_Z + \tfrac{1}{2}iGmM_Z^{-1}\bar{\mu}\gamma_5\mu = 0,$$

$$(3.28)\quad (\{\partial_\alpha - ie(A_\alpha - \cot\theta Z_\alpha)\}^2 + \xi^{-1}M^2)\chi^+ - e^2\cot 2\theta W^{+\alpha}(W^+_\alpha\chi^- - -W^-_\alpha\chi^+) + ie\cot^2\theta W^{+\alpha}\partial_\alpha\chi_Z - ie\cot^2\theta M^{-1}M_Z^2 W^{+\alpha}Z_\alpha - -\frac{ie}{\sqrt{2}}\operatorname{cosec}\theta\, m_\mu M^{-1}\bar{\mu}\nu_L = 0.$$

We carry through canonical quantization with the vector field commutators

$$[W^+_\alpha(x), W^-_\beta(y)] = -i(g_{\alpha\beta} + M^2\partial_\alpha\partial_\beta)\Delta(x-y, M^2) + +iM^{-2}\partial_\alpha\partial_\beta\Delta(x-y, \xi^{-1}M^2),$$

$$(3.29)\quad [Z_\alpha(x), Z_\beta(y)] = -i(g_{\alpha\beta} + M_Z^2\partial_\alpha\partial_\beta)\Delta(x-y, M_Z^2) + +iM_Z^{-2}\partial_\alpha\partial_\beta\Delta(x-y, \eta^{-1}M_Z^2),$$

$$[A_\alpha(x), A_\beta(y)] = -ig_{\alpha\beta}\Delta(x-y; 0) + i(1-\zeta^{-1})\partial_\alpha\partial_\beta E(x-y),$$

$$E(x-y) = -\frac{\partial}{\partial(M^2)}\Delta(x-y, M^2)|_{M^2=0}.$$

The anticommutators of the lepton fields is standard.

The scattering amplitude is given by the integral over paths

$$\mathscr{A} = \int \exp\left\{i\int d^4x(\mathscr{L}_{\mathrm{I}}+\mathscr{L}_{\mathrm{II}}+\mathscr{L}_{\mathrm{III}})\right\} d\{\Phi, \chi\},$$

$$(3.30)\quad \{\Phi\} = \{W_\alpha^\pm, Z_\beta, A_\lambda, \mu, \bar{\mu}, \nu, \bar{\nu}\},$$

$$\{\chi\} = \{\chi_A, \chi_Z, \chi^\pm\}.$$

'Carrying out the integration' over χ we get the primitive amplitude

$$(3.31)\quad \mathscr{A} = \int \exp\left\{i\int d^4x(\mathscr{L}'_{\mathrm{I}}+\mathscr{L}'_{\mathrm{II}}+\mathscr{L}'_{\mathrm{III}})\right\} d\{\Phi\}.$$

This amplitude is not unitary but contains the unitarity violating contributions from the spin 0 parts in the intermediate states. We must remove the contribution from these to get the unitary amplitude. As a first step we note that χ_A obeys a free field equation and may therefore be ignored; in contrast χ_Z obeys a linear inhomogeneous equation with only a lepton source. So we may write

$$(3.32)\quad \mathscr{L}(\chi_Z) = \tfrac{1}{2}\chi_Z(\Box+\eta^{-1}M_Z^2)\chi_Z - 2ie\operatorname{cosec}2\theta m_\mu M_Z^{-1}\bar{\mu}\gamma_5\mu\chi_Z.$$

The extra amplitude due to χ_Z can therefore be written

$$(3.33)\quad X_Z = \int \exp\left\{i\int d^4x\mathscr{L}(\chi_Z)\right\} d\{\chi_Z\}$$

$$= \exp\left\{2ie^2\operatorname{cosec}^2 2\theta m_\mu^2 M_Z^{-2}\iint \bar{\mu}\gamma_5\mu(x)\times\right.$$

$$\left.\times(\Box+\eta^{-1}M_Z^2)^{-1}\bar{\mu}\gamma_5\mu(y)d^4xd^4y\right\}$$

$$= \exp\left\{2ie^2\operatorname{cosec}^2 2\theta m_\mu^2 M_Z^{-2}\iint \bar{\mu}\gamma_5\mu(x)\times\right.$$

$$\left.\times\, G(x-y)\bar{\mu}(y)\gamma_5\mu(y)d^4xd^4y\right\}$$

where

$$(\Box+\eta^{-1}M_Z^2)G(x-y) = \delta^4(x-y).$$

In the equation of motion (3.28) for $\chi^\pm$ there is a term proportional to χ_Z in which we may write

$$\chi_Z(x) = \int G(x-y)(-ie\operatorname{cosec}2\theta M_Z^{-1}\bar{\mu}(y)\gamma_5\mu(y))d^4y.$$

The effective Lagrangian for the Lagrange multiplier fields $\chi^\pm$ can now be written down:

$$\text{(3.34)}\quad \mathscr{L}(\chi^\pm) = \chi^- Q\chi^+ - \tfrac{1}{2}e^2 \operatorname{cosec}^2\theta \{W^{+\alpha}W^+_\alpha(\chi^-)^2 + W^{-\alpha}W^-_\alpha(\chi^+)^2\} + \{S^+\chi^- + S^-\chi^+\}$$

where

$$Q = \{\partial_\alpha - ie(A_\alpha - \cot\theta Z_\alpha)\}^2 + e^2\operatorname{cosec}^2\theta W^{+\alpha}W^-_\alpha,$$

$$S^+ = \frac{1}{2}e^2\cot^2\theta m_\mu M_Z^{-1}W^{+\alpha}(\Box + \eta^{-1}M_Z^2)\,\partial_\alpha\bar{\mu}\gamma_5\mu - ieM_Z^2 M^{-1}W^{+\alpha}Z_\alpha - \frac{ie}{\sqrt{2}}\operatorname{cosec}\theta M^{-1}\bar{\mu}\nu = (S^-)^+.$$

The contribution of $\chi^\pm$ to the unitarity violation is given by the amplitude

$$\text{(3.35)}\quad X = \int \exp\left\{i\int \mathscr{L}(\chi^\pm)\,d^4x\right\} d\{\chi^+, \chi^-\}.$$

The physical unitary amplitude is given by

$$\text{(3.36)}\quad F = \int X_Z^{-1}X^{-1}\exp\left\{i\int d^4x(\mathscr{L}'_{\mathrm{I}} + \mathscr{L}'_{\mathrm{II}} + \mathscr{L}'_{\mathrm{III}})\right\} d\{\varphi\}.$$

As before we can rewrite X_Z^{-1} in terms of a fictitious scalar field obeying Bose statistics:

$$\text{(3.37)}\quad X_Z^{-1} = \int \exp\left\{i\int d^4x(\tfrac{1}{2}\partial_\lambda B\partial^\lambda B - \tfrac{1}{2}\eta^{-1}M_Z^2B^2) + im_\mu M_Z^{-1}e\operatorname{cosec}2\theta\bar{\mu}\gamma_5\mu B)\right\} d\{B\}$$

The factor χ^{-1} could be written

$$X^{-1} = \det\{\tfrac{1}{4}e^4\operatorname{cosec}^4\theta W^{+\alpha}W^+_\alpha W^{-\beta}W^-_\beta - \tfrac{1}{4}\overleftrightarrow{QQ}\}\,Y,$$

$$\text{(3.38)}\quad \begin{aligned} Y &= \int \exp\left\{i\int d^4x\mathscr{L}(C^+, C^-)\right\} d\{C^+, C^-\},\\ \mathscr{L}(C^+, C^-) &= -C^-QC^+ - (S^+C^- + S^-C^+) - \tfrac{1}{2}e^2\operatorname{cosec}^2\theta(W^{+\alpha}W^+_\alpha C^-C^- + W^{-\alpha}W^-_\alpha C^+C^+)\end{aligned}$$

where C is a fictitious field obeying Bose statistics.

The Feynman diagram rules can be worked out for this model and they are as follows: (These rules are for 1-loop approximation only, and at this level, the constraints associated with the fictitious Lagrangian could be ignored. The circle at one end of the F-line denotes that at least one end of the F-line must be attached to the leptons.)

$$2ie^2 g_{\alpha\beta}$$

$$-ie^2 g_{\alpha\beta}$$

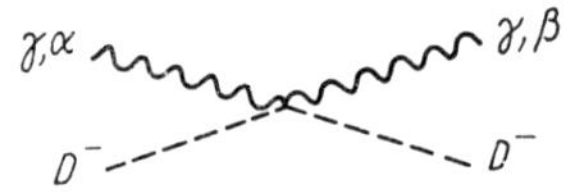

$$2ie^2 g_{\alpha\beta}$$

$$iG\cos^2\theta M_Z^2 g_{\alpha\beta}/M$$

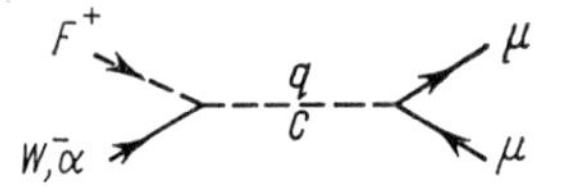

$$i\left(\frac{G^2\cos^4\theta m_\mu}{2M_Z}\right)\frac{q_\alpha}{q^2-M_Z^2/\eta}\,\bar{\mu}\gamma_5\mu$$

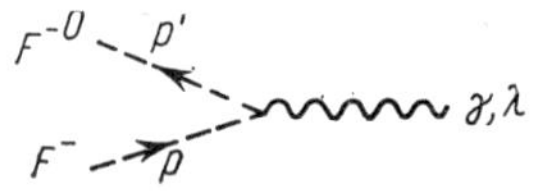

$$-ie(p+p')_\lambda$$

$$-ie\bar{\mu}\gamma_\lambda\mu$$

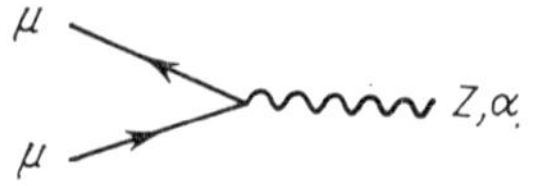

$$iG\mu\left[\frac{\cos 2}{2}\left(\frac{1-\gamma_5}{2}\right)-\sin^2\theta\left(\frac{1+\gamma_5}{2}\right)\right]\gamma_\alpha\mu$$

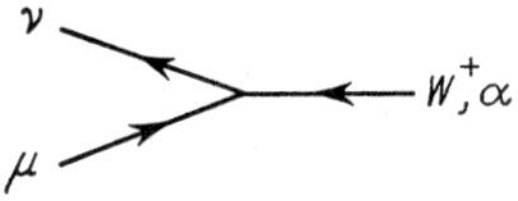

$$\frac{-iG\cos\theta}{\sqrt{2}}\,\bar{\nu}\gamma_\alpha\left(\frac{1+\gamma_5}{2}\right)\mu$$

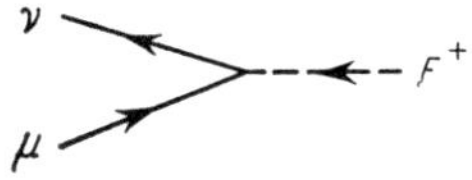

$$-i\left(\frac{Gm_\mu\cos\theta}{\sqrt{2}M}\right)\bar{\nu}\,\frac{1-\gamma_5}{2}\Bigg)\mu$$

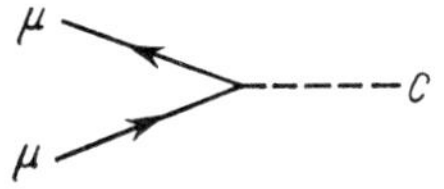

$$\frac{m_\mu G}{2M_Z}\,\bar{\mu}\mu\gamma_5\mu$$

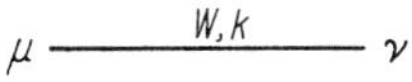

$$-i[g_{\mu\nu}-(1-\xi^{-1})k_\mu k_\nu(k^2-$$
$$-M^2/\xi)^{-1}]/(k^2-M^2)$$

$$-i[g_{\mu\nu}-(1-\eta^{-1})k_\mu k_\nu(k^2-$$
$$-M_Z^2/\eta)^{-1}]/(k^2-M_Z^2)$$

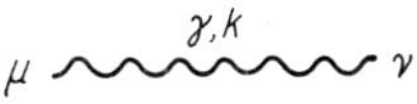

$$-i[g_{\mu\nu}-(1-\zeta^{-1})k_\mu k_\nu k^{-2}]/k^2$$

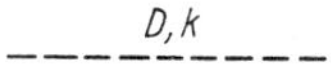

$$i/(k^2-M^2/\xi)$$

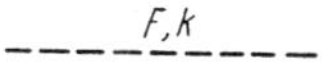

$$i/(k^2-M^2/\xi)$$

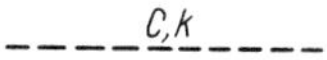

$$i/(k^2-M_Z^2/\eta)$$

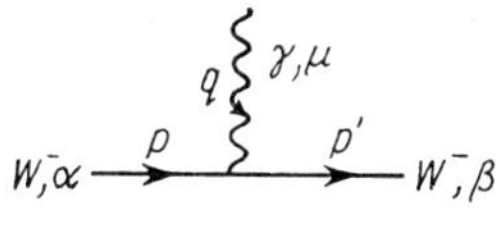

$$ie\{(p+p')_\mu g_{\alpha\beta}+[2q_\beta+(\xi-1)p'_\beta]g_{\alpha\mu}-$$
$$-[2q_\alpha-(\xi-1)p_\alpha]g_{\beta\mu}\}$$

$$-ie^2[2g_{\alpha\beta}g_{\mu\nu}+(\xi-1)g_{\alpha\mu}g_{\beta\nu}+(\xi-1)g_{\alpha\nu}g_{\beta\mu}]$$

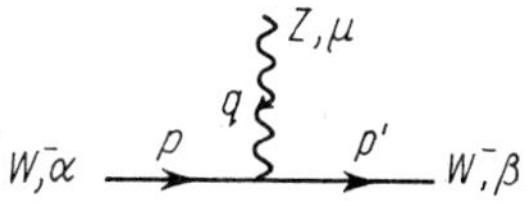

$$-iG\cos^2\theta\{(p+p')_\mu g_{\alpha\beta}+[2q_\beta+(\xi-$$
$$-1)p'_\beta]g_{\alpha\mu}-[2q_\alpha-(\xi-1)p_\alpha]g_{\beta\mu}\}$$

$$-iG^2\cos^4\theta[2g_{\alpha\beta}g_{\mu\nu}+(\xi-1)g_{\alpha\mu}g_{\beta\nu}+$$
$$+(\xi-1)g_{\alpha\nu}g_{\beta\mu}]$$

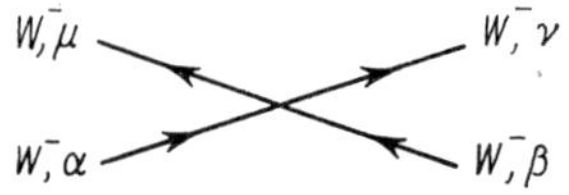

$$-i(e^2+G^2\cos^4\theta)(g_{\mu\alpha}g_{\nu\beta}+g_{\mu\beta}g_{\nu\alpha}-2g_{\mu\nu}g_{\alpha\beta})$$

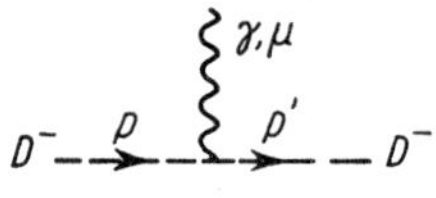

$$-ie(p+p')_\mu$$

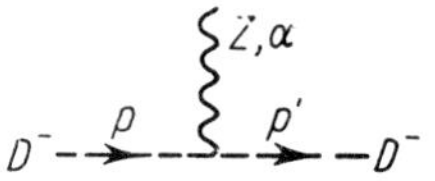

$$2iG\cos^2\theta P_\alpha$$

$$2iG^2\cos^4\theta g_{\alpha\beta}$$

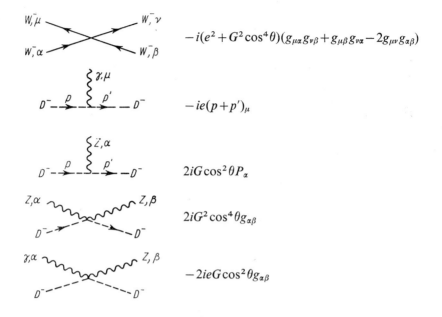

$$-2ieG\cos^2\theta g_{\alpha\beta}$$

A specially simple situation obtains for $\cot\theta = \pm 1$ since then both the A_α and the Z_α fields are coupled with the same coupling constant. For this special case the A, $W^\pm$, Z are coupled to leptons through

$$\text{(3.39)} \quad \pm e(\bar{\nu}_L \underset{\wedge}{W}{}^+ \mu + \bar{\mu} \underset{\wedge}{W}{}^- \nu_L) \pm e\bar{\nu}\underset{\wedge}{Z}\nu_L,$$

$$e \pm \bar{\mu}\underset{\wedge}{Z}\frac{1-\gamma_5}{2}\mu - e\bar{\mu}\underset{\wedge}{A}\mu$$

and to each other through

$$W^+_{\alpha\beta} = \partial_\alpha W^+_\beta - \partial_\beta W^+_\alpha + ie\{W^+_\alpha(A_\beta \pm Z_\beta) - W^+_\beta(A_\alpha \pm Z_\alpha)\}.$$

For this particular value of θ we have therefore genuine universality. If the conventional four-fermion coupling is denoted by G_F then the $W^\pm$ meson mass is determined from the relation

$$\frac{e^2}{4M^2} = \frac{G_F}{\sqrt{2}}$$

which yields

(3.40) $M = 52 \cdot 4$ GeV.

In this theory the mass M_z of the Z field is still undetermined and should be determined from data on neutral currents.

IV. A Model of Universal Unified Interactions

The above model can be thought of as a realistic model for purely leptonic sources for weak and electromagnetic interactions. But both these interactions apply equally well for hadrons. Yet we have so far not introduced hadrons. The simplest method of introducing hadrons is to have a set of quark fields which are coupled to $W^{\pm}$, A, and Z. The number and multiplicity of the quark fields depends to a certain degree on our choice; but if we use the $SU(2)$ group structure we can handle only $SU(2)$ quarks. Of course using quark fields rather than the observed hadrons is a formal device at this stage, but it does have the advantage that we do not have to deal with the electromagnetism of vector mesons and spin 3/2 baryons. Since the construction of this model is only a straight forward extension of what we have described in the last section, we shall not discuss this model in detail.

We have already alluded to the advantages of having a self-coupled vector meson multiplet mediating strong interactions. Such a theory could be asymptotically free and could have relatively simple scaling behaviour at high energies. But if we do that we are immediately led to the problem of making a charged vector boson theory renormalizable. *We no longer have the option of choosing a coupling scheme with the photon as a member of the multiplet.* The mechanism that is left to us is, then, to rely on the mass mixing between the strong and the weak bosons. This scheme is included in our general theory developed in Section II, but we do it more explicitly here.

For convenience of manipulation let us consider strong interactions invariant under the 'strong' chiral $SU(2) \times SU(2) \times U(1)$ and a unification of weak and electromagnetic interactions in terms of a weak $SU(2) \times U(1)$. We will choose, as typical fermion sources, the (μ, ν) doublet of leptons and the (n, p) doublet of four-component 'quarks'. We have the 'strong' triplet of vector mesons V_a^{α}, the 'strong' triplet of axialvector mesons U_a^α, the 'strong' singlet V_0^α, the 'weak' triplet of vector mesons W_a^α and singlet W_0^α. The Lagrangian density may then be written in the form

$$\mathcal{L} = \mathcal{L}_{\mathrm{I}}+\mathcal{L}_{\mathrm{II}}+\mathcal{L}_{\mathrm{III}}+\mathcal{L}_{\mathrm{IV}}, \tag{4.1}$$

$$\begin{aligned}\mathcal{L}_{\mathrm{I}} = & -\tfrac{1}{4}\{\partial^\alpha V_a^\beta-\partial^\beta V_a^\alpha-if\Gamma_{abc}(V_b^\alpha V_c^\beta+U_b^\alpha U_c^\beta)\}^2-\\ & -\tfrac{1}{4}\{\partial^\alpha U_a^\beta-\partial^\beta U_a^\alpha-if\Gamma_{abc}(V_b^\alpha U_c^\beta+U_b^\alpha U_c^\beta)\}^2-\\ & -\tfrac{1}{4}\{\partial^\alpha V_0^\beta-\partial^\beta V_0^\alpha\}^2-\tfrac{1}{2}u_{ab}U_a^\alpha U_{b\alpha}+\tfrac{1}{2}v_{ab}V_a^\alpha V_{b\alpha}+\\ & +u_{30}V_3^\alpha V_{0\alpha}+\tfrac{1}{2}u_{00}V_0^\alpha V_{0\alpha}-\\ & -\tfrac{1}{4}\{\partial^\alpha W_a^\beta-\partial^\beta W_a^\alpha-ig\Gamma_{abc}W_b^\alpha W_c^\beta\}^2+M_{30}W_3^\alpha W_{0\alpha}-\\ & -\tfrac{1}{4}\{\partial^\alpha W_0^\beta-\partial^\beta W_0^\alpha\}^2+\tfrac{1}{2}M_{ab}W_a^\alpha W_{b\alpha}+\tfrac{1}{2}M_{00}W_0^\alpha W_{0\alpha}.\end{aligned} \tag{4.2}$$

$$\begin{aligned}\mathcal{L}_{\mathrm{II}} = & \bar{q}\{i\hat{\partial}-\lambda-\frac{i}{2}f\tau_a\cdot(\hat{V}_a+\gamma_5\hat{U}_a)\}q+\\ & +\bar{l}\left\{i\hat{\partial}-k-\frac{i}{2}g\tau_a\cdot\hat{W}_a\frac{1+\gamma_5}{2}\right\}l+\\ & +\bar{q}\left\{\frac{1}{2}f_0\hat{V}_0\right\}q+\bar{l}\left\{\frac{1}{2}g_0\frac{1+\gamma_5}{2}+\frac{1+\tau_3}{2}g_0\frac{1-\gamma_5}{2}\right\}\hat{W}_0 l.\end{aligned} \tag{4.3}$$

$$\begin{aligned}\mathcal{L}_{\mathrm{III}} = & \chi_0\partial_\alpha W_0^\alpha+\chi_a(\partial_\alpha\delta_{ab}-ig\Gamma_{aQb}W_{Q\alpha})W_b^\alpha+\\ & +\varphi_0\partial_\alpha V_0^\alpha+\varphi_a(\partial_\alpha\gamma_{ab}-if\Gamma_{aQb}V_{Q\alpha})V_b^\alpha+\\ & +\psi_a(\delta_{ab}\partial_\alpha-if\Gamma_{aQb}V_{Q\alpha})U_b^\alpha+\\ & +\tfrac{1}{2}n_{ab}\chi_a\chi_b+n_{30}\chi_3\chi_0+\tfrac{1}{2}n_{00}\chi_0^2+\\ & +\tfrac{1}{2}l_{ab}\varphi_a\varphi_b+l_{30}\varphi_3\varphi_0+\tfrac{1}{2}l_{00}\varphi_0^2+\tfrac{1}{2}m_{ab}\psi_a\psi_b.\end{aligned} \tag{4.4}$$

$$\mathcal{L}_{\mathrm{IV}} = p_{ab}U_a^\alpha W_{b\alpha}+qV_a^\alpha W_{b\alpha}+\gamma_{ab}U_a^\alpha V_{b\alpha}+\gamma_{30}U_3^\alpha V_{0\alpha}. \tag{4.5}$$

Here $\Gamma_{abc} = i\varepsilon_{abc}$ and χ, φ, ψ are the Lagrange multiplier fields associated with W, V, U. The quark and the lepton mass matrices are given by λ and k. The quantities f, f_0 are the strong coupling constants and g, g_0 the weak coupling constant.

If we temporarily ignore $\mathcal{L}_{\mathrm{IV}}$ we can make use of the general technique to diagonalize the neutral vector mesons. In this fashion we find two combinations of V_3 and V_0 as mass eigenstates. We write these in the form

$$(4.6)\qquad \begin{aligned} \mathscr{V} &= V_3 \cos\varphi - V_0 \sin\varphi, \\ \mathscr{V}^1 &= V_3 \sin\varphi + V_0 \cos\varphi. \end{aligned}$$

This value of φ is determined by the mass matrix with elements U_{33}, U_{30}, U_{30}, U_{00}. We choose this matrix such that $\mathscr{V}$ is coupled to the charge operator for the quark doublet. (In particular if the average charge were zero we could have chosen $\varphi = 0$.) For the weak vector bosons we similarly define

$$(4.7)\qquad \begin{aligned} \mathscr{W} &= W_3 \cos\theta - W_0 \sin\theta, \\ \mathscr{W}' &= W_3 \sin\theta + W_0 \cos\theta, \end{aligned}$$

again choosing θ so as to have $\mathscr{W}$ be coupled via the charge operator of the leptons. This implies that

$$(4.8)\qquad \begin{aligned} \tan\varphi &= f_0/f, \\ \tan\theta &= g_0/f. \end{aligned}$$

We now make use of the isotopic triplet nature of the vector part of the weak interactions and the electromagnetic interactions. We can incorporate it by demanding that $\mathscr{L}_{\text{IV}}$ contain only such mass mixings of $\mathscr{V}$ and $\mathscr{W}$, V^+ and $W^\pm$ which can be removed by a rediagonalization of this mass matrix which involves rotation of all the three pairs of vector fields through the same angle $\otimes$ such that

$$\tan\otimes = g/f = \sqrt{g^2+g_0^2}/\sqrt{f^2+f^2}$$

or, equally

$$(4.9)\qquad \tan^2\theta = \tan^2\varphi.$$

The lepton and the quarks should therefore couple to their respective vector mesons in the same manner apart from an overall scale change. The vector meson matrices need not be proportional but they should be diagonalizable by the same rotation. We now define:

$$(4.10)\qquad \begin{aligned} A &= \cos\otimes\mathscr{W} - \sin\otimes\mathscr{V}, \\ B &= \sin\otimes\mathscr{W} + \cos\otimes\mathscr{V}. \end{aligned}$$

We demand that the mass matrix be such as to make A have zero mass so that we may identify it with the photon; and B with a vector meson of mass the average of ω and ϱ meson masses. We identify B with a linear superposition of ω and ϱ:

$$(4.11)\quad B = \frac{1}{\sqrt{2}}(\omega+\varrho).$$

The remaining components $\mathscr{V}'$ and $\mathscr{W}'$ also mix in some manner: we write

$$(4.12)\quad \begin{aligned} Z &= \cos\otimes'\mathscr{W}-\sin\otimes'\mathscr{V}', \\ C &= \sin\otimes'\mathscr{W}'+\cos\otimes'\mathscr{V}' \end{aligned}$$

which are both neutral vector mesons. Z may now be identified with a physical 'weak' heavy neutral boson and C with the conjugate superposition

$$(4.13)\quad C = \frac{1}{\sqrt{2}}(\omega-\varrho).$$

The ω–ϱ mass splitting can be generated by a small mass mixing term $B^{\alpha}C_{\alpha}$ in $\mathscr{L}_{\text{IV}}$. But we ignore this detail.

Let us now get back to electromagnetic interactions. The photon field A is coupled to the p quark and μ lepton by

$$(4.14)\quad -\bar{p}\hat{A}pf\sin\otimes+\bar{\mu}\hat{A}\mu g\cos\otimes = e(\bar{p}\hat{A}p-\bar{\mu}\hat{A}\mu).$$

Provided we choose

$$e = -f\sin\otimes = -\sqrt{f^2+g^2}\sin\otimes\cos\otimes = -g\sin\otimes.$$

Similarly the charged weak boson is coupled via the interaction

$$(4.15)\quad \frac{f\cos\otimes}{2\sin\theta}\bar{p}\hat{W}^{+}n+\frac{g\cos\otimes}{2\sin\theta}\bar{\nu}_L\hat{W}^{+}\mu$$

$$= -\frac{e}{2\sin\theta}(\bar{p}\hat{W}^{+}n+\bar{\nu}_L\hat{W}^{+}\mu).$$

Comparing with the weak four-fermion coupling constant, we have

$$\frac{e^2}{8M^2\sin^2\theta} = \frac{G_F}{\sqrt{2}},$$

$$M = \left(\frac{e^2}{2\sqrt{2G_F}}\right)^{1/2}\frac{1}{\sqrt{2}\sin\theta} = \frac{52\cdot 4}{\sqrt{2}\sin\theta}\ \text{GeV}$$

for the charged boson mass. In addition to this 'almost local' four-fermion interaction mediated by the 50 GeV weak boson there is a contribution to hadron semileptonic weak interactions generated by $p^{\pm}$ with the same form factor as the $\omega+\varrho$ contribution to the electromagnetic form factor.

A. EXTENSION TO $SU(3)$ AND APPLICATIONS TO WEAK INTERACTIONS

To discuss strong interactions the hadrons should be associated with $SU(3)\times SU(3)$ and the decay of strange particles constitute a major portion of weak interaction phenomena. The extension to $SU(3)$ from $SU(2)$ is rather straight forward: V_Q would now become $\frac{1}{2}(V_3+\sqrt{3}V_8)$ in place of V_3 and the structure constants Γ_{abc} now become appropriate to $SU(3)$. Now the mass mixings can become more elaborate; thus providing at once with more possibilities to accommodate the data and more arbitrariness.

We recall the following unusual features: (1) We have both $\Delta S=0$ and $\Delta S=\pm 1$ weak decays. (2) We have both leptonic and nonleptonic decays. (3) The neutrino in $\Delta S=0$ decays $(\pi\to\mu\nu)$ and in $\Delta S=1$ $(K\to\mu\nu)$ seem to be the same. (4) There are both leptonic and nonleptonic decays. (5) The nonleptonic decays seem to be predominately obeying $\Delta I=\frac{1}{2}$. (6) The neutral $\Delta S=1$ decays seem to be totally absent. (7) The semileptonic $\Delta Q=1$, $\Delta S=1$ decays seem to be suppressed by one order of magnitude from the decays. (8) There is a CP violating $\Delta S=1$ interaction which appears to satisfy $\Delta I=\frac{1}{2}$.

To realize an $SU(3)$ multiplet of leptons we have to have at least three leptons. We may choose them so as to have one charged and two neutral leptons (μ,ν_1,ν_2). However only one neutrino may be observed. We choose accordingly the mass matrix

$$k=\begin{bmatrix} m_\mu & 0 & 0 \\ 0 & m_{11} & m_{12} \\ 0 & m_{21} & m_{22}\end{bmatrix}=\begin{bmatrix} m_\mu & 0 & 0 \\ 0 & m_0^{S2} & -m_0^{S} \\ 0 & -m^{CS} & m_0^{C2}\end{bmatrix}.$$

Then for a very large value of m_0 only the mass eigenstate ν with zero mass comes in. If we write

$$C=\cos\varphi,\qquad S=\sin\varphi.$$

then the effective coupling to the $\mu\nu$ channel is proportional to cos for $\Delta S=0$ and proportional to $\sin\varphi$ for $\Delta S=1$ decays. So φ is the Cabibbo angle θ_C in this theory, provided we assume that the intermediate bosons with $Y=0$ and $Y=1$ and $Q=\pm 1$ have the same mass and the weak bosons are coupled directly to the quarks. More specifically

$$\tan\theta_C = \frac{M'^2}{M^2}\tan\varphi, \tag{4.16}$$

where M' and M are the weak boson masses. If we so choose we may take $\varphi = \pi/A$ and then θ_C measures the relative masses of the weak bosons.

The mass mixings of the strange weak and strong bosons lead to the hadronic decays in the theory involving strong vector bosons. These mass mixings must therefore be smaller for the charged strange mesons compared to that for the charged nonstrange mesons; and such mass mixings seem to be totally absent for the neutral strange mesons. A small mass mixing between the strange strong mesons and nonstrange weak mesons could be constructed to obtain a small *CP*-violating term.

The nonleptonic decays could be brought about by one of two mechanisms. In the first, we could choose mass mixings which violate strangeness (but conserve charge) between strong and weak mesons could, in second order, lead to nonleptonic decays. This mechanism is unsatisfactory since it appears quite difficult to obtain a $\Delta I = \frac{1}{2}$ dominance with the circumstance that the neutral $\Delta S = 1$ leptonic decays are absent. This is, of course, the old difficulty in a current-current theory appearing in a new garb.

We must, therefore, invoke another model. In this we have direct mass mixings of the strong bosons which generate the semileptonic decays. Now these interactions could be chosen to be $\Delta S = \frac{1}{2}$ dominantly. The only embarrassment here is that those weak couplings are now divorced from the leptonic weak interactions. We not only unify: we also diversify!

A more detailed examination of the applications to weak interactions is beyond the scope of this report.

B. COMPARISON WITH SPONTANEOUSLY BROKEN GAUGE THEORIES

The models that we have discussed parallel the models discussed by a number of authors for weak interactions. The strong boson-weak boson theory is a revised form of the Theory of Primary Interactions

[13], [14]. The central point of this class of theories is that the weak and electromagnetic properties of the hadrons may be thought of as induced by the strong bosons.

The search for renormalizable interactions lead us from four-fermion interactions to weak boson mediated interactions; this in turn provides the hope of unifying weak and electromagnetic interactions. But with charged bosons the electromagnetism becomes renormalizable only with a unified theory with a highly symmetric coupling. It has been found that if the mass is generated by spontaneous symmetry breaking leading to nonzero expectation values for certain auxiliary fields the massive theory promises to be renormalizable. With specific models of the Weinberg type or the Georgi–Glashow type one can then proceed to calculate higher order effects in such a theory.

We have outlined an entirely different method of generating unitary renormalizable theories.[2] In these theories the interaction structures and the kinematic terms are highly symmetric but the mass terms violate this symmetry. The mass terms may be chosen more or less at will, subject to certain general principles. The negative metric contributions in the primitive Lagrangian can be removed by division by a computable factor. We generate thus a theory with explicit computational rules. To the extent that we have been able to calculate (and we believe we have gone to the same degree of complication as other workers have done) the theory is unitary.

The application to realistic situations looks promising. In particular, since it is known that gauge theories with non-Abelian groups and their attendant nonlinear interactions lead to asymptotic freedom and scaling behaviour, we have endeavoured to show that a theory of gauge vector mesons for strong interactions can be obtained along with a universal and unified theory.

References

* Work supported in part by the USAEC (40–1) 3992.

[1] Many authors seem to have had the same idea. See, for example, [12].

[2] The problem of constraints associated with the fictitious Lagrangian could be avoided by introducing more unobservable fields in the Lagrangian. In this way, one could have the conventional Feynman rules, i.e., without changing the statistical weights of the diagrams in which 'different' unphysical particles related by the constraints

appears simultaneously in the intermediate states. Also, the resultant theory is unitary and renormalizable by power counting (See. J. P. Hsu and E. C. G. Sudarshan, *Phys. Lett.* **51B** (1974) 349; *Nucl. Phys.* **B91** (1975) 477; J. P. Hsu and J. A. Underwood, *Phys. Rev.* **D12** (1975) 620).

Bibliography

[1] J. P. Hsu, E. Mac and E. C. G. Sudarshan, 'The Failure of the Usual Gauge Formalism in a Class of Gauge Conditions', *Phys. Lett.* (in press); N. G. Deshpande, D. A. Dicus, J. P. Hsu and E. C. G. Sudarshan, *Unified Theory of Weak and Electromagnetic Interactions without Higgs Phenomena*, CPT-217 (1974).

[2] R. P. Feynman, *Acta Phys. Polon.* **27** (1963) 697.

[3] R. P. Feynman, *Magic without Magic* (edited by J. R. Klauder), Freeman and Co. (1972) 355–408.

[4] G. Guralnik, C. R. Hagen and T. W. B. Kibble, *Phys. Rev. Letters* **13** (1964) 585.

[5] P. Higgs, *Phys. Lettres* **12** (1964) 132.

[6] P. Higgs, *Phys. Rev.* **145** (1966) 1156.

[7] J. P. Hsu and E. C. G. Sudarshan, *Phys. Rev.* **D9** (1974) 1678.

[8] J. P. Hsu and E. C. G. Sudarshan, *Unified Theory of Weak and Electromagnetic Interactions without Spontaneously Broken Gauge Symmetry, I*, CPT-218 (1974); *Nucl. Phys.* **B91** (1975) 477.

[9] J. P. Hsu and E. C. G. Sudarshan, *Unified Theory of Weak and Electromagnetic Interactions with One Single Coupling Constant*, CPT-225 (1974); *Lett. Nuovo Cimento* (in press); *Violation of Unitarity in Weinberg's Unified Theory with Bilinear Gauge Conditions*, ORO-235 (1975).

[10] T. D. Lee and C. N. Yang, *Phys. Rev.* **119** (1960) 1410.

[11] R. E. Marshak and E. C. G. Sudarshan, *Introduction to Elementary Particle Physics*, Interscience (John Wiley) Inc., New York 1961.

[12] A. Salam, *Proc. of the Eighth Nobel Symposium*, John Wiley and Sons, New York 1968.

[13] E. C. G. Sudarshan, *Nature* **216** (1967) 979.

[14] E. C. G. Sudarshan, 'The Nature of Universal Primary Interactions of Particles', *Proc. Roy. Soc.* **305A** (1968) 319.

[15] E. C. G. Sudarshan and R. E. Marshak, *Proc. International Conference on Elementary Particles, Padua–Venice*, **1957**; reprinted in: *Development of the Theory of Weak Interactions* (edited by P. K. Kabir), Gordon and Breach, New York 1964.

[16] S. Weinberg, *Phys. Rev. Lettres* **19** (1967) 1264.

[17] H. Weyl, *Raum. Zeit, Materie*, fifth edition, Springer, Berlin 1923.

[18] H. Weyl, *The Principle of Relativity*, Dover Publications, Inc. (1923) 201–216.

[19] C. N. Yang and R. L. Mills, *Phys. Rev.* **96** (1954) 191.

PART FIVE

QUANTUM STATISTICAL PHYSICS

PHASE TRANSITIONS IN OPEN SYSTEMS FAR FROM THERMAL EQUILIBRIUM

K. HEPP

Eidgenössische Technische Hochschule, Zürich, Switzerland

Abstract

We have reviewed recent progress (in collaboration with E. H. Lieb) on the thermodynamic limit for 'small' quantum systems coupled to quantum reservoirs in KMS states with different temperatures. If the reservoirs act without memory and if the interaction H_s in the 'small' system is of mean field type, then a multitude of different dissipative structures can be exhibited. These rigorous results have been published in [2], [3], and reviewed in [1], [4], [5].

References

[1] K. Hepp, 'On Phase Transitions in Open Systems Far from Thermal Equilibrium', *Acta Physica Austriaca, Suppl. XI* (1973) 475, and *Proceedings of the International Conference in Mathematical Physics,* Moscow 1972.

[2] K. Hepp and E. H. Lieb, 'Phase Transitions in Reservoir-Driven Open Systems with Applications to Lasers and Superconductors', *Helv. Phys. Acta* **46** (1973) 573.

[3] K. Hepp and E. H. Lieb, 'Constructive Macroscopic Quantum Electrodynamics', in: *Constructive Quantum Field Theory* (eidted by G. Velo and A. S. Wightman), Lecture Notes in Physics Vol. 25, Springer Verlag, Berlin, Heidelberg, New York 1973.

[4] K. Hepp and E. H. Lieb, 'Battelle Seattle Rencontres 1974', to appear in Lecture Notes in Physics, Springer Verlag, Berlin, Heidelberg, New York 1975.

[5] E. H. Lieb, 'Exactly Soluble Models', *Physica* **73** (1974) 226.

SUPER-GAUGE GROUPS*

C. FRONSDAL**

International Centre for Theoretical Physics, Trieste, Italy,
Istituto di Fisica Teorica dell'Universita di Trieste, Trieste, Italy.

Abstract

Following a recent suggestion by Wess and Zumino, and applying methods due to Salam and Strathdee, we study some spinorial transformation groups; that is, groups that transform bosons into fermions and vice versa. Using chiral spinor variables, we give a super-field wave equation that conforms to the Fierz–Pauli principle. Neutrino gauge transformations are defined in close analogy with electromagnetic gauge transformations and a minimal principle for interactions with neutrinos is proposed.

Introduction

Schwinger[1] seems to have been the first to introduce spinorial transformations; that is, groups of transformations whose generators transform like spinors of odd rank with respect to Lorentz transformations. For an anti-commuting fermion field $\psi(x)$, Schwinger considered transformations of the type $\psi(x) \to \psi(x)+\eta(x)$, where $\eta(x)$ is an infinitesimal, anti-commuting c-number, in connection with his variational principles; without, however, investigating the associated abstract group. Flato and Hillion[2] considered both the abstract group and the theory of representations associated with 'translations' of neutrino fields, of the type $\psi(x) \to \psi(x)+\theta F$, where θ is a constant spinor and F is a 'suitable fixed operator', scalar with respect to Lorentz transformations. The components of θ are ordinary commuting numbers, while F presumably anticommutes with $\psi(x)$. The representations of this group, combined with Lorentz transformations into a semi-direct product, are studied by means of the symmetric space isomorphic to θ space; that is, θ is treated as an internal variable.

Recently Zumino and Wess[3] have drawn attention to some exciting new possibilities concerning spinorial groups. We believe that their ideas may contain the key to an understanding of neutrinos and (at least the purely leptonic) weak interactions. An important contribution by Salam and Strathdee[4] has simplified the study of such groups, and facilitated the invention of alternatives. Like Flato and Hillion,[2] they introduce an internal spinorial variable θ, but with the difference that the components of θ are constant anti-commuting numbers ('a-numbers'). Here we consider two further examples. The principal innovation is a proposal to introduce interactions with neutrinos through a concept of neutrino gauge invariance and an associated minimal coupling.

First example. The super field[4] $\Phi(x,\theta)$ depends on x^μ and on the left-handed anti-commuting spinor θ_L. The general form is

$$\Phi = a\varphi + b(\bar\theta_L\psi_R + \bar\chi_R\theta_L) + \frac{1}{a}\bar\theta_L\gamma_\mu\theta_L\varphi_\mu + \frac{A}{2}\bar\theta_R\theta_L + \frac{B}{2}\bar\theta_L\theta_R +$$

$$+\frac{1}{2b}[(\bar\theta_L\theta_R)(\bar\theta_R\psi_L) + (\bar\theta_R\theta_L)(\bar\chi_L\theta_R)] + \frac{a}{4}(\bar\theta_R\theta_L)(\bar\theta_L\theta_R)F.$$

The right-handed spinor θ_R is the charge conjugate of θ_L, so that $\theta_L+\theta_R$ is a Majorana spinor. The rearrangement rules are very simple:

$$\theta_L\bar\theta_R = -\tfrac{1}{2}(\bar\theta_R\theta_L), \quad \theta_R\bar\theta_L = -\tfrac{1}{2}(\bar\theta_L\theta_R),$$

$$\theta_L\bar\theta_L = -\tfrac{1}{2}(\bar\theta_L\gamma_\mu\theta_L)\gamma_\mu, \quad \theta_R\bar\theta_R = -\tfrac{1}{2}(\bar\theta_R\gamma_\mu\theta_R)\gamma_\mu.$$

Clearly, the right-hand sides of these relations should have factors of $(1\pm\gamma_5)/2$; but used as substitution rules in context they are adequate. The coefficients φ, ψ, $\bar\chi$, A, B, F are scalar or spinor fields that depend only on x^μ. All spinors anti-commute. No derivatives have been introduced in Φ since all differential relations should be contained in the wave equation (Fierz–Pauli principle). The coefficients a, b are real numbers to be adjusted later.

Following Wess and Zumino[3], we consider the transformations, with constant anti-commuting parameters α_L and $\bar\alpha_L$:

$$\theta_L \to \theta_L + \alpha_L, \quad \bar\theta_L \to \bar\theta_L + \bar\alpha_L,$$

$$x^\mu \to x^\mu + i\bar\theta_L\gamma^\mu\alpha_L + i\bar\alpha_L\gamma_\mu\theta_L.$$

Actually this differs from the rules of Wess and Zumino in one important detail: the sign of the last term in δx^μ. This has the following consequences: the group is Abelian and the multiplet structure is affected in a way that will appear shortly.

The action of the algebra on Φ is given by

$$\delta\Phi = i[\bar{\alpha}_L \bar{\partial}_L + \alpha_L \partial_L + i(\bar{\theta}_L \gamma^\mu \alpha_L + \bar{\alpha}_L \gamma^\mu \theta_L)\partial_\mu]\Phi,$$

where $\bar{\partial}_L$, ∂_L, ∂_μ stand for differentiation with respect to $\bar{\theta}_L, \theta_L$ and x^μ. In component form this reads

$$\delta\varphi = i\bar{\alpha}_L \psi_R + i\bar{\chi}_R \alpha_L, \qquad \delta F = \bar{\alpha}_L \gamma\partial\psi_L + \bar{\chi}_L \gamma\overleftarrow{\partial}\alpha_L,$$

$$\delta\varphi_\mu = \tfrac{1}{2}\bar{\alpha}_L(\gamma\partial\gamma_\mu\psi_R + i\gamma_\mu\psi_L) + \tfrac{1}{2}(\bar{\chi}_R\gamma_\mu\gamma\overleftarrow{\partial} + i\bar{\chi}_L\gamma_\mu)\alpha_L,$$

$$\delta\psi_R = (i\varphi_\mu - \partial_\mu\varphi)\gamma_\mu\alpha_L + iB\alpha_R,$$

$$\delta\bar{\chi}_R = \bar{\alpha}_L\gamma_\mu(i\varphi_\mu - \partial_\mu\varphi) + i\bar{\alpha}_R A,$$

$$\delta\psi_L = (iF - \gamma_\mu\gamma\partial\varphi_\mu)\alpha_L + \partial_\mu B\gamma_\mu\alpha_R,$$

$$\delta\bar{\chi}_L = \bar{\alpha}_L(iF - \gamma\partial\gamma_\mu\varphi_\mu) + \bar{\alpha}_R\gamma_\mu\partial_\mu A,$$

$$\delta A = (\bar{\chi}_R\gamma\overleftarrow{\partial} - i\bar{\chi}_L)\alpha_R, \qquad \delta B = \bar{\alpha}_R(\gamma\partial\psi_R - i\psi_L).$$

Here we have taken $a = b = 1$.

An invariant wave equation is

$$L\Phi = 0,$$

$$L \equiv \bar{\theta}_L\gamma\partial\theta_L + \partial_L\gamma\partial\bar{\partial}_L - (\theta_L\partial_L - \bar{\theta}_L\bar{\partial}_L)^2 + 2.$$

In component form,

$$\varphi = F, \qquad A = B = 0,$$

$$i\partial_\mu\varphi_\mu + \varphi = 0, \qquad i\partial_\mu\varphi + \varphi_\mu = 0,$$

$$(i\gamma\partial + 1)\psi = 0, \qquad \bar{\chi}(i\gamma\overleftarrow{\partial} + 1) = 0.$$

Here $\psi = \psi_L + \psi_R$, $\bar{\chi} = \bar{\chi}_L + \bar{\chi}_R$. Invariance of these equations can easily be verified directly. The difference between the Wess-Zumino model and ours is that we have two fermions without any vector mesons. The generators are non-hermitian, in the sense that $\int d^4x\Phi^+\Phi$ is not invariant, and we cannot add a constant to L without destroying the invariance of the equations. The masses of the multiplet are not only required to be equal: the value is fixed in terms of the scale introduced by the generators.

As Lagrangian, one may take either the full $\mathscr{L} = \int d^4x\Phi^+ L\Phi$ or, as suggested by Salam and Strathdee[4], the coefficient of the highest power of θ in this expression; the wave equations obtained by independent variation of all the components of Φ^+ are the same in either case.

Wess and Zumino[3] allow a certain limited x-dependence of the parameters: $\alpha \to \alpha(x) = \alpha_1 + \gamma x \alpha_2$. Following this prescription, one loses the commutativity of the algebra; the commutators define new transformations that resemble Lorentz transformations, just as Wess and Zumino obtain generators with a superficial similarity[5] to conformal transformations. Our wave equation is not invariant under this larger algebra. Perhaps it is more interesting to limit the x dependence of α by requiring $\gamma \partial \alpha = 0$. Then one can hope to restore invariance by introducing interactions with neutrinos transforming according to the gauge rule $\delta \nu = \alpha$.

Neutrino Gauge Invariance

By an x-space gauge transformation we mean

$$\Phi \to \exp[i\lambda(x)]\Phi,$$

$$i\partial_\mu \Phi \to i\partial_\mu \exp[i\lambda(x)]\Phi = \exp[i\lambda(x)]\big(i\partial_\mu - \lambda_{,\mu}(x)\big)\Phi.$$

Invariance is achieved by replacing $i\partial_\mu$ by $i\partial_\mu + eA_\mu$ and postulating the associated transformation of $A_\mu(x)$: A_μ: $(x) \to A_\mu(x) + \lambda_{,\mu}(x)$. Similarly, a θ-space gauge transformation could be defined by

$$\Phi \to \exp[i(\bar{\alpha}_R \theta_L + \bar{\theta}_L \alpha_R)]\Phi = \exp[i\bar{\alpha}\theta]\Phi,$$

$$i\bar{\partial}_L \Phi \to i\bar{\partial}_L \exp[i\bar{\alpha}\theta]\Phi = \exp[i\bar{\alpha}\theta]\,(i\bar{\partial}_L - \alpha_R)\Phi.$$

Invariance can be obtained by replacing $i\bar{\partial}_L \to i\bar{\partial}_L + \nu_R$ and $i\partial_L \to i\partial_L - \bar{\nu}_R$ and postulating the associated transformations

$$\nu_R \to \nu_R + \alpha_R, \qquad \bar{\nu}_R \to \bar{\nu}_R + \bar{\alpha}_R.$$

More general possibilities soon come to mind.

Second example. We use the same super field as before, but ignore the components A, B and F. Consider the transformation

$$\delta\Phi = g(\bar{\alpha}_R \theta_L + \bar{\theta}_L \alpha_R)\Phi,$$

$$\delta \nu_R = \alpha_R, \qquad \delta \bar{\nu}_R = \bar{\alpha}_R.$$

This is the first-order term in

$$\exp[g(\bar{\alpha}_R \theta_L + \bar{\theta}_L \alpha_R)]\Phi,$$

which is a unitary (if g is imaginary) neutrino gauge transformation. In terms of the components

$$\delta\varphi = 0, \quad \delta\varphi_\mu = -g(ab/2)(\bar{\chi}_R\gamma_\mu\alpha_R+\bar{\alpha}_R\gamma_\mu\psi_R),$$
$$\delta\psi_L = g(b/a)\varphi_\mu\gamma_\mu\alpha_R, \quad \delta\bar{\chi}_L = g(b/a)\bar{\alpha}_R\gamma_\mu\varphi_\mu,$$
$$\delta\psi_R = g(a/b)\varphi\alpha_R, \quad \delta\bar{\chi}_R = g(a/b)\varphi\bar{\alpha}_R.$$

It is not difficult to include the components A, B, F and evaluate the effect of $\exp[g\bar{\alpha}\theta]$ to all orders (i.e. to fourth order) in g. To obtain equations that are invariant to all orders, one should start with the super-wave equation and introduce the neutrinos by the minimal substitution. It seems, however, hardly worthwhile to carry out this programme at this stage, since the model is unrealistic. Instead, we simply try to introduce interactions by hand, adjusting the coupling so as to obtain invariance to lowest order in g. It is remarkable that this weak requirement is just as restrictive as invariance under the Wess–Zumino group. The masses must be equal and the normalization chosen so that $b^{-2}-a^{-2} = \frac{1}{2}g^*/g = \pm\frac{1}{2}$. If the interaction is free of derivatives, then the unique set of equations is:

$$i\partial_\mu\varphi+\varphi_\mu+\tilde{g}(\bar{\nu}\gamma_\mu\psi+\bar{\chi}\gamma_\mu\nu) = 0,$$
$$i\partial_\mu\varphi_\mu+\varphi-\tilde{g}(\bar{\nu}\psi+\bar{\chi}\nu) = 0,$$
$$(i\gamma\partial+1)\psi+\tilde{g}^*(\varphi_\mu\gamma_\mu-\varphi)\nu = 0,$$
$$\bar{\chi}(i\gamma\overleftarrow{\partial}+1)+\tilde{g}^*\bar{\nu}(\gamma_\mu\varphi_\mu-\varphi) = 0,$$
$$i\gamma\partial\nu+g(\varphi_\mu^\dagger\gamma_\mu\psi_R-\varphi^\dagger\psi_L)+g^*(\varphi_\mu\gamma_\mu\chi_R-\varphi\chi_L).$$

Here $\tilde{g} = gab/2$, $\nu = \nu_R$, $\psi = \psi_R+\psi_L$, $\bar{\chi} = \bar{\chi}_R+\bar{\chi}_L$. These equations have a familiar aspect.[6] The first-order transition amplitudes vanish, although there is an effective interaction of order g^2. The reason is that the invariance extends to the local gauge group (with x^μ-dependent parameters) provided $\gamma\partial\alpha = 0$; hence all external neutrino fields can be eliminated. If additional interaction is included, so as to extend the invariance to second order in g, then it appears that external neutrino fields can be eliminated to that order.

So far this model is unrealistic insofar as it contains only one type of neutrino. If it should turn out to be possible to implement the idea of a minimal neutrino coupling in a more realistic model, then it seems reasonable to expect that neutrino amplitudes will show a tendency to cancel. Subsequent breaking of the ψ, χ, φ mass degeneracy would destroy the cancellations except so far as the lowest order is concerned. In this case it would be possible to identify the fields ψ, χ, φ of the multiplet with the particles e^-, μ^+ and π^-, and g^2 with the Fermi interaction constant.

Comments

The appearance of the 'anti-commuting parameters' α in the transformations seems to suggest that one is dealing with an entirely new type of groups. In fact, however, one quickly ascertains that the spinorial groups considered so far are Lie groups.[7] Consequently one can approach the problem of representation by unitary operators in the space of states, just as one does routinely, in fact, with fermion field operators. This might be one way to obtain soft-neutrino theorems but probably not the simplest. What is really new is not the abstract algebraic structures but the idea of representations in the form of transformations among fermion and boson fields. This may be seen as an extension to fermions of the idea of non-linear representations.

The very revolutionary suggestion of extending the principle of linear superposition to include polynomials in the anti-commuting numbers α also seems to be due to Schwinger,[1] who talks of extending the number system to include anti-commuting elements. This idea is also implicit in the work of Wess and Zumino[3] and in a paper by Salam and Strathdee.[8] For example, in the Wess–Zumino model one finds that a certain commutator has the value $\bar{\alpha}_1 \gamma_\mu \alpha_2 \partial/\partial x^\mu$. This operator is treated as a type of translation, $\xi^\mu P_\mu = i\xi^\mu \partial/\partial x^\mu$. The two can easily be distinguished since the numerical vector ξ^μ enjoys properties that set it quite apart from the form $\bar{\alpha}_1 \gamma_\mu \alpha_2$, and their respective images under a unitary transformation must normally be quite distinct – unless one extends the superposition principle.

Setting aside, nevertheless, that exciting possibility, we note that the type of extended relativistic symmetry achieved by the spinorial groups is of the usual type: a semi-direct product of the Poincaré group with an internal symmetry group, with the latter as invariant factor. The unitary representations are infinite-dimensional and involve infinite degeneracy, as would be expected on consideration of the O'Raifeartaigh theorem or the Goldstone theorem. Neutrinos must play a fundamental role.[5]

Acknowledgements

I am grateful to Professor L. Fonda, Professor Abdus Salam and Dr. J. Strathdee for useful conversations, and wish to thank Professor Abdus Salam, the International Atomic Energy Agency and UNESCO for hospitality at the International Centre for Theoretical Physics, Trieste.

References

* Work supported in part by Istituto Nazionale di Fisica Nucleare.
** On leave of absence from the Univ. of California, Los Angeles, Cal., U.S.A.
[1] J. Schwinger, *Phys. Rev.* **92** (1953) 1283.
[2] M. Flato and P. Hillion, *Phys. Rev.* **D1** (1970) 1667.
[3] J. Wess and B. Zumino, *Nucl. Phys.* **B70** (1974) 39.
[4] Abdus Salam and J. Strathdee, *Nucl. Phys.* **B76** (1974) 477.
[5] See discussion at the end.
[6] To the author, that is; see C. Fronsdal, *Phys. Rev.* **136** (1964) B1190, *Phys. Rev.* **176** (1968) 1846.
[7] The order r of the Lie algebra depends on the dimension d of the linear vector space of the parameters. If d is finite then r is finite. Every subalgebra of finite order is nilpotent. Note that d is not the number of components of the spinor α, but depends on their realization.
[8] Abdus Salam and J. Strathdee, *Nucl. Phys.* **B76** (1974) 477. This paper constructs representations of the Clifford algebra obtained by separating out the anticommuting α's.
[9] The relevance of neutrinos has been stressed by Abdus Salam and J. Strathdee, *Nucl. Phys.* **B80** (1974) 499.

RELATIVISTIC QUANTUM STATISTICAL MECHANICS

RAPHAEL HØEGH-KROHN

Institute of Mathematics, University of Oslo, Oslo, Norway

1. Introduction

Although the study of the statistical mechanics for quantum systems has made good progress in the last decade [8], the progress has been best for the discrete systems or the lattice systems. A major difficulty in connection with the continuous systems has been that the group of time automorphisms α_t for the Schrödinger particles is non local. The consequence of this non locality is that the infinite system of interacting Schrödinger particles do not agree well with the generally accepted picture of a quantum statistical mechanics described in terms of a local C^*-algebra or a C^*-algebra of local operators, on which the time acts as a group α_t of C^*-automorphisms. Hence we get a somewhat discouraging situation, that the only realistic model of a statistical quantum mechanics, namely the system of a dilute gas of Schrödinger particles, does not conform to the well developed abstract theory of quantum statistical mechanics.

For this very reason the question of studying relativistic particles instead of Schrödinger particles comes up quite natural, since in any relativistic theory there should be an upper bound for the propagation speed and this would force the group of time automorphisms α_t to be local. This is the motivation for this paper.

Interacting relativistic particles or interacting quantum fields are by now reasonably well understood in the case of two space-time dimensions. In the case of weak polynomial and exponential interactions in two space-time dimensions one also has a very clear picture of what happens with the vacuum in the infinite volume limit, or as we would like to say it here, one has a very clear picture of the thermodynamic limit in the case of temperature zero. For the weak polynomial interactions this was done by Glimm, Jaffe and Spencer [4], and in the case of exponential interaction by Albeverio and Høegh-Krohn [1]. Hence good candidates for a quantum statistical mechanics of interacting relativistic particles are the polynomial and exponential interactions in two space-time dimensions.

In this paper I study the thermodynamic limit of the positive temperature Gibbs-state for the polynomial and exponential interactions in two space-time dimensions.

The method used is strongly influenced by the Markov field approach, initiated by Nelson [7], and developed by Guerra, Rosen and Simon [5]. In this paper we use the Markov field approach to transform the problem about the thermodynamic limit for the Gibbs-state at temperature $1/\beta$ for the relativistic quantum statistical system into the problem of the uniqueness of the vacuum for the system in a periodic box of length β.

In fact it turns out that for any of the interactions we consider, namely the polynomial and exponential interaction, the Markov fields for the Gibbs-state at temperature $1/\beta$ is the Markov field on a cylinder $S_\beta \times R$, where S_β is the circle of length β, which correspond to the Markov field for the vacuum in the plane $R \times R$ for the same interaction, and this last Markov field is the limit of the first as the temperature $1/\beta$ goes to zero.

Using this method it is here proved that the thermodynamic limit exists for the Gibbs-state for all positive temperatures $1/\beta$ and for all interactions, i.e. for the strong exponentials as well as for the strong polynomial interactions.

We see that this is in strong contrast to the vacuum or temperature zero case for the polynomial interactions, where Glimm and Spencer were only able to prove the existence of the infinite volume limit in the case of weak interactions. In the case of even polynomials we know from Dobrushin and Minlos [2] that this is best possible; in fact for any even polynomial interaction in two space-time dimensions they get that the thermodynamic limit is not unique in the temperature zero case for strong enough interactions. The reason for this remarkable difference is the above mentioned fact that while for the temperature zero case we have a Markov field in the plane $R \times R$ so the problem is two-dimensional, we have for positive temperature a Markov field on a cylinder $S_\beta \times R$ so that the problem is essentially one-dimensional, and therefore in a sense much simpler.

The Gibbs-state at positive temperature $1/\beta$ is of course not invariant under the Lorentz group since it is given in terms of the energy operator.

There is however a Lorentz invariant analogy of the Gibbs-state at positive temperature $1/\beta$. But this Lorentz invariant Gibbs-state is only to be found in the closed universe, the so-called De Sitter universe. The problems of the positive temperature universe will not be dealt with in this paper, and for those interested we refer to [3].

2. The Gibbs-State for the Anharmonic Oscillator with a Finite Number of Degrees of Freedom

Consider the self-adjoint operator

$$(2.1)\quad H_0 = -\tfrac{1}{2}\Delta + \tfrac{1}{2}(x, A^2 x) - \tfrac{1}{2}\operatorname{tr} A$$

on the Hilbert space $\mathscr{H} = L_2(R^N)$, where $\Delta = \sum_{i=1}^{N} \frac{\partial^2}{\partial x_i^2}$ and A is a real symmetric $N \times N$ matrix bounded below by a positive constant, $A \geqslant cI$, $c > 0$, $x \in R^N$ and $(\ ,\ )$ is the natural inner product in $\mathscr{H}$.

Let $\lambda_1, \ldots, \lambda_N$ be the eigenvalues of A. It is well known that H_0 has discrete spectrum consisting of the points of the form $\sum_{k=1}^{N} \lambda_{i_k}$ and zero. Hence for $\beta > 0$, $e^{-\beta H_0}$ is of trace class and we get

$$(2.2)\quad \operatorname{tr} e^{-\beta H_0} = \sum_{n_i \geqslant 0} e^{-\beta \sum_{i=1}^{N} n_i \lambda_i} = \prod_{i=1}^{N} (1 - e^{-\beta \lambda_i})^{-1}$$

so that

$$(2.3)\quad \operatorname{tr} e^{-\beta H_0} = |1 - e^{-\beta A}|^{-1}$$

where $|1 - e^{-\beta A}|$ is the determinant of the matrix $1 - e^{-\beta A}$.

Let $V(x) \geqslant -b$ be a real measurable function bounded below such that

$$(2.4)\quad H = H_0 + V(x)$$

is essentially self-adjoint. We say that H is the Hamiltonian for the anharmonic oscillator. From $V \geqslant -b$ we get that $H \geqslant H_0 - b$, which gives us that H has discrete spectrum and that $e^{-\beta H}$ is of trace class. We may therefore form the normal state ω_β on the von Neumann algebra $B(\mathscr{H})$, given by

$$(2.5)\quad \omega_\beta(B) = (\operatorname{tr} e^{-\beta H})^{-1} \operatorname{tr}(B e^{-\beta H}).$$

ω_β is called the *Gibbs-state for the anharmonic oscillator*.

By the Feynmann–Kac formula we know that the kernel $e^{-\beta H}(x, y)$ of the operator $e^{-\beta H}$ is given by

$$\text{(2.6)} \quad e^{-\beta H}(x, y) = E^{\beta}_{(x,y)}\left[\exp\left[-\int_0^{\beta} U(x(\tau))\, d\tau\right]\right]$$

with $U(x) = \frac{1}{2}(x, A^2x) - \frac{1}{2}\operatorname{tr} A + V(x)$ and $E^{\beta}_{(x,y)}$ is the conditional expectaction with respect to the Brownian motion in R^N given that $x(0) = x$ and $x(\beta) = y$. Hence

$$\text{(2.7)} \quad \operatorname{tr} e^{-\beta H} = \int_{R^N} E^{\beta}_{(x,x)}\left[\exp\left[-\int_0^{\beta} U(x(\tau))\, d\tau\right]\right] dx$$

$$= C_1 \int_{R^N} E^{\beta}_{(x,x)}\left[\exp\left[-\tfrac{1}{2}\int_0^{\beta} (x(\tau), A^2 x(\tau))\, d\tau\right] \times\right.$$

$$\left.\times \exp\left[-\int_0^{\beta} V(x(\tau))\, d\tau\right]\right] dx.$$

One may now verify (see [6], Section 2) that (2.7) is equal to

$$\text{(2.8)} \quad CE^{\beta}\left[\exp\left[-\int_0^{\beta} V(x(\tau))\, d\tau\right]\right]$$

where C is a normalization constant and E^{β} is the expectation with respect to the normal distribution indexed by the real Hilbert space g_{β} of continuous periodic functions from $[0, \beta]$ into R^N with norm square given by

$$\text{(2.9)} \quad \int_0^{\beta} \left[\left(\frac{dx(\tau)}{d\tau}, \frac{dx(\tau)}{d\tau}\right) + (x(\tau), A^2 x(\tau))\right] d\tau,$$

which is the same as the expectation with respect to the homogeneous Gaussian process on the circle S_{β} of length β and covariance function given by

$$\text{(2.10)} \quad E^{\beta}(x_i(0)x_j(t)) = (2A(1 - e^{-\beta A}))^{-1}[e^{-tA} + e^{-(\beta - t)A}]$$

for $0 \leqslant t \leqslant \beta$

Setting $V = 0$ in (2.7) we get $C = \operatorname{tr} e^{-\beta H_0} = |1 - e^{-\beta A}|^{-1}$ hence we get

$$\text{(2.11)} \quad \operatorname{tr} e^{-\beta H} = |1 - e^{-\beta A}|^{-1} E^{\beta}\left[\exp\left[-\int_0^{\beta} V(x(\tau))\, d\tau\right]\right].$$

For more details and proof of the following theorem see [6], Section 2.

THEOREM 2.1. Let $F_i \in B(\mathscr{H})$ be multiplication operators by bounded continuous functions $F_i(x)$, let $0 = s_0 \leqslant \ldots \leqslant s_n = \beta$ and let H be the Hamiltonian for the anharmonic oscillator, then

$$\operatorname{tr}(F_0 e^{-s_1 H} F_1 e^{-(s_2-s_1)H} \ldots F_{n-1} e^{-(\beta-s_{n-1})H})$$

$$= |1-e^{-\beta A}|^{-1} E^\beta \Big[\exp\Big[-\int_0^\beta V(x(\tau))\,d\tau\Big] \prod_{i=0}^{n-1} F_i(x(s_i))\Big],$$

where $|1-e^{-\beta A}|$ is the determinant of the matrix $1-e^{-\beta A}$ and E^β is the expectation with respect to the homogeneous Gaussian process on the circle S_β of length β with mean zero and covariance function given by

$$E^\beta\big(x_i(0)x_j(t)\big) = \big(2A(1-e^{-\beta A})\big)^{-1}[e^{-tA}+e^{-(\beta-t)A}]$$

for $0 \leqslant t \leqslant \beta$.

Let α_t be the C^*-automorphism of $B(\mathscr{H})$ defined by

(2.12) $\alpha_t(B) = e^{-itH} B e^{itH}$,

then

(2.13) $\operatorname{tr}\big(B\alpha_t(C)e^{-\beta H}\big) = \operatorname{tr}(Ce^{-(\beta-it)H}Be^{-itH})$

is analytic in t in the strip $-\beta < \operatorname{Im} t < 0$ with boundary values at real t equal to $\operatorname{tr}\big(B\alpha_t(C)e^{-\beta H}\big)$ and at $t-i\beta$ equal to $\operatorname{tr}\big(C\alpha_{-t}(B)e^{-\beta H}\big)$. Moreover

(2.14) $\operatorname{tr}(F_0 e^{-s_1 H} F_1 e^{-(s_2-s_1)H} \ldots F_{n-1} e^{-(\beta-s_{n-1})H})$

is analytic in the domain $0 < \operatorname{Re} s_1 < \ldots < \operatorname{Re} s_{n-1} < \beta$ with boundary values at $\operatorname{Re} s_i = 0$ which are continuous and uniformly bounded and for $s_k = it_k$ given by

(2.15) $\operatorname{tr}\big(F_0 \alpha_{t_1}(F_1)\alpha_{t_2}(F_2) \ldots \alpha_{t_{n-1}}(F_{n-1})e^{-\beta H}\big)$.

LEMMA 2.1. Let $t_i \in R$ and F_i be bounded continuous functions on R^N, then $B(\mathscr{H})$ is the smallest strongly closed linear space of operators that contains all operators of the form

$$\alpha_1(F_1)\alpha_2(F_2) \ldots \alpha_n(F_n).$$

For the proof of this lemma and also of the following theorem see [6], Section 2.

THEOREM 2.2. Let B and C be in $B(\mathcal{H})$, then

$$\omega_\beta(B\alpha_t(C)) = \omega_\beta(\alpha_{-t}(B)C)$$

is analytic in the strip $-\beta < \operatorname{Im} t < 0$ and continuous and uniformly bounded in $-\beta \leqslant \operatorname{Im} t \leqslant 0$. The boundary values satisfy the KMS condition

$$\omega_\beta(B\alpha_{t-i\beta}(C)) = \omega_\beta(C\alpha_{-t}(B)).$$

Moreover, any operator $B \in B(\mathcal{H})$ may be approximated strongly by the linear combinations of operators of the form $\alpha_{t_1}(F_1) \ldots \alpha_{t_n}(F_n)$ where $F_1, \ldots, F_n$ are multiplication operators by continuous functions $F_1(x), \ldots$ $\ldots, F_n(x)$. Furthermore $\omega_\beta(F_0\alpha_{t_1}(F_1) \ldots \alpha_{t_n}(F_n))$ is analytic in $0 > \operatorname{Im} t_1$ $> \ldots > \operatorname{Im} t_n > -\beta$ and its value for $t_k = -is_k$ with $0 \leqslant s_1 \leqslant \ldots$ $\leqslant s_n \leqslant \beta$ is given by

$$\omega_\beta(F_0\alpha_{-is_1}(F_1) \ldots \alpha_{-is_n}(F_n))$$
$$= \left(E^\beta\left[\exp\left[-\int_0^\beta V(x(\tau))\,d\tau\right]\right]\right)^{-1} E^\beta\left[\prod_{i=0}^n F_i(x(s_i)) \times\right.$$
$$\left.\times \exp\left[-\int_0^\beta V(x(\tau))\,d\tau\right]\right]$$

where E^β is the expectation given in Theorem 2.1.

3. The Gibbs-State for the Anharmonic Oscillator with an Infinite Number of Degrees of Freedom

Let h be a real separable Hilbert space, and let A be a positive self-adjoint operator on h bounded below by a positive constant, i.e. $A \geqslant cI$, $c > 0$. The Hamiltonian for the harmonic oscillator with the harmonic potential $\frac{1}{2}(x, A^2x)$ where $(\, , \,)$ is the inner product in h and $x \in h$ is a self-adjoint operator which we denote by

$$(3.1) \quad H_0 = -\tfrac{1}{2}\Delta + \tfrac{1}{2}(x, A^2x) - \tfrac{1}{2}\operatorname{tr} A,$$

where Δ is the Laplacian on h and $(\, , \,)$ the inner product in h. The definition of (3.1) is usually given in terms of Fock spaces, but for our convenience we shall here use another definition which is due to Nelson (see [7]).

Let g be the real Hilbert space of continuous functions from R into the domain of A, $D_A \subset h$, such that the norm square

$$(3.2)\quad \int_R \left[\left(\frac{dx(\tau)}{d\tau}, \frac{dx(\tau)}{d\tau}\right) + \left(x(\tau), A^2 x(\tau)\right)\right] d\tau$$

is finite. Let E be the expectation with respect to the normal distribution indexed by g, which is the same as the expectation with respect to the homogeneous Gaussian process with values in h with mean zero and covariance function given by

$$(3.3)\quad E[(y_1, x(s))\,(y_2, x(t))] = \left(y_1, (2A)^{-1} e^{-|t-s|A} y_2\right),$$

where (,) is the inner product in h. Let $\mathscr{B}_0$ be the subalgebra of measurable sets of this process which is generated by the functions of the form $(y, x(0))$ when y runs over h. Let E be the conditional expectation with respect to $\mathscr{B}_0$, and $\mathscr{H}$ the L_2-space of square integrable functions which are measurable with respect to $\mathscr{B}_0$. E_0 is then an orthogonal projection onto in the L_2-space of the whole process. It follows from (3.3) that the process $x(t)$ is a homogeneous Markov process with invariant measure. Nelson proved that one may define H as the infinitesimal generator of this Markov process.

If we identify the function $(y, x(0))$ with the linear function (y, x) on h we see that $\mathscr{H}$ is generated by linear functions on h. It follows from (3.3) for $s = t = 0$ that $\mathscr{H}$ is in fact the L_2-space with respect to the normal distribution with covariance given by

$$(3.4)\quad E_0\left((y_1, x)(y_2, x)\right) = \left(y_1, (2A)^{-1} y_2\right)$$

for y_1 and y_2 in h. Let F be a measurable function of the process. $T_s F$ is then the function induced by the transformation $x(t) \to x(t+s)$.

We than have that if F is in $\mathscr{H}$ then

$$(3.5)\quad e^{-tH_0} F = E_0 T_t F$$

or

$$(3.6)\quad e^{-tH_0} = E_0 T_t E_0 .$$

We shall here take (3.5) or (3.6) as the definition of (3.1). Since $\mathscr{H}$ is generated in the sense of measurable functions by linear functions (y, x) on h, we shall use the notation $F(x)$ for functions in $\mathscr{H}$, although the Gaussian measure given by (3.4) does not necessarily have support in h

but rather on some completion of h. We shall also write $F(x(t))$ for $T_t F$ and the identification of $(y, x(0))$ with (y, x) corresponds then to the identification of $F(x(0))$ with $F(x)$.

Let $V(x)$ be a $\mathscr{B}_0$ measurable function which is bounded below and finite almost everywhere, such that

$$(3.7) \quad H = H_0 + V(x)$$

is essentially self-adjoint. We shall call H the Hamiltonian for the anharmonic oscillator. Nelson then proved the following version of the Feynmann–Kac formula (see [7])

$$(3.8) \quad e^{-tH}F = E_0 \exp\left[-\int_0^t V(x(\tau))\,d\tau\right] F.$$

Let us now assume that $e^{-\beta A}$ is of trace class, we can then prove that $e^{-\beta H}$ is also of trace class and we have the following theorem:

THEOREM 3.1. Let $F_i \in B(\mathscr{H})$ be multiplication operators by bounded measurable functions, let $0 \leqslant s_0 \leqslant s_1 \leqslant \ldots \leqslant s_{n-1} \leqslant \beta$, then

$$\begin{aligned} &\operatorname{tr}(F_0 e^{-s_1 H} F_1 e^{-(s_2 - s_1)H} \ldots F_{n-1} e^{-(\beta - s_{n-1})H}) \\ &= |1 - e^{-\beta A}|^{-1} E^{\beta}\left[\exp\left[-\int_0^{\beta} V(x(\tau))\,d\tau\right] \prod_{i=0}^{n-1} F(x(s_i))\right], \end{aligned}$$

where $|1-e^{-\beta A}|^{-1}$ is the determinant of the operator $1-e^{-\beta A}$ and E^{β} is the expectation with respect to the homogeneous Gaussian process on the circle S_β of length β with mean zero and covariance function given by

$$\begin{aligned} &E^{\beta}[(y_1, x(0))(y_2, x(t))] \\ &= ((y_1, 2A(1-e^{-\beta A}))^{-1}[e^{-tA} + e^{-(\beta-t)A}]y_2) \end{aligned}$$

for $0 \leqslant t \leqslant \beta$. This is the same as the expectation with respect to the normal distribution indexed by the real Hilbert space g_β of continuous periodic functions from $[0, \beta]$ into the domain of A such that the norm square

$$\int_0^{\beta} \left[\left(\frac{dx(\tau)}{d\tau}, \frac{dx(\tau)}{d\tau}\right) + (x(\tau), A^2 x(\tau))\right] d\tau$$

is finite.

For the proof of this theorem and also of the following theorem see [6], Section 3.

We now define ω_β and α_t as in the case of a finite number of degrees of freedom.

THEOREM 3.2. Lemma 2.1 and Theorem 2.2 hold also for the case of an infinite number of degrees of freedom under the assumption that $e^{-\beta A}$ is of trace class.

4. The Gibbs-State for a Gas of Relativistic Scalar Bose Particles without Interaction

Let $\Lambda \subset R^n$ be a bounded domain in R^n with regular boundary $\partial\Lambda$. Let $A_\Lambda^2 = -\Delta + m^2$ where Δ is the Laplace operator in $L^2(\Lambda)$ with some self-adjoint boundary conditions on $\partial\Lambda$. If the constant function satisfies the boundary conditions we shall assume $m > 0$ if not only that $m \geqslant 0$ so that in any case A_Λ^2 is a self-adjoint operator on the real Hilbert space $h_\Lambda = L_2^R(\Lambda)$ and $A_\Lambda^2 \geqslant cI$, $c > 0$.

The Hamiltonian for a system of relativistic scalar Bose particles of mass m without interaction in Λ is given by the Hamiltonian $H_0(\Lambda)$ for the free scalar field in Λ with mass m, which in the notation of the previous section is given by

$$(4.1) \quad H_0(\Lambda) = -\tfrac{1}{2}\Delta_\Lambda + \tfrac{1}{2}(x_1 A_\Lambda^2 x) - \tfrac{1}{2}\operatorname{tr} A_\Lambda$$

where Δ_Λ denotes the Laplacian on h_Λ and (,) is the inner product in h_Λ. The definition of (4.1) was given in the previous section.

It is well known that A_Λ has discrete spectrum and that $e^{-\beta A_\Lambda}$ is of trace class for all $\beta > 0$ so that the Fredholm determinant $|1 - e^{-\beta A_\Lambda}|$ exists. Hence the result of the previous section holds with $H = H_0(\Lambda)$. Let $\omega_\beta^0(\Lambda)$ and $\alpha_t^0(\Lambda)$ be the ω_β and α_t of the previous section with $H = H_0(\Lambda)$. Now if $F \in B(\mathscr{H}_\Lambda)$ is a multiplication operator by a bounded $\mathscr{B}_0$ measurable function, then F is simply a bounded function of the time zero fields $\varphi(x)$ since in the notation of the previous section $\varphi(f) = \int_{R^n} \varphi(x) f(x)\,dx$ with $\operatorname{supp} f \subset \Lambda$ is simply the $\mathscr{B}_0$ measurable linear function $(f, x(0))$, where (,) is the inner product in h_Λ.

Let $\mathcal{O} \subset R^n$ be a bounded open set in R^n and let $\mathcal{O} \subset \Lambda$. Let $\mathscr{F}(\mathcal{O})$ be the subalgebra of $B(\mathscr{H}_\Lambda)$ generated by the time zero fields $\varphi(f)$ with $\operatorname{supp} f \subset \mathcal{O}$. Let $\mathscr{A}_0(\mathcal{O})$ be the local algebra on $\mathcal{O}$, i.e. the C^*-algebra generated by the fields and the canonical conjugate fields in $\mathcal{O}$. It is easy to see that $\mathscr{A}_0(\mathcal{O})$ has equivalent representations in $B(\mathscr{H}_{\Lambda_1})$ and $B(\mathscr{H}_{\Lambda_2})$ as soon as $\mathcal{O} \subset \Lambda_1$ and $\mathcal{O} \subset \Lambda_2$. $\mathscr{H}_\Lambda$ is the Hilbert space $\mathscr{H}$ of the previous section with $H = H_0(\Lambda)$. Due to the equivalence of the representations the strong closure $\bar{\mathscr{A}}(\mathcal{O})$ of $\mathscr{A}_0(\mathcal{O})$ is independent of Λ as soon as $\mathcal{O} \subset \Lambda$. We obviously have that $\bar{\mathscr{A}}(\mathcal{O}_1) \subseteq \bar{\mathscr{A}}(\mathcal{O}_2)$ if $\mathcal{O}_1 \subseteq \mathcal{O}_2$. Let $\bar{\mathscr{A}}$ be the smallest C^*-algebra containing all $\bar{\mathscr{A}}(\mathcal{O})$. Due to the finite propagation speed for relativistic particles we have that if $B \in \bar{\mathscr{A}}(\mathcal{O})$, then $\alpha_t^0(\Lambda)(B)$ is independent of Λ for Λ large enough. Let

$$(4.2) \quad \alpha_t^0(B) = \lim_{\Lambda \to R^n} \alpha_t^0(\Lambda)(B),$$

α_t^0 then defines a group of C^*-automorphism of $\bar{\mathscr{A}}$.

The quasi-local algebra for the free field $\mathscr{A}_0$ is the smallest norm closed subalgebra of $\bar{\mathscr{A}}$ containing $\alpha_t^0(F)$ for all t and $F \in \mathscr{F}(\mathcal{O})$ for some $\mathcal{O}$. It is now possible to prove that the limit as $\Lambda \to R^n$ of $\omega_\beta^0(\Lambda)$ on $\mathscr{A}_0$ exists. We have in fact the following theorem. For the proof of this theorem and for more details see [6], Section 3.

THEOREM 4.1. Let $\mathscr{A}_0$ be the local algebra for the free field, then α_t^0 defines a group of C^*-automorphisms of $\mathscr{A}_0$. There is a state ω_β^0 on $\mathscr{A}_0$ which is invariant under α_t^0, i.e.

$$\omega_\beta^0(B \cdot \alpha_t^0(C)) = \omega_\beta^0(\alpha_{-t}^0(B) \cdot C),$$

such that $\omega_\beta^0(B\alpha_t^0(C))$ is analytic in the strip $-\beta < \operatorname{Im} t < 0$ and uniformly bounded and continuous in $-\beta \leqslant \operatorname{Im} t < 0$, and satisfies the KMS conditions on the boundary, i.e.

$$\omega_\beta^0(B \cdot \alpha_{t-i\beta}^0(C)) = \omega_\beta^0(C\alpha_{-t}^0(B))$$

for real t.

Moreover, if $F_0, \ldots, F_m$ is in the subalgebra of $\mathscr{A}_0$ generated by the fields at time zero then $\omega_\beta^0(F_0\alpha_{t_1}^0(F_1) \ldots \alpha_{t_m}^0(F_m))$ is analytic in $0 > \operatorname{Im} t_1 > \ldots > \operatorname{Im} t_m > -\beta$ and continuous and uniformly bounded in $0 \geqslant \operatorname{Im} t_1 \geqslant \ldots \geqslant \operatorname{Im} t_m \geqslant -\beta$ and its value at the imaginary points $t_k = -is_k$ with $0 = s_0 \leqslant s_1 \leqslant \ldots \leqslant s_m \leqslant \beta$ is given by

$$\omega_\beta^0(F_0\alpha_{-is_1}^0(F_1) \ldots \alpha_{-is_m}^0(F_m)) = E^\beta\left[\prod_{k=0}^m F_k^{s_k}\right]$$

where E^β is the expectation with respect to the generalized Gaussian random field on $S_\beta \times R^n$ with covariance function $G^\beta(x-y, 0-t)$ which is the Green's function on $S_\beta \times R^n$ for the self-adjoint operator

$$-\frac{\partial^2}{\partial t^2} - \sum_{i=1}^{m} \frac{\partial^2}{\partial x_i^2} + m^2$$

on $L_2(S^\beta \times R^n)$ where S^β is the circle of length β, and $F_k^{s_k}$ is the translation by the action of the circle group S^β on $S^\beta \times R^n$ of the function F_k by the amount s_k.

Furthermore, if $B \in \mathscr{A}_0$ is in $\mathscr{A}(\mathscr{O})$ for some bounded $\mathscr{O}$ then

$$\omega_\beta^0(B) = \lim_{\Lambda} \omega_\beta^0(\Lambda)(B)$$

as $\Lambda \to R^n$ in the sense that Λ finally contains any fixed bounded set.

If we introduce $\hat{G}^\beta(p, s) = \int_{R^n} G^\beta(x, t) e^{ixp} dx$ then for $0 \leqslant s \leqslant \beta$

(4.3) $\hat{G}^\beta(p, s) = (2\omega(1-e^{-\beta\omega}))^{-1}(e^{-s\omega} + e^{-(\beta-s)\omega})$

with $\omega = \omega(p) = \sqrt{p^2+m^2}$.

The free field at time t is given by $\varphi_t(x) = \alpha_t^0(\varphi(x))$, and the propagator is given by

(4.4) $\Delta^\beta(x, t) = G^\beta(x, -i|t|)$

so that

(4.5) $\hat{\Delta}^\beta(p, t) = (2\omega(1-e^{-\beta\omega}))^{-1}(e^{i|t|\omega} + e^{-\beta\omega}e^{-i|t|\omega})$.

We have the following formula for computing the expectation of a product of fields

(4.6) $\omega_\beta^0(\varphi(x_1, t_1) \ldots \varphi(x_n, t_n))$

$$= \begin{cases} \sum \Delta^\beta(x_{i_1}-x_{i_2}, t_{i_1}-t_{i_2}) \ldots \Delta^\beta(x_{i_{n-1}}-x_{i_n}, t_{i_{n-1}}-t_{i_n}), \\ 0 \quad \text{for } n \text{ odd.} \end{cases}$$

If we define the pressure for the gas of free relativistic bose particles at temperature $1/\beta$ in the usual way by

(4.7) $p_\beta^0 = \beta^{-1} \lim_{\Lambda \to R^n} |\Lambda|^{-1} \log(\mathrm{tr}\, e^{-\beta H_0(\Lambda)})$

where $|\Lambda|$ is the volume of Λ, then we get that this limit always exists and is given by

(4.8) $p_\beta^0 = -(2\pi)^{-n}\beta^{-1} \int_{R^n} \log(1-e^{-\beta\omega(p)})\, dp$.

Similarly we find that the density of particles with momentum p exists and is given by

$$(4.9)\quad \varrho_\beta^0(p) = (2\pi)^{-n}\frac{e^{-\beta\omega(p)}}{1-e^{-\beta\omega(p)}}\,.$$

5. The Gibbs-State for a Gas of Relativistic Scalar Bose Particles with Relativistic Invariant Interaction in Two Space-Time Dimensions

In the case of two space-time dimensions or equivalently one space dimension, the relativistic interacting scalar Bose particles are relatively well understood in the case of polynomial interactions [4] and exponential interactions [1]. It was proved by Glimm–Jaffe–Spencer that the thermodynamic limit for the temperature zero ($\beta = \infty$) state existed and is unique for weak polynomial interactions. Later Nelson has established the existence of the thermodynamic limit also for strong polynomial interactions with Dirichlet boundary conditions. Nelson's method which depends strongly on the Dirichlet boundary conditions, leads to the question of whether this limit is independent of the boundary conditions, and in fact Dobrushin and Minlos [2] have announced the result that for any even polynomial there is a certain value for the interaction strength above which the limit does depend on the boundary conditions. For the strong exponential interactions the existence of the thermodynamic limit for the temperature zero state was proved by Albeverio and Høegh-Krohn [1] in the case of even interactions.

From what is said above we see that the thermodynamic behaviour of the temperature zero state is quite complex already in one space dimension. This is in contrast with what we usually have in statistical mechanics, namely that for one-dimensional system there are usually no phase transitions. We shall see that in one space dimension the existence of phase transitions is something that happens only in the temperature zero case. We have namely that for positive temperature there is no phase transitions and the thermodynamic limit always exists.

So let now

$$(5.1)\quad H_l = H_0 + \int_{-l}^{l} :V(\varphi(x)):dx$$

where $\varphi(x)$ is the time zero free field and H_0 is the Hamiltonian for the free relativistic scalar bose particles, or equivalently for the free relativistic scalar field, and V is either a polynomial which is bounded below or an exponential function, i.e.

$$V(s) = \int e^{\alpha s}\, d\mu(\alpha)$$

where $d\mu$ is a positive measure of compact support in the interval $(-\sqrt{2\pi}, \sqrt{2\pi})$. The Wick ordering : : in (5.1) is taken with respect to the free vacuum.

THEOREM 5.1. Let $\xi(x, s)$ be the generalized Gaussian random field on $R \times S_\beta$ of the previous section and let E^β be the expectation with respect to this random field, and dP_0^β the corresponding measure.

$$dP_l^\beta = \left\{E^\beta\left(\exp\left[-\int_0^\beta \int_{-l}^{l} :V(\xi(x, s)): dx\, ds\right]\right)\right\}^{-1} \times$$

$$\times \exp\left[-\int_0^\beta \int_{-l}^{l} :V(\xi(x, s)): dx\, ds\right] dP_0^\beta$$

is a probability measure that is absolutely continuous with respect to dP_0^β. Then the weak limit dP_∞^β of dP_l^β as $l \to \infty$ exists and is locally equivalent to dP_0^β. Moreover, dP_∞^β is translation invariant on the cylinder $R \times S_\beta$ and it is strongly mixing with respect to space translations, i.e.

$$\lim_{x\to\infty} E_\infty^\beta(e^{i(\varphi_1, \xi)}\, e^{i(\varphi_2^x, \xi)}) = E_\infty^\beta(e^{i(\varphi_1, \xi)}) \cdot E_\infty^\beta(e^{i(\varphi_2, \xi)})$$

where $\varphi^x(t, y) = \varphi(t, y - x)$ and E_∞^β is the expectation with respect to dP_∞^β.

For the proof of this theorem and the following theorems see [6], Section 4.

Let α_t^l be the group of C^*-automorphisms on $\bar{\mathscr{A}}$ induced by H_l. By the finite propagation speed the limit of α_t^l as $l \to \infty$ exists and defines a group of C^*-automorphisms α_t on $\bar{\mathscr{A}}$. We have now the following theorem:

THEOREM 5.2. Let $\mathscr{A}$ be the local algebra for the interacting field, i.e. the smallest norm closed C^*-algebra in $\bar{\mathscr{A}}$ containing $\alpha_t(F)$ for all real t and $F \in \mathscr{F}(\mathcal{O})$ for any bounded $\mathcal{O}$ in R.

Then there exists a state ω_β on $\mathscr{A}$ such that ω_β is invariant under α_t, i.e.

$$\omega_\beta(B \cdot \alpha_t(C)) = \omega_\beta(\alpha_{-t}(B) \cdot C)$$

for any B and C in $\mathscr{A}$. $\omega_\beta(B \cdot \alpha_t(C))$ is analytic in the strip $-\beta < \operatorname{Im} t < 0$ and uniformly bounded and continuous in $-\beta \leqslant \operatorname{Im} t \leqslant 0$, and satisfies the KMS conditions on the boundary

$$\omega_\beta(B \cdot \alpha_{t-i\beta}(C)) = \omega_\beta(C \cdot \alpha_{-t}(B))$$

for real t. ω_β is invariant under space translations

$$\omega_\beta(B_x) = \omega_\beta(B)$$

and has the cluster property

$$\lim_{x \to \infty} \omega_\beta(B_x \cdot C) = \omega_\beta(B) \cdot \omega_\beta(C).$$

ω_β is locally Fock.

Moreover, if $F_0, \ldots, F_n$ is in $\mathscr{F}(\mathcal{O})$ for some bounded $\mathcal{O}$ then $\omega_\beta(F_0 \alpha_t(F_1) \ldots \alpha_{t_n}(F_n))$ is analytic in $0 > \operatorname{Im} t_1 > \ldots > \operatorname{Im} t_n > -\beta$ and continuous and uniformly bounded in $0 \geqslant \operatorname{Im} t_1 \geqslant \ldots \geqslant \operatorname{Im} t_n \geqslant -\beta$ and its values at the imaginary points $t_k = -is_k$ with $0 = s_0 \leqslant \ldots \leqslant s_n \leqslant \beta$ is given by

$$\omega_\beta(F_0 \alpha_{-is_1}(F_1) \ldots \alpha_{-is_n}(F_n)) = E_\infty^\beta \left[\prod_{k=0}^{\infty} F_k^{s_k} \right]$$

where E_∞^β is the expectation with respect to the generalized homogeneous process on $S_\beta \times R$ given by the Theorem 5.1, and $F_k^{s_k}$ is the translated of the function F_k by the amount s_k in the action of the circle group S_β on $S_\beta \times R$.

Furthermore, if B is in the subalgebra generated by $\alpha_t(F)$ for t in a fixed interval $[-a, a]$ and $F \in \mathscr{F}(\mathcal{O})$ for a fixed bounded $\mathcal{O}$ then

$$\omega_\beta(B) = \lim_{l \to \infty} \lim_{\Lambda \to \infty} \omega_\beta^l(\Lambda)(B)$$

where $\omega_\beta^l(\Lambda)$ is the Gibbs-state for the operator

$$H_0(\Lambda) + \int_{-l}^{l} :V(\varphi(x)): dx.$$

THEOREM 5.3 (The duality theorem). Let $W_\infty^\beta(x_1, t_1, \ldots, x_n, t_n)$ be the Wightman functions at temperature $1/\beta$ for the infinite volume interaction, and let W_β^∞ be the usual Wightman functions at temperature zero ($\beta = \infty$) for the interacting field in a periodic box of length l. Let S_∞^β and S_β^∞ be the corresponding Schwinger functions, i.e. the Wightman functions at imaginary times. Then $W_\infty^\beta(x_1, t_1, \ldots x_n, t_n)$ is analytic in $0 > \mathrm{Im}\, t_1 > \ldots > \mathrm{Im}\, t_n > -\beta$ and W_β^∞ is analytic in $\mathrm{Im}\, t_1 > \ldots > \mathrm{Im}\, t_n$ and for the corresponding Schwinger functions we have

$$S_\infty^\beta(x_1, s_1, \ldots, x_n, s_n) = S_\beta^\infty(s_1, x_1, \ldots, s_n, x_n).$$

Moreover, the difference between the pressure for the free and the interacting field at temperature $1/\beta$ is equal to

$$p_\beta(0) - p_\beta(V) = \beta^{-1} e_\beta(V)$$

where $e_\beta(V)$ is the lowest eigenvalue for the interacting Hamiltonian in the periodic box of length β.

Remark. It is understood that the interaction terms in the Hamiltonian for the dual Wightman functions W_∞^β and W_β^∞ are defined by mutually consistent Wick ordering, i.e. if the Wick ordering for the interaction of W_∞^β is with respect to the free vacuum respectively the free Gibbs-state (at temperature $1/\beta$), then the Wick ordering for the interaction of W_β^∞ is with respect to the vacuum for H_0 respectively the vacuum for H_β.

For proof of these theorems see [6], Section 4.

References

[1] S. Albeverio and R. Høegh-Krohn, 'The Wightman-Axioms and the Mass Gap for Strong Interactions of Exponential Type in Two-Dimensional Space Time', *J. Funct. Anal.* **16** (1974) 39–82.

[2] R. L. Dobrushin and R. A. Minlos, 'Construction of a One-Dimensional Quantum Field via a Continuous Markov Field', *Funct. Anal. and Its Appl.* **7** (1973) 324–325 (English transl.).

[3] E. Figari, R. Høegh-Krohn and C. R. Nappi, 'Interacting Relativistic Boson Fields in the De Sitter Universe with Two Space-Time Dimensions', *Commun. Math. Phys.* **44** (1975) 265–278.

[4] J. Glimm, A. Jaffe and T. Spencer, 'The Wightman Axioms and the Particle Structure in the $P(\varphi)_2$ Quantum Field Model', *Ann. Math.* **100** (1974) 585–632.

[5] F. Guerra, L. Rosen and B. Simon, 'The $P(\varphi)_2$ Euclidean Field Theory as a Classical Statistical Mechanics', *Ann. Math.* **101** (1975) 111–259.

[6] R. Høegh-Krohn, 'Relativistic Quantum Statistical Mechanics in Two-Dimensional Space-Time', *Commun. Math. Phys.* **38** (1974) 195–224.

[7] E. Nelson, J. Funct. Anal. **12** (1973) 211–227.

[8] D. Ruelle, *Statistical Mechanics, Rigorous Results*, Benjamin, New York 1969.

INDEX